16.99

Edex
GC
higher
tics

Keith Pled

Gareth Cole

Peter Jolly

Graham Newman

Joe Petran

Sue Bright

D1471594

www.heinemann.co.uk

✓ Free online support
✓ Useful weblinks
✓ 24 hour online ordering

01865 888058

Heinemann
Inspiring generations

Heinemann Educational Publishers
Halley Court, Jordan Hill, Oxford OX2 8EJ
Part of Harcourt Education Limited

Heinemann is the registered trademark of
Harcourt Education Limited

© Harcourt Education Ltd, 2006

First published 2006

10 09 08 07 06
10 9 8 7 6 5 4 3 2 1

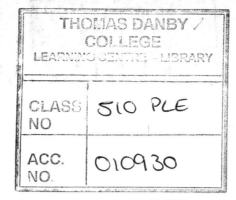

British Library Cataloguing in Publication Data is available from the British Library on request.

10-digit ISBN: 0 435 53409 2
13-digit ISBN: 978 0 435534 09 7

Typeset by Tech-Set Ltd, Gateshead, Tyne and Wear
Original illustrations © Harcourt Education Limited, 2006
Illustrated by Adrian Barclay and Mark Ruffle
Cover design by mccdesign
Printed in the UK by Pindar
Cover photo: Digital Vision ©

Acknowledgements
Harcourt Education Ltd would like to thank those schools who helped in the development and trialling of this course.

This high quality material is endorsed by Edexcel and has been through a rigorous quality assurance programme to ensure that it is a suitable companion to the specification for both learners and teachers. This does not mean that its contents will be used verbatim when setting examinations nor is it to be read as being the official specification – a copy of which is available at www.edexcel.org.uk

The publisher's and authors' thanks are due to Edexcel Limited for permission to reproduce questions from past examination papers. These are marked with an [E]. The answers have been provided by the authors and are not the responsibility of Edexcel Limited.

The authors and publisher would like to thank the following individuals and organisations for permission to reproduce photographs:

Harcourt Education Ltd / Peter Morris p1; Getty Images / PhotoDisc pp18, 78, 156, 185; Corbis pp35 left, 214, 275, 320, 343, 385, 458; Alamy Images pp35 right, 216, 464; Empics pp62, 81, 380, 440; Getty Images p68; iStockPhoto.com / Dan Cooper p88; Digital Vision p110; Getty Images / News & Sport p115; Alvey & Towers p144; Science Photo Library pp152, 154, 340; Guillaume Dargaud p184; iStockPhoto.com p212; iStockPhoto.com / Gisele Wright p291; Photos.com pp338, 371, 376, 388; Alamy Images / Elmtree Images p338; Action+ Images p346; Brand X Photos p403; Lonely Planet Images p429

Every effort has been made to contact copyright holders of material reproduced in this book. Any omissions will be rectified in subsequent printings if notice is given to the publishers.

Publishing team

Editorial	James Orr, Lindsey Besley, Evan Curnow, Katherine Pate, Nick Sample, Alex Sharpe, Laurice Suess, Elizabeth Bowden, Ian Crane
Design	Phil Leafe
Production	Siobhan Snowden
Picture research	Chrissie Martin

Websites
There are links to relevant websites in this book. In order to ensure that the links are up-to-date, that the links work, and that the sites aren't inadvertently linked to sites that could be considered offensive, we have made the links available on the Heinemann website at www.heinemann.co.uk/hotlinks. When you access the site, the express code is **4092P**.

Tel: 01865 888058   www.heinemann.co.uk

Quick reference to chapters

1 Exploring numbers 1   1

2 Essential algebra   16

3 Shapes   29

4 Fractions and decimals   52

5 Collecting and recording data   71

6 Solving equations and inequalities   91

7 Transformations and loci   104

8 Using basic number skills   133

9 Functions, lines, simultaneous equations and regions   160

10 Presenting and analysing data 1   184

11 Estimation and approximation   214

12 Sequences and formulae   225

13 Measure and mensuration   238

14 Simplifying algebraic expressions   255

15 Pythagoras' theorem   276

16 Basic trigonometry   291

17 Graphs and equations   310

18 Proportion   328

19 Quadratic equations   346

20 Presenting and analysing data 2   369

21 Advanced trigonometry   399

22 Advanced mensuration   412

23 Exploring numbers 2   432

24 Probability   446

25 Transformations of graphs   460

26 Circle theorems   485

27 Vectors   501

28 Introducing modelling   519

29 Conditional probability   534

Examination practice papers   542
Formulae sheet   553
Answers   554
Index   601

Introduction

Introduction

This revised and updated edition has been carefully matched to the new two-tier specification for GCSE Maths. It covers everything you need to know to achieve success in your exam, up to and including Grade A*. The author team is made up of Senior Examiners, a Chair of Examiners and Senior Moderators, all experienced teachers with an excellent understanding of the requirements of the Edexcel specification.

Key features

- **Chapters** are divided into **sections**, each with a simple explanation followed by clear examples or a worked exam question. These show you how to tackle questions. Each section also contains practice exercises to develop your understanding and help you consolidate your learning.

- **Key points** are highlighted throughout, like this:

  > If $x \times x = A$, then x is the **square root** of A, written $\sqrt{A}$

  Each chapter ends with a summary of key points you need to remember.

- **Hint boxes** are used to make explanations clearer. They may also remind you of previously learned facts or tell you where in the book to find more information.

  > Remember:
  > $(x + 3)^2 = (x + 3)(x + 3)$

- **Mixed exercises** are designed to test your understanding across each chapter. They include past exam questions which are marked with an [E]. You will find a mixed exercise at the end of almost every chapter.

- **Examination practice papers** are included to help you prepare for the exam at the end of your course.

- **Answers** are provided at the back of the book to use as your teacher directs.

Quick reference and detailed Contents pages

- Use the thumb spots on the edges of the **Quick reference** pages to help you turn to the right chapter quickly.

- Use the detailed **Contents** to help you find a section on a particular topic. The summary and reference codes on the right show your teacher the part(s) of the specification covered by each section in the book. (For example, NA 3h refers to Number and Algebra, section 3 Calculations, subsection h.)

Use of a calculator or a computer

These symbols show you where you must, or must not, use a calculator. Sometimes you may wish to use a spreadsheet package on a computer. There are also links to websites and suggested activities that require an internet search.

Coursework

A Coursework Guide is available online at www.zebramaths.co.uk

Contents

Note: *Italic* indicates paragraphs in the Higher tier specification that have been drawn from the Foundation tier.

1 Exploring numbers 1

1.1	**Number facts**	1–4	Prime factors, HCF and LCM	NA2a
1.2	**Square numbers and cube numbers**	4–5	Squares and cubes	NA2b
1.3	**Finding squares and square roots, cubes and cube roots**	5–7	Understanding and using $\sqrt{}$ and $\sqrt[3]{}$	NA2b
1.4	**Index numbers**	7–8	Simple integer powers and the general form a^n	NA2b
1.5	**The index laws**	8–10	Rules for multiplication and division; negative and zero powers; powers of powers	NA2b/3a
1.6	**Fractional indices**	10–11	Positive and negative fractional powers with exact answers	NA3a
1.7	**Powers of 2 and 10**	12–13	Including solving equations of the type $2^{2n-1} = 32$	NA6a
	Summary of key points	15		

2 Essential algebra

2.1	**Evaluating algebraic expressions**	16–17	Substitution	NA5d
2.2	**Evaluating algebraic expressions involving squares**	17–18	Further substitution	NA5d
2.3	**Indices**	18–19	Index notation and the index laws	NA5d
2.4	**Removing brackets from algebraic expressions**	19–20	Multiplying a single term over a bracket	NA5b
2.5	**Simplifying algebraic expressions**	20–21	Collecting like terms	NA5b
2.6	**Factorising**	22	Taking out common factors	NA5b
2.7	**Multiplying bracketed expressions**	23–24	Multiplying out two linear expressions	NA5b/c
2.8	**Multiplying an expression by itself**	25	Squaring simple expressions, $(a + b)^2$	NA5b
2.9	**Solving simple equations**	25–26	Solving simple equations with the unknown on one side only	NA5e
	Summary of key points	27–28		

3 Shapes

3.1	**Angles**	29–30	Angles at a point and on a line; alternate angles and corresponding angles	SSM2a
3.2	**2-D shapes**	31–34	Triangles, quadrilaterals, polygons and their angles	SSM2a/b/c/d
3.3	**Demonstration and proof**	34–35	Sum of angles in a triangle equals 180°; exterior angle equals the sum of interior opposite angles	SSM/2a

3.4	3-D shapes	35–38	Names and nets; plan and elevation	SSM2i
3.5	Congruent shapes	38–41	Proving congruence using formal arguments	SSM2e
3.6	Similar shapes	42–45	Identifying similar shapes; scale factor	SSM2g/3c/d
3.7	Planes of symmetry	45–46	Reflection symmetry of 3-D shapes	SSM3b/4b
	Summary of key points	49–51		

4 Fractions and decimals

4.1	Equivalent fractions	52–53	Finding equivalent fractions	NA2c
4.2	Ordering fractions	54	Ordering fractions by writing them with a common denominator	NA2c
4.3	Ordering decimals	54–56	Ordering decimals by comparing digits with the same place value	NA2d
4.4	Mental calculations with decimals	56–57	The four rules with simple decimals	NA3i/k
4.5	Written methods for dividing decimals	57–58	Long division by a decimal	NA3i/k
4.6	Changing fractions to decimals	58–60	Converting fractions to decimals and decimals to fractions; recurring decimals	NA3c
4.7	Adding and subtracting fractions	60–63	Adding and subtracting fractions by writing them with a common denominator	NA3c
4.8	Multiplying fractions	63–65	Multiplying a fraction by an integer or a fraction	NA3d
4.9	Dividing fractions	65–66	Dividing a fraction by an integer or a fraction	NA3d
4.10	Solving problems involving fractions	67–68	Word problems involving fractions	NA3c/4a
	Summary of key points	69–70		

5 Collecting and recording data

5.1	Different types of data	71–72	Qualitative, quantitative, discrete and continuous data; primary and secondary data	HD2d/e
5.2	Sampling	73–76	Sampling techniques to minimise bias	HD2c/d
5.3	Collecting primary data	76–80	Questionnaires, observation and measurement	HD2c/3a
5.4	Secondary data	80–82	Sources of secondary data	HD3b
5.5	Recording and presenting grouped data	82–85	Tables and histograms for grouped data	HD2d/3c/4a
5.6	Frequency polygons	86–87	Drawing frequency polygons	HD4a
	Summary of key points	89–90		

6 Solving equations and inequalities

6.1	Solving simple equations	91	A reminder of the balancing method	NA5e
6.2	When the x term includes a fraction	91–92	Equations where the x coefficient is a fraction	NA5e/f
6.3	When the 'unknown' x is on both sides	92	Equations with x on both sides	NA5f

6.4	Unknown on both sides and negative coefficients	93	Equations with a negative coefficient and x on both sides	NA5f
6.5	Solving equations involving brackets	93–94	Equations with brackets	NA5f
6.6	Equations involving algebraic fractions	95	Equations with algebraic fractions	NA5f
6.7	Using equations to help solve problems	95–96	Setting up and solving equations	NA5a/e/f
6.8	Showing inequalities on a number line	97	Representing inequalities on a number line	NA5j
6.9	Solving inequalities	98–100	Solving inequalities in one variable	NA5j
6.10	Solving inequalities involving brackets	100–101	Solving inequalities	NA5j
	Summary of key points	103		

7 Transformations and loci

7.1	Transformations	104–113	Translation, reflection, rotation and enlargement	SSM3a/b/c/d/f
7.2	Combined transformations	113–115	Combining two transformations	SSM3b
7.3	Scale drawings and scale models	115–117	Drawing and interpreting scale diagrams and maps	SSM3d
7.4	Constructions	117–121	Using compasses to construct triangles, perpendiculars and bisectors	SSM2h/4c/d
7.5	Locus of a point	122–126	Constructing loci	SSM4e
7.6	Bearings	126–128	Drawing diagrams and calculating bearings	SSM4a
	Summary of key points	131–132		

8 Using basic number skills

8.1	Percentage increases	133–134	Increasing by a percentage	NA2e/*e*/3j
8.2	Percentage decreases	134–136	Decreasing by a percentage	NA2e/*e*/3j
8.3	Finding percentage increases	136–137	Finding an increase as a percentage	NA2e/*e*/3j
8.4	Finding percentage decreases	137–138	Finding a decrease as a percentage	NA2e/*e*/3j
8.5	Mixing percentage increases and decreases	138–139	Problems involving consecutive percentage changes	NA3e/j
8.6	More percentage calculations	139–140	Reverse percentages, e.g. working out a pre-sale price from a given sale price and percentage discount	NA3e/j/s/4a
8.7	Compound interest	140–143	Calculating compound interest and depreciation	NA3k/4a
8.8	Compound measures	143–147	Average speed, density and rates of flow	NA4a/SSM4a
8.9	Ratios	148–150	Sharing quantities in a given ratio	NA2f/3f/n/4a
8.10	More ratios	150–151	Further ratio problems	NA2f/3f/n/4a
8.11	Standard form	152–155	Understanding and using standard index form	NA2b/3i/m/r
8.12	Calculations in standard form	155–157	Calculations using numbers in standard form	NA2b/3i/m/r
	Summary of key points	159		

9 Functions, lines, simultaneous equations and regions

9.1	Functions	160	Explaining what a function is	NA5a/6a
9.2	Graphs of linear functions with positive whole number inputs	161–162	Graphing sets of discrete points	NA6b/d
9.3	Graphs of linear functions with continuous positive inputs	162–164	Drawing linear graphs for continuous variables	NA6b/d
9.4	Lines and parallel lines	165–167	Finding gradient and y-intercept	NA6b/c
9.5	$y = mx + c$	167–170	The role of m as gradient and c as intercept on the y-axis	NA6c
9.6	Perpendicular lines	170–172	The gradient and equation of a line perpendicular to a given line	NA6c
9.7	The mid-point of a line segment AB	172	Finding mid-points from pairs of coordinates	SSM2a/3e
9.8	Simultaneous equations – graphical solutions	172–174	Solving simultaneous equations by finding the point of intersection of their graphs	NA5i
9.9	Simultaneous equations – algebraic solutions	174–177	Solving simultaneous equations using algebraic methods	NA5i
9.10	Solving problems using simultaneous equations	177–178	Setting up and solving simultaneous equations	NA5i
9.11	Regions	178–181	Finding and sketching regions that satisfy inequalities in two variables	NA5j
	Summary of key points	183		

10 Presenting and analysing data 1

10.1	Averages for discrete data	184–185	Mean, median and mode	HD4e
10.2	Stem and leaf diagrams	185–186	Drawing and using stem and leaf diagrams	HD4a/5b
10.3	Using appropriate averages	186–188	Choosing the right average	HD4e/5a/d
10.4	The quartiles	188–190	Upper and lower quartiles	HD4e
10.5	Measures of spread	191	Range and interquartile range	HD4e
10.6	Averages from frequency distributions	192–194	Finding the mode, median and mean from a frequency table (distribution)	HD4e
10.7	Spread from frequency distributions	194–195	Finding the range and interquartile range from a frequency table (distribution)	HD4e
10.8	Averages from grouped data	195–198	Estimating averages from grouped data	HD4e
10.9	Drawing conclusions from data	199–201	Interpreting averages and spread	HD5a/d
10.10	Moving averages	201–203	Time series, moving averages and trends	HD4a/f/5b/k
10.11	Scatter diagrams	203–205	Drawing scatter graphs	HD2e/4a
10.12	Correlation and lines of best fit	206–209	Using scatter graphs to identify relationships	HD4a/i/5b/c/f
	Summary of key points	212–213		

11 Estimation and approximation

11.1	Rounding to a number of decimal places	214–215	Rounding to a given number of decimal places	NA2a
11.2	Rounding to a number of significant figures	216–217	Rounding to a given number of significant figures	NA3h
11.3	Checking and estimating	218–219	Using estimation to check calculations	NA4b
11.4	Describing the accuracy of a measurement	219–220	Understanding the true value can lie in a range from $-\frac{1}{2}$ a unit to $+\frac{1}{2}$ a unit	SSM4a/NA4b
11.5	Upper and lower bounds	220–223	Calculating upper and lower bounds of numbers expressed to a given degree of accuracy	SSM4a/NA3q/4b

Summary of key points   224

12 Sequences and formulae

12.1	Finding terms of a sequence	225–226	Using a formula to generate terms of a sequence	NA6a
12.2	Finding the formula for a sequence	227–228	Finding formulae for arithmetic sequences	NA6a
12.3	More about formulae	228–230	Evaluating by substitution	NA5g
12.4	Manipulating formulae	230–232	Finding the value of a variable that is not the subject of the formula	NA5g
12.5	Changing the subject of a formula	232–234	Rearranging a formula to change the subject	NA5g
12.6	Substituting one formula in another	235–236	Combining formulae	NA5g

Summary of key points   237

13 Measure and mensuration

13.1	Continuous and discrete data	238	Understanding that accuracy depends on the accuracy of the measuring instrument	SSM4a
13.2	The measurement of continuous data	238–239	Measurements to the nearest unit may be inaccurate by up to 0.5 in either direction	SSM4a
13.3	Perimeter and area of triangles and quadrilaterals	239–243	Calculating perimeters and areas, using formulae as appropriate	SSM2e/4f
13.4	Circumference and area of circles	243–244	Using formulae; includes finding the radius	SSM4d
13.5	Volume and surface area of 3–D shapes	244–248	Volume and surface area of cuboids and prisms	SSM2i/4d
13.6	Coordinates in 3-D	248–250	Using coordinates in three dimensions	SSM3e
13.7	Dimension theory	250–252	Identifying formulae for length, area and volume	SSM3d

Summary of key points   254

14 Simplifying algebraic expressions

14.1	Index notation	255	A reminder of index notation	NA5d
14.2	Multiplying expressions involving indices	255–256	Multiplication rule for indices	NA5d
14.3	Dividing expressions involving indices	257	Division rule for indices	NA5d
14.4	Zero, negative and fractional indices	257–259	Extending understanding of indices	NA3g/5d
14.5	Combining indices	259–260	Powers of indices, $(x^m)^n$	NA3g/5d
14.6	Equations involving indices	260	Solving an equation where an 'unknown' is an index	NA5d
14.7	Algebraic fractions	261–262	Simplifying, multiplying and dividing algebraic fractions	NA5b
14.8	Adding and subtracting algebraic fractions	262–265	Adding and subtracting algebraic fractions	NA5b
14.9	Factorising	265–266	Factorising by taking out the HCF	NA5b
14.10	Factorising by pairing	266–267	Further factorising	NA5b
14.11	Factorising quadratic expressions	267–270	Factorising quadratics	NA5b
14.12	The difference of two squares	271–272	Factorising expressions of the form $a^2 - b^2$	NA5b
14.13	Further fractional algebraic expressions	272–273	Simplifying by factorising and cancelling common factors	NA5b
	Summary of key points	275		

15 Pythagoras' theorem

15.1	Pythagoras' theorem	276–277	Introducing the theorem	SSM2f
15.2	Finding lengths using Pythagoras' theorem	277–279	Using $a^2 + b^2 = c^2$	SSM2f
15.3	Using Pythagoras' theorem to find one of the shorter sides of a triangle	279–281	Rearranging the formula	SSM2f
15.4	Pythagoras' theorem applied twice	281	Finding two lengths – plane figures only	SSM2f
15.5	The distance between two points	282–283	Finding the distance between coordinates	SSM2f/3e
15.6	Non-right-angled triangles	283–284	Using Pythagoras' theorem to identify obtuse- and acute-angled triangles	SSM2f
15.7	Pythagorean triples	284–285	Integer sets satisfying $a^2 + b^2 = c^2$	SSM2f
15.8	Pythagoras' theorem in three dimensions	285–286	Using Pythagoras' theorem to find lengths in 3-D shapes	SSM2f
15.9	The equation of a circle	286–287	Finding the equation of a circle with centre the origin	NA6h
	Summary of key points	289–290		

16 Basic trigonometry

16.1	Right-angled triangles and the trigonometric ratios	291–294	Opposite, adjacent, hypotenuse and the sin, cos and tan formulae	SSM2g

16.2	Using a calculator	294–295	Obtaining sin, cos and tan, and their inverses	NA3o
16.3	Using trigonometric ratios to find angles	295–296	With any two sides given	SSM2g
16.4	Using trigonometric ratios to find the lengths of sides	296–298	With one angle and any side	SSM2g
16.5	More about trigonometric functions	298–299	For angles greater than 90°	SSM2g
16.6	The graph of sin x	300	For 0°–360° and then extended	SSM2g
16.7	The graph of cos x	301	For 0°–360° and then extended	SSM2g
16.8	The graph of tan x	302–304	For 0°–90° and 90°–270°, and then extended	SSM2g
16.9	Solving a trigonometric equation	304–306	Finding one solution algebraically and others from the symmetries of the graph	SSM2g
	Summary of key points	308–309		

17 Graphs and equations

17.1	Graphs of quadratic functions	310–311	Drawing graphs of quadratic functions	NA6e
17.2	Graphs of cubic functions	311–312	Drawing graphs of cubic functions	NA6f
17.3	Graphs of reciprocal functions	312–314	Drawing graphs of reciprocal functions	NA6f
17.4	Graphs of the form $y = ax^3 + bx^2 + cx + d + \dfrac{e}{x}$	314–315	Drawing graphs of cubic functions with up to three terms	NA6f
17.5	Using graphs to solve equations	315–316	Finding solutions to a quadratic equation from its graph	NA6e/f
17.6	Trial and improvement methods for solving equations	317–318	Using trial and improvement to solve equations	NA5m
17.7	Solving problems by trial and improvement	318–319	Using trial and improvement to solve problems	NA5m
17.8	Graphs that describe real-life situations	320–325	Drawing and interpreting graphs, both accurate and sketched	NA6d
	Summary of key points	327		

18 Proportion

18.1	Direct proportion	328	Direct proportion and the ∝ symbol	NA5h
18.2	Graphs that show direct proportion	329–331	Straight line graphs showing direct proportion	NA5h
18.3	Using ratios to find proportionality rules	331–333	Finding the rule connecting quantities, using ratio	NA5h
18.4	Writing proportionality formulae	333–334	Connecting variables in direct proportion with a formula	NA3l/5h
18.5	Square and cubic proportionality	335–338	Extending to square and cubic relationships	NA3l/5h
18.6	Inverse proportion	339–342	Solving problems involving inverse proportion	NA3l/4a/5h
	Summary of key points	344–345		

19 Quadratic equations

19.1	Solving quadratic equations by factorising	346–348	Rearranging and then factorising	NA5k
19.2	Completing the square	349	Completing the square for a quadratic expression	NA5k
19.3	Solving quadratic equations by completing the square	350	Using the method of completing the square	NA5k
19.4	Solving quadratic equations by using a formula	351–353	Using the quadratic formula to solve quadratic equations	NA5k
19.5	Equations involving algebraic fractions	354–355	Rearranging into quadratic equations and then solving	NA5b/k
19.6	Problems leading to quadratic equations	355–358	Setting up and solving quadratic equations	NA5k
19.7	Solving linear and quadratic equations simultaneously	358–360	Using algebraic methods and geometric interpretation	NA5l
19.8	The intersection of a line and a circle	361–362	Using algebraic methods and geometric interpretation; includes tangents to a circle	NA5l/6e/h
19.9	Solving equations graphically	363–366	Rearranging equations and solving graphically	NA6e/f
Summary of key points		368		

20 Presenting and analysing data 2

20.1	Upper class boundaries	369–370	Upper boundaries for class intervals	SSM4a/HD2d
20.2	Cumulative frequency	370–372	Calculating cumulative frequencies	HD4a
20.3	Cumulative frequency graphs	372–373	Drawing cumulative frequency graphs	HD4a
20.4	Using cumulative frequency graphs	374–377	Estimating the median and quartiles	HD4a/e
20.5	Using cumulative frequency graphs to solve problems	377–379	Estimating percentages from cumulative frequency graphs	HD4e
20.6	Box plots	379–381	Drawing box plots	HD4a/e
20.7	Comparing sets of data with box plots	382–383	Comparing distributions	HD4a/4e/5d
20.8	Plotting data in a histogram	383–388	Understanding frequency density	HD4a/5d
20.9	Interpreting histograms	389	Interpreting histograms	HD5b/d
Summary of key points		398		

21 Advanced trigonometry

21.1	The area of a triangle	399–401	Calculating area of a triangle using $A = \frac{1}{2}ab \sin C$	SSM2g
21.2	The sine rule	402–405	The sine rule	SSM2g
21.3	The cosine rule	405–407	The cosine rule	SSM2g
21.4	Trigonometry and Pythagoras' theorem in three dimensions	407–409	Finding lengths and angles in 3-D problems	SSM2g
Summary of key points		411		

22 Advanced mensuration

22.1	Finding the length of an arc of a circle	412–414	Calculating arcs given the angle and vice versa	SSM4d
22.2	Finding the area of a sector of a circle	414–415	Calculating areas of sectors	SSM4d
22.3	Finding the area of a segment of a circle	416–418	Calculating areas of segments	SSM2i
22.4	Finding volumes and surface areas	418–421	Volume and surface area of prisms, including cylinders	SSM2i
22.5	The volume of a pyramid	421–423	Volume of pyramids, including cones	SSM2i
22.6	The surface area and volume of a sphere	423–424	Surface area and volume of spheres	SSM2i
22.7	Areas and volumes of similar shapes	424–427	Using area and volume scale factors	SSM3d
22.8	Compound solids	427–429	Volume of combined shapes; includes truncated solids	SSM2i
	Summary of key points	431		

23 Exploring numbers 2

23.1	Terminating and recurring decimals	432–434	Converting fractions into decimals that terminate or recur	NA2d/3c
23.2	Finding a fraction equivalent to a recurring decimal	434–435	Converting recurring decimals to fractions	NA2d/3c
23.3	Surds	435–438	Manipulating expressions with surds	NA3n
23.4	Rounding	438	Rounding measurements	NA2a
23.5	Upper bounds and lower bounds	439–440	Finding bounds of numbers expressed to a given degree of accuracy	NA3q
23.6	Calculations involving upper and lower bounds – addition and multiplication	440–441	Calculating the bounds in problems involving addition or multiplication	NA3q
23.7	Calculations involving upper and lower bounds – subtraction and division	441–443	Calculating the bounds in problems involving subtraction or division	NA3q
	Summary of key points	444–445		

24 Probability

24.1	Finding probabilities	446–448	Calculating theoretical probabilities	HD4b
24.2	Mutually exclusive events	448–449	Calculating probabilities of mutually exclusive events	HD4d/g
24.3	Lists and tables	449–452	Listing outcomes systematically	HD4c
24.4	Relative frequency	452–453	Estimating probability using relative frequency	HD4b/5h/i
24.5	Estimating from experience	453	Estimating and justifying probabilities	HD4b
24.6	Independent events	453–454	Calculating probabilities of independent events	HD4g
24.7	Probability trees	455–456	Using probability trees to calculate probabilities	HD4g/h
	Summary of key points	459		

25 Transformations of graphs

25.1	Function notation	460–462	Using f(x) notation	NA6a/e
25.2	Applying vertical translations to graphs	463–466	Using transformations of the form $y = f(x) + a$	NA6g
25.3	Applying horizontal translations to graphs	467–470	Using transformations of the form $y = f(x + a)$	NA6g
25.4	Applying double translations to graphs	470–472	Combining horizontal and vertical translations	NA6g
25.5	Applying reflections to graphs	472–474	Reflections in the x and y axes	NA6g
25.6	Applying combined transformations	474–475	Combining reflections and translations	NA6g
25.7	Applying stretches to graphs	475–479	Stretches parallel to the axes, $y = af(x)$ and $y = f(ax)$	NA6g
25.8	Applying transformations to trigonometric functions	479–480	Transformations applied to the sine and cosine functions	NA6g
	Summary of key points	483–484		

26 Circle theorems

26.1	Circle theorems	485–487	Tangents from a point to a circle; angle between radius and tangent; perpendicular from centre to chord	SSM2h
26.2	More circle theorems	487–492	All the recognised circle theorems	SSM2h
26.3	Proving circle and geometrical theorems	492–496	Proofs of key circle theorems	SSM2h
	Summary of key points	500		

27 Vectors

27.1	Translations	501–503	Using column vectors to describe translations	SSM3f
27.2	Vectors	503–506	Vector addition and subtraction by calculation and graphically	SSM3f
27.3	Vector algebra	506–508	Applying rules of algebra to vectors	SSM3f
27.4	Finding the magnitude of a vector	508–509	Using Pythagoras' theorem	SSM3f
27.5	Linear combinations	509–511	Combinations of vectors of the form $p\mathbf{a} + q\mathbf{b}$	SSM3f
27.6	Position vectors	511–513	Vectors related to the origin	SSM3f
27.7	Proving geometrical results	514–515	Using properties of parallel vectors	SSM3f
	Summary of key points	518		

28 Introducing modelling

28.1	Modelling using exponential functions	519–521	Using the exponential function	NA3t/6f
28.2	Modelling using trigonometric functions	522–525	Using trigonometric functions	NA6f/SSM2g

28.3 Using a line of best fit   525–527   Using lines of best fit   NA6c/HD1g/4i
      to obtain a relationship
28.4 Reducing equations to   527–531   Rewriting relationships in linear form   NA6c/e
      linear form   and reading values from the graph
28.5 Finding the constants   531–533   Matching a graph or coordinates   NA6f
      in an exponential   to $y = pq^x$
      relationship

      Summary of key points   533

29 Conditional probability

29.1 Dependent and   534–536   Extending ideas of probability   HD4g
      independent events
29.2 Paths through tree   536–538   Calculating conditional probabilities   HD4h
      diagrams
29.3 Probability and human   538–540   Conditional probabilities   HD4h/5g
      behaviour

      Summary of key points   541

① Exploring numbers 1

You will need to use and apply numbers in a wide variety of situations throughout your GCSE course and in everyday life.

This chapter deals with some of the terms that describe numbers and their properties.

You should remember these key points from earlier work:

A **factor** of a number is a whole number that divides exactly into the number. The factors include 1 and the number itself.

A **multiple** of a number is the result of multiplying the number by a positive whole number.

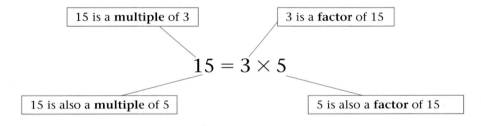

| 15 is a **multiple** of 3 | | 3 is a **factor** of 15 |

$$15 = 3 \times 5$$

| 15 is also a **multiple** of 5 | | 5 is also a **factor** of 15 |

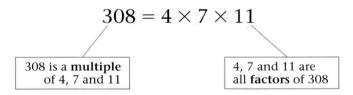

$$308 = 4 \times 7 \times 11$$

| 308 is a **multiple** of 4, 7 and 11 | | 4, 7 and 11 are all **factors** of 308 |

Codes based on prime numbers are used to provide the highest level of security for banks and computer systems.

A **prime number** is a number greater than 1 which has only two factors: itself and 1. For example, 7 is a prime number because the only factors of 7 are 7 and 1.

1.1 Number facts

Prime factors

Any factor of a number that is a prime number is a **prime factor**. For example, 2 and 3 are the prime factors of 6.

You can write any whole number as the product of its prime factors.

Example 1

Write 720 as the product of its prime factors.

Method 1   Split 720 into factors in stages, each time finding the smallest possible prime number that is a factor:

$$720 = 2 \times 360$$
$$= 2 \times 2 \times 180$$
$$= 2 \times 2 \times 2 \times 90$$
$$= 2 \times 2 \times 2 \times 2 \times 45$$
$$= 2 \times 2 \times 2 \times 2 \times 3 \times 15$$
$$= 2 \times 2 \times 2 \times 2 \times 3 \times 3 \times 5$$

This can be written as: $720 = 2^4 \times 3^2 \times 5$

> Use powers to write $2 \times 2 \times 2 \times 2$ in the shorter form 2^4.

Here 720 is in **prime factor form**   2, 3 and 5 are **prime factors** of 720

Method 2   Use a factor tree to help find the prime factors:

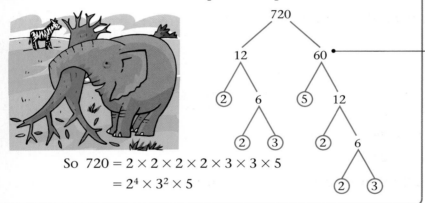

So $720 = 2 \times 2 \times 2 \times 2 \times 3 \times 3 \times 5$
$$= 2^4 \times 3^2 \times 5$$

> 12 and 60 is not the only way to split 720. You could use 72 and 10 or 80 and 9, for example. However you start, the factor tree will always end with the same prime factors.

> The circled numbers at the ends of the branches are the prime factors.

Writing a number as the product of its prime factors is called writing it in **prime factor form**.

Finding the highest common factor (HCF)

Sometimes you will need to find the largest factor that two numbers have in common; this is called the **highest common factor** or **HCF**.

For example, to find the highest common factor of 720 and 84:

First write each number in prime factor form:

$720 =$ 2 $\times$ 2 $\times$ 2 $\times$ 2 $\times$ 3 $\times$ 3 $\times$ 5

$84\ \ =$ 2 $\times$ 2 $\times$ 3 $\times$ 7

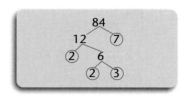

Then pick out the common factors: those that appear in *both* numbers. These are:

$2 \times 2 \times 3$

The highest common factor of 720 and 84 is $2 \times 2 \times 3 = 12$.

> The **highest common factor** (HCF) of two whole numbers is the highest factor that is common to both of them.

what about the HCM ?

Finding the lowest common multiple (LCM)

Sometimes you will need to find the lowest multiple that two numbers have in common; this is called the **lowest common multiple** or **LCM**.

The multiples of 6 are: 6   12   18   24   30   36   42   48   54   60   66   72 ...

The multiples of 8 are: 8   16   24   32   40   48   56   64   72   80   88   96 ...

is 72 the HCM ? make it clear for me

Lowest common multiple

6 and 8 have some common multiples.

The lowest common multiple (LCM) of 6 and 8 is 24.

> The **lowest common multiple (LCM)** of two whole numbers is the lowest number that is a multiple of them both.

You can find the LCM by looking at the prime factor forms of the two numbers.

___

Example 2

Find the LCM of 6 and 8.

First write the numbers in prime factor form:

$$8 = 2 \times 2 \times 2 \quad \text{and} \quad 6 = 2 \times 3$$

The LCM must contain all the prime factors of 8 *and* all the prime factors of 6:

$$\begin{array}{c} 8 \\ \overbrace{2 \times 2 \times 2} \times 3 \\ \underbrace{}_{6} \end{array}$$

8 and 6 have a common prime factor of 2 which is only counted once.

The LCM of 6 and 8 is $2 \times 2 \times 2 \times 3 = 24$.

___

Exercise 1A

1   Find all the factors of:

   (a)   24   (b)   36   (c)   308

   (d)   1001   (e)   1400   (f)   53

2   Find all the prime numbers between 1 and 100.

3   Can the sum of two prime numbers be a prime number? Explain your answer.

> Try adding some pairs of prime numbers.

4 Write each number in prime factor form.

(a) 24 (b) 32 (c) 18 (d) 13

(e) 72 (f) 50 (g) 99 (h) 120

5 Find the highest common factor (HCF) of

(a) 4 and 8 (b) 9 and 12 (c) 18 and 24

(d) 36 and 48 (e) 720 and 252 (f) 19 600 and 756

6 Find the lowest common multiple (LCM) of

(a) 3 and 4 (b) 4 and 6 (c) 7 and 14

(d) 6 and 15 (e) 36 and 16 (f) 85 and 50

7 A ship is at anchor between two lighthouses L and H.
The light from L shines on the ship every 30 seconds.
The light from H shines on the ship every 40 seconds.
Both lights started at the same moment.
How often do both lights shine on
the ship at once?

> Find the LCM of
> the light times.

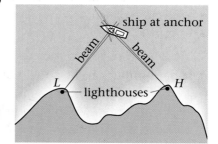

ship at anchor

beam   beam

L H

lighthouses

8 **The Sigma Function: an investigation**

In this question, $f(n)$ means the sum of the factors of n.

So $f(36) = 1 + 2 + 3 + 4 + 6 + 9 + 12 + 18 + 36 = 91$

(a) Find $f(n)$ when n is

(i) 24 (ii) 48 (iii) 100 (iv) 256 (v) 32

(b) Write down an expression for $f(p)$, where p is any prime number.

(c) Show that $f(2^n) = 1 + 2 + 2^2 + 2^3 + \cdots + 2^n$

(d) Find an expression for $f(p^n)$, where p is any prime number.

1.2 Square numbers and cube numbers

These numbers sometimes occur in number patterns in investigations.

> A **square number** is the result of multiplying a whole number by itself.

Square numbers:

$1 \times 1 = 1$ ———— 1st square number

$2 \times 2 = 4$ ———— 2nd square number

$3 \times 3 = 9$ ———— 3rd square number

$4 \times 4 = 16$ ———— 4th square number

$x \times x = x^2$ ———— the square of any number x

4×4, for example, can also be written as

- the square of 4
- 4 squared
- 4^2

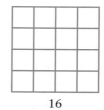

1   4   9   16

A square number can be shown as a square pattern.

A **cube number** is the result of multiplying a whole number by itself then multiplying by the number again.

Cube numbers:

$1 \times 1 \times 1 = 1$ ———— 1st cube number
$2 \times 2 \times 2 = 8$ ———— 2nd cube number
$3 \times 3 \times 3 = 27$ ———— 3rd cube number
$4 \times 4 \times 4 = 64$ ———— 4th cube number
$x \times x \times x = x^3$ ———— the cube of any number x

$4 \times 4 \times 4$, for example, can also be written as
- the cube of 4
- 4 cubed
- 4^3

A cube number can be shown as a cube pattern.

 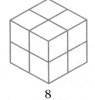 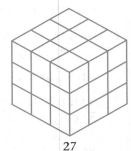

1   8   27

Example 3

1, 3, 4, 6, 8, 13, 16, 18, 24, 27, 30

From the list write down

(a)   the square numbers   (b)   the cube numbers.

(a)   1, 4, 16   (b)   1, 8, 27

Exercise 1B

1   Find
   (a)   the 8th square number   (b)   the 7th cube number
   (c)   the 12th square number   (d)   the 10th cube number
   (e)   the first 12 square numbers (f)   the first 8 cube numbers.

2   From each list write down all the numbers which are
   (i)   square numbers   **(ii)** cube numbers.
   (a)   50, 20, 64, 30, 1, 80, 8, 49, 9
   (b)   10, 21, 57, 4, 60, 125, 7, 27, 48, 16, 90, 35
   (c)   137, 150, 75, 110, 50, 125, 64, 81, 144
   (d)   90, 180, 216, 100, 81, 75, 140, 169, 125

1.3 Finding squares and square roots, cubes and cube roots

Squares

To find the **square** of any number, multiply the number by itself.
The square of $3.7 = 3.7^2 = 3.7 \times 3.7 = 13.69$
The square of $-6.2 = -6.2^2 = -6.2 \times -6.2 = 38.44$

Square roots

$4 \times 4 = 16$, so we say that 4 is a **square root** of 16. It is a number which multiplied by itself gives 16.

You can write the square root of 16 as $\sqrt{16}$

$$1.5 \times 1.5 = 2.25$$

So 1.5 is a square root of 2.25, written $\sqrt{2.25}$

In general, for any number A:

> If $x \times x = A$, then x is the **square root** of A, written $\sqrt{A}$

> Notice that
> $-4 \times -4 = 16$
> so -4 is also a square root of 16.
> $\sqrt{16} = \pm 4$

> Remember:
> $\sqrt{2.25}$ can also be -1.5 if that is a sensible answer to the problem you are solving.

You can find the square root of any positive number, though the answer may not be a whole number. Most calculators have a function key that finds the square root of a number. You often need to round the answer on the calculator display. For example

$$\sqrt{18} = 4.242\ 640\ 6\ldots = 4.24 \text{ (to 2 d.p.)}$$

Cubes

> To find the **cube** of any number, multiply the number by itself then multiply by the number again.
> The cube of $5.3 = 5.3^3 = 5.3 \times 5.3 \times 5.3 = 148.877$
> The cube of $-2.1 = (-2.1)^3 = -2.1 \times -2.1 \times -2.1 = -9.261$

Cube roots

$2 \times 2 \times 2 = 8$, so we say that 2 is the **cube root** of 8. It is a number which multiplied by itself, then multiplied by itself again, gives 8.

You can write the cube root of 8 as $\sqrt[3]{8}$.

$$3.4 \times 3.4 \times 3.4 = 39.304$$

So 3.4 is the cube root of 39.304, written $\sqrt[3]{39.304}$

In general, for any number A:

> If $y \times y \times y = A$, then y is the **cube root** of A, written $\sqrt[3]{A}$

You can find the cube root of any positive or negative number, though the answer may not be a whole number. Some calculators have a cube root function key to find the cube root of numbers. As with square roots, you often have to round the answer. For example

$$\sqrt[3]{18} = 2.620\ 741\ 3\ldots = 2.62 \text{ (to 2 d.p.)}$$

> Notice that
> $\sqrt[3]{-64} = -4$ because
> $-4 \times -4 \times -4 = -64$

Exercise 1C

Use your calculator to work out

Sometimes you are asked to leave an answer as a square root or cube root.
e.g. $x^2 = 5$
$x = \sqrt{5}$
This is called writing your answer in **surd form**.

1 (a) 13^2 (b) 3.5^2 (c) 40^2

    (d) 8.7^2 (e) 19.6^2 (f) $(-57.4)^2$

2 (a) 6^3 (b) 2.4^3 (c) 20^3

    (d) $(-1.3)^3$ (e) 13.4^3 (f) 36.2^3

3 (a) $\sqrt{121}$ (b) $\sqrt{225}$ (c) $\sqrt{16\,900}$

    (d) $\sqrt{2.89}$ (e) $\sqrt{0.49}$ (f) $\sqrt{33.64}$

4 In this question give your answers correct to 2 d.p.
    (a) $\sqrt{253}$ (b) $\sqrt{2341}$ (c) $\sqrt{18.4}$

    (d) $\sqrt{8476}$ (e) $\sqrt{29.44}$ (f) $\sqrt{1\,825\,963}$

5 In this question give your answers correct to 3 s.f.
    (a) $\sqrt[3]{68}$ (b) $\sqrt[3]{3654}$ (c) $\sqrt[3]{26.5}$

    (d) $\sqrt[3]{-9.2}$ (e) $\sqrt[3]{882.5}$ (f) $\sqrt[3]{6\,547\,812}$

1.4 Index numbers

In Section 1.2 you saw that

$3 \times 3 = 3^2 = 9$ is a **square number**

and

$4 \times 4 \times 4 = 4^3 = 64$ is a **cube number**

because they can be shown as square and cube patterns.

You cannot draw patterns which go beyond three dimensions, but you can extend the notation for multiplying numbers like this:

$3 \times 3 \times 3 \times 3 = 3^4 = 81$

You say that 81 is **3 raised to the power 4**, or 81 is **3 to the fourth**.

Also:

$2 \times 2 \times 2 \times 2 \times 2 = 2^5 = 32$

You say that 32 is **2 raised to the power 5**, or 32 is **2 to the fifth**, or 32 is **2 to the power 5**.

The ^{4} and the ^{5} are examples of **index numbers**.

32 written in **index form** is 2^5.

The power a number is raised to is called the **index** (plural **indices**).

General form for index numbers

You can use algebra to write about index numbers in general. For any number a, three lots of a multiplied together can be written:

$$a \times a \times a = a^3$$

and n lots of a multiplied together can be written:

$$a \times a \times a \times \cdots \times a = a^n$$
$$\longleftarrow n \text{ lots of } a \longrightarrow$$

A number written in the form a^n is written in **index form**.

Example 4

(a) Calculate 4^7. (b) Write 243 in index form.

(a) $4^7 = 4 \times 4 \times 4 \times 4 \times 4 \times 4 \times 4$
$= 16\,384$

(b) $243 = 3 \times 81$
$= 3 \times 9 \times 9$
$= 3 \times 3 \times 3 \times 3 \times 3$
$= 3^5$

Exercise 1D

1 Find the values of these powers.
(a) 2^6 (b) 10^5 (c) 5^4 (d) 7^3 (e) 12^4 (f) $(0.9)^5$

2 Write these numbers in index form.
(a) 1000 (b) 125 (c) 512
(d) 2401 (e) 625 (f) 19 683

3 Show that 64 can be written as either 2^6 or 4^3.
Explain why this is the case.

4 Which is the greater, and by how much, 2^5 or 5^2?

1.5 The index laws

There are rules to help you calculate with numbers written in index form.

Multiplication

$$3^2 \times 3^3 = (3 \times 3) \times (3 \times 3 \times 3) = 3 \times 3 \times 3 \times 3 \times 3$$
$$= 3^5$$

So $\qquad\qquad 3^2 \times 3^3 = 3^5 = 3^{2+3}$

To multiply two powers of the same number add the indices.

$$a^n \times a^m = a^{n+m}$$

Division

$$2^5 \div 2^3 = \frac{2 \times 2 \times \cancel{2} \times \cancel{2} \times \cancel{2}}{\cancel{2} \times \cancel{2} \times \cancel{2}}$$

$$= 2 \times 2$$
$$= 2^2$$

So $2^5 \div 2^3 = 2^2 = 2^{5-3}$

To divide two powers of the same number subtract the indices.

$$a^n \div a^m = a^{n-m} \quad \text{or} \quad \frac{a^n}{a^m} = a^{n-m}$$

Raising to the power 0

Notice that

$$1 = \frac{8}{8} = \frac{2^3}{2^3} = 2^{3-3} = 2^0$$

The 2 could be replaced by any other non-zero number a. The result would be the same.

$$a^0 = 1 \text{ (where } a \neq 0)$$

Negative powers

Sometimes numbers are raised to a negative power.
For example

$$2^2 \div 2^5 = 2^{2-5} = 2^{-3}$$

and $2^2 \div 2^5 = \dfrac{\cancel{2} \times \cancel{2}}{\cancel{2} \times \cancel{2} \times 2 \times 2 \times 2} = \dfrac{1}{2 \times 2 \times 2} = \dfrac{1}{2^3}$

So $2^{-3} = \dfrac{1}{2^3} = \dfrac{1}{2 \times 2 \times 2} = \dfrac{1}{8}$

$$a^{-n} = \frac{1}{a^n} \quad \text{(where } a \neq 0)$$

Powers raised to further powers

$$(2^3)^2 = (2 \times 2 \times 2)^2$$
$$= (2 \times 2 \times 2) \times (2 \times 2 \times 2)$$
$$= 2 \times 2 \times 2 \times 2 \times 2 \times 2 = 2^6$$

So $(2^3)^2 = 2^{3 \times 2} = 2^6$

To raise a power to a further power multiply the indices together.

$$(a^n)^m = a^{n \times m}$$

Example 5

Calculate each of the following using the index laws. Give your answers both in index form and without using indices (where possible).

(a) $2^5 \times 2^3$ (b) $3^7 \div 3^5$ (c) $(5^3)^2$ (d) 2^{-3} (e) 117^0 (f) $2^2 + 2^3$

(a) $2^5 \times 2^3 = 2^{5+3} = 2^8 = 256$

(b) $3^7 \div 3^5 = 3^{7-5} = 3^2 = 9$

(c) $(5^3)^2 = 5^{3 \times 2} = 5^6 = 15\,625$

(d) $2^{-3} = \dfrac{1}{2^3} = \dfrac{1}{2 \times 2 \times 2} = \dfrac{1}{8} = 0.125$

(e) $117^0 = 1$

(f) $2^2 + 2^3 = 4 + 8 = 12$

Exercise 1E

1 Work out each of the following. Give your answer both in index form and without using indices (where possible).

(a) $5^2 \times 5^4$ (b) $5^{10} \div 5^2$ (c) $10^3 \div 10$

(d) $12^2 \times 12$ (e) $3^3 \times 3^4$ (f) 6^0

(g) $(1.3)^0$ (h) $\dfrac{1}{25}$ (i) $6^{10} \div 6^9$

(j) $5^3 + 5^2$ (k) $2^1 + 2^2 + 2^3$ (l) $(4^3)^2$

(m) $(5^3)^3$ (n) $(2^5)^2$ (o) $\dfrac{3^{-2}}{3^2}$

(p) $\dfrac{4^3}{4^{-2}}$ (q) $\dfrac{6}{6^2}$ (r) $5^2 \div 5^{-3}$

(s) $\dfrac{2^3 \times 2^5}{(2^2)^3}$ (t) $\dfrac{(4^2)^3}{4^2 \times 4^3}$ (u) $(5^{-2})^3$

(v) $(3^2)^{-2}$ (w) $\dfrac{2^3 \times 2^{-3}}{(2^2)^2}$ (x) $\dfrac{(3^{-2})^3}{3^{-2} \times 3^{-6}}$

2 Investigate to find possible solutions of $n^m = m^n$.

1.6 Fractional indices

You should know that $\quad \sqrt{9} = 3$

So: $\quad \sqrt{9} \times \sqrt{9} = 3 \times 3 = 9 = 9^1$

We can use this to evaluate $9^{\frac{1}{2}}$.

Using the index law for multiplication:

$$9^{\frac{1}{2}} \times 9^{\frac{1}{2}} = 9^{\frac{1}{2} + \frac{1}{2}} = 9^1$$

But $\qquad \sqrt{9} \times \sqrt{9} = 9^1 \qquad$ so $\quad 9^{\frac{1}{2}} = \sqrt{9} \qquad$ or $\quad 9^{\frac{1}{2}} = 3$

Similarly, $\qquad 27^{\frac{1}{3}} \times 27^{\frac{1}{3}} \times 27^{\frac{1}{3}} = 27^{\frac{1}{3} + \frac{1}{3} + \frac{1}{3}} = 27^1$

But $\qquad \sqrt[3]{27} \times \sqrt[3]{27} \times \sqrt[3]{27} = 27 \quad$ so $\quad 27^{\frac{1}{3}} = \sqrt[3]{27} \quad$ or $\quad 27^{\frac{1}{3}} = 3$

$$a^{\frac{1}{n}} = \sqrt[n]{a}$$

Cube roots

The cube root of 27 is 3 because

$$3 \times 3 \times 3 = 27$$

The cube root is written

$$\sqrt[3]{27}$$

Example 6

Work out (a) $8^{\frac{1}{3}}$ (b) $8^{\frac{2}{3}}$ (c) $8^{-\frac{1}{3}}$

(a) $8^{\frac{1}{3}} = \sqrt[3]{8} = 2$ because $8 = 2 \times 2 \times 2$

(b) $8^{\frac{2}{3}} = 8^{\frac{1}{3}+\frac{1}{3}} = 8^{\frac{1}{3}} \times 8^{\frac{1}{3}} = 2 \times 2 = 4$

(c) $8^{-\frac{1}{3}} = \dfrac{1}{8^{\frac{1}{3}}} = \dfrac{1}{2}$

Example 7

Work out (a) $64^{\frac{1}{3}}$ (b) $64^{-\frac{2}{3}}$

(a) $64^{\frac{1}{3}} = \sqrt[3]{64} = 4$ because $4 \times 4 \times 4 = 64$

(b) $64^{-\frac{2}{3}} = 64^{-\frac{1}{3}} \times 64^{-\frac{1}{3}} = \dfrac{1}{64^{\frac{1}{3}}} \times \dfrac{1}{64^{\frac{1}{3}}}$

$= \dfrac{1}{4} \times \dfrac{1}{4} = \dfrac{1}{16}$

So $64^{-\frac{2}{3}} = \dfrac{1}{64^{\frac{2}{3}}} = \dfrac{1}{16}$

Exercise 1F

1   Work out

    (a) $4^{\frac{1}{2}}$ (b) $4^{-\frac{1}{2}}$ (c) $4^{\frac{3}{2}}$

    (d) $125^{-\frac{1}{3}}$ (e) $16^{\frac{1}{4}}$ (f) $16^{-\frac{1}{4}}$

    (g) $16^{-\frac{3}{4}}$ (h) $25^{-\frac{1}{2}}$ (i) $100^{-\frac{1}{2}}$

    (j) $1000^{\frac{1}{3}}$ (k) $25^{\frac{3}{2}}$ (l) $8^{-\frac{2}{3}}$

2   Work out the following. Write your answers both in index form
    and without using indices (where possible).

    (a) $2^3 \times 2^7$ (b) $3^4 \div 3^2$ (c) 12^0

    (d) $\dfrac{(5^2)^3}{5^6}$ (e) $3^0 + 3 + 3^2 + 3^3$ (f) $(4^2)^{-3}$

    (g) $\dfrac{10^3}{10^5}$ (h) $(5^{-2})^{-2}$

3   Work out

    (a) $27^{\frac{1}{3}}$ (b) $25^{\frac{1}{2}}$ (c) $49^{-\frac{1}{2}}$

    (d) $(8^{-\frac{1}{3}})^{-2}$ (e) $16^{\frac{1}{4}} \times 32^{\frac{2}{5}}$ (f) $1000^{-\frac{2}{3}}$

1.7 Powers of 2 and 10

You should be able to recall very quickly the powers of 2 and the powers of 10.

Remember that $2^3 = 2 \times 2 \times 2 = 8$ and $2^{-3} = \dfrac{1}{2^3} = \dfrac{1}{8}$

So the powers from 2^{-5} to 2^5 are:

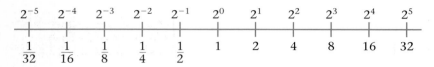

2^{-5}	2^{-4}	2^{-3}	2^{-2}	2^{-1}	2^0	2^1	2^2	2^3	2^4	2^5
$\dfrac{1}{32}$	$\dfrac{1}{16}$	$\dfrac{1}{8}$	$\dfrac{1}{4}$	$\dfrac{1}{2}$	1	2	4	8	16	32

Similarly,

$$10^3 = 10 \times 10 \times 10 = 1000 \quad \text{and} \quad 10^{-3} = \frac{1}{10^3} = \frac{1}{1000} = 0.001$$

So the powers from 10^{-5} to 10^5 are

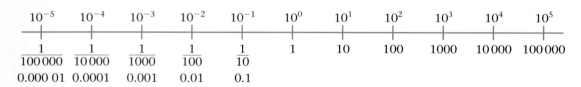

10^{-5}	10^{-4}	10^{-3}	10^{-2}	10^{-1}	10^0	10^1	10^2	10^3	10^4	10^5
$\dfrac{1}{100\,000}$	$\dfrac{1}{10\,000}$	$\dfrac{1}{1000}$	$\dfrac{1}{100}$	$\dfrac{1}{10}$	1	10	100	1000	10 000	100 000
0.000 01	0.0001	0.001	0.01	0.1						

Example 8

Work out the value of

(a) 2^6 (b) 2^{10} (c) 2^{-10} (d) 10^7 (e) 10^{-7} (f) $2^2 \times 10^3$

(a) $2^6 = 2 \times 2 \times 2 \times 2 \times 2 \times 2$ (b) $2^{10} = 2 \times 2 \times 2 \times 2 \times 2 \times 2 \times 2 \times 2 \times 2 \times 2 = 1024$
       $= 64$

(c) $2^{-10} = \dfrac{1}{2^{10}} = \dfrac{1}{1024}$ (d) $10^7 = 10 \times 10 \times 10 \times 10 \times 10 \times 10 \times 10 = 10\,000\,000$

(e) $10^{-7} = \dfrac{1}{10^7} = \dfrac{1}{10\,000\,000}$ (f) $2^2 \times 10^3 = (2 \times 2) \times (10 \times 10 \times 10)$
           $= 0.000\,000\,1$ $= 4 \times 1000$
                                                       $= 4000$

Example 9

Solve the equation $2^{3n-1} = 64$

First recognise that $64 = 2^6$ so
     $2^{3n-1} = 2^6$

Hence, equating the indices:
     $3n - 1 = 6$
         $3n = 7$ ──────────────── Add 1 to both sides.
           $n = \tfrac{7}{3}$ or $n = 2\tfrac{1}{3}$ ──────── Divide both sides by 3

Example 10 (Summing a series of powers)

Find an expression for the sum S of the series

$$1 + 10 + 10^2 + 10^3 + \cdots + 10^n$$

Write $S = 1 + 10 + 10^2 + 10^3 + \cdots + 10^n$

then multiply by 10:

$$10 \times S = 10(1 + 10 + 10^2 + \ldots + 10^n)$$

So $\quad 10S = 10 + 10^2 + 10^3 + \cdots + 10^{n+1}$

and $\quad S = 1 + 10 + 10^2 + \cdots + 10^n$

Subtract:

$$10S = \qquad 10 + 10^2 + 10^3 + \cdots + 10^n + 10^{n+1}$$
$$\underline{-S = -1 - 10 - 10^2 - \cdots - 10^n}$$
$$9S = 10^{n+1} - 1$$

So $\quad S = \dfrac{10^{n+1} - 1}{9}$

> You can use this method for other similar series.

Exercise 1G

1   Work out the value of

(a) 2^3 (b) 2^{-2}

(c) 10^{-4} (d) 10^{10}

(e) $2^3 \times 10^5$ (f) $2^2 \times 10^2$

(g) $2^3 \times 10^{-3}$ (h) $2^5 \times 10^5$

(i) 10^{-6} (j) 2^{-8}

2   Round 2^{10} to the nearest 100.

3   For values of x from -3 to 5, draw the graphs of:

(a) $y = 2^x$ (b) $y = 10^x$

> Draw up a table of values.

4   Solve these equations

(a) $10^{5n-2} = 100$

(b) $2^{5-2n} = 8$

(c) $10^{2n+3} = 0.000\,01$

5   Find an expression for the sum of the series

$$1 + 2 + 2^2 + 2^3 + \cdots + 2^n$$

Mixed exercise 1

1   Write 156 in prime factor form.

2   Find the HCF and LCM of   (a) 6 and 12   (b) 18 and 30

3   Work out   (a) $\dfrac{2^4 \times 2^3}{(2^2)^2}$ (b) $(5^2)^{-3}$

4   Work out   (a) $81^{\frac{1}{2}}$ (b) $216^{\frac{1}{3}}$ (c) $36^{-\frac{1}{2}}$

5   For any positive whole number n its **Tau Function** $\tau(n)$ is defined as the number of positive whole number factors of n.
7 is a prime number. It has two factors, 1 and 7, so $\tau(7) = 2$.
   (a)   Show that if p is *any* prime number then $\tau(p) = 2$.
   (b)   For any prime number p and any positive value of n find an expression for $\tau(p^n)$.
   (c)   **An investigation**
        $6 \times 7 = 42$
        The factors of 6 are 1, 2, 3, 6 so $\tau(6) = 4$
        The factors of 42 are 1, 2, 3, 6, 7, 14, 21 and 42,
        so 42 has 8 factors and $\tau(42) = 8$.
        $6 \times 7 = 42$ and $\tau(42) = 4 \times 2 = \tau(6) \times \tau(7)$,
        so $\tau(6 \times 7) = \tau(6) \times \tau(7)$.
        Investigate to see whether or not $\tau(n \times m) = \tau(n) \times \tau(m)$ for other values of n and m.

> Remember that $\tau(7) = 2$ (from above).

6   One way of making the number 5 by adding 1s and 3s is
$5 = 3 + 1 + 1$.
Another different way is $5 = 1 + 3 + 1$.
Investigate the number of different ways of making any number by adding 1s and 3s.

7   $1^3 + 2^3 + 3^3 \;=\; 1 + 8 + 27 \;=\; 36 \;=\; 6^2 \;=\; (1 + 2 + 3)^2$
$1^3 + 2^3 + 3^3 + 4^3 \;=\; 1 + 8 + 27 + 64 \;=\; 100 \;=\; 10^2$
$ =\; (1 + 2 + 3 + 4)^2$
Investigate whether this is true for different sets of consecutive numbers.

8   The **last digit** of 146 is 6. This is written as LD(146) = 6.
What comments can you make about
   (a)   LD(any square number)   (b)   LD($n \times m$)?
   (c)   Show that $10n + 7$ can never be a square number for any positive whole number value of n.

9   Solve each of these equations
   (a) $2^{5n-2} = 1024$ (b) $2^{3n+1} = 0.125$
   (c) $10^{4n+3} = 0.000\,000\,1$

10   Find an expression for the sum of the series
        $1 + 5 + 25 + 125 + 625 + \cdots + 5^n$

Summary of key points

1 A **factor** of a number is a whole number that divides exactly into the number. The factors include 1 and the number itself.

2 A **multiple** of a number is the result of multiplying the number by a positive whole number.

3 A **prime number** is a number greater than 1 which has only two factors: itself and 1.

4 Any factor of a number that is a prime number is a **prime factor**.

5 Writing a number as the product of its prime factors is called writing it in **prime factor form**.

6 The **highest common factor** (HCF) of two whole numbers is the highest factor that is common to both of them.

7 The **lowest common multiple** (LCM) of two whole numbers is the lowest number that is a multiple of them both.

8 A **square number** is the result of multiplying a whole number by itself.

9 A **cube number** is the result of multiplying a whole number by itself, then multiplying by the number again.

10 To find the **square** of any number, multiply the number by itself.

11 If $x \times x = A$, then x is the **square root** of A, written $\sqrt{A}$

12 To find the **cube** of any number, multiply the number by itself, then multiply by the number again.

13 If $y \times y \times y = A$, then y is the cube root of A, written $\sqrt[3]{A}$

14 The power a number is raised to is called the **index** (plural **indices**).

15 A number written in the form a^n is written in **index form**, where n is the index.

16 The **index laws**:
  - $a^n \times a^m = a^{n+m}$
  - $a^n \div a^m = a^{n-m}$ or $\dfrac{a^n}{a^m} = a^{n-m}$
  - $a^0 = 1$ (where $a \neq 0$)
  - $a^{-n} = \dfrac{1}{a^n}$ (where $a \neq 0$)
  - $(a^n)^m = a^{n \times m}$

17 Fractional indices: $a^{\frac{1}{n}} = \sqrt[n]{a}$

2 Essential algebra

Algebra is the branch of mathematics in which letters are used to represent numbers. This chapter shows you ways of simplifying algebraic expressions and how to solve simple equations.

You should already know that:

> $a + a = 2a$ $\qquad a \times b = ab$ $\qquad ab = ba$ $\qquad a \times a = a^2$

An **algebraic expression** is a collection of letters, numbers and symbols.

A **function** is a rule which shows how one set of quantities relates to another. A function can be written like this: $n \rightarrow 2n + 3$.

2.1 Evaluating algebraic expressions

You need to be able to **evaluate** expressions such as $2n + 3$, $3(n - 2)$ and $3n^2 + 1$ for different values of n.

> **Remember:**
>
> - Work out the parts in brackets first. $\longrightarrow$ $2(n + 3)$ means add 3 to n then multiply by 2.
>
> - In an expression such as $\dfrac{n + 3}{2}$ the line acts like a bracket. $\longrightarrow$ $\dfrac{n + 3}{2}$ means add 3 to n then divide by 2.
>
> - $2n$ means $2 \times n$ and $\dfrac{n}{2}$ means $n \div 2$.
>
> - Do multiplication and division before addition and subtraction. $\longrightarrow$ $2n + 1$ means multiply n by 2 then add 1.
> $\dfrac{n}{2} - 3$ means divide by 2 then subtract 3.

An algebraic expression can have letters other than n. It can also have more than one letter, for example $a + 3b - 2c$.

Example 1

Evaluate the expression $\dfrac{2a + b}{c}$ when $a = 3$, $b = 4$ and $c = 2$.

$$\frac{2a + b}{c} = \frac{2 \times 3 + 4}{2} \qquad \text{Work out the part above the line first.}$$

$$= \frac{6 + 4}{2}$$

$$= \frac{10}{2}$$

$$= 5$$

Exercise 2A

Work out the value of each algebraic expression using the values given.

1 (a) $4a + 1$ when $a = 3$

(b) $3b + c$ when $b = 5$, $c = 2$

(c) $2f - g$ when $f = 1.5$, $g = 4$

(d) $hg - 2$ when $h = 1.5$, $g = 3$

(e) $10 + 3x$ when $x = -2$

(f) $2x - 3y$ when $x = 4$, $y = -2$

(g) $2x + 3$ when $x = \frac{1}{5}$

(h) $3ab$ when $a = \frac{1}{4}$, $b = 2$

2 (a) $2(a + 3)$ when $a = 5$

(b) $3(s - 2)$ when $s = 7$

(c) $4(p + q)$ when $p = 5$, $q = 3$

(d) $r(8 - s)$ when $r = 3$, $s = 5$

(e) $3(b + 7)$ when $b = -2$

(f) $2(3 - c)$ when $c = -4$

3 (a) $5(a + b)$ when $a = 3$, $b = 4$

(b) $4(x + y)$ when $x = 5$, $y = -3$

(c) $\dfrac{a}{4} + 3$ when $a = 12$

(d) $\dfrac{a}{b} + 5$ when $a = 20$, $b = 4$

(e) $\dfrac{m - 4}{2}$ when $m = 12$

(f) $\dfrac{7 - x}{y}$ when $x = -3$, $y = -2$

4 (a) $\dfrac{m + n}{r}$ when $m = 8$, $n = 7$, $r = 5$

(b) $\dfrac{4q + r}{6}$ when $q = 5$, $r = 4$

(c) $\dfrac{3s - r}{t}$ when $s = 8$, $r = 6$, $t = 3$

(d) $\dfrac{3s}{4} - r$ when $s = 8$, $r = 3$

(e) $x - \dfrac{3y}{6}$ when $x = 3$, $y = -4$

2.2 Evaluating algebraic expressions involving squares

Remember:

- n^2 is called n squared or n to the power of 2. $\longrightarrow$ n^2 means n is multiplied by itself: $n \times n$.

- Squaring is done before multiplying. $\longrightarrow$ $2n^2 + 1$ means work out n^2, multiply by 2, then add 1.

A good way to remember the order in which to evaluate an algebraic expression is to use the nonsense word **BIDMAS**.

Use **BIDMAS** to help you remember the order of mathematical operations.

Follow this order:

Brackets
Indices
Divide
Multiply
Add
Subtract

Remember:
The index in x^2 is ^{2}.
The plural of *index* is *indices*.

Example 2

Work out the value of $\dfrac{4r^2 - t}{5}$ when $r = 3$ and $t = 1$.

$$\frac{4r^2 - t}{5} = \frac{4 \times 3^2 - 1}{5}$$

$$= \frac{4 \times 9 - 1}{5} \quad\text{—— Indices}$$

$$= \frac{36 - 1}{5} \quad\text{—— Multiply}$$

$$= \frac{35}{5} \quad\text{—— Subtract}$$

$$= 7 \quad\text{—— Divide}$$

Work out the top part first.

Exercise 2B

Work out the value of each algebraic expression using the values given.

1 (a) $4m^2 + 1$ when $m = 5$ (b) $3m^2 - 1$ when $m = 4$

(c) $2p^2 - c$ when $p = 3$, $c = 4$ (d) $4x^2 + d$ when $x = 6$, $d = 3$

(e) $2p^2 + 5$ when $p = -3$ (f) $10 - q^2$ when $q = -2$

2 (a) $ar^2 + t$ when $r = 3$, $a = 2$, $t = 1$ (b) $3m^2 + 2m$ when $m = 3$

(c) $p^2q - r$ when $p = 4$, $q = 2$, $r = 5$ (d) $2p^2 - 3p$ when $p = 4$

(e) $x^2 - y^2$ when $x = -5$, $y = 3$ (f) $a^2 - b^2$ when $a = 4$, $b = -3$

3 (a) $4p^2 + ar$ when $p = 3$, $a = 2$, $r = 4$ (b) $\dfrac{p^2 + 4}{2}$ when $p = 4$

(c) $\dfrac{p^2}{3} + 4$ when $p = 6$ (d) $\dfrac{p^2}{2} + \dfrac{q^2}{3}$ when $p = 4$, $q = 6$

(e) $2a(b^2 + c^2)$ when $a = 3$, $b = -2$, $c = 1$ (f) $\dfrac{2r^2 - t}{8}$ when $r = -3$, $t = -6$

2.3 Indices

A short way of writing a number multiplied by itself is to use an index. You should be used to seeing a^2 (a squared or a to the power 2). In the same way,

$$a \times a \times a = a^3 \quad\text{and}\quad a \times a \times a \times a = a^4.$$

The number of times a is multiplied by itself is the index (or power).

Example 3

Simplify $y^2 \times y^3$.

Using the index law for multiplication: $a^n \times a^m = a^{n+m}$

$$y^2 \times y^3 = (y \times y) \times (y \times y \times y) = y^5$$

If the length of each small square on the board is a, then the area of each small square is a^2. The total area of all the small squares on the board is $64a^2$.

Example 4

Simplify $3a^2b \times 4a^4b^3$.

$$3a^2b \times 4a^4b^3 = (3 \times a^2 \times b) \times (4 \times a^4 \times b^3) = (3 \times 4) \times (a^2 \times a^4) \times (b \times b^3) = 12a^6b^4$$

Exercise 2C

Simplify these algebraic expressions.

1 (a) $a \times a^3$ (b) $b^3 \times b^2$
   (c) $c^2 \times c \times c^3$ (d) $y \times y \times y^2$

2 (a) $2ab \times 3ab^3$ (b) $5p^2q^3 \times 2p^4q$
   (c) $3x^2y^2 \times 3x^2y^2$ (d) $st^4 \times 3st$

2.4 Removing brackets from algebraic expressions

To remove brackets from an algebraic expression, multiply each term inside the brackets by the term outside.

This is sometimes called **expanding the brackets**.

You can also use the areas of rectangles like these to show the same results:

$$\text{total area} = \text{length} \times \text{width}$$
$$= 2(n + 1)$$
$$\text{sum of separate areas} = 2n + 2$$
so
$$2(n + 1) = 2n + 2$$

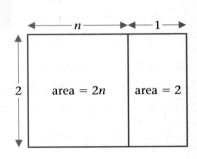

Example 5

Remove the brackets from the expression $4(2n - 3)$.

Multiply each term inside the brackets by 4

$$4(2n - 3) = 4 \times 2n + 4 \times -3$$
$$= 8n - 12$$

Or, using rectangles

$$\text{area of left rectangle} = \text{whole area} - \text{area of right rectangle}$$
$$4(2n - 3) = 4 \times 2n - 4 \times 3$$
$$= 8n - 12$$

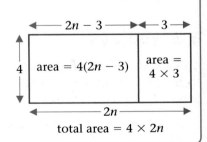

Exercise 2D

1 Draw rectangles to find expressions without brackets that are equivalent to these expressions:

   (a) $2(n + 3)$ (b) $4(3p + 2)$ (c) $y(2y + 3)$
   (d) $2x(x + 5)$ (e) $3(n - 2)$ (f) $1(2p - 3)$

2 Write these expressions without brackets:

(a) $4(n + 3)$ (b) $2(p - 4)$ (c) $5(a + 1)$

(d) $2(3x - 1)$ (e) $4(5q - 3)$ (f) $3(2r + s)$

(g) $a(b + 2)$ (h) $p(p + q)$ (i) $p(2p - 1)$

3 Write these expressions without brackets:

(a) $3(x + 6)$ (b) $4(y - 2)$ (c) $7(t + 1)$

(d) $4(3x + 2)$ (e) $5(3y - 4)$ (f) $2(3p + q)$

(g) $x(x + 3)$ (h) $x(x + y)$ (i) $y(3y - 2)$

4 Expand the brackets:

(a) $2(m + 5)$ (b) $9(n - 4)$ (c) $8(u - 1)$

(d) $5(7x + 3)$ (e) $4y(y - 6)$ (f) $3x^2(2x^2 + 5)$

(g) $4t^3(5t^2 - 3)$ (h) $4x^2(3x^3 + 7y)$ (i) $5y(3y^3 - 4x^2)$

2.5 Simplifying algebraic expressions

Sometimes you can simplify an algebraic expression by expanding the brackets and collecting **like terms** together.

Terms with the same power of the same letter(s) are called **like terms**.

Example 6

Simplify the expression $3(3a + 2b) + 2(a - b)$.

$3(3a + 2b) + 2(a - b)$

$= 9a + 6b + 2a - 2b$ ——— First expand the brackets.

$= 9a + 2a + 6b - 2b$ ——— Rearrange by collecting like terms together:
9a and 2a are like terms and
6b and −2b are like terms.

$= 11a + 4b$

Example 7

Simplify the expression $\dfrac{4n + 2}{2}$

$\dfrac{4n + 2}{2}$

$= \frac{1}{2}(4n + 2)$

$= \frac{1}{2} \times 4n + \frac{1}{2} \times 2$ ——— Expand the brackets.

$= 2n + 1$

> Remember:
> The line acts like a bracket.

Example 8

Simplify the expression $x(x + 3y) + 2x(x + y)$.

$x(x + 3y) + 2x(x + y)$

$= x^2 + 3xy + 2x^2 + 2xy$ ——— Expand the brackets.

$= x^2 + 2x^2 + 3xy + 2xy$ ——— Collect like terms.

$= 3x^2 + 5xy$

Exercise 2E

Expand and simplify the expressions in questions **1–4**.

1 (a) $2(3r + 1) + 2r$ (b) $3(4p - 2) - 10p$
  (c) $4(2x + y) + 2x - 3y$ (d) $2m - 3n + 3(m + n)$

2 (a) $4(2a + b) + 2(3a - b)$ (b) $3(p + q) + 2(p - q)$
  (c) $2(3r + s) + 4(r - s)$ (d) $3(2p - q) + 3(p + q)$

3 (a) $3(x + y) + 4(x - y)$ (b) $4(m - n) + 3(2m - n)$
  (c) $2(p - q) + 2(p - q)$ (d) $3(p - q) + 3(p + q)$

4 (a) $3(b - a) + 1(6a - 6b)$ (b) $r(r + 2) - r$
  (c) $m(n + m) + m(n - m)$ (d) $m(3 + n) + n(1 - m)$

5 Write these expressions in a simpler form.

  (a) $\dfrac{6n + 4}{2}$ (b) $\dfrac{9n - 6}{3}$ (c) $\dfrac{10n - 20}{5}$

  (d) $\dfrac{6p + 10r}{2}$ (e) $\dfrac{4r - 8q}{4}$ (f) $\dfrac{6r - 3s}{3}$

Dealing with negative numbers

The expression $-2(3a - 2)$ has a negative number outside the brackets. You expand the brackets by multiplying each term inside the brackets by -2:

$$-2 \times 3a = -6a$$
$$-2(3a - 2) = -6a + 4$$
$$-2 \times -2 = +4$$

> $-2(3a - 2)$ means
> $-2 \times (3a - 2)$
> $= -6a + 4$

Example 9

Simplify the expression $4(2a + b) - 2(a - b)$.

$4(2a + b) - 2(a - b)$
$= 8a + 4b - 2a + 2b$ ———— Expand the brackets.
$= 8a - 2a + 4b + 2b$ ———— Collect like terms.
$= 6a + 6b$

Exercise 2F

Simplify these expressions.

1 (a) $4a - 3(a - b)$ (b) $2(2p + q) - 3(p - q)$
  (c) $3(2n - 3) - 4(n - 2)$ (d) $2(s - r) - 2(r - s)$
  (e) $3(x + y) - 3(x - y)$ (f) $r(s - t) - r(3 - t)$

2 (a) $3x - 2(x + y)$ (b) $2(6x - y) - (x + y)$
  (c) $x(4x + 5) - 2x(x + 3)$ (d) $5y(4y - 3) - 2(4y - 3)$
  (e) $x(2x - 5) - 2(2x - 5)$ (f) $2y^3(2y - 3) + 3y^2(4y^2 + 5y)$

2.6 Factorising

Factorising is the opposite of expanding. You write algebraic expressions in a shorter form using brackets. It also helps you solve some algebra problems.

> **Factorising** is the opposite of expanding brackets.

Here are some algebraic expressions before and after factorising:

Before	After
$6x + 18$	$6(x + 3)$
$21x - 28$	$7(3x - 4)$
$x^2 - 9x$	$x(x - 9)$
$ax + 4a$	$a(x + 4)$
$x^2y - xy^2$	$xy(x - y)$
$3x^2 + 12x$	$3x(x + 4)$

> To factorise an expression completely, the highest common factor (HCF) must appear outside the brackets.

In $3x^2 + 12x$ the HCF of $3x^2$ and $12x$ is $3x$. The HCF is the largest term which is a factor of $3x^2$ and $12x$.

$3(x^2 + 4x)$ and $x(3x + 12)$ are both ways of factorising $3x^2 + 12x$ but they are *not factorised completely* because the HCF $3x$ does not appear outside the brackets. Factorising completely, $3x^2 + 12x = 3x(x + 4)$.

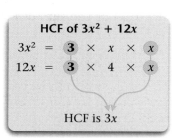

HCF of $3x^2 + 12x$

$3x^2 = \mathbf{3} \times x \times x$

$12x = \mathbf{3} \times 4 \times x$

HCF is $3x$

____ **Example 10** _____

Factorise $8x^2 - 12x$ completely.

  $8x^2 - 12x = 4x(x - 3)$

> $4x$ is the HCF of $8x^2$ and $-12x$

Exercise 2G

Factorise these expressions completely.

1 $5x - 20$ **2** $8x + 24$ **3** $12x + 18$

4 $20x - 25$ **5** $8x^2 - 24$ **6** $9x^2 + 36$

7 $10x^2 - 15$ **8** $x^2 + 6x$ **9** $x^2 - x$

10 $px - 3p$ **11** $qx + q^2$ **12** $5x^2 - 15x$

13 $6x^2 + 9x$ **14** $15x^2 - 35x$ **15** $8x - 12x^2$

16 $ax^2 + ax$ **17** $4ax - 6a$ **18** $6ax^2 + 15ax$

2.7 Multiplying bracketed expressions

Sometimes you will need to multiply bracketed expressions by each other, for example $(e + f)(g + h)$.

This means $(e + f)$ multiplied by $(g + h)$, or $(e + f) \times (g + h)$.

Look at the rectangles on the right.

The area of the whole rectangle is $(e + f)(g + h)$.

It is the same as the sum of the **four** separate areas so:

$$(e + f)(g + h) = eg + eh + fg + fh$$

	g	h
e	area = eg	area = eh
f	area = fg	area = fh

Notice that each term in the first bracket is multiplied by each term in the second bracket:

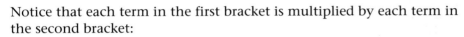

You can also think of the area of the rectangle as the sum of **two** separate parts:

$$(e + f)(g + h) = e(g + h) + f(g + h)$$

	$(g + h)$
e	area = $e(g + h)$
f	area = $f(g + h)$

Think of multiplying each term in the first bracket by the whole of the second bracket.

These are two ways of thinking about the same process. The end result is the same.

This is called **multiplying out** the brackets.

Example 11

Multiply out the brackets in the expression $(2a + 2)(3a - 4)$.

$$(2a + 2)(3a - 4) = 2a(3a - 4) + 2(3a - 4)$$
$$= 6a^2 - 8a + 6a - 8$$
$$= 6a^2 - 2a - 8$$

Equations such as

$$(e + f)(g + h) = eg + eh + fg + fh$$

and $(e + f)(g + h) = e(g + h) + f(g + h)$

which are true for all values of e, f, g and h are also known as **identities**.

Exercise 2H

1 Expand and simplify these products, drawing rectangles to illustrate your answers.

   (a) $(x + 2)(x + 1)$ (b) $(x + 8)(x + 2)$

   (c) $(x + 3)^2$ (d) $(x + a)(x + b)$

In questions **2–12** multiply out the brackets then simplify your expressions where possible.

2 (a) $(a + 4)(b + 3)$ (b) $(c + 5)(d + 4)$

   (c) $(x + 3)(y + 6)$ (d) $(a + 3)(a + 8)$

   (e) $(b + 7)(b + 4)$ (f) $(x + 6)(x + 2)$

3 (a) $(2a + 3)(b + 4)$ (b) $(3b + 2)(c + 3)$

   (c) $(4c + 3)(d + 6)$ (d) $(2a + 5)(a + 3)$

   (e) $(3b + 4)(b + 2)$ (f) $(4c + 5)(c + 2)$

4 (a) $(a - 4)(b - 3)$ (b) $(c - 5)(d - 4)$

   (c) $(x - 3)(y - 6)$ (d) $(a - 3)(a - 8)$

   (e) $(b - 7)(b - 4)$ (f) $(x - 6)(x - 2)$

5 (a) $(a - 4)(b + 3)$ (b) $(c + 5)(d - 4)$

   (c) $(x - 3)(y + 6)$ (d) $(a - 3)(a + 8)$

   (e) $(b - 7)(b + 4)$ (f) $(x + 6)(x - 2)$

6 (a) $(2a - 5)(b + 5)$ (b) $(3b + 4)(b - 6)$

   (c) $(3c - 2)(c - 3)$ (d) $(3a - 4)(a - 5)$

   (e) $(3x - 4) (x + 5)$ (f) $(2a + 3)(a - 6)$

7 (a) $(x + y)(2x + 3y)$ (b) $(x + y)(3x - 4y)$

   (c) $(x - y)(5x - 3y)$ (d) $(2x + 5y)(3x + 4y)$

   (e) $(3x + 5y)(2x - 5y)$ (f) $(6x - 5y)(2x - 3y)$

8 (a) $(x + 3)^2$ (b) $(y - 5)^2$

   (c) $(2x + 1)^2$ (d) $(3y - 4)^2$

   (e) $(3x + 5y)^2$ (f) $(6x - 7y)^2$

> Remember:
> $(x + 3)^2 = (x + 3)(x + 3)$

9 (a) $(3a + 1)(a - 2)$ (b) $(3p + 2)(4p - 1)$

   (c) $(3a + 2)(a + 1)$ (d) $(4p + 2)(3p + 2)$

   (e) $(2b - 3)(b - 1)$ (f) $(4b - 1)(2b + 1)$

10 (a) $(a + 1)(b + 2) - ab$ (b) $(2p + 1)(q - 2) + 3p - pq$

   (c) $(3r - s)(r - 1) + s - 2r$ (d) $(2p + 3a)(2p - 3a) + 9a^2$

11 (a) $(3x + 1)(3x - 1)$ (b) $(1 - 4x)(1 + 4x)$

   (c) $(2x - 5)(2x + 5)$ (d) $(7x + 2)(7x - 2)$

12 (a) $5(x + 2)(x - 6)$ (b) $4(x + 1)(x - 1)$

   (c) $2(3x + 1)(5x - 1)$ (d) $3(4x - 3)(3x + 2)$

Wot, no letter z?

2.8 Multiplying an expression by itself

You need to be able to square simple expressions.

Example 12

Expand and simplify   (a) $(x + 6)^2$ (b) $(x - 8)^2$

(a) $(x + 6)^2 = (x + 6)(x + 6)$
$= x(x + 6) + 6(x + 6)$
$= x^2 + 6x + 6x + 36$
$= x^2 + 12x + 36$

(b) $(x - 8)^2 = (x - 8)(x - 8)$
$= x(x - 8) - 8(x - 8)$
$= x^2 - 8x - 8x + 64$
$= x^2 - 16x + 64$

In general: $(x + a)^2 = x^2 + 2ax + a^2$
$(x - b)^2 = x^2 - 2bx + b^2$

Exercise 21

Expand and simplify the expressions in questions **1–9**.

1 $(x + 5)^2$ **2** $(x - 3)^2$ **3** $(x + 9)^2$

4 $(x - 2)^2$ **5** $(p + q)^2$ **6** $(p - q)^2$

7 $(3 + x)^2$ **8** $(7 - x)^2$ **9** $(a - x)^2$

10   Copy and complete these expressions by writing an appropriate number in each box.

(a) $(x + \square)^2 = x^2 + \square x + 16$
(b) $(x - \square)^2 = x^2 - 20x + \square$
(c) $(x + \square)^2 = x^2 + 2x + \square$
(d) $(x - \square)^2 = x^2 - \square x + 144$

2.9 Solving simple equations

Sometimes you will need to **solve** equations such as $4x + 1 = 7$.
Solving means finding the value or values of x that make the
equation true.

To solve an equation like this, rearrange it so that the x term is
on its own on one side of the equation.
The x term in $4x + 1 = 7$ is $4x$.

To rearrange an equation you can use the **balancing method**:
- add the same quantity to both sides
- subtract the same quantity from both sides
- multiply both sides by the same quantity
- divide both sides by the same quantity.

Whatever you do to one side of an equation you must also do to the other side.

Example 13

Solve $4x + 1 = 7$.

$$4x + 1 = 7$$
$$4x = 7 - 1 \qquad \text{Subtract 1 from both sides.}$$
$$4x = 6$$
$$x = 1\tfrac{1}{2} \qquad \text{Divide both sides by 4.}$$

Check by substituting $1\tfrac{1}{2}$ in place of x in $4x + 1 = 7$:

$$(4 \times 1\tfrac{1}{2}) + 1 = 6 + 1 = 7$$

So $1\tfrac{1}{2}$ is the solution.

Example 14

Solve $5 - 3x = 7$.

$$5 - 3x = 7$$
$$-3x = 2 \qquad \text{Subtract 5 from both sides.}$$
$$x = -\tfrac{2}{3} \qquad \text{Divide both sides by } -3.$$

Check by substituting $-\tfrac{2}{3}$ in place of x in $5 - 3x = 7$:

$$5 - (3 \times -\tfrac{2}{3}) = 5 + 2 = 7$$

Exercise 2J

Solve these equations and check your answers.

1	$3x + 2 = 17$	**2**	$5x - 2 = 18$	**3**	$2x + 5 = 10$
4	$5x - 1 = 12$	**5**	$7x = 3$	**6**	$4x + 3 = 3$
7	$5 + 6x = 10$	**8**	$9x - 4 = 3$	**9**	$2x + 7 = 1$
10	$3a + 8 = 2$	**11**	$4b + 17 = 7$	**12**	$5c - 1 = -13$
13	$8d + 7 = 2$	**14**	$8 + 10e = 3$	**15**	$10f + 7 = 3$
16	$6 - x = 5$	**17**	$8 - x = 10$	**18**	$7 - 2y = 3$
19	$4 - 3y = 16$	**20**	$-2a = 12$	**21**	$-5b = 13$
22	$5 - 2c = 2$	**23**	$2 - 6d = 1$	**24**	$5 - 8e = 7$
25	$3 - 5f = 12$				

Don't forget to check your answers by substitution!

Mixed exercise 2

1   In each of the following expressions $a = 2$, $b = 9$, $c = 5$.
Work out the value of each.

   (a) $2a + 3c$ **(b)** $a^2 + c$ **(c)** $3ab$

   (d) abc **(e)** $(2c)^2$ **(f)** $3b^2$

   (g) $a\sqrt{b}$ **(h)** $(a + b)(b + c)$ **(i)** $\dfrac{2b + a}{c}$

   (j) $\dfrac{2bc}{5a^2}$

2   Simplify

   (a) $3a + 5b + 6a - 2b$ **(b)** $2ab + 3ba$

   (c) $a^2 + 3a - 4a^2 + a$ **(d)** $7p - 4q - 2p - q$

   (e) $8c + 5d - 9c - d$

3   Expand and simplify

   (a) $5(3p + 2) - 2(5p - 3)$ **(b)** $2(3p + 4) + 3(2p - 1)$

   (c) $4(3a - 3b) + 5(b - a)$ **(d)** $3(2q - 5) - 2(3q - 2)$

4   Expand $p(q - p^2)$.   [E]

5   Expand

   (a) $(y + 5)(y - 3)$ **(b)** $(3x + 2)(4x + 1)$

   (c) $(5y - 3)^2$ **(d)** $(3a - 2b)^2$

6   Solve

   (a) $x - 2 = 22$ **(b)** $10 - 2x = 2$ **(c)** $\dfrac{x}{3} = 4$

   (d) $5x = 3$ **(e)** $8 = 3 - x$ **(f)** $4 + 3x = 19$

7   Factorise completely

   (a) $4p - 6$ **(b)** $9q + 3$ **(c)** $6a + 4b$

   (d) $ab + a^2$ **(e)** $2ab + 2ac$ **(f)** $8a^2 - 6a$

Summary of key points

1 $a + a = 2a$ $a \times b = ab$ $ab = ba$ $a \times a = a^2$

2   An **algebraic expression** is a collection of letters, numbers and symbols.

3   A **function** is a rule which shows how one set of quantities relates to another.
A function can be written like this: $n \rightarrow 2n + 3$.

4   Use **BIDMAS** to help you remember the order of mathematical operations.

   Follow this order:

   | **B**rackets
   | **I**ndices
   | **D**ivide
   | **M**ultiply
   | **A**dd
   ▼ **S**ubtract

5 To remove brackets from an algebraic expression, multiply each term inside the brackets by the term outside.

This is sometimes called **expanding the brackets**.

6 Terms with the same power of the same letter(s) are called **like terms**.

7 **Factorising** is the opposite of removing brackets.

8 To factorise an expression completely, the highest common factor (HCF) must appear outside the brackets.

9 Equations such as

$$(e + f)(g + h) = eg + eh + fg + fh$$
$$\text{and } (e + f)(g + h) = e + (g + h) + f(g + h)$$

which are true for all values of e, f, g and h are also known as **identities**.

10 In general: $(x + a)^2 = x^2 + 2ax + a^2$
$(x - b)^2 = x^2 - 2bx + b^2$

11 To rearrange an equation you can use the **balancing method**:
- add the same quantity to both sides
- subtract the same quantity from both sides
- multiply both sides by the same quantity
- divide both sides by the same quantity.

Whatever you do to one side of an equation you must also do to the other side.

3 Shapes

Euclid was a Greek mathematician who lived at Alexandria in Egypt around 300 BC. In his books *The Elements*, he explained everything that was known about 2-D (plane) and 3-D (solid) shapes, like these:

This chapter investigates the properties of 2-D and 3-D shapes.

3.1 Angles

You should remember these key points from earlier work:

Names of types of angles

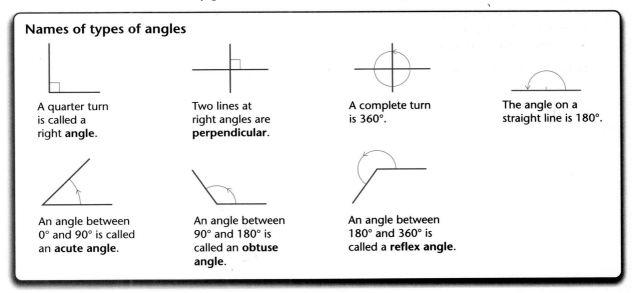

A quarter turn is called a right **angle**.

Two lines at right angles are **perpendicular**.

A complete turn is 360°.

The angle on a straight line is 180°.

An angle between 0° and 90° is called an **acute angle**.

An angle between 90° and 180° is called an **obtuse angle**.

An angle between 180° and 360° is called a **reflex angle**.

You need to know these properties of angles:

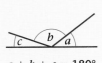

$a + b + c = 180°$

$a + b = 180°$

$a + b + c + d = 360°$

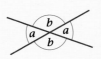

Angles which meet at a point on a straight line add up to 180°.

A pair of angles that add together to make 180° are called **supplementary angles**.

Angles which meet at a point add up to 360°.

Two straight lines which cross at a point form two pairs of **vertically opposite angles**. Vertically opposite angles are **equal**.

Angles formed when a straight line crosses a pair of parallel lines have the following properties:

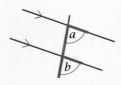 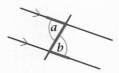 

Corresponding angles are equal. So $a = b$. You can find them by looking for an F shape.

Alternate angles are equal. So $a = b$. Look for a Z shape.

Co-interior angles x and y are supplementary as $x + y = 180°$. Co-interior angles are sometimes called **allied angles**. Look for a C shape.

Example 1 ─────────

Find angles a and b, giving reasons for your answers.

a and 25° are corresponding angles, so $a = 25°$
80° and b are co-interior angles, so:

$$80° + b = 180°$$
$$b = 180° - 80°$$
$$= 100°$$

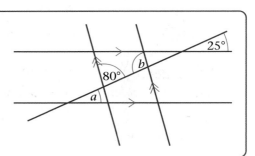

Exercise 3A

1 Calculate the size of each lettered angle.

(a)
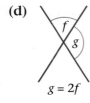

(b)
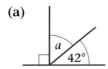

(c)

(d)
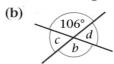

$g = 2f$

(e)
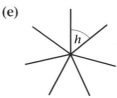

(all angles are equal)

2 Calculate each of the lettered angles.

(a)
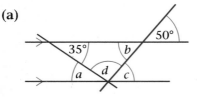

(b)
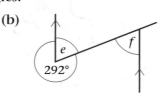

(c)
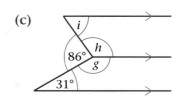

3.2 2-D shapes

Triangles

Remember these key points from earlier work:

The **interior angles** in a triangle add up to 180°.

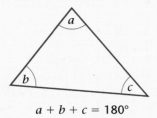

$$a + b + c = 180°$$

An **equilateral triangle** has:
• 3 sides equal
• 3 interior angles equal (60°).

An **isosceles triangle** has:
• 2 sides equal
• base angles equal.

A **right-angled triangle** has:
• 1 angle of 90°.

Triangles without any of these properties are called **scalene triangles**.

Example 2

Calculate the sizes of the lettered angles, giving reasons for your answers.

$$34° + 92° + a = 180° \text{ (interior angles of a triangle)}$$
$$a = 180° − 126° = 54°$$

$$a + b = 180° \text{ (angles on a straight line)}$$
$$b = 180° − 54° = 126°$$

Exercise 3B

Calculate the size of each lettered angle.
Give reasons for your answers.

1

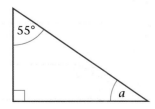

2

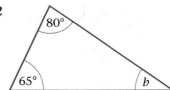

3

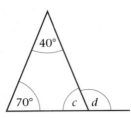

4

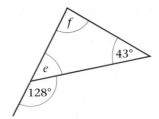

5

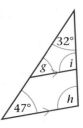

Quadrilaterals

Some quadrilaterals have special names and properties:

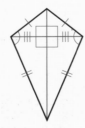

A **square** has:
• all sides equal
• opposite sides parallel
• all interior angles 90°
• diagonals that bisect
  at 90°.

A **rectangle** has:
• opposite sides equal
• opposite sides parallel
• all interior angles 90°
• diagonals that bisect
  each other.

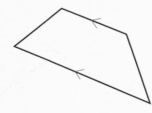

A **parallelogram** has:
• opposite sides equal
• opposite sides parallel
• diagonally opposite
  angles equal
• adjacent angles are
  supplementary angles
• diagonals that bisect
  each other.

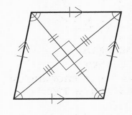

A **rhombus** has:
• all sides equal
• opposite sides parallel
• opposite angles equal
• adjacent angles are
  supplementary angles
• diagonals that bisect
  at 90°.

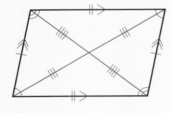

A **trapezium** has:
• 1 pair of parallel lines.

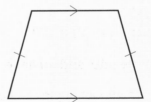

An **isosceles trapezium**
has:
• 1 pair of parallel lines
• 2 sides equal
• 1 line of symmetry.

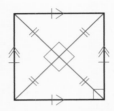

A **kite** has:
• 2 pairs of adjacent sides
  equal
• 1 pair of opposite angles
  equal
• diagonals cut at 90°
• 1 line of symmetry.

An **arrowhead** has:
• 2 pairs of adjacent sides
  equal
• 1 line of symmetry

Each special name means the shape has *all* the properties listed.

The sum of the interior angles of a quadrilateral

> The **interior angles** in a quadrilateral add up to 360°.

You can see this by dividing it into two triangles:

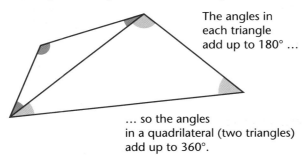

The angles in
each triangle
add up to 180° …

… so the angles
in a quadrilateral (two triangles)
add up to 360°.

The interior angles of any polygon

The sum of the interior angles of any polygon can be found by
dividing the shape into triangles from one vertex:

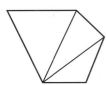

Pentagon
3 triangles
sum of interior angles:
$3 \times 180° = 540°$

Hexagon
4 triangles
sum of interior angles:
$4 \times 180° = 720°$

Octagon
6 triangles
sum of interior angles:
$6 \times 180° = 1080°$

The number of triangles is 2 less than the number of sides of the
polygon.

> For a polygon with n sides:
>   sum of interior angles $= (n - 2) \times 180°$

A **regular polygon** has all its sides and angles equal.

> If a polygon is **regular** each interior angle can be calculated from
>   interior angle $= \dfrac{(n - 2) \times 180°}{n}$

Sometimes it is simpler to calculate the exterior angle.

> The sum of the **exterior angles** of any polygon is 360°.

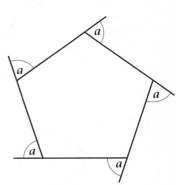

$a + a + a + a + a = 5a = 360°$

So for a regular polygon with n sides:

$$\textbf{exterior angle} = \frac{360°}{n}$$

and   **interior angle** $= 180° -$ **exterior angle**

> For a regular pentagon:
> exterior angle $= 72°$
> interior angle $= 180° - 72°$
> $= 108°$

LOGO can be used to draw polygons. The exterior angle is used in instructions like these:

```
to pentagon
repeat 5[fd 100 rt 72]
end
```

Exercise 3C

1 Write down the special names of these shapes
   (a) a triangle with two sides equal
   (b) a quadrilateral with opposite sides equal
   (c) a quadrilateral with one pair of opposite sides parallel and equal
   (d) a quadrilateral with diagonals equal and intersecting at 90°.

2 (a) Work out the sum of the interior angles of a ten-sided shape (decagon).
   (b) Work out the interior angle of a regular decagon.

3.3 Demonstration and proof

You need to understand the difference between demonstration and proof.

Demonstration

You can **demonstrate** the result that the interior angles of a triangle add up to 180° by drawing a triangle, tearing off the angles b and c and placing them next to angle a to make a straight line as shown:

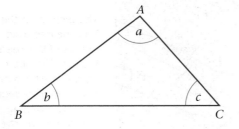 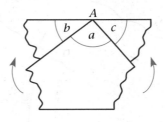

Proof

For any triangle you can draw a straight line PQ through A, parallel to BC:

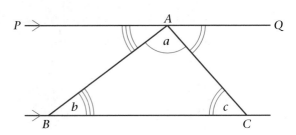

$$P\hat{A}B = b \text{ (alternate angles)}$$
$$C\hat{A}Q = c \text{ (alternate angles)}$$
$$a + b + c = 180° \text{ (angles on a straight line)}$$

So, angles in a triangle add up to 180°.

> The proof applies to **any** triangle.

Exterior angle of a triangle

Here is another result you should know:

The **exterior angle** of a triangle is equal to the sum of the interior opposite angles.

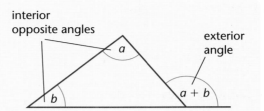

To prove this result first draw line *CD* parallel to *BA*:

$A\hat{C}D = a$ (alternate angles)

$D\hat{C}X = b$ (corresponding angles)

$A\hat{C}X = A\hat{C}D + D\hat{C}X = a + b$

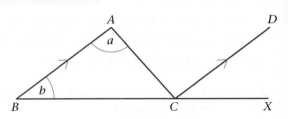

Hence,

exterior angle = the sum of the interior opposite angles.

Exercise 3D

Draw a sketch to demonstrate that the angles of a quadrilateral add up to 360°.

3.4 3-D shapes

A solid shape with plane faces and straight edges is called a **polyhedron** (plural: polyhedra).

The picture on the left is of the geodesic domes of the Eden Project. The outer layer of each dome is made up of hexagons and a few pentagons. The picture on the right shows the glass pyramid at the Louvre.

Some polyhedra have special names:

cube

cuboid

square-based pyramid

triangle-based pyramid

A **pyramid** takes its name from the shape of its base.

A **prism** is a polyhedron that has the same shape or cross-section wherever you slice it along its length. A prism takes its name from the shape of its cross-section.

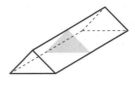

triangular prism

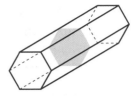

hexagonal prism

These solids have some circular edges so they are *not* polyhedra:

circular prism
(usually called a cylinder)

circle-based pyramid
(usually called a cone)

sphere

Nets: solid shapes unfolded

The **net** of a 3-D shape is a 2-D shape that can be folded to make the 3-D shape.

To fold up correctly, edges which will touch must be the same length.

Both these nets will make the cuboid shown.

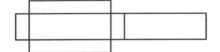

Plan and elevation

You can show all the details of a 3-D shape by drawing three different 2-D views of it.

- The **plan** of a solid is the view from above.
- The **front elevation** is the view from the front.
- The **side elevation** is the view from the (right-hand) side.

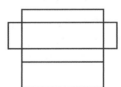

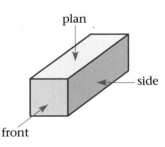

Example 3

Draw to scale the plan and elevations of this 3-D shape.

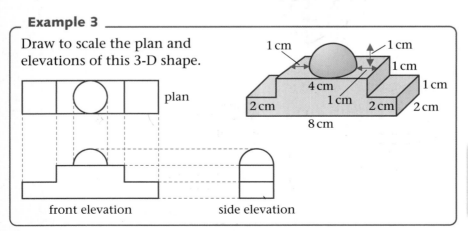

You should link your plans and elevations with dotted lines to show how the different drawings match up.

Example 4

Draw the plan and elevations of this
house and garage. Use a scale of 1 cm : 4 m.

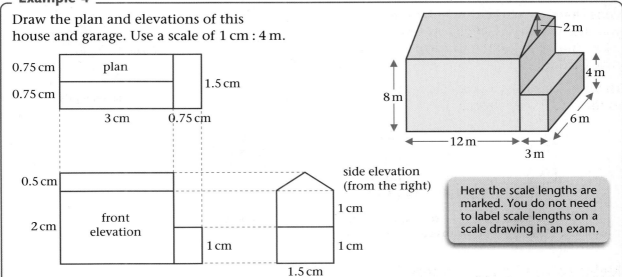

Here the scale lengths are
marked. You do not need
to label scale lengths on a
scale drawing in an exam.

Exercise 3E

1   Sketch six different nets which would make a cube.

2   Sketch a net to make a square-based pyramid.

3   Sketch the net to make the prism
    shown on the right.
    Label the length of each side.

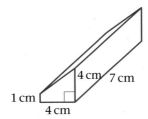

4   Sketch a plan, front elevation and side elevation for each of
    these solids.

    (a)    **(b)**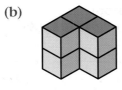

5   Sketch the plans and elevations of these shapes:

    (a) 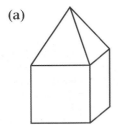   **(b)** 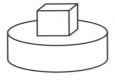   **(c)**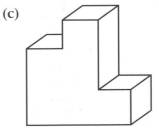

6   This regular pyramid has a triangular base.
Each edge of the pyramid is 5 cm long.
Sketch the plan and front elevation.

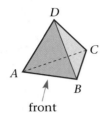

7   Here are the plan and the front elevation of a prism.
The front elevation shows the cross-section of the prism.

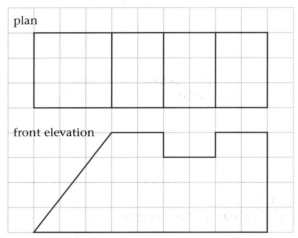

(a)   Draw a side elevation of the prism.

(b)   Draw a 3-D sketch of the prism.

8   The diagram shows a scale drawing.
Draw to scale the plan and the
front and side elevations from the
direction of the arrows.

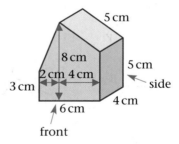

3.5 Congruent shapes

When 2-D shapes are exactly the same shape and size they
are **congruent**.

One shape may have to be rotated or turned over to see the match.
Corresponding sides and corresponding angles are identical.

These keys are congruent.

Triangles are congruent if one of these facts is true:

Three pairs of sides are equal (remember this with the abbreviation **SSS** for three sides).

 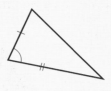

Two pairs of sides are equal and the angles between them (the included angle) are equal (**SAS**).

Two pairs of angles are equal and the sides between them are equal (**ASA**).

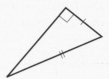 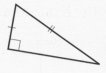

Both triangles have a right angle, the hypotenuses are equal and one pair of corresponding sides is equal (**RHS**).

Example 5

State whether these pairs of shapes are congruent.
List the vertices in corresponding order and give reasons for congruency.

(a)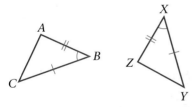

(b)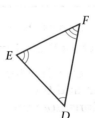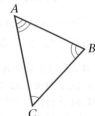

(c)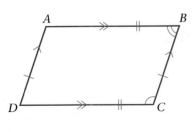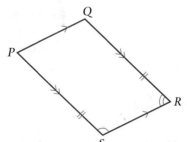

(a)   Yes. *ABC* is congruent to *ZXY* (SAS).

(b)   No. Only the angles are equal. The corresponding sides may not be the same length.

(c)   No. Parallelogram *ABCD* is not congruent with *QRSP*. It is not clear whether *AD* = *PQ* or *BC* = *RS*.

Example 6

In the diagram, *ACYX* and *BCQP* are squares.

Prove that triangles *ACQ* and *BCY*
are congruent.

$AC = CY$ (sides of square *ACYX*)

$CQ = BC$ (sides of square *BCQP*)

$\angle ACQ = 90° + \angle ACB$

$\angle BCY = 90° + \angle ACB$

So $\angle ACQ = \angle BCY$

So triangles *ACQ* and *YCB* are congruent (SAS).

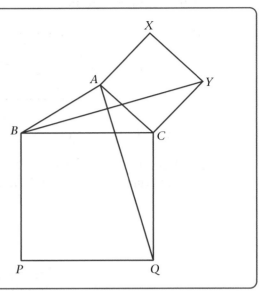

Exercise 3F

1 State whether these pairs of shapes are congruent. List the
vertices in corresponding order and give reasons for congruency.

(a)
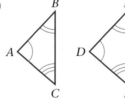

(b)
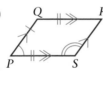
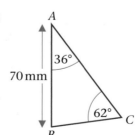

(c)
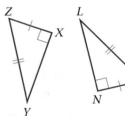

2 Which of these pairs of triangles are congruent? Give the vertices
in corresponding order and the reason for congruency.

(a)
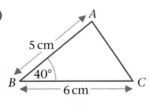

(b)

(c)
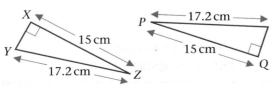

3 ABCD is a parallelogram.

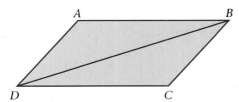

Prove that *ABD* is congruent to *CDB*.

4 In the diagram, *AC = AD* and *BD = CE*.
Prove that triangles *ABC* and *ADE* are congruent.

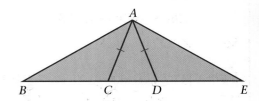

5

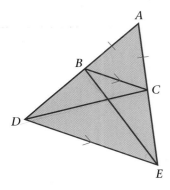

In the diagram, *ABC* is an isosceles triangle with *AB = AC*.

Prove that triangles *ACD* and *ABE* are congruent.

6 In the diagram,

$$AB = BE$$
$$BD = BC$$
$$\angle AEB = \angle BDC$$

Prove that triangles *ABD* and *EBC* are congruent.

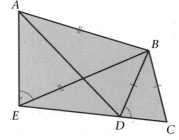

7 State whether these triangles are congruent. Give reasons for congruency.

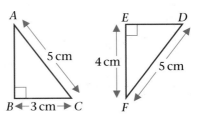

Activity
To prove that a statement is incorrect, you need only one **counter example**.
Draw triangles to demonstrate that SSA triangles might not be congruent.

3.6 Similar shapes

Shapes are **similar** if one shape is an enlargement of the other.

Polygons are similar if all corresponding angles are *equal* and the ratio of object length to image length is the same for all sides.

The **scale factor** of an enlargement is the ratio

$$\frac{\text{length of a side on one shape}}{\text{length of corresponding side on other shape}}$$

Example 7

These rectangles are similar. *B* is an enlargement of *A*.

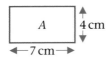

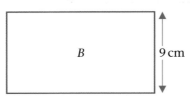

(a) Find the scale factor of the enlargement.

(b) Find the missing length on rectangle *B*.

(a) Using corresponding sides, the scale factor of the enlargement is $\frac{9}{4} = 2.25$

(b) So the length of *B* is $7 \times 2.25 = 15.75$ cm

To decide whether two triangles are similar, you need to check that all the corresponding angles are equal, or that all the corresponding sides are in the same ratio.

Triangles are similar if one of these facts is true:

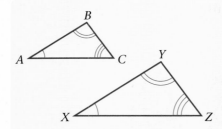

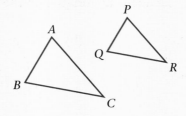

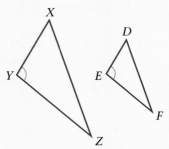

All corresponding angles are equal:

$\hat{A} = \hat{X} \quad \hat{B} = \hat{Y} \quad \hat{C} = \hat{Z}$

All corresponding sides are in the same ratio:

$$\frac{PQ}{AB} = \frac{QR}{BC} = \frac{PR}{AC} = \text{scale factor}$$

Two pairs of corresponding sides are in the same ratio and the included angles are equal:

$$\frac{XY}{DE} = \frac{YZ}{EF} \quad \hat{Y} = \hat{E}$$

Example 8

Find the length of the side marked x.

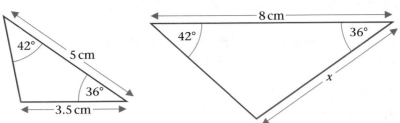

Two pairs of angles are equal, so the third pair must be equal. The triangles are similar.
The scale factor of the enlargement is $\frac{8}{5} = 1.6$

So $x = 3.5 \times 1.6 = 5.6\,\text{cm}$

Exercise 3G

1 Each pair of shapes is similar. Calculate each length marked by
   a letter.

   (a)

   (b)

   (c)

   (d)

2 Each pair of shapes is similar. Calculate the lengths marked by
   letters.

   (a)

   (b)

   (c)
   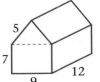

3   Match up pairs of similar rectangles.

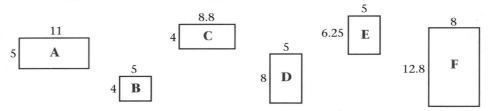

4   Which of these shape families are always similar?
- (a)   all squares
- (b)   all rectangles
- (c)   all parallelograms
- (d)   all circles
- (e)   all equilateral triangles
- (f)   all isosceles triangles
- (g)   all regular hexagons
- (h)   all trapeziums

5   Which of these solid shape families are always similar?
- (a)   cubes
- (b)   cuboids
- (c)   spheres
- (d)   tetrahedrons
- (e)   pyramids
- (f)   hexagonal prisms
- (g)   octahedrons
- (h)   cones

6   Each group of three triangles has two similar and one 'different' triangle. Which triangle is 'different'?

(a)   (i)   (ii)   (iii)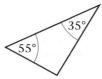

(b)   (i)    (ii)    (iii)

(c)   (i)   (ii) 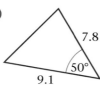   (iii)

7   Write down why these pairs of triangles are similar. Calculate the length of each side marked by a letter.

(a) 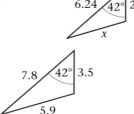   (b) 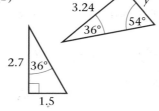   (c)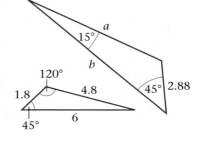

8 (a) Explain why the two triangles in this diagram are similar.

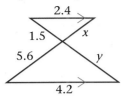

(b) Calculate the lengths of x and y.

3.7 Planes of symmetry

Some 3-D shapes can be divided by a plane to produce two identical solid shapes. The plane is called a **plane of symmetry**.

A cuboid has three planes of symmetry:

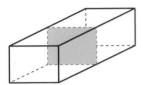 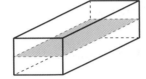 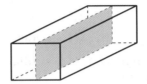

___ **Example 9** _____

Copy the house shape. Draw all its planes of symmetry on separate diagrams.

There are two planes of symmetry:

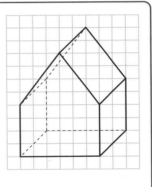

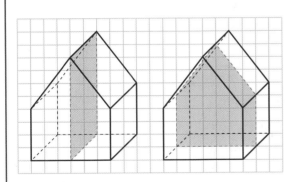

Exercise 3H

1 Write down the number of planes of symmetry for each of these solids.

(a) 　　(b) 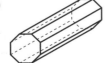　　(c)

2 Write down how many planes of symmetry a regular tetrahedron has. Draw sketches to show them all.

3 Draw sketches to show the planes of symmetry of

(a) a cube

(b) a regular octahedron.

What do you notice?

> A regular tetrahedron is a solid shape whose faces are equilateral triangles.

Mixed exercise 3

1 (a) *MNOP* is a trapezium.
       MN = *MP*
       $N\hat{P}O = 29°$
       Find (i) $M\hat{N}P$ (ii) $P\hat{M}N$

(b) *ABCD* is a square.
Triangle *PDC* is equilateral.
Calculate: (i) $P\hat{C}B$ (ii) $C\hat{P}B$
       (iii) $A\hat{P}D$ (iv) $A\hat{P}B$

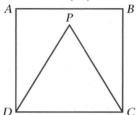

2 *V* = number of vertices   *F* = number of faces   *E* = number of edges

(a) Copy and complete this table for each solid.

Solid	*V*	*F*	*E*
Cube	8	6	12
Cuboid			
Triangular prism			
Hexagonal prism			
Octagonal prism			
Triangle-based pyramid			
Square-based pyramid			

> This relationship is known as Euler's theorem after the Swiss mathematician who discovered it.

(b) Write down a relationship between *V*, *F* and *E*.

3 A regular polyhedron is a solid whose faces are regular polygons. The same number of these polygons meet at each vertex.

tetrahedron   cube   octahedron   dodecahedron   icosahedron

> These five regular solids were known to the Greek mathematician and philosopher Plato, and are often called the **Platonic solids**. A tetrahedron is the same as a triangle-based pyramid.

(a) Check that Euler's theorem (from question **2**) also works for these polyhedra:

Solid	V	F	E
Tetrahedron			
Cube			
Octahedron			
Dodecahedron			
Icosahedron			

(b) Copy and complete this table:

Solid	Number of triangular faces meeting at a vertex
Tetrahedron	
Octahedron	
Icosahedron	

(c) Explain why a regular polyhedron could not have six triangular faces meeting at a vertex.

(d) Explain why a regular polyhedron cannot be made with regular hexagons as faces.

(e) Construct these five 3-D solids.

4 Which pairs of shapes are congruent?

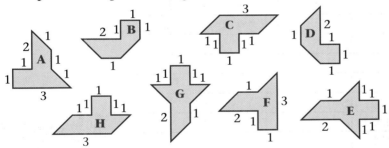

5 Each pair of shapes is similar. Calculate each length marked by a letter.

(a)

3.5   5.25
1.8   y

(b)

3
4.2   4.2
5

q
r   r
7

(c)

9
12   8   x

(d)

5   9   14   p

6 (a) Explain why the two triangles in this diagram are similar.

(b) Calculate the lengths of *x* and *y*.

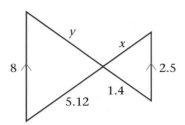

7 (a) Name the similar triangles.

(b) Explain why they are similar.

(c) Calculate the length *AB*.

(d) Calculate the length *AY*.

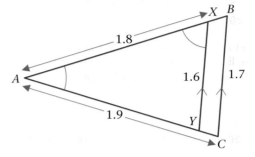

8 *PQRS* is a kite.

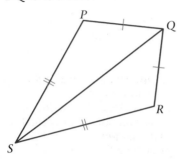

Prove that triangle *PQS* is congruent to triangle *RSQ*.

9 The diagram shows triangles *ABC* and *ACD*. *BCD* is a straight line. The perpendicular distance from *A* to the line *BCD* is *h* cm.

Explain why $\dfrac{\text{area of triangle } ABC}{\text{area of triangle } ACD} = \dfrac{BC}{CD}$

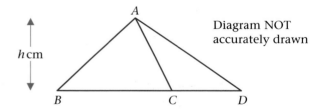

Diagram NOT accurately drawn

10

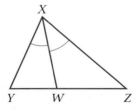

Diagram NOT accurately drawn

The diagram shows triangle *XYZ*.

W is the point on *YZ* such that angle *YXW* = angle *WXZ*.

Using expressions for the area of triangle *YXW* and the area of triangle *WXZ*, or otherwise, show that

$$\frac{XY}{XZ} = \frac{YW}{WZ}$$

Summary of key points

1 Properties of angles:

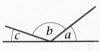

$a + b + c = 180°$

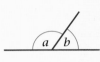

$a + b = 180°$

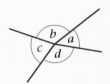

$a + b + c + d = 360°$

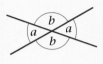

Angles which meet at a point on a straight line add up to 180°.

A pair of angles that add together to make 180° are called **supplementary angles**.

Angles which meet at a point add up to 360°.

Two straight lines which cross at a point form two pairs of **vertically opposite angles**. Vertically opposite angles are **equal**.

2 Angles formed when a straight line crosses a pair of parallel lines have the following properties:

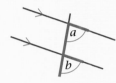

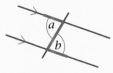

Corresponding angles are equal. So $a = b$. You can find them by looking for an F shape.

Alternate angles are equal. So $a = b$. Look for a Z shape.

Co-interior angles x and y are supplementary as $x + y = 180°$. Co-interior angles are sometimes called **allied angles**. Look for a C shape.

3 The **interior angles** in a triangle add up to 180°.

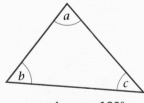

$a + b + c = 180°$

4

An **equilateral triangle** has:
- 3 sides equal
- 3 interior angles equal (60°).

An **isosceles triangle** has:
- 2 sides equal
- base angles equal.

A **right-angled triangle** has:
- 1 angle of 90°.

Triangles without any of these properties are called **scalene triangles**.

5   Some quadrilaterals have special names and properties.

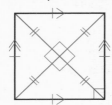

A **square** has:
- all sides equal
- opposite sides parallel
- all interior angles 90°
- diagonals that bisect at 90°.

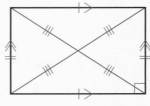

A **rectangle** has:
- opposite sides equal
- opposite sides parallel
- all interior angles 90°
- diagonals that bisect each other.

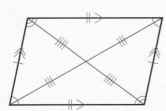

A **parallelogram** has:
- opposite sides equal
- opposite sides parallel
- diagonally opposite angles equal
- adjacent angles are supplementary angles
- diagonals that bisect each other.

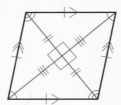

A **rhombus** has:
- all sides equal
- opposite sides parallel
- opposite angles equal
- adjacent angles are supplementary angles
- diagonals that bisect at 90°.

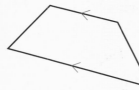

A **trapezium** has:
- 1 pair of parallel lines.

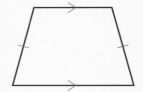

An **isosceles trapezium** has:
- 1 pair of parallel lines
- 2 sides equal
- 1 line of symmetry.

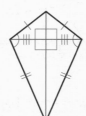

A **kite** has:
- 2 pairs of adjacent sides equal
- 1 pair of opposite angles equal
- diagonals cut at 90°
- 1 line of symmetry.

An **arrowhead** has:
- 2 pairs of adjacent sides equal
- 1 line of symmetry

Each special name means the shape has all the properties listed.

6   The **interior angles** in a quadrilateral add up to 360°.

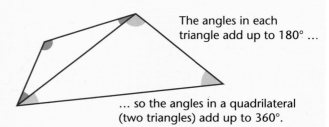

The angles in each triangle add up to 180° …

… so the angles in a quadrilateral (two triangles) add up to 360°.

7   For a polygon with n sides:

sum of interior angles $= (n - 2) \times 180°$

8   If a polygon is **regular** each interior angle can be calculated from

$$\text{interior angle} = \frac{(n - 2) \times 180°}{n}$$

9   The sum of the **exterior angles** of any polygon is 360°.

10 The **exterior angle** of a triangle is equal to the sum of the interior opposite angles.

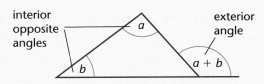

interior opposite angles

exterior angle

$a + b$

11 The **net** of a 3-D shape is a 2-D shape that can be folded to make the 3-D shape.

12 When 2-D shapes are exactly the same shape and size they are **congruent**.

13 Triangles are congruent if one of these facts is true:

Three pairs of sides are equal (remember this with the abbreviation **SSS** for three sides).

Two pairs of sides are equal and the angles between them (the included angle) are equal (**SAS**).

Two pairs of angles are equal and the sides between them are equal (**ASA**).

Both triangles have a right angle, the hypotenuses are equal and one pair of corresponding sides is equal (**RHS**).

14 Shapes are **similar** if one shape is an enlargement of the other.

15 The **scale factor** of an enlargement is the ratio $\dfrac{\text{length of a side on one shape}}{\text{length of corresponding side on other shape}}$

16

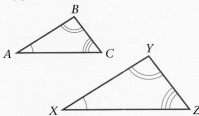   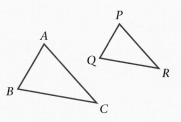   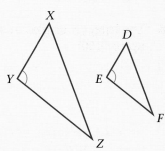

All corresponding angles are equal:

$\hat{A} = \hat{X}$ $\hat{B} = \hat{Y}$ $\hat{C} = \hat{Z}$

All corresponding sides are in the same ratio:

$\dfrac{PQ}{AB} = \dfrac{QR}{BC} = \dfrac{PR}{AC} = \text{scale factor}$

Two pairs of corresponding sides are in the same ratio and the included angles are equal:

$\dfrac{XY}{DE} = \dfrac{YZ}{EF}$ $\hat{Y} = \hat{E}$

17 Some 3-D shapes can be divided by a plane to produce two identical 2-D shapes. The plane is called a **plane of symmetry**.

④ Fractions and decimals

All these things can be divided into parts called **fractions**:

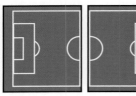

A football pitch has two halves. Each part is $\frac{1}{2}$ the pitch.

This CD has eight equal sectors. Each sector is $\frac{1}{8}$ of the disk.

A chessboard has 64 equal squares. Each square is $\frac{1}{64}$ of the board.

A fraction is a package of information. For example, if $\frac{3}{8}$ of my computer disk is full of data:

The top number shows how many parts are full of data. ———— **3** ———— The top number is called the **numerator**.

The bottom number shows how many parts the disk is divided into. ——— **8** ——— The bottom number is called the **denominator**.

In your GCSE exam, fractions will usually appear in the context of a number problem, or in questions on probability, areas or volumes.

This chapter shows you how to add, subtract, multiply and divide fractions, to order fractions and decimals, and to use mental and written methods.

You should remember these key points from earlier work:

$\frac{5}{2}$ is called an **improper fraction**. The numerator is greater than the denominator.

$2\frac{1}{2}$ is called a **mixed number**. It contains a whole number part and a fractional part smaller than 1.

Mixed numbers can be written as improper fractions.
Improper fractions can be written as mixed numbers.

4.1 Equivalent fractions

You can find **equivalent fractions** by multiplying (or dividing) the top and bottom of a fraction by the same number.

Here are three examples:

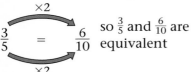

so $\frac{3}{5}$ and $\frac{6}{10}$ are equivalent

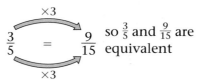

so $\frac{3}{5}$ and $\frac{9}{15}$ are equivalent

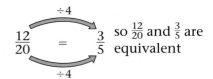

so $\frac{12}{20}$ and $\frac{3}{5}$ are equivalent

So $\frac{3}{5}, \frac{6}{10}, \frac{9}{15}$ and $\frac{12}{20}$ are all equivalent fractions.

Example 1

Find the first five equivalent fractions of $\frac{5}{7}$.

$$\frac{5}{7} = \frac{10}{14} = \frac{15}{21} = \frac{20}{28} = \frac{25}{35} = \frac{30}{42}$$

$\times 3$
$\times 2$
$\times 2$
$\times 3$

Example 2

Complete these equivalent fractions:　(a) $\frac{3}{8} = \frac{?}{16}$　(b) $\frac{4}{5} = \frac{?}{10} = \frac{?}{25}$

(a) $\frac{3}{8} = \frac{?}{16}$　so　$\frac{3}{8} = \frac{6}{16}$
$\times 2$

(b) $\frac{4}{5} = \frac{?}{10} = \frac{?}{25}$　so　$\frac{4}{5} = \frac{8}{10} = \frac{20}{25}$
$\times 2$　$\times 5$
$\times 5$
$\times 2$

Exercise 4A

1　Copy and complete these sets of equivalent fractions:

(a) $\frac{3}{4} = \frac{}{8} = \frac{}{12} = \frac{}{16} = \frac{}{20} = \frac{}{24}$

(b) $\frac{2}{7} = \frac{}{14} = \frac{}{21} = \frac{}{28} = \frac{}{35} = \frac{}{42}$

(c) $\frac{4}{5} = \frac{}{10} = \frac{}{15} = \frac{}{20} = \frac{}{25} = \frac{}{30}$

(d) $\frac{3}{9} = \frac{}{18} = \frac{}{27} = \frac{}{36} = \frac{}{45} = \frac{}{54}$

In questions **2** and **3**, copy and complete the pairs of equivalent fractions.

2　(a) $\frac{1}{6} = \frac{}{18}$　　(b) $\frac{3}{7} = \frac{}{14}$

(c) $\frac{3}{8} = \frac{}{48}$　　(d) $\frac{4}{7} = \frac{}{21}$

(e) $\frac{5}{6} = \frac{}{36}$　　(f) $\frac{2}{3} = \frac{6}{}$

(g) $\frac{4}{9} = \frac{24}{}$　　(h) $\frac{5}{7} = \frac{}{56}$

3　(a) $\frac{9}{10} = \frac{90}{}$　　(b) $\frac{3}{5} = \frac{}{15}$

(c) $\frac{2}{5} = \frac{}{20}$　　(d) $\frac{5}{8} = \frac{}{40}$

(e) $\frac{8}{9} = \frac{40}{}$　　(f) $\frac{7}{12} = \frac{84}{}$

(g) $\frac{7}{8} = \frac{49}{}$　　(h) $\frac{2}{9} = \frac{}{81}$

4.2 Ordering fractions

You can **order** fractions by rewriting them as equivalent fractions with a common denominator.

Example 3

Put the following fractions in order of size, smallest first:

$$\frac{2}{3}, \frac{5}{8}, \frac{7}{12}$$

The smallest number that is a multiple of 3, 8 and 12 is 24. Write them as equivalent fractions with denominator 24:

$$\frac{2}{3} = \frac{2 \times 8}{3 \times 8} = \frac{16}{24}$$

$$\frac{5}{8} = \frac{5 \times 3}{8 \times 3} = \frac{15}{24}$$

$$\frac{7}{12} = \frac{7 \times 2}{12 \times 2} = \frac{14}{24}$$

So, in order of size, smallest first: $\frac{14}{24}, \frac{15}{24}, \frac{16}{24} = \frac{7}{12}, \frac{5}{8}, \frac{2}{3}$

> 24 is the LCM of 3, 8 and 12.

> Compare the three fractions.

Exercise 4B

In each question, put the fractions in order of size, smallest first.

1 $\frac{1}{2}, \frac{2}{3}, \frac{5}{12}$ **2** $\frac{2}{5}, \frac{5}{8}, \frac{3}{7}$ **3** $\frac{3}{4}, \frac{7}{8}, \frac{8}{12}$ **4** $\frac{2}{15}, \frac{1}{5}, \frac{1}{3}$ **5** $\frac{13}{16}, \frac{7}{8}, \frac{5}{6}$

4.3 Ordering decimals

Decimals are used for parts of a number that are smaller than 1.

Hundreds	Tens	Units	.	tenths	hundredths	thousandths	
1.12			1	.	1	2		This equals $1 + \frac{1}{10} + \frac{2}{100}$
15.8		1	5	.	8			This equals $15 + \frac{8}{10}$
3.576			3	.	5	7	6	This equals $3 + \frac{5}{10} + \frac{7}{100} + \frac{6}{1000}$

> A place value table is a useful way of relating decimal numbers to fractions.

> This is a decimal point. It separates the whole numbers from the part that is smaller than 1.

Example 4

Put the number 7.3208 into a place value table and write it in its separate parts.

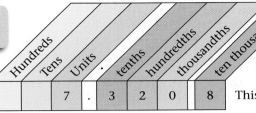

You say 'seven point three, two, zero, eight'.

Hundreds	Tens	Units	.	tenths	hundredths	thousandths	ten thousandths
		7	.	3	2	0	8

7.3208

Separate parts

This equals $7 + \frac{3}{10} + \frac{2}{100} + \frac{8}{10\,000}$

You will often need to sort measurements and other numbers into size order.
You need to be able to sort decimals as well as whole numbers.

Example 5

Rearrange these numbers in order of size, largest first:

1.07, 1.7, $1\frac{7}{8}$, 2, 1.085

First write any fractions as decimals. Then compare whole numbers, then digits in the tenths place, then digits in the hundredths place, and so on.

$1\frac{7}{8} = 1.875$

$7 \div 8 = 0.875$

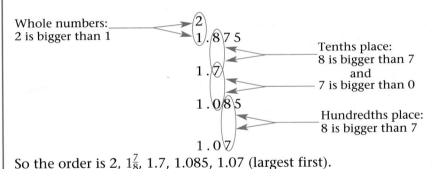

Whole numbers:
2 is bigger than 1

Tenths place:
8 is bigger than 7
and
7 is bigger than 0

Hundredths place:
8 is bigger than 7

So the order is 2, $1\frac{7}{8}$, 1.7, 1.085, 1.07 (largest first).

Exercise 4C

1 Make a place value table with
these headings:

You will need room for
16 answers.

Put each of the numbers into
the table and write them
in their separate parts.

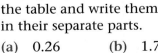

| Tens | Units | . | tenths | hundredths | thousandths | ten thousandths | hundred thousandths |

(a) 0.26   (b) 1.79   (c) 38.24   (d) 14.8

(e) 0.0101   (f) 0.033   (g) 2.645   (h) 80.934

(i) 4.398 52   (j) 0.0011   (k) 5.107   (l) 9.8302

(m) 7.457   (n) 13.0306   (o) 54.705   (p) 10.503 01

2   Rearrange each set of measurements in order of size, largest first.

(a)   4.101 m, 4.009 m, 4.0059 m

(b) $\frac{1}{8}$ kg, 0.55 kg, 0.525 kg, 1.25 kg

(c)   5.202 km, 5.305 km, 5.306 km, 5.204 km

(d) $9\frac{9}{10}$ tonnes, 9.904 tonnes, 9.804 tonnes, 9.99 tonnes

(e)   Times taken in an experiment for a model car to roll down
a slope: $5\frac{3}{4}$ s, 6.556 s, 5.623 s, 6.554 s

(f)   The capacities of four containers:
2.02 cl, $\frac{1}{5}$ cl,   2.0 cl,   0.022 cl

(g)   The distances Barry runs before he is out of breath:
6.202 km, 6.306 km, 6.204 km, 6.305 km

(h)   The weights of four lorries on a weighbridge:
4.512 tonnes, 6.443 tonnes, 6.448 tonnes, 6.643 tonnes

(i)   The widths of four machine parts:
1.08 cm, 9.8 cm, 9.08 cm, $10\frac{4}{5}$ cm

(j)   The lengths of four tubes:
8.701 m, 8.8 m, 8.88 m, 8.801 m

4.4 Mental calculations with decimals

You need to be able to calculate with decimals without using a
calculator.

Example 6

Work out 4.5 + 3.8

$$3.8 = 3 + 0.8$$

So $4.5 + 3.8 = 4.5 + 3 + 0.8$

$$= 5.3 + 3 \text{———————Adding 0.8 to 4.5}$$

$$= 8.3 \text{————————Adding 3 to 5.3}$$

Split into whole number
and decimal parts.

You should be able to do
this is in your head.

Example 7

Work out 54.6 − 6.7

$$54.6 − 6.7 = 54.6 − 6 − 0.7$$

$$= 53.9 − 6 \text{———— Subtracting 0.7 from 54.6}$$

$$= 47.9 \text{————— Subtracting 6 from 53.9}$$

You should be able to do
this in your head.

Example 8

Work out 3.8×2

Rewrite without the decimal point:

$38 \times 2 = 76$

Count the number of decimal places in the question: one.

The answer will have one decimal place, 7.6

Example 9

Work out $3.2 \div 4$

Rewrite without the decimal point:

$32 \div 4 = 8$

Count the number of decimal places in the question: one.

So the answer is 0.8

Exercise 4D

Write down the answers to the following calculations.

1	$9.7 + 3.4$	**2**	$8.4 + 11.9 + 7.6$	**3**	$11.2 - 9.5$	**4**	$56.3 - 27.5$
5	$3.8 + 2.7 - 1.6$	**6**	$11.5 + 7.6 - 12.7$	**7**	5.3×2	**8**	6.9×4
9	$4.8 \div 4$	**10**	$7.8 \div 6$	**11**	$12.6 \div 2$	**12**	$25.5 \div 5$

13 Use the information

$73 \times 154 = 11\,242$

to write down the value of

(a) 7.3×1.54 (b) $112\,420 \div 0.73$

4.5 Written methods for dividing decimals

You need to be able to divide by decimals. It is easier to divide by a whole number, so first multiply the dividing number by a power of ten so that it becomes a whole number. Then multiply the number to be divided by the same power of ten.

Example 10

Work out $8.704 \div 0.17$ without using a calculator.

First multiply 0.17 by 100 to give 17. Then multiply 8.704 by 100 to give 870.4.

Here are three ways of setting out the calculation:

Method 1   Long division

Follow these steps:

$17\overline{)870.4}$

17 divides into 87
5 times remainder 2

$\begin{array}{r} 5 \\ 17\overline{)870.4} \\ -85 \\ \hline 2 \end{array}$

17 divides into 20
1 time remainder 3

$\begin{array}{r} 51 \\ 17\overline{)870.4} \\ -85\downarrow \\ \hline 20 \\ -\ 17 \\ \hline 3 \end{array}$

17 divides into 34
2 times exactly

$\begin{array}{r} 51.2 \\ 17\overline{)870.4} \\ -85 \\ \hline 20 \\ -\ 17\downarrow \\ \hline 3\ 4 \\ 3\ 4 \\ \hline 0 \end{array}$

So 17 divides into 870.4 **51.2** times.

Method 2   Short division

This is a shorter way of setting out the steps in method 1:

$\begin{array}{r} 5\ 1.\ 2 \\ 1\overline{)787^20.^34} \end{array}$

Method 3   Repeated subtraction

Here you need to multiply both numbers by 1000 to remove the decimals:

$8704 - 170 = 8534$

$\qquad 8534 - 170 = 8364$

You would have to do this 512 times – not a practical method – and then sort out the decimal point by dividing by 10.

Exercise 4E

Work these out, showing all your working.

1 (a) $4 \div 0.5$ (b) $12 \div 0.3$ (c) $1.2 \div 0.5$

2 (a) $3.42 \div 0.2$ (b) $14.6 \div 0.5$ (c) $18.2 \div 0.7$

3 (a) $39 \div 1.5$ (b) $76.85 \div 5.3$ (c) $356.9 \div 0.86$
   (d) $813.6 \div 0.18$ (e) $239.66 \div 0.46$ (f) $129.26 \div 2.3$

4.6 Changing fractions to decimals

A fraction such as $\frac{3}{4}$ can be thought of as $3 \div 4$. You can change fractions into decimals easily using a calculator:

$3 \div 4 = 0.75$

You can check this:

0.75 is equal to $\dfrac{75}{100}$

$\dfrac{75}{100}$ simplifies to the equivalent fraction $\dfrac{3}{4}$

So the decimal 0.75 is equivalent to the fraction $\frac{3}{4}$.

> You can convert a fraction to a decimal by dividing the numerator by the denominator.

Example 11

Convert these to decimals.

(a) $\frac{3}{8}$ (b) $4\frac{1}{5}$

(a) $\frac{3}{8} = 3 \div 8 = 0.375$ (b) $4\frac{1}{5} = 4 + \frac{1}{5}$
$\qquad\qquad\qquad\qquad\qquad\qquad\frac{1}{5} = 1 \div 5 = 0.2$
$\qquad\qquad\qquad\qquad\text{So } 4\frac{1}{5} = 4 + 0.2 = 4.2$

Exercise 4F

Write these as decimals.

1 (a) $\frac{4}{5}$ (b) $\frac{1}{4}$ (c) $1\frac{1}{8}$ (d) $\frac{19}{100}$
   (e) $3\frac{3}{5}$ (f) $\frac{13}{25}$ (g) $\frac{5}{8}$ (h) $3\frac{17}{40}$

2 (a) $\frac{7}{50}$ (b) $4\frac{3}{16}$ (c) $3\frac{3}{20}$ (d) $4\frac{5}{16}$
   (e) $\frac{7}{1000}$ (f) $1\frac{7}{25}$ (g) $15\frac{15}{16}$ (h) $2\frac{7}{20}$

You can convert a decimal to a fraction.

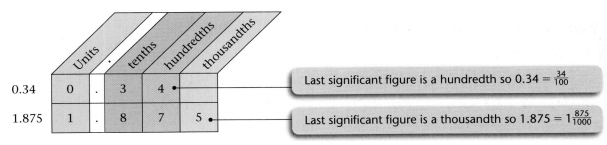

Last significant figure is a hundredth so $0.34 = \frac{34}{100}$

Last significant figure is a thousandth so $1.875 = 1\frac{875}{1000}$

You can write 0.34 as $\frac{34}{100}$ which simplifies to $\frac{17}{50}$.

1.875 is $1\frac{875}{1000}$ which simplifies to $1\frac{7}{8}$.

Example 12

Write these as fractions in their simplest form.

(a)   0.204   (b)   1.17   (c)   2.4375

(a) $0.204 = \frac{204}{1000} = \frac{51}{250}$ (b) $1.17 = \frac{117}{100}$

cannot be simplified further

(c) $2.4375 = 2\frac{4375}{10\,000} = 2\frac{175}{400} = 2\frac{7}{16}$

Exercise 4G

Write these as fractions in their simplest form.

1   (a)   0.48   (b)   0.25   (c)   1.7   (d)   3.406   (e)   4.003

2   (a)   2.025   (b)   0.049   (c)   4.875   (d)   3.75   (e)   10.101

3   (a)   0.625   (b)   2.512   (c)   0.8125   (d)   14.14   (e)   9.1875

Recurring decimals

Not all fractions have an exact equivalent decimal.

The fraction $\frac{2}{3}$ is $2 \div 3 = 0.6666666...$. This is called a **recurring decimal** since one of the digits recurs (repeats).

You usually put a dot over the digits that repeat:

$\frac{2}{3} = 0.6666666... = 0.\dot{6}$ $\frac{5}{12} = 0.4166666... = 0.41\dot{6}$

$\frac{1}{3} = 0.3333333... = 0.\dot{3}$ $\frac{3}{11} = 0.2727272... = 0.\dot{2}\dot{7}$

$\frac{7}{9} = 0.7777777... = 0.\dot{7}$

$\frac{1}{7} = 0.142857142\,857142857... = 0.\dot{1}4285\dot{7}$

The two dots show that all the digits from 1 to 7 repeat.

Recurring decimal notation:
$0.\dot{3}$ means 0.3333333... recurring and
$0.\dot{1}\dot{7}$ means 0.17171717... recurring.

| 0.6666667 |

If you work out $2 \div 3$ or $\frac{2}{3}$ on a calculator the result on the display could be 0.6666667. The result has been corrected to 7 s.f. by the calculator.

You need two dots here since both the 2 and the 7 repeat.

Example 13

Write these as recurring decimals. (a) $\frac{6}{11}$ (b) $3\frac{8}{9}$

Write your answers

(i) as shown on the calculator display
(ii) using recurring decimal notation.

(a) (i) $\frac{6}{11} = 6 \div 11 = 0.5454545$ (ii) $0.\dot{5}\dot{4}$

(b) (i) $3\frac{8}{9} = 3 + (8 \div 9) = 3.8888888$ (*or* 3.8888889) (ii) $3.\dot{8}$

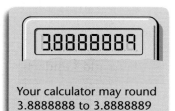

Your calculator may round 3.8888888 to 3.8888889

Exercise 4H

Convert these fractions to recurring decimals.

Write your answers

(i) as shown on the calculator display
(ii) using recurring decimal notation.

1 $\frac{5}{6}$ **2** $1\frac{2}{9}$ **3** $3\frac{1}{6}$ **4** $\frac{11}{12}$ **5** $5\frac{5}{9}$

6 $4\frac{9}{11}$ **7** $\frac{3}{44}$ **8** $2\frac{7}{11}$ **9** $9\frac{21}{22}$ **10** $\frac{25}{30}$

4.7 Adding and subtracting fractions

Adding fractions with the same denominator

It is easy to add fractions when they have the same denominator. You add the numerators. You must not add the denominators.

Example 14

Work out: (a) $\frac{4}{9} + \frac{1}{9}$ (b) $\frac{7}{8} + \frac{5}{8}$ (c) $1\frac{1}{5} + 2\frac{3}{5}$

Add the top numbers only

(a) $\frac{4}{9} + \frac{1}{9} = \frac{4+1}{9} = \frac{5}{9}$

(b) $\frac{7}{8} + \frac{5}{8} = \frac{7+5}{8} = \frac{12}{8} = \frac{3}{2} = 1\frac{1}{2}$ ← Change to a mixed number

Simplify

(c) $1\frac{1}{5} + 2\frac{3}{5} = 3 + \frac{1}{5} + \frac{3}{5} = 3 + \frac{1+3}{5} = 3\frac{4}{5}$

Add the whole numbers first

Why you don't add denominators

$\frac{1}{8} + \frac{3}{8} = \frac{4}{8}$

not $\frac{4}{16}$!

Adding fractions with different denominators

$$\frac{3}{4} + \frac{1}{8} = ?$$

The denominators of these fractions are different.

Example 15

Work out   (a) $1\frac{3}{4} + 2\frac{1}{8}$ (b) $\frac{2}{3} + \frac{1}{5}$ (c) $\frac{5}{6} + \frac{3}{4}$

Add the whole numbers first

(a) $1\frac{3}{4} + 2\frac{1}{8} = 3 + \frac{3}{4} + \frac{1}{8}$

To add $\frac{3}{4}$ and $\frac{1}{8}$ find equivalent fractions with the same denominator.

$\frac{3}{4}$ is equivalent to $\frac{6}{8}$

$\frac{3}{4} + \frac{1}{8} = \frac{6}{8} + \frac{1}{8} = \frac{7}{8}$ •————————

> Write $\frac{3}{4}$ as an equivalent fraction with denominator 8.

So $1\frac{3}{4} + 2\frac{1}{8} = 3\frac{7}{8}$

(b) $\frac{2}{3} + \frac{1}{5}$ •————————

> The smallest number that is a multiple of both 3 and 5 is 15, so change both the $\frac{2}{3}$ and the $\frac{1}{5}$ to equivalent fractions with denominator 15.

$\frac{2}{3}$ is equivalent to $\frac{10}{15}$, and $\frac{1}{5}$ is equivalent to $\frac{3}{15}$.

So $\frac{2}{3} + \frac{1}{5} = \frac{10}{15} + \frac{3}{15} = \frac{13}{15}$

(c) $\frac{5}{6} + \frac{3}{4} = \frac{10}{12} + \frac{9}{12} = \frac{19}{12} = 1\frac{7}{12}$ •————————

> The smallest number that is a multiple of both 6 and 4 is 12, so change both fractions to equivalent fractions with denominator 12.
> You could use any multiple of 12 as the denominator.

To **add** fractions, change them to equivalent fractions that have the same denominator (bottom). Then add the numerators.

Exercise 4I

Work out the additions in questions **1–3**.

1 (a) $\frac{1}{8} + \frac{3}{8}$ (b) $\frac{2}{7} + \frac{4}{7}$ (c) $\frac{2}{5} + \frac{4}{5}$ (d) $\frac{9}{10} + \frac{7}{10}$

   (e) $\frac{7}{9} + 2\frac{4}{9}$ (f) $\frac{5}{6} + 1\frac{5}{6}$ (g) $\frac{3}{4} + \frac{3}{4} + \frac{1}{4}$ (h) $\frac{3}{8} + \frac{5}{8} + \frac{7}{8}$

2 (a) $\frac{5}{8} + \frac{1}{4}$ (b) $1\frac{1}{2} + 2\frac{1}{8}$ (c) $\frac{1}{2} + \frac{7}{8}$ (d) $2\frac{3}{4} + 3\frac{7}{8}$

   (e) $1\frac{3}{4} + 2\frac{5}{16}$ (f) $\frac{3}{4} + 3\frac{5}{8}$ (g) $\frac{3}{8} + \frac{11}{16}$ (h) $2\frac{9}{16} + 1\frac{5}{8}$

3 (a) $\frac{1}{5} + \frac{3}{8}$ (b) $\frac{1}{5} + \frac{1}{6}$ (c) $1\frac{3}{10} + 1\frac{2}{3}$ (d) $\frac{2}{3} + \frac{2}{7}$

   (e) $3\frac{1}{6} + \frac{2}{7}$ (f) $2\frac{5}{6} + 1\frac{1}{7}$ (g) $3\frac{2}{5} + 2\frac{7}{15}$ (h) $1\frac{1}{3} + 1\frac{2}{9}$

4 Becky cycled $2\frac{3}{4}$ miles to one village then a further $4\frac{1}{3}$ miles to her home. What is the total distance Becky cycled?

5 Work out the perimeter of this photograph.

6 Two pieces of wood are fixed together, one on top of the other. One piece has thickness $2\frac{3}{8}$ inches and the other has thickness $1\frac{5}{16}$ inches.
What is the total thickness of the two pieces of wood?

7 In a class, $\frac{1}{6}$ of the pupils own one pet, and $\frac{2}{5}$ of the pupils own more than one pet.
What total fraction of the pupils own at least one pet?

8 A bag weighs $\frac{3}{7}$ lb. The contents weigh $1\frac{1}{5}$ lb.
What is the total weight of the bag and its contents?

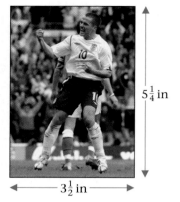

$5\frac{1}{4}$ in

$3\frac{1}{2}$ in

Subtracting fractions

It is easy to subtract fractions when they have the same denominator.
You subtract the numerators. You must not subtract the denominators.

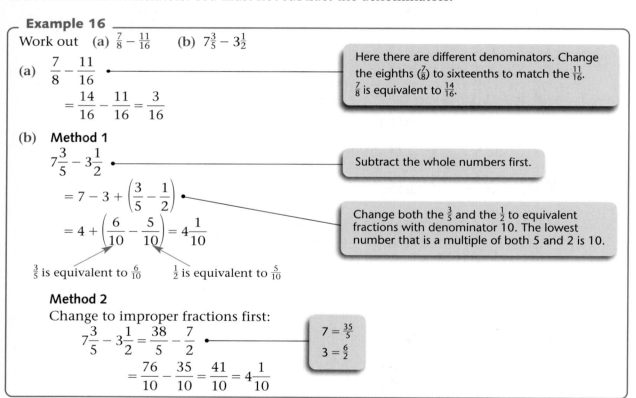

Example 16

Work out   (a) $\frac{7}{8} - \frac{11}{16}$ (b) $7\frac{3}{5} - 3\frac{1}{2}$

(a) $\dfrac{7}{8} - \dfrac{11}{16}$

$= \dfrac{14}{16} - \dfrac{11}{16} = \dfrac{3}{16}$

Here there are different denominators. Change the eighths ($\frac{7}{8}$) to sixteenths to match the $\frac{11}{16}$. $\frac{7}{8}$ is equivalent to $\frac{14}{16}$.

(b) **Method 1**

$7\dfrac{3}{5} - 3\dfrac{1}{2}$

Subtract the whole numbers first.

$= 7 - 3 + \left(\dfrac{3}{5} - \dfrac{1}{2}\right)$

$= 4 + \left(\dfrac{6}{10} - \dfrac{5}{10}\right) = 4\dfrac{1}{10}$

Change both the $\frac{3}{5}$ and the $\frac{1}{2}$ to equivalent fractions with denominator 10. The lowest number that is a multiple of both 5 and 2 is 10.

$\frac{3}{5}$ is equivalent to $\frac{6}{10}$ $\frac{1}{2}$ is equivalent to $\frac{5}{10}$

Method 2

Change to improper fractions first:

$7\dfrac{3}{5} - 3\dfrac{1}{2} = \dfrac{38}{5} - \dfrac{7}{2}$

$7 = \frac{35}{5}$
$3 = \frac{6}{2}$

$= \dfrac{76}{10} - \dfrac{35}{10} = \dfrac{41}{10} = 4\dfrac{1}{10}$

To subtract fractions, change them into equivalent fractions that have the same denominator. Then subtract the numerators.

Exercise 4J

Work out the subtractions in questions **1–4**.

1 (a) $\frac{3}{4} - \frac{1}{4}$ (b) $\frac{5}{8} - \frac{3}{8}$ (c) $\frac{15}{16} - \frac{7}{16}$ (d) $\frac{6}{7} - \frac{3}{7}$

2 (a) $\frac{1}{2} - \frac{3}{8}$ (b) $\frac{7}{8} - \frac{1}{2}$ (c) $\frac{7}{8} - \frac{3}{4}$ (d) $4\frac{5}{8} - 2\frac{1}{4}$

3 (a) $2\frac{3}{4} - \frac{3}{5}$ (b) $4\frac{4}{5} - 2\frac{2}{7}$ (c) $8\frac{2}{3} - 3\frac{2}{5}$ (d) $4\frac{5}{8} - 1\frac{2}{9}$

4 (a) $4\frac{7}{8} - 1\frac{2}{3}$ (b) $5\frac{7}{9} - 3\frac{1}{3}$ (c) $3\frac{4}{5} - \frac{3}{8}$ (d) $7\frac{4}{7} - 4\frac{2}{5}$

5 A box containing vegetables has a total weight of $5\frac{7}{8}$ kg.
   The empty box has a weight of $1\frac{1}{4}$ kg.
   What is the weight of the vegetables?

6 A tin contains $7\frac{1}{2}$ pints of oil. Julie pours out $4\frac{3}{8}$ pints from the tin.
   How much oil remains?

7 A plank of wood is $6\frac{1}{2}$ feet long. A $4\frac{3}{8}$ foot length is cut from one
   end of the plank. What length of wood remains?

8 Carol spends $\frac{2}{3}$ of her salary on food for her family. She spends
   $\frac{1}{4}$ of her salary on bills. What fraction of her salary is left?

9 Amarjit travels $5\frac{3}{8}$ km due North in his car, then $3\frac{1}{4}$ km due South.
   How far is he from his starting point?

4.8 Multiplying fractions

How to multiply a fraction by a whole number

To work out $\frac{7}{10} \times 4$ first write 4 as an improper fraction: $\frac{4}{1}$.

$$\frac{7}{10} \times \frac{4}{1} \quad \text{is} \quad \frac{7 \times \overset{2}{\cancel{4}}}{\underset{5}{\cancel{10}} \times 1} = \frac{7 \times 2}{5 \times 1} = \frac{14}{5}$$

So $\frac{7}{10} \times 4 = \frac{14}{5}$ which as a mixed number is $2\frac{4}{5}$.

> To multiply a fraction by a whole number, write the whole number as
> an improper fraction, then multiply the numerators and denominators.

Example 17

Find $\frac{2}{5}$ of 65 metres.

$$\frac{2}{5} \text{ of 65 m is } \frac{2}{5} \times \frac{65}{1} = \frac{2 \times 65}{5 \times 1}$$

Simplify by dividing
top and bottom by 5

$$= \frac{2 \times \overset{13}{\cancel{65}}}{\underset{1}{\cancel{5}} \times 1} = 26 \text{ m}$$

Worked examination question

Work out $\frac{3}{8}$ of 6 metres.

$$\frac{3}{8} \times \frac{6}{1} = \frac{3 \times \overset{3}{\cancel{6}}}{\underset{4}{\cancel{8}} \times 1} = \frac{9}{4} = 2\frac{1}{4} \text{ or 2.25 metres}$$

How to multiply two fractions

Finding $\frac{2}{3}$ of $\frac{6}{7}$ is equivalent to $\frac{2}{3} \times \frac{6}{7}$.

The blue part represents $\frac{6}{7}$

The pink part represents $\frac{2}{3}$ of $\frac{6}{7}$

So $\frac{2}{3}$ of $\frac{6}{7} = \frac{4}{7}$.

In this context the word **of** means **multiplied by**.

Multiply the numerators (top).

$$\frac{2}{3} \times \frac{6}{7} = \frac{12}{21} = \frac{4}{7}$$

Multiply the denominators (bottom).

It can be useful to simplify before multiplying:

$$\frac{2}{3} \times \frac{6}{7} = \frac{2 \times \overset{2}{6}}{\underset{1}{3} \times 7}$$ Simplify by dividing top and bottom by 3

$$= \frac{4}{7}$$

To **multiply** two fractions, multiply the numerators (top) then multiply the denominators (bottom).

To multiply or divide mixed numbers, change them to improper (top-heavy) fractions first.

Example 18

(a) Find the weight of a quarter of a $\frac{3}{5}$ kg packet of biscuits.

(b) Work out $2\frac{1}{2} \times 1\frac{2}{5}$

> Multiplying by $\frac{1}{4}$ is the same as dividing by 4.
> $\frac{1}{4}$ is called the **reciprocal** of 4.
> The reciprocal of a is $1 \div a$.

(a) $\frac{1}{4}$ of $\frac{3}{5}$ kg is $\frac{1}{4} \times \frac{3}{5}$ kg $= \frac{3}{20}$ kg

First multiply the numerators, then multiply the denominators.

(b) $2\frac{1}{2} \times 1\frac{2}{5} = \frac{5}{2} \times \frac{7}{5}$ ← First change mixed numbers to improper fractions.

$= \frac{\overset{1}{5} \times 7}{2 \times \underset{1}{5}}$ ← Simplify by dividing top and bottom by 5

$= \frac{7}{2} = 3\frac{1}{2}$ ← Always write the answer in its simplest form.

Exercise 4K

Work out the quantities in questions **1–4**.

1 (a) $\frac{1}{8}$ of 72   (b) $\frac{5}{6}$ of 36   (c) $\frac{4}{9}$ of 63 litres   (d) $\frac{7}{16}$ of 48 pints

2 (a) $\frac{1}{12}$ of £180   (b) $\frac{4}{5}$ of 4.63 kg   (c) $\frac{2}{3}$ of £3.96   (d) $\frac{17}{20}$ of 73 litres

3 (a) $\frac{3}{4}$ of 36 kg (b) $\frac{2}{5}$ of £5.55 (c) $\frac{11}{12}$ of 714 km (d) $\frac{4}{7}$ of 490 people

4 (a) $\frac{4}{9}$ of £38.25 (b) $\frac{4}{15}$ of 630 toys (c) $\frac{4}{9}$ of 1620 cars (d) $\frac{5}{7}$ of £52.01

5 During a 28-week holiday season, $\frac{2}{7}$ of the days were wet. How many dry days were there?

6 Jomo delivers 56 newspapers on his round. On Fridays $\frac{3}{8}$ of the newspapers have a magazine supplement. How many supplements does he deliver?

Work out the multiplications in questions **7–9**.

7 (a) $\frac{1}{2} \times \frac{3}{8}$ (b) $\frac{4}{5} \times \frac{2}{3}$ (c) $\frac{4}{7} \times \frac{1}{3}$ (d) $\frac{2}{3} \times \frac{2}{5}$

   (e) $\frac{2}{7} \times \frac{1}{5}$ (f) $\frac{2}{3} \times \frac{5}{7}$ (g) $\frac{1}{2} \times \frac{3}{4}$ (h) $\frac{3}{5} \times \frac{1}{3}$

8 (a) $\frac{1}{3} \times \frac{6}{7}$ (b) $\frac{6}{7} \times \frac{5}{12}$ (c) $\frac{1}{2} \times \frac{4}{5}$ (d) $\frac{2}{3} \times \frac{1}{4}$

   (e) $\frac{3}{7} \times \frac{2}{6}$ (f) $\frac{6}{5} \times \frac{1}{3}$ (g) $5 \times \frac{7}{10}$ (h) $\frac{9}{10} \times \frac{13}{18}$

9 (a) $\frac{2}{3} \times 1\frac{1}{3}$ (b) $\frac{2}{5} \times 2\frac{1}{3}$ (c) $1\frac{1}{2} \times \frac{1}{4}$ (d) $1\frac{1}{2} \times 2\frac{1}{2}$

   (e) $2\frac{1}{2} \times \frac{1}{4}$ (f) $1\frac{1}{5} \times 1\frac{1}{3}$ (g) $6 \times 2\frac{2}{3}$ (h) $2\frac{1}{7} \times 1\frac{2}{5}$

10 Barry earns £130.60 in one week. He pays $\frac{1}{4}$ of this in tax. How much money does he pay in tax each week?

11 Sara receives £2.70 pocket money each week. She saves $\frac{2}{5}$ of it. How much money does she save each week?

12 A melon weighs $2\frac{1}{2}$ lb. Work out the total weight of $8\frac{1}{4}$ melons.

13 Kieran takes $2\frac{1}{4}$ minutes to complete one lap at the Go Kart Centre. How long will it take him to complete $6\frac{1}{2}$ laps?

14 Find the area of this rectangle. Leave your answer as a fraction.

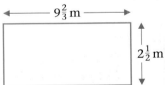

$9\frac{2}{3}$ m

$2\frac{1}{2}$ m

4.9 Dividing fractions

Here is a problem that you can solve if you divide by a fraction:

> The length of a relay race is 2 laps of a running track.
> Each member of a team runs $\frac{1}{5}$ of a lap.
> How many runners are there in a team?

You can easily work out in your head that there are 10 runners in a team. To show your method you need to find how many fifths of a lap there are in 2 whole laps.

You could work this out by dividing 2 by $\frac{1}{5}$.

To get the answer 10 you could work out $\frac{2}{1} \times \frac{5}{1}$.

You get this expression from $\frac{2}{1} \div \frac{1}{5}$ if you invert the second fraction (turn it upside down) and change the $\div$ sign to a $\times$ sign.

Why inverting works

A fraction like $\frac{3}{4}$ is the same as

$$3 \div 4 \quad \text{or} \quad 3 \times \frac{1}{4}$$

But $\frac{1}{4}$ is $\frac{4}{1}$ inverted.

So dividing is the same as multiplying by the inverted number:

$$12 \div 3 = 4$$
$$12 \times \frac{1}{3} = 4 \text{ too}$$

Example 19

$\frac{1}{4} \div \frac{3}{5} = ?$ — Turn the fraction you are dividing by upside down. This is called inverting the fraction.

$\frac{1}{4} \times \frac{5}{3} = \frac{5}{12}$ — Change the ÷ sign to a × sign.

To **divide** by a fraction, invert the dividing fraction (turn it upside down) and change the division sign to multiplication.

To divide mixed numbers, change to improper fractions first.

Example 20

Work out (a) $\frac{2}{5} \div \frac{1}{2}$ (b) $\frac{4}{9} \div 6$ (c) $3\frac{1}{2} \div 2\frac{2}{3}$

(a) $\frac{2}{5} \div \frac{1}{2} = \frac{2}{5} \times \frac{2}{1} = \frac{4}{5}$ — The $\frac{1}{2}$ becomes $\frac{2}{1}$. The ÷ becomes ×

(b) $\frac{4}{9} \div 6 = \frac{4}{9} \div \frac{6}{1}$ — The 6 can be written $\frac{6}{1}$, which becomes $\frac{1}{6}$ when you invert it.

$= \frac{4}{9} \times \frac{1}{6} = \frac{4}{54} = \frac{2}{27}$

(c) $3\frac{1}{2} \div 2\frac{2}{3} = \frac{7}{2} \div \frac{8}{3}$ — First write the mixed numbers as improper fractions.

$= \frac{7}{2} \times \frac{3}{8} = \frac{21}{16} = 1\frac{5}{16}$ — Remember to simplify the final answer.

Exercise 4L

Work out the divisions in questions **1–3**.

1 (a) $\frac{2}{9} \div \frac{1}{2}$ (b) $\frac{2}{5} \div \frac{3}{4}$ (c) $\frac{3}{8} \div \frac{2}{3}$ (d) $\frac{1}{2} \div \frac{1}{4}$

2 (a) $\frac{8}{9} \div 4$ (b) $\frac{2}{3} \div 6$ (c) $4\frac{2}{3} \div 4$ (d) $5\frac{1}{4} \div 3$

3 (a) $1\frac{1}{3} \div 1\frac{1}{2}$ (b) $2\frac{1}{2} \div \frac{1}{3}$ (c) $4\frac{3}{5} \div \frac{2}{3}$ (d) $2\frac{1}{5} \div 1\frac{1}{3}$

   (e) $1\frac{1}{3} \div 2\frac{2}{9}$ (f) $2\frac{2}{3} \div 2\frac{2}{5}$ (g) $3\frac{1}{3} \div 7\frac{1}{2}$ (h) $4\frac{4}{5} \div 5\frac{1}{3}$

4 A tin holds $10\frac{2}{3}$ litres of methylated spirit for a lamp. How many times will it fill a lamp holding $\frac{2}{3}$ litre?

5 A metal rod is $10\frac{4}{5}$ metres long. How many short rods $\frac{3}{10}$ metre long can be cut from the longer rod?

6 Tar & Stone Ltd can resurface $2\frac{1}{5}$ km of road in a day. How many days will it take them to resurface a road of length $24\frac{3}{5}$ km?

4.10 Solving problems involving fractions

Fractions are used in many different situations. The problems in this section are some of the types that are included in GCSE exams.

Example 21

A bag of flour weighs 2.25 kg. More flour is added and the weight of the bag of flour is increased by three fifths.
What is the new weight of the bag of flour?

$$\frac{3}{5} \text{ of } 2.25 \text{ kg is } \frac{3}{5} \times 2.25 = \frac{3 \times 2.25}{5} = \frac{6.75}{5} = 1.35 \text{ kg}$$

The new weight is 2.25 kg + 1.35 kg = 3.6 kg

Example 22

A 5 litre tin of paint is filled with blue and yellow paint to make a shade of green. The tin contains $1\frac{1}{2}$ litres of blue paint.
What fraction of the paint in the 5 litre tin is blue?

The fraction of paint that is blue is

$$\frac{\text{blue paint}}{\text{whole tin}} = \frac{1\frac{1}{2}}{5}$$

$$= 1\frac{1}{2} \div \frac{5}{1} = \frac{3}{2} \times \frac{1}{5} = \frac{3}{10}$$

Exercise 4M

1   A loaded lorry has a total weight of 13.2 tonnes. This weight is decreased by five eighths when the load is removed.
Find the weight of the lorry without the load.

2   Last year 204 cars were imported by a garage. This year the number of cars imported has increased by five twelfths.
How many cars have been imported this year?

3   Of 144 rail passengers surveyed, 32 claimed their train was regularly late.
What fraction of the total number of passengers was this?

4   There are 225 houses on an estate. Of these houses, 85 have no garage.
What fraction of houses have no garage?

5   Find the difference between $\frac{3}{5}$ of 36 miles and $\frac{2}{3}$ of 30 miles.

6   A tin contains approximately 440 beans. The manufacturer increases the volume of the tin by three eighths.
Approximately how many beans would you expect to find in the larger tin?

7   A newspaper has 14 columns of photographs and 18 columns of
     advertisements.
     What fraction of the paper is advertisements?

8   144 men, 80 women and 216 children went on the rollercoaster.
     What fraction of the total number is made up of children?

Mixed exercise 4

1   Complete these sets of equivalent fractions:

   (a) $\dfrac{5}{9} = \dfrac{}{18} = \dfrac{}{27} = \dfrac{}{36} = \dfrac{}{45}$ (b) $\dfrac{2}{5} = \dfrac{}{10}$

   (c) $\dfrac{3}{} = \dfrac{24}{64}$ (d) $\dfrac{}{7} = \dfrac{40}{35}$

2   Order these fractions, starting with the smallest.

     $\dfrac{3}{4}, \quad \dfrac{7}{8}, \quad \dfrac{2}{5}, \quad \dfrac{7}{10}$

3   Express these as decimals.

   (a) $\dfrac{7}{8}$ (b) $2\dfrac{3}{4}$ (c) $4\dfrac{9}{25}$ (d) $\dfrac{7}{9}$ (e) $2\dfrac{5}{6}$

4   Express these as fractions.

   (a) 0.04   (b) 2.725   (c) 5.68   (d) 0.3   (e) 0.45

5   Find the difference between $\dfrac{2}{3}$ of 15 kg and $\dfrac{3}{5}$ of 18 kg.

6   Two pipes of length $4\dfrac{5}{8}$ ft and $3\dfrac{3}{4}$ ft are put end to end.
     What is their total length?

7   Work out

   (a) 11.7 + 73.4   (b) 4.9 × 5
   (c) 28.5 ÷ 5   (d) 4.8 − 5.9 + 2.3
   (e) 15 ÷ 0.2   (f) 106.8 ÷ 0.15

8   A length of wire is $5\dfrac{7}{8}$ feet long. A length of $1\dfrac{3}{16}$ feet is cut off.
     What length of wire is left?

9   There are 70 ounces of sweets in a jar.
     How many $4\dfrac{1}{4}$ ounce bags can be filled from the jar?

10   Derek spent £2 on motor oil and £16 on petrol.
     What fraction of the total amount did he spend on petrol?

11 Two rods are fixed together to make a part
for a motor.

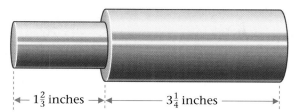

Work out the total length of the two rods.
Write your answer as a mixed number in its
simplest form.   [E]

12 A personal stereo was priced at £48. In a sale it was reduced to
£42. By what fraction was the original price reduced?   [E]

13 Work out $\frac{3}{4}$ of £24.   [E]

14 Six distances have been recorded:

   3.451 km, 3.506 km, 3.9 km, 3.008 km, 3.671 km, 3.91 km

Rewrite these distances in order, starting with the largest.

Summary of key points

1 $\frac{5}{2}$ is called an **improper fraction**. The numerator is greater than the denominator.

2 $2\frac{1}{2}$ is called a **mixed number**. It contains a whole number part and a fractional part smaller
than 1.

3 Mixed numbers can be written as improper fractions.
Improper fractions can be written as mixed numbers.

4 You can find **equivalent fractions** by
multiplying (or dividing) the top and
bottom of a fraction by the same number.

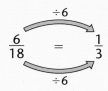

5 You can **order** fractions by rewriting them as equivalent fractions with a common denominator.

6 **Decimals** are used for parts of a number that are smaller than 1.

7 You can convert a fraction to a decimal by dividing the numerator by the denominator:
   $\frac{4}{5} = 4 \div 5 = 0.8$

8 0.6666666... is called a **recurring decimal** since one of the digits recurs (repeats).
Recurring decimal notation:
   0.$\dot{3}$ means 0.3333333... recurring
   0.$\dot{1}\dot{7}$ means 0.17171717... recurring
   0.$\dot{2}15\dot{4}$ means 0.215421542154... recurring.

9 To **add** fractions, change them to equivalent fractions that have the same denominator (bottom). Then add the numerators:

$$\frac{1}{2} + \frac{2}{6} = \frac{3}{6} + \frac{2}{6} = \frac{3+2}{6} = \frac{5}{6}$$

equivalent fractions

10 To **multiply** two fractions, multiply the numerators (top) then multiply the denominators (bottom):

$$\frac{3}{4} \times \frac{1}{7} = \frac{3 \times 1}{4 \times 7} = \frac{3}{28}$$

11 To multiply or divide **mixed numbers**, change them to improper (top-heavy) fractions first.

12 To **divide** by a fraction, invert the dividing fraction (turn it upside down) and change the division sign to multiplication:

$$\frac{1}{4} \div \frac{2}{5} = \frac{1}{4} \times \frac{5}{2} = \frac{1 \times 5}{4 \times 2} = \frac{5}{8}$$

5 Collecting and recording data

Every day you are surrounded by a sea of information.

This chapter is about the ways that information can be collected.

The information collected is called **data**.

5.1 Different types of data

Data is either qualitative or quantitative.

> **Qualitative data** is described using words.
> Place names such as North Wales, activities such as mountain biking, and colours such as red and green are **qualitative data**.

> **Quantitative data** consists of numbers.
> Data that can be counted or measured, such as the number of days in a holiday, temperatures and weights, are **quantitative data**.

Quantitative data can be **discrete** or **continuous**.

> **Discrete data** can only take particular values. For example, you can buy shoes in these sizes:
>
> $6 \qquad 6\frac{1}{2} \qquad 7 \qquad 7\frac{1}{2} \qquad 8$
>
> These values are **discrete** (meaning separate).
>
> There are *no* values in between them. Discrete data has an *exact* value.

> **Continuous data** can take any value. For example, your foot could be:
>
>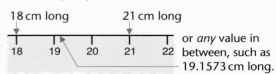
>
> 18 cm long   21 cm long
>
> or *any* value in between, such as 19.1573 cm long.
>
> Continuous data *cannot be measured exactly*. The accuracy of a measurement depends on the accuracy of the measuring device.

Exercise 5A

For each type of data below, write down whether it is qualitative, quantitative and discrete, or quantitative and continuous.

(a) length of a nail

(b) colour of a flower

(c) time taken to play a CD

(d) weight of a packet of sugar

(e) number of goals scored in a hockey match

(f) age of a tree

(g) taste of a drink

(h) smell of a scent

(i) number of buses per day

(j) score on a dice

(k) temperature of an oven

Reasons for collecting data

There are lots of reasons for collecting data. For example:

- market research prior to launching a new product
- quality control to make sure a product is up to standard
- collecting information for reference.

Market research

A company is about to launch a new product, for example a new soap powder.

It wants to know what to call its product and which type of advertising campaign is most likely to be successful.

It conducts a survey using interviews and a questionnaire.

Quality control

A fireworks company needs to be sure that its products will perform well.

Every day they take a batch of 20 of each type of firework and test them. The fireworks are graded 'excellent', 'good' or 'poor', depending on the number in the sample that perform well.

The production manager studies the results and decides what action to take.

> Information collected directly at first hand is called **primary data**.

Information and reference

A tourist office in a holiday resort keeps daily records of sunshine, rainfall and maximum and minimum temperatures.

Hotels in the resort may use this information in their brochures.

> Information taken from existing records is called **secondary data**.

5.2 Sampling

When collecting data you may have to restrict the number of items you survey. This could be because of the time available, the cost, or because taking a large survey is not practical.

The **population** is the complete set of items under consideration.

A **sample** is part of the population selected for surveying.

Selecting a suitable sample is crucial for reliable survey results. You would hardly expect to get a fair picture if you only asked your friends what they thought.

A **biased sample** is one which would be likely to have an undue influence on the outcome.

For example, a boys-only sample might well give a totally different picture to a girls-only sample.

In a **random sample** every member of the population has an equal chance of being chosen.

A **random process** must be unpredictable.

The sample size should be small enough to be manageable, but large enough to be representative of the whole population. In most cases a random sample of around 5% to 10% of the population should be sufficient.

Random sampling

If there are 600 students in your school, a manageable and representative sample size would be 30 students.

It is important that your sample is random. If you only ask Radiohead fans about their favourite band, your sample's views will not be representative of those of the whole population.

To take your random sample of 30:

1   List all 600 students in any order and label their names from 000 to 599.

000	C.J. Adams
001	P.E. Ash
002	C.R. Ashbee

2 Choose a way of generating random numbers from 000 to 599.
Here are two possible methods:

> Random numbers are generated by a random process – they cannot be predicted.

Method 1

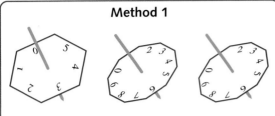

Use one 6-sided and two 10-sided spinners to generate the three digits.

Method 2

Use the **RND** function on a scientific calculator.

Stratified sampling

In a school survey on favourite bands it is likely that the students' age will influence their choices. Some bands have more appeal to younger students, some to older ones.

You might want to look at the popularity of the various bands according to the ages of students. To do this you can put the students in groups called **strata**, according to their ages.

> A **stratified sample** is one in which the population is divided into groups called strata and each of the strata is randomly sampled.

A convenient way of doing this in a school would be to use the year groups of the students as the strata.

For example, the table shows the numbers in each year group in a school of 600 students.

There are more students in Year 7 than in Year 10, so the sample should include more students from Year 7 than Year 10. Year 7 students are $\frac{150}{600} = \frac{1}{4}$ of the whole school, so they should make up $\frac{1}{4}$ of the sample.

For example you could take a 6% sample of each of Years 7 to 11.

A 6% sample is in the range 5%–10% and gives whole numbers of students in each of the strata:

Year group	Number of students
7	150
8	150
9	100
10	100
11	100

Year group	Number of students	Number in 6% sample
7	150	9
8	150	9
9	100	6
10	100	6
11	100	6

> Sample size = 36
> Year 7 = $\frac{9}{36} = \frac{1}{4}$ of sample
> Year 10 = $\frac{6}{36} = \frac{1}{6}$ of sample

For each of the strata the sample you take must be random.

Systematic sampling

Systematic sampling is often used in quality control, where items such as light bulbs, batteries, computer chips and tinned food need to be tested at the factory to ensure that they have been made to satisfactory standards.

A **systematic sample** is one in which every nth item is chosen.

To take a 5% sample you select 1 in every 20 items.

$5\% = \frac{5}{100} = \frac{1}{20}$

You could take a 5% systematic sample of the students at your school. To do this you would list all the names. Randomly select any number from 1 to 20, for example 8. Then from your list choose the 8th, 28th, 48th, 68th... student names.

Example 1

There are 1200 students at Lucea High School. This table shows how students are distributed by year group and gender.

Jamilla is conducting a survey about the students' favourite school subject. She decides to use a stratified random sample of 100 students according to year group and gender.

Year group	Number of boys	Number of girls
7	144	156
8	150	150
9	87	113
10	102	96
11	100	102

How many Year 7 girls should there be in her sample?

The number of girls in Year 7 is 156.

As a proportion of the total of 1200 students, this is $\frac{156}{1200} = 0.13$

Jamilla's total sample is to be 100 so she should sample

$0.13 \times 100 = 13$ girls from Year 7

Exercise 5B

1   For each method, say whether it would give a random sample or whether it is likely to be biased.

　　(a)　Take every 10th name from a list.

　　(b)　Interview all the people in a launderette.

　　(c)　Put all the names into a hat and pick without looking.

　　(d)　Test the first article produced each hour by a machine.

　　(e)　Number all the items and then select by using random number tables.

Random number tables give lists of numbers generated at random by a computer.

2 Use the data in Example 1 on page 75 for this question.

(a) Work out the number of students Jamilla should sample from
(i) Year 8   (ii) boys in Year 10.

(b) Explain why the correct answer to (ii) above will create a problem for Jamilla. How could Jamilla resolve this problem?

(c) Work out the number of students Jamilla should sample from boys in Year 9.

(d) Comment on your answers to (b) and (c).

3 Just before a local election a market research company makes a survey of the voting intentions of the electorate. There

Age range (years)	Number of males	Number of females
Under 30	5800	6200
30 or over	9500	8500

are 30 000 people who can vote in the election. They are categorised according to age and gender as in the table above.

The market research company will try to obtain the views of a 2000-person sample.

Work out the number in the sample who should be

(a) aged under 30   (b) male   (c) female and aged 30 or more.

4 Explain how you could take a systematic sample of 2% of the names in a telephone directory.

5.3 Collecting primary data

Questionnaires

One way of collecting data is to use a **questionnaire**.

> When you are writing questions for a **questionnaire**:
> - be clear about what you want to find out, and what data you need
> - ask short, simple questions.

Here are some good examples:

Are you:

Male ☐

Female ☐

———— This has a clear choice of two answers.

What age are you?

Under 17 ☐

17–20 ☐

21–30 ☐

Over 30 ☐

Which of these styles of music do you like best?

Metal ☐

House ☐

Grunge ☐

None of them ☐

These both offer four choices.

Avoid questions that are vague, personal or **biased** (which may influence the answer).

How often do you go swimming?

Sometimes ☐   Occasionally ☐   Often ☐

Sometimes, occasionally and *often* may mean different things to different people.

Have you ever stolen anything from a shop?

Yes ☐   No ☐

Even a hardened criminal is unlikely to answer this question honestly!

Do you agree that the UK should have the euro?

Yes ☐   No ☐

This question suggests that the right answer is Yes. It is **biased**.

Test your questionnaire on a few people first to see if it works or needs to be improved. This is called a **pilot survey**.

___ **Example 2** _____

The Governors of Lucea High School want to find out whether parents think that girls attending the school should wear ties. Design a question to obtain parents' views on this which could be used in a questionnaire.

Several designs are possible. Here is one example:

Please tick the box which most accurately reflects your view on this statement:

Girls attending Lucea High School should wear ties.

☐ Strongly   ☐ Agree   ☐ Not sure   ☐ Disagree   ☐ Strongly
   agree   disagree

You could use only the middle three boxes. Using any more than the five boxes shown here would make it harder to find a pattern in the responses.

Avoiding bias

When you are collecting data you need to make sure that your survey or experiment is **fair** and avoids **bias**. A bias is anything that might make the data unrepresentative.

For example, if you want to find out which sport is most popular amongst 14–16 year olds in the UK, you will not be able to ask everyone in this age group. You will have to ask a **sample** of 14–16 year olds and treat their views as being representative of all people of their age.

If you only asked 14–16 year old boys at a football match, your survey would be biased because you would be collecting the views of male football enthusiasts.

The question 'Do you agree that running is good for you?' is biased too, because it suggests that a particular answer is correct.

Exercise 5C

1 Look at these pairs of questions. Choose whether Question X or Question Y is better to do the job in the *To find out* column.

	To find out:	Question X	Question Y
(a)	if people like the Labour party	Do you like the Labour party? Yes/No	Do you agree that New Labour is the best party? Yes/No
(b)	which age group a person is in	How old are you?	Are you under 21, 21 to 40, 41 to 60, over 60?
(c)	the most popular soap on TV	Do you watch *Eastenders*? Yes/No	Do you watch *Eastenders/ Coronation Street/ Emmerdale/Hollyoaks/ Other*?
(d)	if they watched the film	Did you see the film *War of the Worlds*? Yes/No	Do you like Steven Spielberg?
(e)	if someone is short sighted	How good is your eyesight?	Can you read that sign without glasses? Yes/No
(f)	if people give money to charity	Do you give money to charity? Yes/No	Everyone gives money to charity, don't they? Yes/No
(g)	if the hotel was satisfactory	Was everything in the hotel all right? Yes/No	Did you enjoy your stay? Yes/No
(h)	if newspapers should be censored	Newspapers should be censored: Agree/Disagree/Don't know	What do you think about censoring newspapers?

2 Here are some questions that are not suitable for a questionnaire. For each one, say why and write a more suitable question.

(a) Do you agree that the UK should have a monarchy?
Yes ☐   No ☐   Don't know ☐

(b) What was the weather like on your holiday?
Terrible ☐   Quite good ☐   OK ☐

(c) Most people approve of corporal punishment. Do you?
Yes ☐   No ☐

(d) Do you still play football?
Yes ☐   No ☐

(e) How many hours of TV do you watch?
1 ☐   2 ☐   3 ☐

(f) Does your local library have wheelchair access?
Yes ☐   No ☐

3 Finn wants to find out about people's mobile phones. He has designed the following questionnaire for his web page.

    (a) How could you improve part **4** of the questionnaire?

    (b) Finn also wants to write a question about how people buy their call time. Design a question he could use. Include tick boxes for a response.

Questionnaire: Mobile phones

1 Do you have a mobile phone?

    Yes ☐   No ☐

2 Who is your service provider?

    Mobile P ☐ H_4 ☐   Five ☐

    Fonavode ☐   Pineapple ☐   Other ☐

3 Do you: Pay monthly ☐

             Pay-As-You-Go ☐

4 How much do you spend on your phone each month?

    £5 ☐   £10 ☐   £15 ☐   £20 ☐

Activity – Holidays

Design a questionnaire to find out about the kind of holidays people had last year.

Test your questions by asking some of your friends.

Observation and measurement

Data can also be collected by observation (e.g. a traffic survey) or by measurement (e.g. weighing bars of chocolate produced). This method is suitable for quality control.

Example 3

You are commissioned to do a survey on traffic flows on a particular road to see whether parking restrictions are required. What factors might affect traffic flow? Design a data capture sheet for this survey.

Things which might make a difference to traffic flow include:

- day of the week
- time of day
- types of vehicle using the road
- direction of flow.

> A data capture sheet can be used to record primary data.

Here is a possible data capture sheet to record results:

Monday		7 am – 10 am	10 am – 1 pm	1 pm – 4 pm	4 pm – 7 pm
North/ South	cars vans lorries others				
South/ North	cars vans lorries other				

Exercise 5D

1 Prepare data capture sheets for surveys to find out two of the following by observation:

(a) the make of people's MP3 player

(b) the colour of people's eyes

(c) the most popular brand of sports shoe of people in your school

(d) the CPU speed and hard disk space on people's computers

(e) the age and gender of people entering a supermarket.

2 Design a suitable data capture sheet for quality control at a fireworks factory. There is a day shift and a night shift and five machines. Since testing means that the firework cannot be sold, you are only allowed to test 1000 fireworks each week.

3 A supermarket wishes to conduct research to follow the sales trends of five items over a period of 3 months. There are 2045 branches, and 200 branches are going to run cut-price promotion offers on the five items.
Plan the data collection.

5.4 Secondary data

Secondary data is information that has already been collected and recorded. It is quick, cheap and ready for you to use. If such data is adequate for your research, it makes sense to use it.

For example, suppose you want to know if there is any evidence to suggest that homes are becoming safer places. You would need to collect records first hand for years before you could make any comment.

However, the *Annual Abstract of Statistics,* available at libraries and online, will give you details of deaths in the home and accidents at home that needed hospital treatment. This provides information going back 20 years or more.

> Data like these are obtained from the National Census, held every ten years.

The internet is a good place for finding secondary data.

Obtaining data from the internet

The internet can be a very useful source of data. But remember that data collected from the internet may be inaccurate or out of date.

When you collect data from the internet make sure that
● the data comes from a reliable source
● the data is accurate – check against other sources.

Activity – Internet search

Use the internet to find the following information.
Give two reliable sources for each answer.

(a) The members of the European Union.

(b) The heights of the five highest mountains in the world.

(c) The average life expectancy of people in the UK.

(d) The number of gold, silver and bronze medals won
by Great Britain in the Paralympic Games in 2004.

Obtaining data from a database

A **database** is an organised collection of information. It can be stored
on paper or on a computer.

Here is a spreadsheet showing part of a database stored on a computer:

	A	B	C	D	E	F	G	H
	Year group	Surname	Years	Months	Gender	Sport	Height (m)	Weight (kg)
2	10	Abejurouge	15	3	Male	Rugby	1.63	60
3	10	Aberdeen	15	0	Male	Rounders	1.75	45
4	11	Ableson	16	6	Female	Table tennis	1.83	60
5	11	Acton	16	3	Female	Basketball	1.67	52
6	10	Adam	15	1	Male	Judo	1.80	49
7	10	Agha	15	7	Male	Cricket	1.66	70

A computer database allows you to obtain information quickly and in
a variety of forms, for example:

- in alphabetical order
- girls' results only
- in numerical order
- male students over a certain height.

Exercise 5E

1   Say whether you would collect primary data or use secondary
data for the following research. Explain your answers.

(a) a survey on train punctuality

(b) research into wildlife in an African National Park

(c) investigating levels of student attendance at
your school

(d) market research into sales at the school tuck shop

(e) compiling a report on the major tourist attractions in the UK

2 (a) Give one example of
     (i) a project where you would use primary data
     (ii) a project where you would use secondary data.
   (b) Give an example of a situation where you could collect primary data and then compare it with secondary data.

Activity – Mayfield School database (Go to www.heinemann.co.uk/hotlinks, insert the express code 4092P and click on this activity.)

Use the Mayfield school database to answer these questions:

(a) How tall is David Hazelwood (Year 9)?

(b) How many KS4 students were born in April?

(c) How many KS3 students said their favourite TV programme is *The Simpsons*? Which one of these said their favourite sport is running?

(d) How many KS3 students are right handed and have dark brown eyes? Which of these said their favourite TV programme is *Eastenders*?

(e) Which female KS4 student has 5 pets and said her favourite subject is PE?

5.5 Recording and presenting grouped data

When quantitative data has a wide range of values, it makes sense to group sets of values together. This makes it easier to record the data and to spot any patterns or trends.

For example, scores out of 50 in a test might be grouped:

0–10,   11–20,   21–30,   31–40,   41–50

It often makes life easier if all the groups are the same size, but they do not have to be. In this case the first group has 11 members and the others have 10.

The groupings 0–10, 11–20 and so on are called **class intervals**.

Class intervals are groupings of quantitative data.

Here is a frequency table for a test taken by some pupils:

The class intervals must not overlap.
Each score can only be in one class interval.

The test scores are discrete data.
You can score 31 or 32, but *not* 31.4

Score	Tally	Frequency
0–10	‖	2
11–20	ЖН ‖‖	8
21–30	ЖН ЖН ‖‖‖	14
31–40	ЖН	5
41–50	‖	1

You can also group continuous data. Here is some data from a traffic study. Kuljit has timed the interval in seconds between successive vehicles passing her school for 50 vehicles. The times are correct to the nearest 0.1 second.

3.3	21.0	23.5	9.4	14.6	46.8	34.3	45.8	56.1	50.4	12.2	7.6	14.4
16.5	34.5	35.7	3.5	2.8	5.7	10.0	45.1	46.2	23.6	17.1	9.7	38.4
3.5	7.3	28.7	32.1	51.7	21.3	6.1	17.4	12.2	43.2	56.3	34.5	37.9
12.0	17.3	3.2	15.3	4.2	24.4	35.6	29.8	7.1	3.7	10.2		

To record this data in a frequency table you need to group it into class intervals. If the class intervals are too narrow there will be too many groups to show up any pattern in the data. If the class intervals are too wide there will be too few groups to show up a pattern.

There is no golden rule for choosing the sizes of the class intervals or the number of groups, but it is usually best to have no fewer than five groups and no more than ten.

Here is a frequency table for the data grouped into class intervals of width 10 seconds:

This means 0 up to but not including 10

This means 50 up to but not including 60

Class interval (seconds)	Tally	Frequency
$0 \leqslant t < 10$	ЖН ЖН IIII	14
$10 \leqslant t < 20$	ЖН ЖН II	12
$20 \leqslant t < 30$	ЖН II	7
$30 \leqslant t < 40$	ЖН III	8
$40 \leqslant t < 50$	ЖН	5
$50 \leqslant t < 60$	IIII	4

Presenting grouped data

Data that is grouped and continuous can be displayed in a **histogram**.

Here is a histogram for the traffic data from the frequency table:

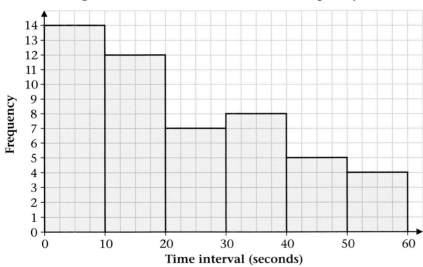

A histogram *looks* like a bar chart but:
- the data is continuous so there can be no gaps between the bars
- the *area* of the bars represents the frequency, so the data must be grouped into class intervals of equal width if you want to use the *lengths* of the bars to represent data values.

There is more about histograms in Chapter 20.

Exercise 5F

1 As part of his geography fieldwork, Tony took measurements of the
steepness of slopes. The steepness was measured as the angle the
slope made with the horizontal. Tony's results are shown below.

      15° 16°   9° 21° 32° 37° 25° 36° 40°   8° 32°

      13° 21° 32° 29 32° 29° 32°   7°   4° 18° 17°

Copy and complete the frequency distribution table below, using
four equal class intervals.

Class interval (steepness in °)	Tally	Frequency
1–10		

                                                                [E]

2 Forty people took part in a clay pigeon shooting competition.
The points they scored are shown below.

    18   24   19   3   24   11   25   10   25   14

    25   14   25   9   16   26   21   27   13   23

     5   26   22   12   27   20   7   28   21   20

    22   16   12   25   7   25   19   17   15   8

  (a) Using the class intervals 1–5, 6–10, 11–15, 16–20, 21–25,
26–30, construct the frequency table.

  (b) Use the same data to construct a frequency table with ten
class intervals.

  (c) Which of your frequency tables do you think is better?
Why?

3 Neville weighed 3 dozen hens' eggs to see if there was much
difference between them. They all came from boxes marked
Large. These are his results:

      63.1 g 65.4 g 61.9 g 62.2 g 63.4 g 65.1 g 61.7 g 62.2 g

      64.9 g 63.5 g 67.1 g 62.8 g 61.9 g 62.6 g 63.1 g 64.3 g

      65.0 g 63.7 g 65.2 g 61.8 g 62.1 g 63.5 g 65.1 g 66.3 g

      64.4 g 62.3 g 61.5 g 63.7 g 64.7 g 65.1 g 62.3 g 62.8 g

      64.7 g 62.2 g 63.6 g 61.4 g

Use class intervals 61.0–61.9, 62.0–62.9, 63.0–63.9, 64.0–64.9,
65.0–65.9, 66.0 and over, to construct a frequency table.

Activity – Hand-span

(a) Measure the hand-span of each person in your class.

(b) Record the data in a frequency table using class intervals of equal width.

(c) Draw a histogram to display your data.

Grouped data in two-way tables

Daniel wants to find out if the amount of pocket money school students receive is related to their age. For each student in his survey he needs to record two variables – age and amount of pockey money.

> A two-way table records two variables.

He records the data in a two-way table:

		Amount of pocket money		
		£0–£3.00	£3.01–£6.00	£6.01–£9.00
Age	$11 \leqslant x < 12$			
	$12 \leqslant x < 13$			
	$13 \leqslant x < 14$			
	$14 \leqslant x < 15$			
	$15 \leqslant x < 16$			

> The variables can be either discrete (e.g. money) or continuous (e.g. age).

Example 4

The tourist office at a holiday resort records the hours of sunshine and the maximum daily temperature for publicity purposes. Design a suitable two-way table for collecting the data.

		Hours of sunshine				
		$0 \leqslant y < 3$	$3 \leqslant y < 6$	$6 \leqslant y < 9$	$9 \leqslant y < 12$	over 12
Maximum daily temperature (°C)	$15 \leqslant x < 20$					
	$20 \leqslant x < 25$					
	$25 \leqslant x < 30$					
	over 30					

Exercise 5G

Design two-way tables to record the data in questions **1–3**.

1 Height (ranging from 160 cm to 200 cm).
   Waist measurement (ranging from 65 cm to 100 cm).

2 Average speed of a journey (ranging from 30 km/h to 60 km/h).
   Length of journey (ranging from 10 km to 100 km).

3 Height of a nest above the ground (ranging from 0 cm to 400 cm).
   Number of eggs in the nest (between 1 and 7 inclusive).

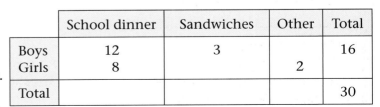

4 Thirty pupils were asked about their lunch one day. The table gives some information about their answers.

   (a) Copy and complete the table.

   (b) How many of the girls had sandwiches?   [E]

	School dinner	Sandwiches	Other	Total
Boys	12	3		16
Girls	8		2	
Total				30

5.6 Frequency polygons

A **frequency polygon** can be used to show the general pattern of data. Plot the frequency for each class interval against the mid-point of that interval.

This table shows the frequency distribution of marks in a test:

Mark	0–9	10–19	20–29	30–39	40–49	50–60
Frequency	16	14	18	9	6	2

This is discrete data grouped into class intervals 0–9, 10–19 and so on.

One way of presenting this data is to plot a graph of frequency against mark. For each class interval, plot the mid-point against the frequency. Join the points with straight lines.

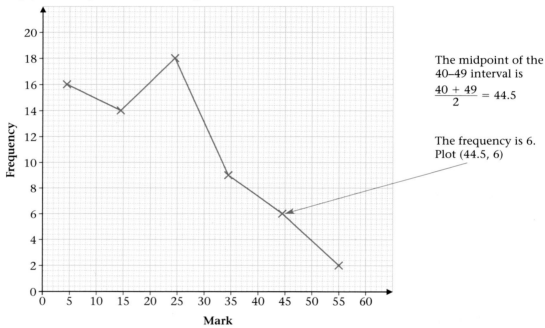

The midpoint of the 40–49 interval is

$$\frac{40 + 49}{2} = 44.5$$

The frequency is 6. Plot (44.5, 6)

The resulting graph is called a **frequency polygon**. Here the mid-point of each class interval is treated as being representative of all the points in that interval.

Exercise 5H

1   James asked people to draw a line 5 cm long using a straight edge without any markings on it. Here are the lengths in centimetres of the lines they drew.

    4.3   3.2   3.9   4.7   5.8   6.1   5.7   6.2   6.5   3.7   4.2
    5.1   6.5   7.2   7.4   3.7   5.8   4.2   4.1   5.0   5.1   4.7
    3.2   3.5   5.2   2.9   2.8   4.3   5.1   4.8

    (a)   Organise these results in a grouped frequency table. Use a class interval of 1 centimetre.

    (b)   Draw a frequency polygon for the data.

2 This frequency polygon shows the times of 25 girls who each had to sew on two buttons with standard lengths of thread.

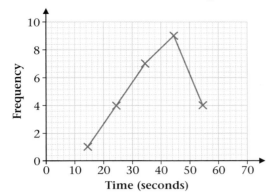

A group of 25 boys was challenged to do the same task. Here are their times, to the nearest second.

    25 32 46 32 57 46 51 28 49 68 55 78 32
    44 89 22 56 67 77 40 41 48 60 *   *

The stars represent two boys who gave up.

(a) How will you deal with the two boys who gave up?

(b) Group the data and draw a frequency polygon.

(c) Compare the two frequency polygons. What conclusions can you draw from the two frequency polygons?

Mixed exercise 5

1 Grace and Gemma were carrying out a survey on the food people eat in the school canteen. Grace wrote the question 'Which food do you eat?' Gemma said this question was too vague.

Write down two ways in which this question could be improved.
[E]

2 Fred is conducting a survey into television viewing habits. One of the questions on his survey is 'How much television do you watch?' His friend, Sheila, tells him that it is not a very good question.
Write down two ways in which Fred could improve this question.
[E]

3 Nazia is going to carry out a survey of the types of DVD her friends have watched.
Draw a suitable data collection sheet that Nazia could use.
[E]

> 'Data collection sheet' is another name for 'data capture sheet'.

4 Rachel and Terry were collecting information on the types of meals bought by students in the school canteen.
Draw a suitable data collection sheet for this information.
[E]

5 There are 1000 students in Nigel and Sonia's school.
Explain how Nigel could take a random sample of students to carry out a survey.
[E]

6   Categorise these sampling methods as random, not random but unlikely to be biased, or biased.
    (a)   Take every 20th name from a telephone directory.
    (b)   Use all those with a birthday in June.
    (c)   Use married men for a survey on football.
    (d)   Give all members of a tennis club a number and use random number tables.

7   Luca is trying to find out which weekly magazines are read most by students in his school. He decides to ask 10 of his friends.
    (a)   Explain why this is not a reliable method.
    (b)   Suggest two steps that Luca could take to get more reliable results.

8   There are 1200 students at Russell High School.

    Maria wishes to take a random sample of 60 of these students for her humanities project.

    Describe at least three different ways in which Maria could take such a sample.

9   The table shows the number of students in each year group at a school.

Year group	Number of students
7	270
8	270
9	240
10	200
11	180

    Mrs Fox is the music teacher. She is carrying out a survey about the students' favourite radio channel. She uses a stratified sample of 40 students according to year group.
    Calculate how many Year 10 students should be in her sample.   [E]

10 Bob carried out a survey of 100 people who buy tea. He asked them about the tea they buy most. The two-way table gives some information about his results.

	Tea bags	Packet tea	Instant tea	Total
50 g	2	0	5	
100 g	35	20		60
200 g	15			
Total		25		100

    Copy and complete the two-way table.   [E]

Sales of Fairtrade tea are increasing in Britain. Fairtrade initiatives should ensure workers earn fair wages.

11 This table gives information on the sales of petrol and diesel in a garage last week.

Day	Number of litres of petrol sold	Number of litres of diesel sold
Monday	400	400
Tuesday	600	200
Wednesday	700	300
Thursday	450	700
Friday	200	900
Saturday	700	800
Sunday	600	700

   (a) Draw two frequency polygons on the same axes to represent this data.

   (b) Comment on the sales of petrol and diesel last week.

12 The table shows the frequency distribution of the ages of members of a sports club in 2000 and 2005.

Age	0–9	10–19	20–29	30–39	40–49	50–59
2000	6	16	22	29	18	9
2005	9	23	27	21	10	7

Draw two frequency polygons on the same axes and compare the ages of members in 2000 and 2005.

Summary of key points

1 There are two types of data:
   - **Qualitative data** is described using words. For example, Sheffield, blue, euro...
   - **Quantitative data** consists of numbers: it is data that can be counted or measured.

2 Quantitative data can be discrete or continuous:
   - **Discrete data** can take only particular values. For example, shoe sizes 6, $6\frac{1}{2}$, 7...
   - **Continuous data** can take any value. For example, lengths 1.5 cm, 2.54 cm, 62.643 cm...

3 Information collected directly at first hand is called **primary data**.
   Information taken from existing records is called **secondary data**.

4 The **population** is the complete set of items under consideration.
   A **sample** is part of the population selected for surveying.
   A **biased sample** is one which would be likely to have an undue influence on the outcome.
   In a **random sample** every member of the population has an equal chance of being chosen.
   A **random process** must be unpredictable.

5 A **stratified sample** is one in which the population is divided into groups called strata and each of the strata is randomly sampled. It must have the same proportion from each of the strata as occurs in the whole population.

6 A **systematic sample** is one in which every nth item is chosen.

7 When you are writing questions for a **questionnaire**:
   - be clear about what you want to find out, and what data you need
   - ask short, simple questions
   - avoid questions that are vague, personal, or **biased** (which may influence the answer).

   Test your questionnaire on a few people first to see if it works or needs to be improved. This is called a **pilot survey**.

8 When you collect data from the internet make sure that:
   - the data comes from a reliable source
   - the data is accurate – check against other sources.

9 A **database** is an organised collection of information. It can be stored on paper or on a computer.

10 **Class intervals** are groupings of quantitative data.

11 Data that is grouped and continuous can be displayed in a **histogram**.

12 A histogram looks like a bar chart but:
   - the data is continuous so there can be no gaps between the bars
   - the *area* of the bars represents the frequency, so the data must be grouped into class intervals of equal width if you want to use the *lengths* of the bars to represent data values.

13 A **frequency polygon** can be used to show the general pattern of data. Plot the frequency for each class interval against the mid-point of that interval.

⑥ Solving equations and inequalities

In this chapter you will learn more about solving equations.

6.1 Solving simple equations

In Section 2.9 you solved simple equations using the balancing method.
Remember this key point from that section:

> Whatever you do to one side of an equation you must also do to the other side.

___ **Example 1** _____

Solve $14 = 9 - 2x$.

$$14 = 9 - 2x \bullet$$
$$14 - 9 = -2x \quad\text{——— Subtract 9 from both sides.}$$
$$5 = -2x$$
$$-2\tfrac{1}{2} = x \quad\text{——— Divide both sides by } -2.$$

Check by substituting $x = -2\tfrac{1}{2}$ into the original equation:
$$14 = 9 - 2(-2\tfrac{1}{2}) = 9 + 5 = 14$$

So $x = -2\tfrac{1}{2}$ is the solution.

> You need to collect the number terms on one side of the equation and the letter terms on the other.

Exercise 6A

Solve these equations. Check your answers by substitution.

1 $2 - 3x = 17$ **2** $4x + 3 = -5$ **3** $7 = 2x + 2$ **4** $16 = 5 - 2x$ **5** $2a + 7 = 1$

6 $11 - 5b = 3$ **7** $9 = 16 + 2c$ **8** $-13 = 5 - 6p$ **9** $14 - 3q = 2$ **10** $23 = 3 - 5s$

6.2 When the x term includes a fraction

Sometimes the coefficient of x is a fraction, for example in $\tfrac{1}{4}x - 5 = 2$.
Here are two ways of solving the equation:

___ **Method 1** _____

$$\tfrac{1}{4}x - 5 = 2$$
$$\tfrac{1}{4}x = 2 + 5 \quad\text{——— Add 5 to both sides.}$$
$$\tfrac{1}{4}x = 7$$
$$x = 28 \quad\text{——— Multiply both sides by 4.}$$

___ **Method 2** _____

$$\tfrac{1}{4}x - 5 = 2$$
$$x - 20 = 8 \quad\text{——— Multiply both sides by 4.}$$
$$x = 8 + 20 \quad\text{——— Add 20 to both sides.}$$
$$x = 28$$

Check by substituting $x = 28$ in $\tfrac{1}{4}x - 5 = 2$: $\tfrac{1}{4} \times 28 - 5 = 7 - 5 = 2$

So $x = 28$ is the solution.

Exercise 6B

Solve these equations and check your answers.

1 $\frac{1}{3}x - 4 = 2$ **2** $\frac{1}{2}x + 7 = 4$

3 $\frac{x}{4} - 5 = 1$ **4** $\frac{3}{4}x + 2 = 9$

5 $\frac{2}{3}x - 4 = -1$ **6** $9 - \frac{x}{2} = 2$

7 $7 - \frac{x}{3} = 9$ **8** $6 - \frac{x}{4} = 9$

9 $5 - \frac{2}{3}x = 1$ **10** $3 - \frac{3x}{4} = 8$

6.3 When the 'unknown' x is on both sides

In some equations, for example in $5x - 4 = 3x - 10$, there is an x term on both sides. To solve this, rearrange the equation so that all the x terms are on one side.

Example 2

Solve $5x - 4 = 3x - 10$.

$$5x - 3x - 4 = -10 \quad\text{————— Subtract } 3x \text{ from both sides.}$$
$$2x - 4 = -10$$
$$2x = -10 + 4 \quad\text{————— Add 4 to both sides.}$$
$$2x = -6$$
$$x = -3 \quad\text{————— Divide both sides by 2.}$$

Check by substituting $x = -3$ in $5x - 4 = 3x - 10$

Left side: $(5 \times -3) - 4 \ = -15 - 4 = -19$

Right side: $(3 \times -3) - 10 = -9 - 10 = -19$

So $x = -3$ is the solution.

For equations with the unknown on both sides, rearrange so the unknowns are on one side of the equation and the numbers are on the other.

Exercise 6C

Solve these equations and check your answers by substitution.

1 $2a + 5 = a + 9$ **2** $6b - 5 = 4b + 1$

3 $6c + 1 = 4c + 2$ **4** $2d + 3 = 7d - 1$

5 $5e + 1 = 2e - 5$ **6** $3f - 2 = 8f + 3$

7 $4g - 5 = 8g - 11$ **8** $7h - 3 = 4h + 8$

9 $2j - 9 = 8j - 1$ **10** $9k + 8 = 4k - 3$

6.4 Unknown on both sides and negative coefficients

Some equations, for example $5x + 4 = 2 - 3x$, have the 'unknown' x on both sides and one of the x terms has a negative coefficient, in this case $-3x$.

To solve this type of equation, rearrange it so that the unknown appears on only one side.

Example 3

Solve $5x + 4 = 2 - 3x$.

$5x + 3x + 4 = 2$ ——————— Add $3x$ to both sides.

$8x + 4 = 2$

$8x = 2 - 4$ ——————— Subtract 4 from both sides.

$8x = -2$

$x = -\frac{2}{8}$ ——————— Divide both sides by 8.

$= -\frac{1}{4}$

Example 4

Solve $1 - 5x = 10 - 2x$.

$1 = 10 - 2x + 5x$ ——————— Add $5x$ to both sides.

$1 = 10 + 3x$

$1 - 10 = 3x$ ——————— Subtract 10 from both sides.

$-9 = 3x$

$-3 = x$ ——————— Divide both sides by 3.

So $x = -3$

> Here both x terms have negative coefficients.

Exercise 6D

Solve these equations. Check your answers by substitution.

1   $4x + 3 = 8 - x$ 2 $3x + 4 = 7 - 3x$ 3 $2 - 5x = 3x + 14$

4   $9 - 2x = 3 - 4x$ 5 $7 - 8a = 4 - 3a$ 6 $5 - 6b = 8 - 2b$

7   $9 - 7c = 3 - 2c$ 8 $5d - 7 = 2 - 4d$ 9 $9 - 4e = 9 + 3e$

10   $8 - 3f = 1 - 6f$

6.5 Solving equations involving brackets

Sometimes you will need to solve equations involving brackets.

> See Section 2.4 to remind yourself how to multiply out brackets.

Example 5

Solve $3(x + 2) = 10$.

$3x + 6 = 10$ ——————— Expand the brackets.

$3x = 4$ ——————— Subtract 6 from both sides.

$x = \frac{4}{3} = 1\frac{1}{3}$ ——————— Divide both sides by 3.

Example 6

Solve $3(x + 8) = 27x - 6$.

You could always multiply out the brackets first and then continue as before. Here, the quicker thing to do is to divide both sides by 3 (as both the terms on the RHS are divisible by 3).

$$x + 8 = 9x - 2 \quad\text{————— Divide both sides by 3.}$$
$$x + 8 + 2 = 9x \quad\text{————— Add 2 to both sides.}$$
$$x + 10 = 9x$$
$$10 = 9x - x \quad\text{————— Subtract } x \text{ from both sides.}$$
$$10 = 8x$$
$$x = 1.25 \quad\text{————— Divide both sides by 8.}$$

Exercise 6E

Solve these equations.

1 $5(x + 2) = 30$

2 $2(x + 7) = 8$

3 $6(x - 2) = 1$

4 $3(x - 1) = 7$

5 $4(x + 1) = 9x$

6 $5(x - 4) = 3x - 7$

7 $3(2h + 1) = h - 2$

8 $2(3p - 3) = 5(p - 1)$

9 $2(2q + 3) = 3(4q - 2)$

10 $4(9 - 3r) = 3(r + 2)$

11 $4 - 2t = 2(8 + t)$

12 $2(13 - 2x) = 9(9 + 2x)$

13 $3(4x - 5) - 5(2x - 3) = 4(x - 1)$

14 $6 + (5x - 7) - 2(3x + 4) = 7(x + 1)$

15 $4(3x - 2) - (7x + 8) = 2(x - 5)$

16 $6(x + 5) - 5(3x - 4) = 11 + 3(2x - 7)$

17 $2(7x - 4) + (5 - 9x) = 1 - 3x$

18 $3(5x + 2) - (9x - 8) = 7(x - 4) + (x + 12)$

19 $8x - 5(2x + 3) - (1 - 6x) = 8 + (4 + 6x)$

20 $5(3x - 7) + 2(9 - 2x) = 8x - (2x + 1) - 16$

21 $3(5x - 4) - 7(2x - 3) = 3(x + 1) + 5$

6.6 Equations involving algebraic fractions

Example 7

Solve the equation $\dfrac{p}{4} - 2 = \dfrac{p + 4}{6}$

> $\dfrac{p + 4}{6}$ is an algebraic fraction.

Multiply both sides of the equation by the LCM of the denominators.

The LCM of 4 and 6 is 12.

$$12\left(\frac{p}{4} - 2\right) = 12\left(\frac{p + 4}{6}\right)$$ —— Multiply both sides by 12.

$$3p - 24 = 2p + 8$$ —— Expand the brackets.

$$p - 24 = 8$$ —— Subtract $2p$ from both sides.

$$p = 32$$ —— Add 24 to both sides.

> $12\left(\dfrac{p}{4} - 2\right) = \dfrac{12 \times p}{4} - 24$
> $= 3p - 24$
> $12\left(\dfrac{p + 4}{6}\right) = 2(p + 4)$
> $= 2p + 8$

Exercise 6F

Solve these equations.

1 $\dfrac{h}{5} + 4 = 16$

2 $\dfrac{p - 4}{8} = 7$

3 $\dfrac{m + 7}{5} = 7$

4 $\dfrac{m}{3} + 2 = \dfrac{m}{4} + 3$

5 $\dfrac{g}{2} - 1 = \dfrac{g}{4} + 1$

6 $\dfrac{n + 3}{3} = \dfrac{n}{5} + 3$

7 $\dfrac{x}{6} + 1 = \dfrac{x - 4}{4}$

8 $3(2p - 10) = \dfrac{4p - 7}{2}$

9 $5\left(\dfrac{r}{3} - 3\right) = 40$

10 $\dfrac{x}{4} - 6 = 1 - 2x$

6.7 Using equations to help solve problems

To solve 'mystery number' problems:

- Use a letter to represent the unknown number.
- Write the problem as an equation.
- Solve the equation.
- Write the answer.

Mystery Number Challenge

Example 8

I add 3 to a mystery number and multiply the result by 5. My answer is 55. Find the mystery number.

Write the problem as an equation with x as the mystery number:

$$(x + 3) \times 5 = 55$$
$$5(x + 3) = 55$$
$$x + 3 = 11$$ —— Divide both sides by 5.
$$x = 11 - 3$$ —— Subtract 3 from both sides.
$$x = 8$$

The mystery number is 8.

Exercise 6G

Solve the problems in questions **1** and **2** by writing and solving an equation.

1 I subtract 9 from a number and multiply the result by 4.
My answer is 36. Find the number.

2 I add 2 to a number and multiply the result by 5. The answer is the same as when I add 8 to the number and multiply the result by 3. Find the number.

3

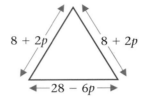

(a) The diagram shows an equilateral triangle.
Write down an equation in p that must be true.

(b) Solve your equation to find the value of p.

(c) What is the length of a side of the triangle?

4 (a) The diagram shows a square.
Write down an equation in x that must be true.

(b) Solve your equation to find the value of x.

(c) What is the length of a side of the square?

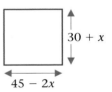

5 Priya has £15 at the start of half-term holiday. During the holiday she goes to the cinema three times. Serena starts with £12 and goes to the cinema twice. Neither of them spends any other money. At the end of the week they have equal amounts of money left. Let c stand for the cost of going to the cinema.

(a) Write down an equation in c that must be true.

(b) Solve your equation to find the cost of going to the cinema.

6 A cafe sells orange juice and apple juice.
At the start of one day there are 70 litres of orange juice and 60 litres of apple juice. During the day the cafe sells 60 glasses of orange juice and 20 glasses of apple juice.
At the end of the day there are equal amounts of orange and apple juice left.
Let g stand for the amount of juice in a glass.

(a) Write down an equation in g that must be true.

(b) Solve your equation to find the amount of juice in a glass.

6.8 Showing inequalities on a number line

An expression in which the left hand side and the right hand side are *not* equal is called an **inequality**. For example $x > 2$

You can use a number line to show an inequality such as $x > 2$:

> | > | greater than |
> | < | less than |
> | ⩾ | greater than or equal to |
> | ⩽ | less than or equal to |

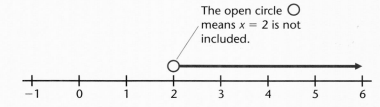

The open circle ○ means $x = 2$ is not included.

Here is $x \leqslant 3$:

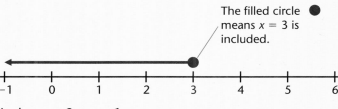

The filled circle ● means $x = 3$ is included.

And this shows $-2 \leqslant x < 1$:

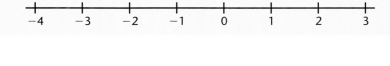

Exercise 6H

Write down the inequalities shown in questions **1–4**.

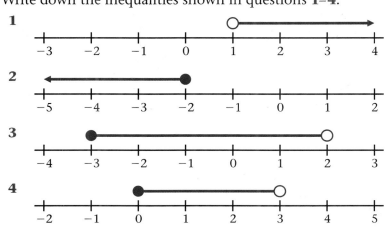

1

2

3

4

Draw number lines to show the inequalities in questions **5–10**.

5 $x < 4$　　　　　**6** $x \geqslant -2$　　　　**7** $x \leqslant -1$

8 $1 < x < 5$　　　**9** $-2 < x \leqslant 0$　　**10** $-3 \leqslant x < -1$

6.9 Solving inequalities

$3 > -6$ is an inequality. It is also a true statement, so $3 > -6$ is a true inequality.

But does the inequality remain true when you do the following things to it?

(a)   Add 4 to both sides: $3 + 4 > -6 + 4$
                                    $7 > -2$ is still true

(b)   Subtract 2 from both sides: $3 - 2 > -6 - 2$
                                    $1 > -8$ is still true

(c)   Multiply both sides by 2: $3 \times 2 > -6 \times 2$
                                    $6 > -12$ is still true

(d)   Divide both sides by 3: $3 \div 3 > -6 \div 3$
                                    $1 > -2$ is still true

(e)   Multiply both sides by -5: $3 \times -5 > -6 \times -5$
                                    $-15 > 30$ is **not true**

(f)   Divide both sides by -3: $3 \div -3 > -6 \div -3$
                                    $-1 > 2$ is **not true**

> You can solve inequalities in the same way as solving equations **except** that you must not multiply or divide by a *negative* number.

To solve an inequality, rearrange it so that x is on one side of the inequality, to give a range of values for x.

To solve an inequality you **can**:
- add the same quantity to both sides
- subtract the same quantity from both sides
- multiply both sides by the same *positive* quantity
- divide both sides by the same *positive* quantity.

But you **must not**:
- multiply both sides by a *negative* quantity
- divide both sides by a *negative* quantity.

Example 9

Solve $3x + 4 > 22$.

$3x > 18$ ———————————— Subtract 4 from both sides.

$x > 6$ ———————————— Divide both sides by 3.

Example 10

Solve $6 - 5x \geqslant 3x + 2$.

$6 \geqslant 3x + 5x + 2$ ———————— Add $5x$ to both sides.

$6 \geqslant 8x + 2$

$6 - 2 \geqslant 8x$ ———————— Subtract 2 from both sides.

$4 \geqslant 8x$

$\frac{1}{2} \geqslant x$ ———————— Divide both sides by 8.

Another way of writing $\frac{1}{2} \geqslant x$ is $x \leqslant \frac{1}{2}$.

Exercise 6I

Solve these inequalities.

1 $3x - 5 \leqslant 4$

2 $5x + 2 > 4$

3 $4x + 7 \geqslant 7x + 2$

4 $7x + 2 < 5x - 4$

5 $8 - 3a \leqslant 5$

6 $1 - 6b > 7$

7 $-5c < 20$

8 $-2d \geqslant -3$

9 $\frac{1}{2}e + 3 \leqslant 1$

10 $5 - \frac{1}{4}f > 12$

11 $2 - \frac{1}{2}g \geqslant 1$

12 $4 - 5h \leqslant 3h$

13 $8 - 3j < 2j + 13$

14 $7 - 5k \leqslant 12 - 8k$

15 $8 - 2m \geqslant 1 - 4m$

16 $4 - 9n > 1 - 11n$

Solving two-sided inequalities

To solve a two-sided inequality, treat it as two inequalities. Solve each inequality separately, then combine the two solutions.

___ **Example 11** ___

Solve $7 \leqslant 3x - 2 < 9$.

Treat this as two inequalities:

$$7 \leqslant 3x - 2 \quad \text{and} \quad 3x - 2 < 9$$
$$9 \leqslant 3x \qquad\qquad 3x < 11$$
$$3 \leqslant x \qquad\qquad x < 3\tfrac{2}{3}$$

Write the answer as $3 \leqslant x < 3\frac{2}{3}$.

Exercise 6J

Solve these inequalities.

1 $7 < 4p + 3 \leqslant 27$

2 $3 < 2q - 7 < 8$

3 $1 \leqslant 3r + 4 \leqslant 10$

4 $-2 < \frac{1}{2}s + 1 < 3$

Giving integer solutions to inequalities

Integers are the positive and negative whole numbers and 0:
$$\dots -3 \quad -2 \quad -1 \quad 0 \quad 1 \quad 2 \quad 3 \dots$$

Some questions will ask you for only the integer solutions of an inequality.

___ **Worked examination question 1** ___

List all the possible integer values of n such that $-3 \leqslant n < 2$.

$$n = -3, \quad -2, \quad -1, \quad 0, \quad 1$$

Worked examination question 2

Given that n is an integer, find the greatest value for n for which $4n + 3 < 18$.

$$4n + 3 < 18$$
$$4n < 15$$
$$n < 3\tfrac{3}{4}$$

The greatest integer value less than $3\tfrac{3}{4}$ is 3. So $n = 3$

Exercise 6K

1 List the possible integer values of n.

    (a) $-2 \leqslant n \leqslant 1$ (b) $0 < n < 5$

    (c) $-3 \leqslant n < 0$ (d) $-1 < n \leqslant 3$

2 Write down an inequality satisfied by the integers listed.

    (a) 2, 3, 4, 5   (b) 0, 1, 2, 3

    (c) $-3, -2, -1$ (d) $-2, -1, 0, 1, 2$

3 Find the greatest integer value of n.

    (a) $3n + 2 < 10$ (b) $8 - 5n \geqslant 2$

    (c) $4n + 7 < 5$ (d) $1 - 3n > 9$

4 Find the lowest integer value of n.

    (a) $4n + 5 \geqslant 1$ (b) $7 - 2n < 4$

    (c) $6n + 7 > 2$ (d) $5 - 4n < -2$

5 Find all the possible integer values of n.

    (a) $5 \leqslant 2n + 1 < 9$ (b) $4 < 5n - 1 \leqslant 14$

    (c) $-4 < 3n + 5 < 12$ (d) $-2 \leqslant 1 - 2n < 3$

6.10 Solving inequalities involving brackets

Remember that inequalities can be solved in the same way as equations except that you must not multiply or divide both sides by a *negative* number.

Example 12

Solve $3(x - 2) \leqslant 20$.

$$3x - 6 \leqslant 20 \text{ ————— Expand the brackets.}$$
$$3x \leqslant 26 \text{ ————— Add 6 to both sides.}$$
$$x \leqslant 8\tfrac{2}{3} \text{ ————— Divide both sides by 3.}$$

Another way of solving $3(x - 2) < 20$ is to divide by 3 first.

You can do this because 3 is positive.

Example 13

Solve $2(x + 2) + 3(2x - 3) > 2$.

$2x + 4 + 6x - 9 > 2$ —————— Expand the brackets.

$8x - 5 > 2$ —————— Simplify.

$8x > 7$ —————— Add 5 to both sides.

$x > \frac{7}{8}$ —————— Divide both sides by 8.

Example 14

Solve $2(3x - 2) + 3(4 - 3x) > 12 - (5x + 7)$.

$6x - 4 + 12 - 9x > 12 - 5x - 7$ —————— Expand the brackets.

$8 - 3x > 5 - 5x$ —————— Simplify.

$8 + 2x > 5$ —————— Add $5x$ to both sides.

$2x > -3$ —————— Subtract 8 from both sides.

$x > -1\frac{1}{2}$ —————— Divide both sides by 2.

Exercise 6L

Solve these inequalities.

1 $3(x - 7) > 14$

2 $5(x + 4) > 40$

> Divide both sides by 5.

3 $4(x + 1) \leqslant 2x - 3$

4 $2(3x - 5) \geqslant 8 - 3x$

5 $3(x + 2) - 5(x - 4) < 27$

6 $5 - (3x + 4) \geqslant 4$

7 $7(x - 2) - (5x - 9) > 0$

8 $3x + 4(2x + 3) < 1$

9 $4(2x - 1) \leqslant 3(x + 4)$

10 $3 - (2x - 3) > 6$

11 $6(3x + 1) - 2(5x - 4) < 5(x + 2)$

12 $3(4x + 5) + (8 - 10x) \leqslant 4(x + 6)$

13 $7(3x - 4) - (12x - 5) > 5(2x - 3)$

14 $8x + 7(2 - x) < 9 - 3(x + 1)$

Mixed exercise 6

In questions **1–14** solve the equations.

1 $6a - 5 = 13$

2 $5b + 2 = 9$

3 $8c + 10 = 2$

4 $4d + 5 = 3d + 9$

5 $7e + 4 = 5e - 8$

6 $6f - 1 = 3f + 1$

7 $8 - g = 10$

8 $-4h = 10$

9 $7 - 5i = 4$

10 $\frac{j}{2} + 3 = 7$

11 $9 - \frac{1}{4}k = 11$

12 $\frac{2m}{3} + 5 = 9$

13 $7n - 6 = 14 - 3n$

14 $8 - 3p = 13 - 5p$

15 I multiply a number by 5 and add 3. The result is the same as when I multiply the number by 8 and subtract 18. Find the number.

16 Pineapples cost six times as much as lemons.
The cost of 20 pineapples and 200 lemons is £86.40.
Find the cost of a lemon.

17 (a)

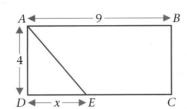

ABCD is a rectangle, measuring 4 cm by 9 cm.
E is a point on DC with $DE = x$ cm.
The area of trapezium ABCE is 26 cm². Find the value of x.

(b)

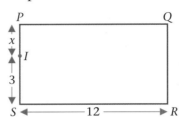

The area of rectangle PQRS is 60 cm². Find the value of x.

18 Draw number lines to show these inequalities:
(a) $x > 5$ (b) $x \leqslant -3$
(c) $3 < x \leqslant 5$ (d) $-4 \leqslant x < 0$

19 Solve these inequalities:
(a) $8x + 3 < 19$ (b) $6 - 5x \geqslant 16$ (c) $3 - 2x < 7 - 5x$

20 Find (i) the greatest and (ii) the lowest integer values of n for which these inequalities are true:
(a) $4 \leqslant 3n - 2 < 10$ (b) $1 < 4n + 5 < 20$

21 The diagram shows the lengths, in cm, of the sides of a rectangle. The perimeter of the rectangle is P cm.
Work out the value of P.   [E]

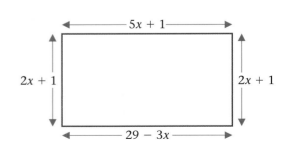

22 Solve the inequality
$6(x + 4) - 7 - (3x + 10) \geqslant 2(4x - 5)$

23 Solve the equation $\dfrac{y + 5}{3} = 6 - 2y$

Summary of key points

1 Whatever you do to one side of an equation you must also do to the other side.

2 For equations with the unknown on both sides, rearrange so the unknowns are on one side of the equation and the numbers are on the other.

3 An expression in which the left hand side and the right hand side are *not* equal is called an **inequality**. For example $x > 2$

4 You can use a number line to show an inequality such as $x > 2$:

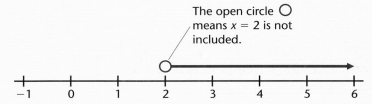

The open circle ○ means $x = 2$ is not included.

Here is $x \leqslant 3$:

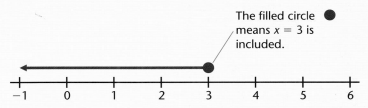

The filled circle ● means $x = 3$ is included.

And this shows $-2 \leqslant x < 1$:

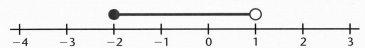

5 To solve an inequality you **can**:
   • add the same quantity to both sides
   • subtract the same quantity from both sides
   • multiply both sides by the same *positive* quantity
   • divide both sides by the same *positive* quantity.
   But you **must not**:
   • multiply both sides by a *negative* quantity
   • divide both sides by a *negative* quantity.

6 **Integers** are the positive and negative whole numbers and 0:

$$\ldots -3 \quad -2 \quad -1 \quad 0 \quad 1 \quad 2 \quad 3 \ldots$$

7 Transformations and loci

This chapter shows you how to translate, rotate, reflect and enlarge shapes on a grid, and how to make accurate constructions and scale drawings.

Translations, rotations and reflections change the position of the object, but not its size or shape. The image and object are **congruent**.

7.1 Transformations

A **transformation** is a change in an object's position or size. You need to know the four kinds of transformation – translation, reflection, rotation and enlargement.

Translation

A **translation** moves every point on a shape the same distance and direction.

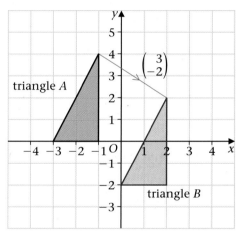

In this diagram triangle A is translated to B. B is the **image** of A. All the points on A are moved $+3$ units parallel to the x-axis followed by -2 units parallel to the y-axis.

This translation is described by the **vector** $\begin{pmatrix} 3 \\ -2 \end{pmatrix}$.

In the vector $\begin{pmatrix} x \\ y \end{pmatrix}$

- x gives the movement parallel to the x-axis
- y gives the movement parallel to the y-axis.

We say that triangle **A** 'maps onto' triangle **B**.

To describe a translation fully you need to give the distance moved and the direction of the movement. You can do this by giving the vector of the translation.

Example 1

Triangle B is a translation of triangle A.
Describe the translation that takes A to B.

Triangle A has moved 2 squares in the x-direction
 1 square in the y-direction.

The translation has vector $\begin{pmatrix} 2 \\ 1 \end{pmatrix}$.

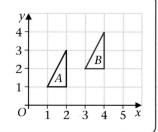

Exercise 7A

1   Write the vectors describing these translations:

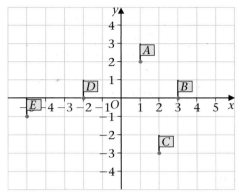

    (a)   flag A to flag B

    (b)   flag B to flag A

    (c)   flag D to flag B

    (d)   flag C to flag E

    (e)   flag A to flag E.

2   Use graph paper or squared paper. Draw a set of axes and label each one from -5 to 5. Use the same scale for each axis. Draw a trapezium A, with vertices at $(-5, 3)$, $(-4, 3)$, $(-3, 2)$ and $(-3, 1)$.

    (a)   Transform A by the translation $\begin{pmatrix} 7 \\ 1 \end{pmatrix}$.

        Label the new trapezium B.

    (b)   Transform B by the translation $\begin{pmatrix} -1 \\ -6 \end{pmatrix}$.

        Label the new trapezium C.

    (c)   Write the single translation required to transform A to C.

    (d)   Write the single translation required to transform C back to A.

3   Use graph paper or squared paper. Draw a set of axes and label each one from -5 to 5. Plot points $A\,(0, 1)$ and $B\,(3, 2)$.

    T_1 describes a translation of $\begin{pmatrix} 3 \\ 2 \end{pmatrix}$. T_2 describes a translation of

    $\begin{pmatrix} -2 \\ 1 \end{pmatrix}$. Write down the coordinates of A and B after these translations:

    (a) T_1 (b) T_1 followed by T_1 again

    (c)   Write down a single translation for **(b)**

    (d) T_1 followed by T_2 (e) T_2 followed by T_1

    (f)   Write down what you notice about the images in **(d)** and **(e)**.

Reflection

Stand directly in front of a mirror. You see your reflected **image**. The image appears to be the same distance away from the mirror as you are. Move closer to the mirror and the image moves closer. The mirror acts as a plane of symmetry.

> A **reflection** in a line produces a mirror image.
> The mirror line is a line of symmetry.

The diagram on the right shows a point A reflected in a line.

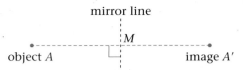

The original shape A and the image A' are the same distance from the line, but on opposite sides.

If you draw a line from A to A' to cross the line of reflection at M. $AM = A'M$ and AA' is at right angles to the line of reflection

Notice that A' is a reflection of A and A is a reflection of A'.

You can draw the reflection of a shape by using these facts to find the images of different points on the shape.

Example 2

Copy the diagram on squared paper.

(a) Draw the reflection of the shape in the line $y = x$.

(b) Reflect the image back in the line $y = x$.
What do you notice?

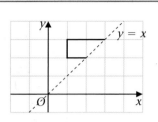

(a) Take each vertex on the object and locate its image.
Imagine a line perpendicular to the line of symmetry.
Join the points to produce the image.

> For a sloping line of reflection, it is easier to draw a reflection if you turn the paper until the line is vertical.

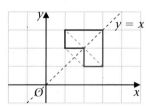

(b) The image reflected in the line $y = x$ returns back to the object position.

A point on the mirror line remains unchanged by reflection.

A **repeated reflection** in the same mirror line returns the image to the original object.

To describe a reflection fully you need to give the equation of the mirror line.

Exercise 7B

1 Draw a coordinate grid with both x- and y-axes going from -4 to 4.

(a) Plot the points $P(-2, 1)$, $Q(0, 1)$, $R(1, 0)$, $S(3, 4)$ and $T(-2, 4)$ and join them in order to form the closed shape $PQRST$.

(b) Draw the image of $PQRST$ after a reflection in the x-axis.

2 Describe fully the transformation that maps shape *A* onto *B*.

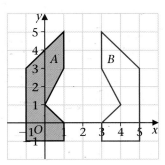

3 Use graph paper or squared paper. Draw a set of axes and label each axis from −5 to 5. Use the same scale for each axis.
Draw a rectangle with vertices (3, 1), (3, 2), (5, 2), (5, 1).
Label the rectangle *A*.

   (a)   **(i)**   Draw the reflection of *A* in the *x*-axis.
   Label the image *B*.

   (ii)   Write the coordinates of the vertices of *B*.

   (iii)   Describe how the coordinates of *A* are changed by the transformation.

   (b)   **(i)**   Draw the reflection of *A* in the *y*-axis. Label the image *C*.

   (ii)   Write the coordinates of the vertices of *C*.

   (iii)   Describe how the coordinates of *A* are changed by the transformation.

   (c)   Draw the reflection of *A* in the line $y = x$. Label the image *D*.

   (d)   Draw the reflection of *D* in the *y*-axis. Label the image *E*.

   (e)   Describe fully the reflection that transforms *E* to *C*.

4   **(a)**   Copy this diagram. Extend the
   x- and *y*-axes to −5.
   Reflect the object in the line $x = 3$.

   (b)   Reflect the image in the line $y = 2$.

   (c)   Reflect the object in the line $y = -x$.

Rotation

An object can be turned about a point.
This point is called the centre of rotation.

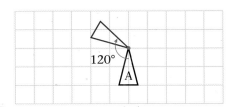

Triangle *A* is turned 120° clockwise about a **centre of rotation**.

To describe a **rotation** fully you need to give the
- centre of rotation
- amount of turn
- direction of turn.

(A clockwise direction may be indicated by a negative sign, and an anticlockwise direction by a positive sign.)

Example 3

Describe fully the transformation which maps shape P onto shape Q.

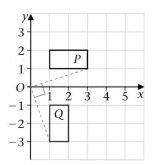

Rotation of 90° clockwise about (0, 0).

Each vertex of the rectangle has been rotated 90° clockwise.

> You can use tracing paper to help you find the angle and centre of rotation.

The transformation in Example 3 could also be described as a rotation of 270° anticlockwise about the origin.

Exercise 7C

1 Use graph paper or squared paper. Draw a set of axes and label each axis from −5 to 5. Use the same scale for each axis. Draw a triangle with vertices (3, 1), (5, 1) and (5, 3). Label the triangle A.

   (a) Rotate A about the origin through 90° anticlockwise. Label the image B. Write the coordinates of the vertices of B.

   (b) Rotate B about the origin through 90° anticlockwise. Label the image C. Write the coordinates of the vertices of C.

   (c) Write a single rotation to transform A to C.

   (d) Rotate A about the origin through 90° clockwise. Label the image D. Write the coordinates of the vertices of D.

   (e) Write a single rotation to transform B to D.

   (f) Rotate A about the point (3, 0) through 90° clockwise. Label the image E.

2 Describe fully the transformation which maps shape A onto shape B.

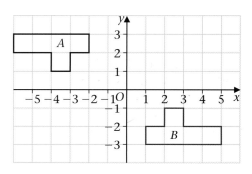

Enlargement

Different sized photographs can be printed from the same negative. In each photograph the dimensions are in the same proportion as in the original. The images are not distorted. Each length is enlarged or reduced by a **scale factor**.

> An **enlargement** changes the size but not the shape of an object. The **scale factor** of the enlargement is the number of times the lengths of the original object have been enlarged.

The sides of shape B are all twice as long as the corresponding sides of shape A. You can say that B is an enlargement of A with scale factor 2. All the corresponding angles are equal so A and B are **similar** (they are the same shape). A and B are the same way up.
Each line of B is parallel to the corresponding line of A.

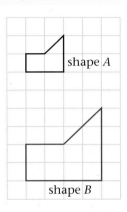

The scale factor can be positive, fractional or negative. In the diagram below, triangle ABC has been enlarged by scale factor 2, using X as the centre of enlargement.

For an enlargement about a centre X:

length $XA' =$ scale factor $\times$ length XA
length $XB' =$ scale factor $\times$ length XB,   etc.

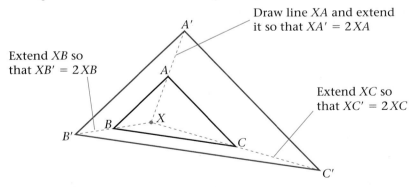

Draw line XA and extend it so that $XA' = 2XA$

Extend XB so that $XB' = 2XB$

Extend XC so that $XC' = 2XC$

The enlarged triangle $A'B'C'$ is similar to triangle ABC. Each side of $A'B'C'$ is twice as long as the corresponding side in ABC. This is because the scale factor of the enlargement is 2. Corresponding angles are identical. Each of the lines in the enlarged shape is parallel to the corresponding original side.

> To describe an enlargement fully you need to give its centre and the scale factor.

Example 4

(a) Enlarge triangle ABC with scale factor $\frac{1}{2}$ using O as the centre of enlargement.

(b) Enlarge triangle ABC with scale factor -1 using O as the centre of enlargement.

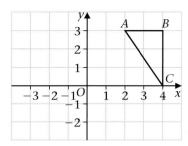

(a) $OA \times \frac{1}{2} = OA'$ $\quad OB \times \frac{1}{2} = OB'$ $\quad OC \times \frac{1}{2} = OC'$
An enlargement of scale factor $\frac{1}{2}$ has reduced each side of the triangle ABC. Triangles ABC and $A'B'C'$ are similar.
The ratio of corresponding sides is $1 : \frac{1}{2}$ or $2 : 1$

(b) $OA \times -1 = OA''$ $\quad OB \times -1 = OB''$ $\quad OC \times -1 = OC''$
A negative scale factor indicates measuring from the centre of the enlargement in the opposite direction.

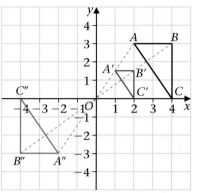

An enlargement by a scale factor smaller than 1 gives a reduced image.

An enlargement by a negative scale factor indicates that the measuring from the centre of the enlargement must be in the opposite direction.

Example 5

Describe fully the transformation that maps $ABCD$ onto $PQRS$.

$PQRS$ is the same shape as $ABCD$. The sides of $PQRS$ are all 2 times the lengths of the corresponding sides of $ABCD$.
So the transformation is an enlargement by a scale factor 2.

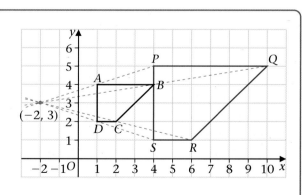

To find the centre of enlargement:
Join P to A and continue the line.
Join Q to B and continue the line.
In the same way, join R to C and S to D and continue the lines.

The lines all pass through the same point $(-2, 3)$, so this is the centre of enlargement.

The transformation is an enlargement by scale factor 2 with centre of enlargement $(-2, 3)$.

Exercise 7D

1   Copy this shape on squared paper and
    then draw these enlargements:

    (a)   centre of enlargement at the origin
          and scale factor of enlargement 2

    (b)   centre of enlargement at point (2, 1)
          and scale factor of enlargement 2

    (c)   centre of enlargement at point (0, 0)
          and scale factor of enlargement $\frac{1}{2}$.

    (d)   centre of enlargement at point (0, 4) and scale factor of
          enlargement $\frac{1}{2}$

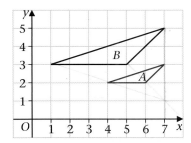

2   Describe fully the transformation that maps shape B onto
    shape A.

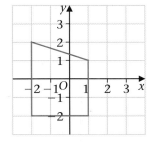

3   Copy this shape on squared paper and then draw the enlargement
    with centre of enlargement at the origin and scale factor of
    enlargement $\frac{1}{2}$.

4   Copy this shape on squared paper
    and then draw the enlargement with
    centre of enlargement at the origin
    and scale factor of enlargement -3.

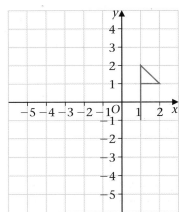

5   State the centre of enlargement and scale
factor of enlargement for

(a)   shape *A* to shape *B*

(b)   shape *A* to shape *C*

(c)   shape *C* to shape *A*

(d)   shape *A* to shape *D*.

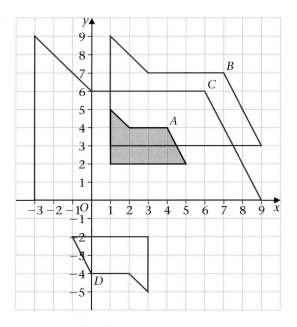

6   These LOGO instructions produce a square:

Alter these instructions to enlarge the square by scale factor 3
with centre of enlargement at the starting position.

```
repeat 4 [fd 15 rt 90]
end
```

7   Describe fully the
enlargement that

(a)   maps *A* onto *B*

(b)   maps *B* onto *A*.

(c)   What do you notice
about the scale factors?

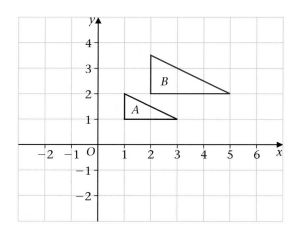

8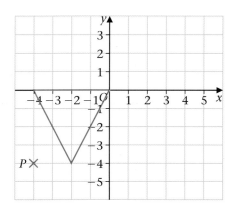

Draw a copy of the diagram
on squared paper.
Enlarge the triangle by a
scale factor $1\frac{1}{2}$, centre *P*.

9 Describe fully the single transformation which maps triangle *P* onto triangle *Q*.

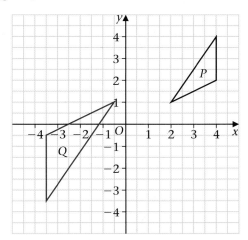

7.2 Combined transformations

Transformations can be combined by performing one transformation and then performing another transformation on the image.

Example 6

(a) Reflect the flag in the *y*-axis.

(b) Reflect the image in the line *y* = *x*.

(c) Describe the single transformation to replace (a) and (b).

(d) Reflect the image from (a) in the line *x* = 2.

(e) Describe the single transformation to replace (a) and (d).

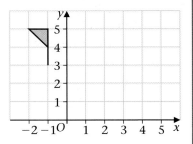

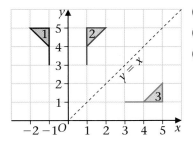

(a) Flag 2 is the image of flag 1.

(b) Flag 3 is the image of flag 2.

(c) Single transformation: rotate flag 1 through 90° clockwise about the origin.

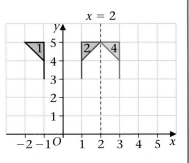

(d) The image from (a) is flag 2. Flag 4 is the image of flag 2.

(e) Single transformation: translate flag 1 by $\begin{pmatrix} 4 \\ 0 \end{pmatrix}$.

A reflection followed by a reflection can be replaced by a single transformation of
- a rotation if the reflection lines are *not* parallel
- a translation if the reflection lines are parallel.

Exercise 7E

1 (a) Reflect the triangle in the *x*-axis.

  (b) Reflect the image in the *y*-axis.

  (c) Describe the single transformation that replaces (a) and (b).

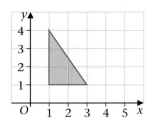

2 (a) Reflect the shape in the line *x* = 3.

  (b) Reflect the image in the line *x* = 6.

  (c) Describe the single transformation that replaces (a) and (b).

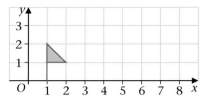

3 (a) Reflect the rectangle in the line *y* = −*x*.

  (b) Rotate the image 90° clockwise about the origin.

  (c) Describe the single transformation that replaces (a) and (b).

4 (a) Start with the same rectangle as in question **3**.
  Rotate the rectangle 90° clockwise about the origin.

  (b) Reflect the image in the line *y* = −*x*.

  (c) Describe the single transformation that replaces (a) and (b).

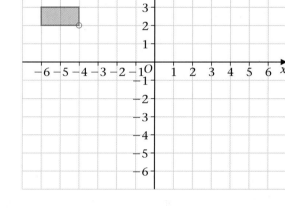

5 (a) Enlarge the shape in question **2** with centre of enlargement at the point (2, 1) and scale factor −1.

  (b) Describe this transformation in another way.

6 (a) Rotate the rectangle 90° anticlockwise (+90°) about the point (1, 0).

  (b) Enlarge the image with centre of enlargement at the point (−1, 3) and scale factor 2.

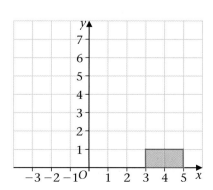

7 **(a)** Describe fully the single transformation that maps shape *P* onto shape *Q*.

(b) On a copy of the diagram, reflect shape *P* in the line *x* = 1.

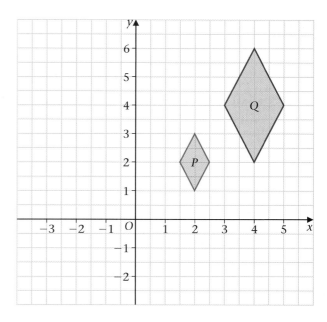

7.3 Scale drawings and scale models

The model of the racing car on the left is a small replica of the real racing car on the right. Each part of the model is a smaller version of the corresponding part on the real car.

Each length of 1 cm on the model represents 18 cm on the real car. The model is built to a scale of 1 : 18. The racing car is an enlargement of the model using a scale factor of 18.

> width of model × 18 = real width
> length of model × 18 = real length

To check the model steering wheel is in the correct scale:

> real steering wheel = 28.8 cm
> model steering wheel = 1.6 cm

Comparing measurements   1.6 : 28.8
                         1 : 18

The model steering wheel is correctly made to a scale of 1 : 18.

> Scale drawings (or models) of objects are enlargements of a real object. The size of enlargement is given by the scale factor.

$28.8 \div 1.6 = 18$

The enlargement has a fractional scale factor.

Example 7

A map has a scale of 1 : 500 000.

(a) Complete this sentence: 1 cm on the map represents _____ km in real life.

(b) The distance between Arken and Blorcastle is 4 cm on the map.
How far is Arken from Blorcastle in real life?

(c) Blorcastle is 12 km from Clarkwood. How far is this on the map?

(a) 1 : 500 000 is shorthand for '1 unit on the map represents 500 000 units in real life'. 1 cm on the map represents 5 km in real life.

(b) Arken and Blorcastle are 20 km apart.

(c) 12 km is represented on the map by 2.4 cm.

Both parts of the ratio 1 : 500 000 have the same units. 500 000 cm = 5 km

$4 \times 500\,000$ cm $= 20$ km
$$\frac{12}{5} \text{ km} = \frac{1\,200\,000}{500\,000} \text{ cm}$$

Exercise 7F

1 A model boat is made using a scale 1 : 10.

(a) The model is 76 cm long. How long is the real boat?

(b) The model is 43 cm wide. How wide is the real boat?

2 A model aeroplane is built using a scale of 1 : 50.

(a) The model is 1.60 m wide. How wide is the real aeroplane?

(b) The real aeroplane is 75 m long. How long is the model?

3 This map of the Norfolk Broads is drawn to a scale of 1 : 600 000.

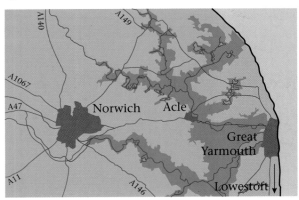

(a) Calculate the approximate distance in kilometres between the centres of Acle and Great Yarmouth.

(b) Calculate the approximate distance in kilometres between the centres of Norwich and Great Yarmouth.

(c) Lowestoft is 16.25 km from Great Yarmouth. How far would this be on the map?

4 This is a scale drawing of Shardia's bedroom: She has drawn it so that 1 cm on the diagram represents 2 m in her bedroom.

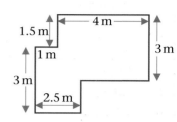

   (a) Write this as a scale.

   (b) Measure the length and width of the bedroom on the plan.

   Hint: 1 : ___ ___ ___

   (c) What are the length and width of the real bedroom?

   (d) How long is the real desk in metres?

5 This is a sketch of Arfan's bedroom. It is *not* drawn to scale. Draw an accurate scale drawing on cm squared paper of Arfan's bedroom. Use a scale of 1 : 50.

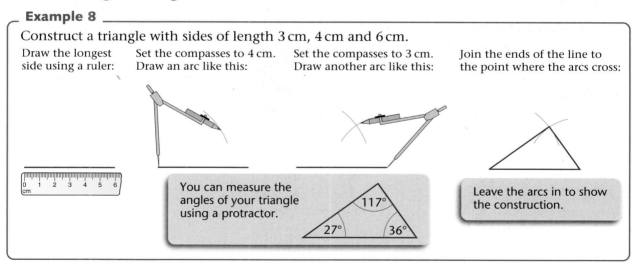

7.4 Constructions

A **construction** is an accurate drawing carried out using a straight edge (ruler), a pencil and a pair of compasses.

Constructing a triangle

Example 8

Construct a triangle with sides of length 3 cm, 4 cm and 6 cm.

Draw the longest side using a ruler:

Set the compasses to 4 cm. Draw an arc like this:

Set the compasses to 3 cm. Draw another arc like this:

Join the ends of the line to the point where the arcs cross:

You can measure the angles of your triangle using a protractor.

117° 27° 36°

Leave the arcs in to show the construction.

Example 9

Use triangle constructions to draw an angle of 60° without using a protractor.

An equilateral triangle has angles of 60°. Construct an equilateral triangle by keeping your compasses set to the same length throughout.

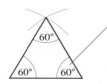

Each angle of the equilateral triangle is 60°.

60° 60° 60°

Constructing a regular hexagon

Example 10

Construct a regular hexagon.
Keep your compasses set to the same length throughout.

| Draw a circle of any radius: | With the point of the compasses on the circumference, draw an arc: | Place your compass point where the first arc crossed the circle. Draw another arc: | Draw four more arcs like this and join them up to make a regular hexagon: |

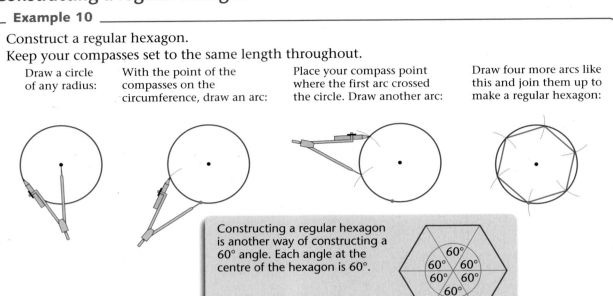

Constructing a regular hexagon is another way of constructing a 60° angle. Each angle at the centre of the hexagon is 60°.

Constructing perpendiculars

Perpendicular lines meet at right angles.
The green line is the perpendicular bisector of the blue one.

Bisect means 'cut in half'.

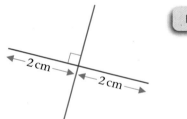

←2 cm→ ←2 cm→

Example 11

Draw a line segment AB of length 8 cm and construct its perpendicular bisector.

You need to set your compasses to more than half the length of the line segment, say 6 cm.

| With the compass point at B, draw a large arc: | Place the compass point at A. Draw this arc: | Join the points where the arcs cross. This is the perpendicular bisector of AB. |

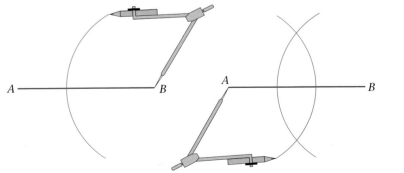

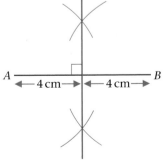

A ————— B   ←4 cm→ ←4 cm→

Example 12

Construct a perpendicular to *AB* that passes
through the point *P*.
This is called constructing the perpendicular
from *P* to *AB*.

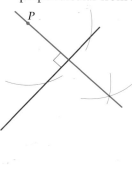

Place your compass point at *P*.
Draw two arcs to cut the line
AB:

With the compass point at
the points where the arcs cut
the line, draw these arcs:

Join the point where the arcs
cross to *P*. This line is the
perpendicular from *P* to *AB*.

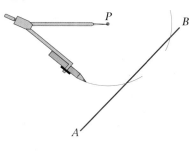

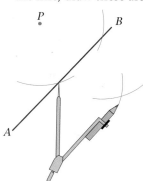

Example 13

Construct a perpendicular to *AB* that passes through the point *P*.
This is called constructing the perpendicular **to *AB* at *P***.

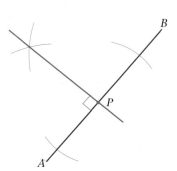

Place your compass point at *P*.
Draw two arcs to cut the line
AB:

Open the compasses more.
With the compass point at
the points where the arcs cut
the line, draw these arcs:

Join the point where the arcs
cross to *P*. This line is the
perpendicular to *AB* at *P*.

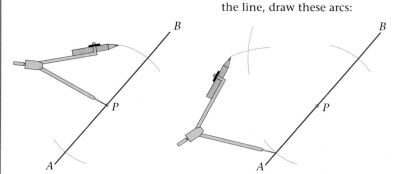

Examples 11, 12 and 13 all show methods of constructing right angles.

You could also construct a right angle using your knowledge of Pythagoras' theorem.

$3^2 + 4^2 = 5^2$, so you can create a right-angle by constructing a triangle with sides of length 3 cm, 4 cm and 5 cm.

> You will learn more about Pythagoras' theorem in Chapter 15.

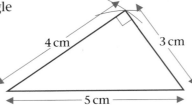

Bisecting an angle

The bisector of an angle is the line that divides the angle into two equal parts. The green line is the bisector of this angle:

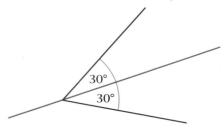

Example 14

Draw a 50° angle and construct its bisector.

Keep your compasses set to the same distance throughout.

Use a protractor to draw an angle of 50°. With your compass point at the vertex of the angle, draw two arcs to cut the sides of the angle:

Place your compass point at the points where the arcs cut the sides of the angle. Draw these arcs:

Join the point where the arcs cross to the vertex of the angle. This line is the bisector of the angle.

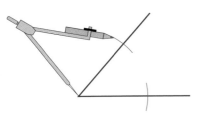

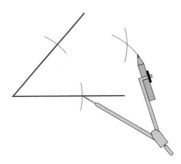

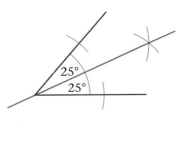

The bisector is a line of symmetry for the angle. The perpendicular distances from the bisector to both sides of the angle are the same.

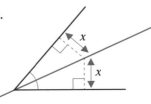

Exercise 7G

1 Using a ruler, a pair of compasses and pencil only, construct triangles with sides of length

    (a)   4 cm, 10 cm and 9 cm   **(b)**   8 cm, 7 cm and 12 cm.

2 Use a ruler and protractor to draw triangles with the following lengths and angles.

    (a)   6 cm, 50°, 7 cm   **(b)**   80°, 5.5 cm, 58°

    (c)   6 cm, 5 cm, 40°

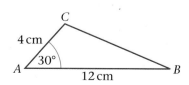

Draw a sketch first.

3 Draw a line segment of length 10 cm.
Using a straight edge, a pair of compasses and pencil only, construct the perpendicular bisector of this line segment.

4 Draw this line accurately.
Construct the perpendicular
to *XY* at *P*.

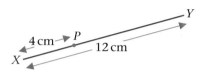

5 Draw a line segment *AB* and a point below it, *Q*.
Construct the perpendicular from *Q* to *AB*.

6 Construct an equilateral triangle with sides of length 8 cm.
What is the size of each angle of your construction?

7 Draw an angle of any size.
Without using any form of angle measurer, construct the bisector of the angle.

8 The diagram shows a construction
of a regular hexagon.
What is the size of

    (a)   angle *x*   **(b)**   angle *y*?

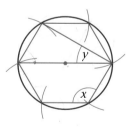

9 Without using any form of angle measurer, construct an angle of

    (a)   90°   **(b)**   45°   **(c)**   135°   **(d)**   60°

    (e)   30°   **(f)**   15°   **(g)**   120°

10 This diagram is a sketch of a
triangle *ABC*.

    (a)   Without using any form of
angle measurer, construct
the triangle *ABC*.

    (b)   Measure the length of *BC*.

7.5 Locus of a point

When Michael throws a cricket ball to his friend Simon, the ball moves along the path shown in the diagram on the right.

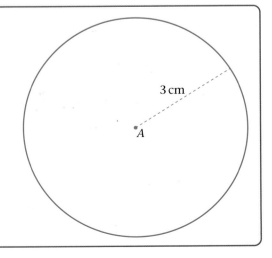

Michael   Simon

As this merry-go-round turns, the horses move up and down.

The mechanism on the merry-go-round makes a point on the tip of a horse's nose (or any point on a horse) obey a particular rule which takes the same path every rotation:

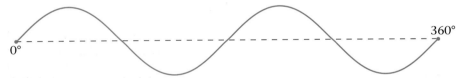

0°   360°

Path of the tip of the horse's nose during one revolution of the merry-go-round.

This path is called the **locus** of the point.

> A **locus** is a set of points which obey a particular rule (plural : loci).

A locus may be produced by something moving (such as a cricket ball or a merry-go-round) according to a set of rules, or by a set of points which follow a mathematical rule.

Locus around a fixed point

Example 15

Draw the locus of a point which moves so that it is always 3 cm from a fixed point.

A is the fixed point. Draw a circle, radius 3 cm and centre at *A*. The locus of points satisfying the rule is the circumference of the circle.

3 cm

A

The locus of a point which moves so that it is always a fixed distance from a point A is a circle, centre A.

Example 16

The diagram shows a rectangular lawn.

A garden sprinkler rotates from a fixed position on the lawn. Water from the sprinkler reaches a maximum distance of 4 m from the sprinkler.

Complete the diagram and shade the parts of the lawn watered by the sprinkler.

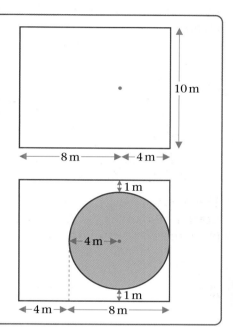

The sprinkler waters the area within a circle, centre the sprinkler, radius 4 m.

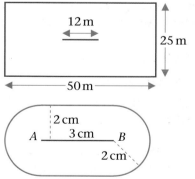

Locus around a fixed line

Example 17

(a) Sketch the locus of a point which moves so that it is always 2 cm from a straight line 3 cm long.

(b) A goat is tethered in a field by a 10 m rope. The other end of the rope is allowed to slide along a horizontal bar 12 m long. Make a scale drawing to show the parts of the field the goat can graze.

(a) The locus is two parallel lines (2 cm either side of AB and the same length as AB) with a semicircle of radius 2 cm drawn at each end.

(b) The goat can graze any of the shaded area.

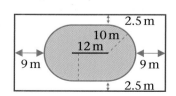

The locus of a point that moves so that it is always the same distance from a straight line is two parallel lines with a semicircle at each end.

Locus of a point the same distance from two lines at an angle

Example 18

Draw the locus of a point which moves so that it is always the same distance from two straight lines.

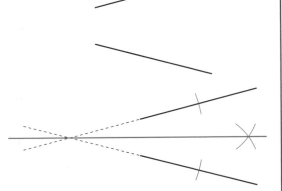

The perpendicular distances from two lines to their angle bisector are always equal.

Continue the two lines and bisect the angle between them.

The locus of a point that moves so that it is always the same distance from two straight lines is the bisector of the angle between the lines.

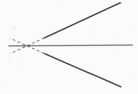

Locus of a point the same distance from two points

Example 19

Draw the locus of a point that moves so that it is always the same distance from points A and B.

Draw the perpendicular bisector of AB.

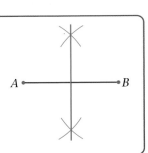

The locus of a point which moves so that it is always the same distance from two points A and B is the perpendicular bisector of the line AB.

Exercise 7H

1 Mark two points A and B roughly 4 cm apart. Draw a path equidistant from A and B.

2 Draw the locus of a point that moves so that it is always 2.5 cm from a line 4 cm long.

3 A heavy cuboid box is moved by tipping it over on each edge. Draw the locus of any point P on the box as it makes a whole turn.

> You could model this using a cube.

4 A running track is designed so that any point on the track is 22.3 m from a fixed line 150 m long.

    (a) Draw the locus of the point.

    (b) Calculate the distance once round the running track.

5 A bicycle has wheels of radius 45 cm. It has a reflector on the front wheel attached 30 cm from the centre.

    (a) Write the shortest distance the reflector can be from the road.

    (b) Draw the locus of the reflector as the bicycle moves forward through one complete wheel turn.

6 Baby Susan is placed inside a rectangular playpen measuring 1.2 m by 0.8 m. She can reach 30 cm outside the playpen. Draw the locus of the points Susan can reach beyond the edge of the playpen.

7 Daniel lives at the point marked X. There are schools at Ayleton, Bankbury and Corley. Each school allows pupils to attend provided they live in the catchment area. This is a particular distance from the school.

    Ayleton

                              Bankbury

              X

                              Corley

Scale: 1 cm = 1 mile

Ayleton accepts pupils who live less than 5 miles away. Bankbury accepts pupils who live less than 4 miles away, and Corley accepts pupils who live less than 3 miles away. Trace the diagram. Shade the catchment area for each school. Write down which school(s) Daniel could attend.

8 Samia enters a rectangular field 40 m by 30 m by a gate in one corner. She crosses the field by taking a path equidistant from the hedges at the side of the gate. Construct the locus of her path accurately.

9 *A* and *B* are two fixed points.
A point *P* moves in the plane such that $A\hat{P}B = 45°$.
Sketch the complete locus of *P* above and below the line *AB*.

10 Repeat question **9** for $A\hat{P}B = 30°$.

7.6 Bearings

Imagine your boat is in the harbour at point *H*. There is a ship out at sea at point *S*, 5 km away. You can describe the position of the ship by giving its bearing and distance from you.

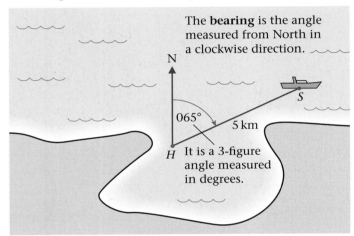

The ship's bearing from you is 065°. It is 5 km from you, on a bearing of 065°.

> A **bearing** is an angle expressed in three digits to indicate direction.

Example 20

Fred lives at *F*. His friend Winston lives 1 km away on a bearing of 210°.

(a) Show this on a diagram.

Fred visits Winston. After a while Fred returns home.

(b) Calculate the bearing Fred takes from Winston's home to arrive back home.

(a)

The bearing
of *W* from *F*
is 210°

(b)

Method 1

W is on a bearing of 210° from F. Fred needs to turn through half a turn (180°) to face home again. The bearing of F from W is

$$210° - 180° = 030°$$

The bearing of F from W is 030°.

Method 2

Draw another North line at F and use alternate angles:

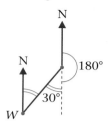

The bearing of F from W is 030°.

A bearing giving the reverse direction is called a **back-bearing**.

Exercise 7I

1 Draw diagrams to show these bearings:
    (a) 016°   **(b)** 200°   **(c)** 325°
    (d) 009°   **(e)** 168°

2 Write
    (a) the bearing of A **(b)** the bearing of B **(c)** the bearing of C
        from O.   from O.   from O.

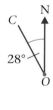

3 Three friends live near each other. Their homes are shown by X, Y and Z. Y is on a bearing of 055° from X. Angle $XYZ = 120°$ as shown on the diagram.

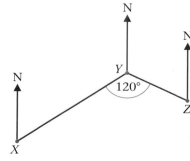

Remember:
Co-interior angles

$x + y = 180°$

Calculate the bearing of
    (a) X from Y **(b)** Z from Y **(c)** Y from Z.

4 The bearing of *B* from *A* is 060°.
What is the bearing of *A* from *B*?

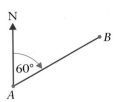

5 The bearing of *E* from *D* is 145°.
What is the bearing of *D* from *E*?

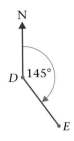

6 Investigate bearings and back-bearings where the bearing of *B* from *A* is

(a) less than 180°

(b) greater than 180°.

Mixed exercise 7

1 A ship, *S*, leaves a harbour, *H*, and travels 40 km due North to reach a marker buoy, *B*.
At *B* the ship turns and travels for a further 30 km on a bearing of 075° from *B* to reach a lighthouse, *L*.
At *L* it turns again and travels back to *H* in a straight line.

(a) Sketch the locus of the ship's path.

(b) Make an accurate scale drawing of the path taken by the ship.

(c) Measure the bearing of *L* from *H*.

(d) Use your scale drawing to find the distance *HL*.

2 Make a copy on squared paper of the diagram below.

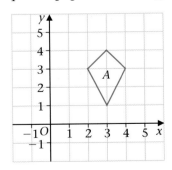

Reflect shape *A* in the *x*-axis to give shape *B*.
Draw and label shape *B*.

Enlarge shape *A* by scale factor 2, centre *O*, to give shape *C*.
Draw and label shape *C*.

3   Make a copy on squared paper of the diagram below.

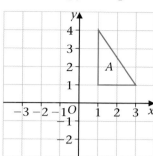

Shape *A* is rotated 90° anticlockwise, centre (0, 1), to shape *B*.

Shape *B* is rotated 90° anticlockwise, centre (0, 1), to shape *C*.

Shape *C* is rotated 90° anticlockwise, centre (0, 1), to shape *D*.

(a)   Mark the positions of shapes *B*, *C* and *D*.

(b)   Describe the single transformation that takes shape *C* to shape *A*.

4   Make a copy on squared paper of the diagram on the right.

(a)   Rotate triangle *A* 180° about *O*. Label your new triangle *B*.

(b)   Enlarge triangle *A* by scale factor $\frac{1}{2}$, centre *O*. Label your new triangle *C*.

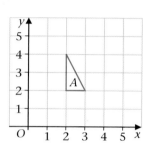

5   Triangle B is a reflection of triangle A.

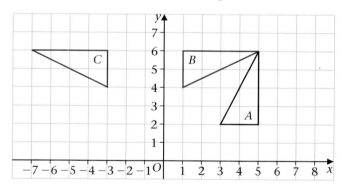

(a)   Copy the diagram on squared paper and draw the line of reflection.

(b)   Write down the equation of the line of reflection.

(c)   Describe fully the single transformation that maps triangle *A* onto triangle *C*.   [E]

6   A map is drawn to scale of 1 : 25 000.

Two schools *A* and *B* are 12 cm apart on the map.

(a)   Work out the actual distance from *A* to *B*. Give your answer in kilometres.

B is due East of *A*.

C is another school. The bearing of *C* from *A* is 064°. The bearing of *C* from *B* is 312°.

(b) On a copy of the scale drawing below, mark, with a cross, the position of the school C.

N

A ———————————————————————————— B

7 Make a copy of the line XY.
Draw the locus of all points which
are 3 cm away from the line XY.

X

Y

8 The diagram represents a triangular garden ABC.

The scale of the diagram is 1 cm to represent 1 m.

A tree is to be planted in the garden so that it is
- nearer to AB than to AC
- within 5 m of point A.

On a copy of the diagram, shade the
region where the tree may be planted.

A

B

C

9 Draw a line and mark the point A at one end.
Use a ruler and a pair of compasses to construct
an angle of 45° at A.

You must show all construction lines.

10 Construct a regular pentagon and a regular octagon of any size.

> Draw a circle and divide
> the angle at the centre by
> the number of sides.

11 The diagram shows the position of each of three buildings in a
town.

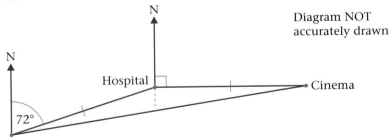

N

Diagram NOT
accurately drawn

N

Hospital

Cinema

72°

Art gallery

The bearing of the hospital from the art gallery is 072°.
The cinema is due East of the hospital.
The distance from the hospital to the art gallery is equal to the
distance from the hospital to the cinema.
Work out the bearing of the cinema from the art gallery.   [E]

Summary of key points

1 A **translation** moves every point on a shape the same distance and direction.

2 To describe a translation fully you need to give the distance moved and the direction of the movement. You can do this by giving the vector of the translation.

3 In the vector $\begin{pmatrix} x \\ y \end{pmatrix}$

- x gives the movement parallel to the x-axis
- y gives the movement parallel to the y-axis.

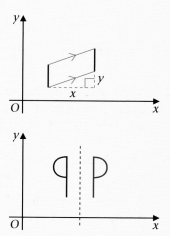

4 A **reflection** in a line produces a mirror image. The mirror line is a line of symmetry.

5 A point on the mirror line remains unchanged by reflection.
A **repeated reflection** in the same mirror line returns the image to the original object.

6 To describe a reflection fully you need to give the equation of the mirror line.

7 To describe a **rotation** fully you need to give the
- centre of rotation
- amount of turn
- direction of turn.
(A clockwise direction may be indicated by a negative sign, and an anticlockwise direction by a positive sign.)

8 An **enlargement** changes the size but not the shape of an object. The **scale factor** of the enlargement is the number of times the lengths of the original object have been enlarged.

9 To describe an enlargement fully you need to give its centre and the scale factor.

10 An enlargement made with a scale factor smaller than 1 gives a reduced image.

11 An enlargement made with a negative scale factor indicates that the measuring from the centre of the enlargement must be in the opposite direction.

12 A reflection followed by a reflection can be replaced by a single transformation of
- a rotation if the reflection lines are *not* parallel
- a translation if the reflection lines are parallel.

13 Scale drawings (or models) of objects are enlargements of a real object. The size of enlargement is given by the scale factor.

14 A **construction** is an accurate drawing carried out using a straight edge (ruler), a pencil and a pair of compasses.

15 A **locus** is a set of points which obey a particular rule (plural : loci).

16 The locus of a point which moves so that it is always a fixed
distance from a point A is a circle, centre A.

17 The locus of a point that moves so that it is always the same distance from a straight line is two
parallel lines with a semicircle at each end.

18 The locus of a point that moves so that it is always the same
distance from two straight lines is the bisector of the angle between
the lines.

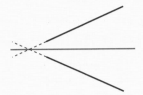

19 The locus of a point which moves so that it is always the same
distance from two points A and B is the perpendicular bisector
of the line AB.

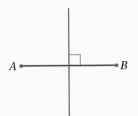

20 The position of a point P with reference to a fixed point O can be
identified by the **bearing** of P (bearings are angles measured
clockwise from the North line and expressed in three digits),
together with the distance OP.

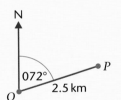

21 A bearing giving the reverse direction is called a **back-bearing**.
It is the outward bearing plus 180°.

8 Using basic number skills

You need to use number skills in a wide range of mathematical calculations. In this chapter you will calculate percentages and rates, and learn the notation for very large or very small numbers.

8.1 Percentage increases

2003 was a 'boom year' for house prices in the UK. Prices went up by an average of 24% during the first quarter.

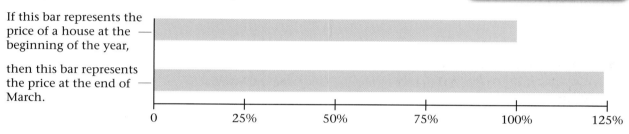

If this bar represents the price of a house at the beginning of the year,

then this bar represents the price at the end of March.

As a percentage the new value is

100% + 24% = 124% of the old value.

Example 1

Jenny earns £26 000 a year. She gets a 12.5% increase. Calculate her new salary.

A 12.5% increase means that her new salary will be (100 + 12.5)% of her old salary.

112.5% as a decimal is 1.125

Her new salary is 1.125 × 26 000 = £29 250

> 1.125 is called the multiplier.

Example 2

Tom gets a 15% rise in pay.

He was earning £9500 a year. What is his new salary?

As a percentage of his old salary his new salary is

(100 + 15)% = 115%

As a decimal, 115% is 1.15

Tom's new salary is 1.15 × £9500 = £10 925

If an amount is increased by x% the new amount is (100 + x)% of the original amount. So the new amount = the old amount × (100 + x)%.

Exercise 8A

1   Write the new value as a percentage of the old value if
(a)   a house increases in value by 12%
(b)   the balance in a bank account increases by 17%
(c)   the weight of a man goes up by 30%
(d)   the cost of a car goes up by 11%
(e)   the cost of a washing machine increases by 8.5%.

2   (a)   If you want to increase a quantity by 24%, what do you multiply by?
(b)   Graham's salary of £10 850 is increased by 24%. What is his new salary?

3   (a)   In order to increase an amount by 40%, what do you multiply by?
(b)   The cost of a theatre ticket is increased by 40% for a special concert. What is the new price if the normal price was £15.40?

4   Katherine buys a woodburning stove for £1298 plus VAT at $17\frac{1}{2}$%. Work out the total price including VAT.

5   Gail wants to work out what her weight will be if it increases by 4%.
What should she multiply her present weight by?

6   (a)   Increase £120 by 20%.
(b)   Increase 56 kg by 25%.
(c)   Increase 2.4 m by 16%.
(d)   Increase £1240 by 10.5%.
(e)   Increase 126 cm by 2%.

8.2 Percentage decreases

During 1993 house prices in the UK fell by an average of 5% during the first quarter.

If this bar represents the price of a house at the beginning of the year,

then this bar represents the price at the end of March.

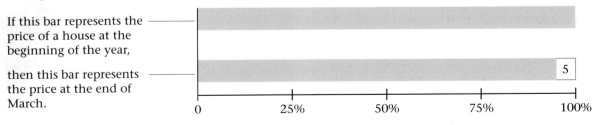

As a percentage the new value is
$(100 - 5)\% = 95\%$ of the old value.

> If an amount is decreased by x% the new amount is $(100 - x)$% of the original amount. So the new amount = the old amount $\times (100 - x)$%.

Exercise 8B

1 Write the new value as a percentage of the old value if
   (a)   a house decreases in value by 16%
   (b)   the balance in a bank account decreases by 11%
   (c)   the weight of a man goes down by 5%
   (d)   the value of a car goes down by 22%.

2 (a)   If you want to decrease a quantity by 34%, what do you multiply by?
   (b)   The value of a company's shares falls by 34%. Before the drop the shares cost £5.25 each. What is the new price?

3 Alan weighs 82 kg before going on a diet. He sets himself a target of losing 5% of his original weight. What is his target weight?

4 (a)   Decrease £120 by 20%.   (b)   Decrease 56 kg by 25%.
   (c)   Decrease 2.4 m by 16%.   (d)   Decrease £1240 by 10.5%.
   (e)   Decrease 126 cm by 2%.

5 Carla invested £100 in a savings account on 1 January 1999.
   The interest paid was 6% per annum.
   (a)   What do you multiply by to find the total amount in her savings account at the end of the first year?
   (b)   How much money did she have in her account at the end of the first year?
   (c)   Copy and complete this table. Give the amounts to the nearest pound.

Date	1/1/99	1/1/00	1/1/01	1/1/02	1/1/03
Amount	£100				

6 A car costing £14 000 depreciates (loses value) by
   • 12% in its first year
   • 18% in its second year
   • 5% each year thereafter.
   (a)   Find the value of the car after 1 year.
   (b)   Copy and complete this table.
         Give your answers to the nearest pound.

New price	£14 000
Value after 1 year	
Value after 2 years	
Value after 3 years	
Value after 4 years	
Value after 5 years	

7 The amount in an investment account increases by 10% in the first year and a further 8% in the second.

   (a) What should you multiply by to find the value after the first year?

   (b) What should you multiply by to find the value after two years?

> Be careful: it is not 1.18.

8.3 Finding percentage increases

> The **percentage increase** is $\dfrac{\text{actual increase}}{\text{original price}} \times 100\%$

Example 3

In 2000 a box of tissues cost 60p.
In 2004 a similar box cost 80p.
What was the percentage increase in price?

The actual increase is $80 - 60 = 20$p.

The fractional increase is $\dfrac{\text{actual increase}}{\text{original price}} = \dfrac{20}{60}$

$\frac{20}{60}$ as a decimal is $0.3\dot{3}$

Percentage increase is $0.3\dot{3} \times 100 = 33\frac{1}{3}\%$

> Remember:
> To change a decimal to a percentage multiply by 100.

Exercise 8C

1 The box of tissues in Example 3 had increased in price from 45p in 1998 to 80p in 2004. Find

   (a) the fractional increase from the 1998 cost of 45p to the 2004 price

   (b) the fractional increase from the 1995 cost of 40p to the 2004 price

   (c) the percentage increase from 1998 to the 2004 price

   (d) the percentage increase from 1995 to the 2004 price.

2 Calculate the percentage increase from

   (a) £24 to £36   (b) 12.5 kg to 20 kg

   (c) 2.45 m to 2.86 m   (d) 50 s to 1 minute.

3 Before fitting a 'fuel-saver' Jamie's car travelled 36 km on 4.5 litres of petrol. After fitting the fuel-saver it travelled 58.3 km on 5.5 litres. Find

   (a) the distance travelled on 1 litre of fuel before the fuel saver was fitted

   (b) the distance travelled on 1 litre of fuel after it was fitted

   (c) the percentage increase in distance travelled.

4 Siamak buys an old car for £8400. He spends £1040 on repairs, then sells the car on eBay for £14 900. Find

    (a) his actual profit

    (b) his percentage profit to the nearest 1%.

5 A mechanic at a garage pours 1.5 litres of oil into a 24 litre oil drum which was exactly $\frac{1}{4}$ full.

    (a) Find the amount of oil now in the drum.

    (b) By what percentage has the amount of oil in the drum increased?

8.4 Finding percentage decreases

The **percentage decrease** is $\dfrac{\text{actual decrease}}{\text{original price}} \times 100\%$

___ **Example 4** ___

Josie invests £800 in a company called Spec-U-Late Ltd. One year later she finds her investment is only worth £680.

Find   **(a)** her actual loss   **(b)** her percentage loss.

(a) Her actual loss is £800 − £680 = £120

(b) Her fractional loss is $\frac{120}{800}$

    $\frac{120}{800}$ as a decimal is 0.15

    So her percentage loss is 0.15 × 100 = 15%

Exercise 8D

1 Josie invested £800. One year later this investment was only worth £680 (see Example 4). After a further year Josie's investment had fallen to £540. Find

    (a) her actual loss over the two years

    (b) her percentage loss over the two years

    (c) her percentage loss between the end of the first year and the end of the second year of her investment.

2 Find the percentage decrease to the nearest 1% of

    (a) a decrease from £48 to £32

    (b) a decrease from 5.2 kg to 3.8 kg

    (c) a decrease from 45 cm to 39.5 cm

    (d) a decrease from 2 min to 110 s

    (e) a decrease from 1 m to 45 cm.

3   This table shows the value of a car after different numbers of years from new.

New price	£12 500
after 1 year	£10 000
after 2 years	£8 600
after 3 years	£7 500
after 4 years	£6 600

   (a)   Calculate the percentage loss from new to the end of each year.

   (b)   Calculate the percentage loss between the second and third years.

8.5 Mixing percentage increases and decreases

Example 5

In 1997, Mr Peck's flat was valued at £60 000.
The value of the flat increased by 20% during 1998.
During 1999 the value of the flat decreased by $12\frac{1}{2}$%.

Work out the value of the flat at the end of 1999.

By the end of 1998, the value of the flat, in pounds, was

          60 000 + 20% of 60 000

or $60\,000 \times (1 + 0.2)$

or $60\,000 \times 1.2$

          = 72 000

By the end of 1999, the value of the flat, again in pounds, had fallen by 12.5% of the £72 000.

So by the end of 1999, the value of the flat was

          72 000 − 12.5% of 72 000

or $72\,000 \times (1 - 0.125)$

or $72\,000 \times 0.875$

          = 63 000

Exercise 8E

1   The price of a holiday in 2004 was £450.
    The same holiday cost an extra 12% in 2005.
    In 2006 the same holiday was reduced by 15% of its price in 2005.
    Work out the price of this holiday in 2006.

2   Jenny bought a flat in September 2002.
    The value of the flat then was £52 000.
    During the first year, the value of the flat decreased by 8%.
    Over the next year the value of the flat increased by 14%.
    Work out the value of the flat in September 2004 to the nearest pound.

3 Erica bought £10 000 worth of shares in May 2002.
By May 2003 the value of the shares had increased by 15%.
By May 2004 the value of the shares had decreased by n% of
their value in May 2003.
The value of the shares in May 2004 was again £10 000.
Work out the value of n.

4 Sanjit bought a house in 1994 for £84 000. In 1999 he sold it to
James for £120 000. In 2004 James had to sell the house for
£80 000. Calculate
   (a) Sanjit's percentage profit
   (b) James's percentage loss
   (c) the percentage change in the value of the house between
   1994 and 2004.

8.6 More percentage calculations

Example 6

The price of a carpet is reduced by 20% in the sale. It now costs
£60. What was the pre-sale price?

The sale price is $(100 - 20)$% = 80% of the pre-sale price
$$= 0.8 \text{ in decimal form}$$

A flow diagram can be used to represent the information in the
question:

$$\text{pre-sale price} \xrightarrow{\times 0.8} £60$$

To solve this type of flow diagram problem, draw a second flow
diagram reversing the direction and using the inverse operation:

$$£75 \xleftarrow{\div 0.8} £60$$

The pre-sale price of the carpet was £75.

> **Remember:**
> In this type of question you are *not* finding 80% of the *given* price, but 80% of the *unknown* price.

Example 7

A gas bill of £47.25 includes VAT at 5%. How much VAT is paid?

Method 1

$$£47.25 = 105\%$$
So $\quad 1\% = 47.25 \div 105$
$$= £0.45$$
So $\quad 5\%$ VAT $= 5 \times 0.45 = £2.25$

Method 2

Use the multiplier method:
Bill without VAT
$$= 47.25 \div 1.05$$
$$= £45$$
So VAT $= 47.25 - 45 = £2.25$

Exercise 8F

1 A television is reduced by 10% in a sale. The sale price is £540.

    (a) What percentage of the pre-sale price is the sale price?

    (b) Draw a flow diagram to show this information.

    (c) Find the pre-sale price.

2 An electricity bill of £58.38 includes VAT at 5%.
How much VAT is paid?

3 A dining room table costs £699, including VAT at $17\frac{1}{2}$%.
How much VAT is paid?

4 The total price of a winter holiday includes VAT at 17.5%.
In the brochure the total price is quoted as £320 per person.

    (a) What percentage of the pre-VAT price is the total price?

    (b) Draw a flow diagram to show this information.

    (c) Find the price before VAT is added.

5 Due to falling orders, Jonson's Electrical Company decreases its workforce by 8%, down to 161 employees. Find

    (a) the number of employees before the decrease

    (b) the actual decrease in the number of employees.

6 After two years the value of a new drilling machine has fallen to £12 000. The depreciation is estimated at 10% per year. Find, to the nearest penny,

    (a) the value of the drilling machine after the first year

    (b) the original price.

8.7 Compound interest

Tessa invests £250 in a bank's internet account. The account pays an annual interest rate of 6% per annum. She wants to know how much her money will be worth in 6 years.

This problem would be easy if the bank paid **simple interest** of 6% of the original amount for each of the 6 years:

    6% interest on £250 for 1 year is £15

    6% interest on £250 for 6 years is £15 × 6 = £90

Tessa's investment after 6 years would then be worth

    £250 + £90 = £340.

But banks usually pay **compound interest**, meaning they pay interest on the interest too. So at the end of the second year Tessa will get 6% interest on both her original investment and the first year's interest.

This table shows how her investment will grow over 6 years.

Year	Amount at start of year	Interest	Total amount at year end
1	£250	250×0.06	£265
2	£265	$265 \times 0.06 = 250 \times 0.06^2$	£280.90
3	£280.90	$280.90 \times 0.06 = 250 \times 0.06^3$	£297.75
4	£297.75	$297.75 \times 0.06 = 250 \times 0.06^4$	£315.62
5	£315.62	$315.62 \times 0.06 = 250 \times 0.06^5$	£334.56
6	£334.56	$334.56 \times 0.06 = 250 \times 0.06^6$	£354.63

After 6 years' compound interest, Tessa's investment will have grown to £354.63 to the nearest penny, compared with £340 if simple interest only had been paid.

Compound interest is interest paid on an amount and on the interest already earned.

To calculate compound interest, find the multiplier:

Amount after n years = original amount $\times$ multiplier^n

Example 8

A student borrows £600 to buy a laptop. Interest is added to the loan each year at the rate of 5% per annum.
How much does the student owe after 3 years if no repayments are made?

Each year the loan is increased by 5% to 105%, which is 1.05 times the previous value.

$£600 \times 1.05^3 = £694.58$

Exercise 8G

Use the table in Section 8.7 for questions **1–4**.

1 (a) How much more interest does Tessa receive over the six years if compound interest is paid, rather than simple interest?

(b) What percentage of the original investment is the extra interest?

2 Copy the table showing Tessa's compound interest and extend it to find how much the original investment would have grown to after ten years.

3 (a) How much more interest does £250 invested at 6% compound interest for 10 years gain than £250 invested at 6% simple interest?

(b) What percentage of the £250 is the extra interest?

(c) What annual simple interest rate over the 10 year investment period would give the same amount of interest as the 6% compound interest rate?

4 How many years will it take £250 to double in value when invested

(a) at 6% compound interest

(b) at 6% simple interest?

5 On 1 January Surjit invests £1000 in an investment account that pays compound interest at a rate of 8.2%.
Calculate the value at

(a) the end of the first year

(b) the end of the fourth year

(c) the end of the eighth year.

6 The Bank of Nirvana pays compound interest at an annual rate such that an investment doubles in value every two years.
Claire thinks that the interest rate paid per annum must be 50%.
Explain why she must be wrong.

7 Colin rents a house at a monthly rent of £620. In the rent agreement it states that the rent will be reviewed annually and will be increased by the annual inflation rate.
Assuming an annual inflation rate of 4%, find

(a) the amount of rent paid in the first year

(b) the rent increase at the end of the first year

(c) the monthly rent paid during the second year

(d) the monthly rent to be paid during the seventh year.

8 A car costs £12 000 when bought new. It depreciates (loses value) at an annual rate of 10%. Find

(a) the value of the car at the end of the first year

(b) the value at the end of the second year

(c) the value at the end of the fifth year

(d) the time it takes for the value of the car to reach £6000.

> The value goes down each year so do the opposite of compound interest: take the amount off each year. The amount each year will be different.

9 By what factor will an investment increase if compound interest is paid at a rate of 12.5% per annum over a period of

(a) 4 years   (b) 6 years   (c) 10 years?

Write your answers as decimals or percentages.

10 For insurance purposes the value of house contents depreciate by 8% each year once purchased. What is the value of the following items for insurance purposes?

(a) A television bought for £650 and 2 years old

(b) A dining room table and chairs bought for £2450 and 4 years old

(c) A three piece suite bought for £1250 and 3 years old

8.8 Compound measures

Compound measures combine measurements of two or more different types. For example, a speed is a measurement of a *distance* and the *time* taken to travel it. You need to know how to use a variety of compound measures.

Average speed (rate of travel)

Lisa travelled from Nottingham to London, a distance of 205 km. She started at 07.00 and arrived in London two and a half hours later.

She did not travel at a constant speed throughout the journey, but her average speed can be found from:

$$\textbf{Average speed} = \frac{\text{distance travelled}}{\text{time taken}}$$

So average speed $= \dfrac{205}{2.5} = 82 \text{ km/h}$

> In travel problems, times are often given in hours and minutes. To make calculations easier, change the minutes into a decimal fraction of an hour.
>
> To do this, divide the minutes by 60.
> For example:
>
> 42 minutes $= 42 \div 60$
> $\qquad\qquad = 0.7$ hour

_____ **Example 9** _____

Jacki takes 36 minutes to walk to school, a distance of 1.8 km. Find her average speed in

(a) metres per minute   (b) km per hour   (c) miles per hour.

(a) Distance travelled = 1800 m
    Time taken = 36 minutes

$$\text{Average speed} = \frac{1800}{36} = 50 \text{ m/min}$$

(b) Distance travelled = 1.8 km
    Time taken = 36 ÷ 60 = 0.6 hour

$$\text{Average speed} = \frac{1.8}{0.6} = 3 \text{ km/h}$$

(c) Distance travelled = 1.8 × 0.625 = 1.125 miles
    Time taken = 36 ÷ 60 = 0.6 hour

$$\text{Average speed} = \frac{1.125}{0.6} = 1.88 \text{ mph}$$

> You can use this triangle to help you remember the formulae.

> Cover the value you wish to find with your thumb: e.g. to find speed cover S. You are left with D over T, that is
>
> $\dfrac{\text{distance}}{\text{time}}$

> 1 km $= \frac{5}{8}$ mile
> 1 km $= 0.625$ mile

Exercise 8H

1   *The Flying Scotsman* took
    7 hours 15 minutes to travel
    from Edinburgh to London,
    a distance of 632 km.

    (a)   Calculate the average speed
          of the train in km per hour.

    (b)   Given that 1 km is
          approximately equal to
          0.62 miles, calculate
          the average speed in
          miles per hour.

The Flying Scotsman first broke the 100 mph speed barrier in 1934.

2   Paul travelled by coach from Derby to Birmingham, a distance of
    45 miles.
    The journey took 1 hour 45 minutes.
    Calculate the average speed of the coach.

3   Change these times into hours, giving your answers correct
    to 2 d.p. where appropriate.

    (a)   75 minutes   (b)   24 minutes   (c)   6 minutes
    (d)   25 minutes   (e)   45 minutes   (f)   10 minutes

4   A car travelled 50 km in 2 hours 42 minutes. How long was the
    journey in

    (a)   minutes   (b)   hours?

    (c)   Calculate the average speed for the journey.

5   Shannon took 1 hour 40 minutes to complete a science experiment.
    Jack took 1.6 hours to complete the same experiment. Work out

    (a)   who took the longer time, and by how much, to complete
          the experiment

    (b)   the average (mean) time taken by the two students.

> To change a decimal
> fraction of an hour into
> minutes, multiply by 60.
> For example:
> 0.6 hour = 0.6 × 60
>   = 36 minutes

6   Change these times into hours and minutes.

    (a)   0.4 hours   (b)   5.3 hours   (c)   3.25 hours
    (d)   6.29 hours   (e)   3.9 hours   (f)   9.88 hours

7   Ron and Garth set out at 1200 hours for a 25 km walk. They
    walked for 2 hours 18 minutes, then had a 30 minute rest before
    completing their walk in a further 2.2 hours. Calculate

    (a)   the time it took to complete the 25 km walk

    (b)   the average speed for the journey.

8   A train travelled at 80 km per hour for 2.25 hours, then at 60 km
    per hour for a further 4 hours 36 minutes. Find

    (a)   how far the train travelled altogether

    (b)   how long the journey took in total

    (c)   the average speed for the journey.

Other rates

Example 10

The average fuel consumption of a car can be measured in kilometres per litre.

A car uses 17.6 litres (*l*) of petrol in travelling 220 km. What is its average fuel consumption?

$$\text{Average fuel consumption} = \frac{220}{17.6} = 12.5 \text{ km per litre or km/}l$$

Exercise 8I

1   What is the average fuel consumption (in km per litre) of a car which travels 464 km on 32 litres of petrol?

2   Mr Robson sets out with a full tank of petrol on a business trip of 588 km. When he arrives, his petrol gauge indicates that his tank is $\frac{1}{8}$ full.

   (a) If a full tank holds 48 litres of petrol, calculate

      (i)   how much petrol it took to complete the journey

      (ii)   the average fuel consumption.

   (b) If the average fuel consumption remains the same, how much further can he travel before he runs out of petrol?

3   Carol fits a *regular* 4.5 V battery into her radio. The battery costs £2.60 and lasts for 320 hours. She replaces it with a *long life* battery which costs £4.68 but lasts for 600 hours.

   (a) Find   (i)   the cost per hour for each battery

             (ii)   the time per pound for each battery.

   (b) Which type of battery is the better buy, and why?

4   Kay found these exchange rates in a newspaper:

       £1 was equivalent to 2.25 Swiss francs
       £1 was equivalent to 1.58 US dollars

   Calculate the exchange rate between the USA and Switzerland in

   (a) francs per dollar   (b) dollars per franc.

5   If this bath is filled from the cold tap only, it takes 5 minutes to fill. If filled from the hot tap only, it takes 8 minutes to fill. The bath holds 920 litres. Calculate

   (a) the flow rate in litres per minute for each tap

   (b) how long it would take to fill the bath if both taps were turned fully on.

6 A lorry travelling at 40 km per hour uses diesel fuel at an average consumption of 6.25 km per litre.

(a) What is the consumption:
    (i) in litres per km (ii) in litres per hour?

(b) (i) How far will the lorry travel in 15 minutes?
    (ii) How much diesel will it use?

Measuring flow

For liquid flowing through a pipe:
volume flowing per second = cross-sectional area × speed

Example 11

Roger is watering his vegetable patch with a hosepipe which has a cross-sectional area of 1.6 cm^2.
Water flows out of the pipe at a speed of 5 cm per second.

Find the amount of water flowing out of the pipe in

(a) cm^3 per second   (b) litres per hour.

(a) Volume flowing = cross-sectional area × speed
$$= 1.6 \times 5$$
$$= 8 \text{ cm}^3/\text{second}$$

In one second a cylinder of water 5 cm long and 1.6 cm^2 in cross-section flows out of the pipe.

(b) If 8 cm^3 flow per second then
$$60 \times 8 \text{ cm}^3 \text{ will flow per minute} = 480 \text{ cm}^3/\text{minute}$$
and $60 \times 480 \text{ cm}^3$ will flow per hour $= 28\,800 \text{ cm}^3/\text{hour}$
$$= \frac{28\,800}{1000}$$
$1000 \text{ cm}^3 = 1 \text{ litre}$ ———— $= 28.8 \text{ litres/hour}$

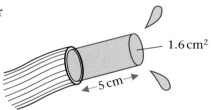

1.6 cm²

5 cm

Exercise 8J

1 A water butt holds 1.5 m^3 when full. It is leaking at a rate of 0.25 litres/minute.
How long will it take in hours and minutes for a full butt of water to leak away?

2 Water flows through a plastic pipe at a rate of 260 cm^3 per second. Find

(a) the amount of water discharged in
    (i) cubic metres per second (ii) litres per minute

(b) the time it takes to discharge one cubic metre of water

(c) the time it would take to fill a water tank holding 800 litres of water.

Density

The density of a substance is its mass per unit volume:

$$\textbf{density} = \frac{\text{mass}}{\text{volume}}$$

Densities are often given in g/cm³ (i.e. grams per cubic centimetre).

Example 12

This block of lead has a volume of 540 cm³ and a mass of 6.156 kg.
Find its density in g/cm³.

$$\text{Density} = \frac{\text{mass}}{\text{volume}} = \frac{6156\,\text{g}}{540\,\text{cm}^3} = 11.4\,\text{g/cm}^3$$

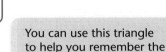

Example 13

A gold bar has a density of 19.3 g/cm³.
Its mass is 57 900 g.
Calculate the volume of the bar.

$$\text{Volume} = \frac{\text{mass}}{\text{density}} = \frac{57\,900}{19.3} = 3000\,\text{cm}^3$$

> You can use this triangle to help you remember the formulae.
>
>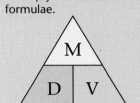

Exercise 8K

1 A cast iron rod has a mass of 240 kg. Its cross-sectional area is 10 cm² and its density is 7.5 g/cm³. Calculate

   (a) the volume of the rod

   (b) the length of the rod in metres.

2 A cast iron rod 2 metres long has a cross-sectional area of 14 cm². The mass of the rod is 210 kg. Calculate

   (a) the volume of metal in the rod

   (b) the density of cast iron in g/cm³.

3 The density of aluminium is 2590 kg/m³. Find

   (a) the mass of a piece of aluminium which has a volume of 2.5 m³

   (b) the volume of a piece of aluminium whose mass is 1200 kg.

4 This is the cross-section of a prism 12 cm long. It has a mass of 3.5 kg. Calculate

   (a) its volume

   (b) the density in g/cm³ of the material from which it is made.

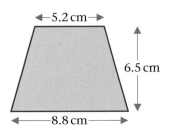

8.9 Ratios

You can use **ratios** to show how things are divided or shared.

Example 14

Tony and Kamiljit bought a pack of five recordable
DVDs. They shared the DVDs and the cost.
Tony took two DVDs and Kamiljit took three.

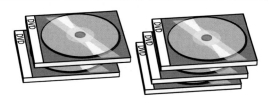

(a)   In what ratio did they share the DVDs?

(b)   What fraction of the DVDs did Tony take?

(c)   The pack cost £3. What was a fair amount for Tony to pay?

(a)   They shared the DVDs in the ratio $2:3$

(b)   Tony took $\frac{2}{5}$ of the DVDs.

(c)   Tony should pay $\frac{2}{5} \times £3 = £1.20$

> ratio: $2:3$
> fraction: $\frac{2}{5}$ $\frac{3}{5}$

Example 15

Barbara and Peter inherit £4000 in their aunt's will. The money is to be shared $3:2$ in favour of
Barbara. How much does each receive?

Method 1

Sharing in the ratio $3:2$ means
that the money is divided into
5 parts, with Barbara getting
3 parts and Peter getting 2 parts.

So Barbara gets $\frac{3}{5} \times £4000 = £2400$
and Peter gets $\frac{2}{5} \times £4000 = £1600$

Method 2

Use the unitary method.
Find the value of one part:

5 parts = £4000
 1 part = £4000 ÷ 5 = £800

So Barbara gets $3 \times £800 = £2400$
and Peter gets $2 = £800 = £1600$

> Check this by adding:
> £2400 + £1600 = £4000

Exercise 8L

1   Errol, Jane and Karl open a bag of chocolate and find it contains
12 bars. They decide to share them in the ratio $1:2:3$. In turn
Errol takes one bar, Jane two and Karl three. They keep doing
this until they have shared out all the bars.

(a)   How many chocolate bars will
(i) Errol have   (ii) Jane have   (iii) Karl have?

(b)   When all the chocolate bars have been shared out, what
fraction of the total will
(i) Errol have   (ii) Jane have   (iii) Karl have?

Write each fraction in its simplest form.

2   Sally, Brian and Mark want to share £40 in the ratio $2:3:5$.

(a)   What fraction of the £40 will each person receive?

(b)   How much will each person receive?

3 Donna and Shirley share £270 in the ratio 3:5, respectively.

    (a) To work out how much each receives, how many parts should £270 be divided into?

    (b) What fraction of £270 does Donna receive?

    (c) How much does Donna receive?

    (d) What fraction does Shirley receive?

    (e) How much does Shirley receive?

4 Danny and Melissa work in a café. Danny works five days a week but Melissa only works two days a week.

   To be fair they agree to share their tips in the ratio of the numbers of days they work.

    (a) In what ratio do they share their tips?

    (b) What fraction of the tips should Danny receive?

    (c) If the tips in one week came to £66.50, how much did Melissa receive?

    (d) The following week the tips came to £84. How much did Danny receive?

    (e) One week Danny received £25 in tips. How much did Melissa receive?

5 In a GCSE maths course, 80 of the final marks are for written exam papers. The other 20 are from coursework assignments.

    (a) Write the ratio of exam marks to coursework marks in its simplest form.

    (b) Daly got a final percentage of 66. The ratio of his exam marks to coursework marks was 8:3. How many coursework marks did he get?

6 Tom sells three types of soap powder in his shop: New Mold, New Drift and New Purr. In one week he sells 476 boxes of soap powder in the ratio 1:2:4, respectively.

    (a) What fraction of the total sales are New Drift?

    (b) How many boxes of New Mold does he sell?

    (c) If he makes 8p profit on each box sold, how much profit does he make altogether from the two best-selling powders?

7 Val and Bill run The Grange Residential Home. They held a fund raising morning and raised £360. Val raised £240 from selling raffle tickets. The rest was raised by Bill selling refreshments.

    (a) How much did Bill raise from the sale of refreshments?

    (b) What is the ratio of the amount raised from selling raffle tickets to the amount raised from refreshments? Write your ratio in its simplest form.

    (c) The following year a second fund raising morning was held and a total of £423 raised. The amounts raised were in the same ratio as in the previous year. How much was raised from selling raffle tickets?

8 Four darts players each threw three rounds of three darts to raise money for charity. The charity received £1 for each point scored. The players scored a total of 1728 between them. Their individual scores were in the ratio 3:4:4:5.
   (a) What score did the best player obtain?
   (b) What must the best player have scored each round to achieve this total score?
   (c) What was the lowest total score?
   (d) What was the difference in score between the highest scorer and the lowest?

> The highest possible score for one dart is 60.

9 In Midgrove School the ratio of pupils to teachers is 17.2 to 1.
   (a) Multiply 17.2 by 5.
   (b) Rewrite the ratio in the form $m:n$ where m and n are both whole numbers.
   (c) What is the smallest possible number of pupils in the college?
   (d) If the actual total of pupils and teachers is 1456, how many teachers are there?

> Use part (a) to help you.

8.10 More ratios

Example 16

In a bottle of Pepsea two secret ingredients are added in the ratio 5:8. If 5.2 ml of the first ingredient is added

(a) how much of the second ingredient is added

(b) how much in total is added?

(a) Of the total amount added, $\frac{5}{13}$ is the first ingredient and $\frac{8}{13}$ the second.

   $\frac{5}{13}$ of the total amount added is 5.2 ml.

   Find 1 part from 5 parts:

   $\frac{1}{13}$ of the total amount added is $5.2 \div 5 = 1.04$ ml

   so $\frac{8}{13}$ of the total amount is $8 \times 1.04 = 8.32$ ml.

   8.32 ml of the second ingredient is added.

(b) There are 5.2 ml of the first ingredient and 8.32 ml of the second so a total of $5.2 + 8.32 = 13.52$ ml of the two ingredients are added.

Exercise 8M

1 At Christmas Darren, Nathan and Joe received Christmas cards in the ratio 4:4:5, respectively. Joe received 75 cards.
   (a) What fraction of the cards did Joe receive?
   (b) What fraction did Darren and Nathan receive between them?
   (c) How many cards did Darren receive?
   (d) How many cards did they receive altogether?

2 A petrol company carried out a fuel consumption test and found that the winter to summer ratio for the same car over the same test track was 3.5 : 4. The winter fuel consumption rate was 8.2 km per litre. Find the summer consumption rate.

5 miles is approximately equal to 8 km.

3 When driving in France the speed limit in towns is 60 km/h.

   (a) Write the distance in miles to the distance in kilometres as a ratio.

   (b) Kevin drives through a French town at 35 mph. Is he breaking the speed limit?

   (c) The maximum speed allowed on British roads is 70 mph. What is this speed in km/h?

4 The average marks of three teaching groups at Midgrove School were in the ratio 4 : 5 : 7. The average mark of the middle group was 62.5. Find

   (a) the lowest average mark

   (b) the difference between the highest and lowest average marks

   (c) the ratio of the lowest average mark to the highest average mark, expressing your answer in the form $1 : n$.

5 Ben and Catherine win the National Lottery and have a motoring holiday in Europe. They visit Paris, Rome, Berlin and Oslo, and spend time in each city in the ratio 4 : 5 : 7 : 9, respectively. They spend 12 days in Paris. Find

   (a) how many days they spend in Oslo

   (b) how many days they spend in Berlin

   (c) how much longer they spend in Berlin than in Rome

   (d) how many days they spend away altogether.

6 During one season East Ham Rovers, Muncaster United and Sheffield Thursday scored a total of 270 goals between them. East Ham Rovers scored twice as many as Muncaster United, who scored 3 times as many as Sheffield Thursday. Find

   (a) the scoring ratio between the three football clubs

   (b) how many goals Muncaster United scored

   (c) the difference between the number of goals scored by East Ham Rovers and by Sheffield Thursday.

7 A gang of four villains robbed a bank but got caught. On passing sentence the judge told the first villain: 'You are the gang leader and will serve the longest sentence'. To the next: 'You were second in charge so you will serve the next longest sentence'. To the fourth robber he said, 'You were led astray by the others so I will give you the shortest sentence. I sentence you all to a total of 122 years in prison to be served in the ratio $1 : \frac{1}{2} : \frac{1}{3} : \frac{1}{5}$.' Work out how long each person's prison sentence was.

8.11 Standard form

Standard form is a convenient kind of shorthand for writing large and small numbers. To use it you need to know how to write powers of 10 in index form.

For example:

$$10 = 10^1$$
$$100 = 10 \times 10 = 10^2$$
$$1000 = 10 \times 10 \times 10 = 10^3$$

This **power** or **index** means multiply 10 by itself 3 times. It does not mean 3×10.

Astronomers use standard form to record large measurements. The Sun's diameter is about 1 392 000 km or 1.392×10^6 km.

Numbers greater than 1

Here are some numbers greater than 1 written in standard form:

number			in standard form
3000	=	3×1000 =	3×10^3
5 000 000	=	$5 \times 1 000 000$ =	5×10^6
7 200 000	=	$7.2 \times 1 000 000$ =	7.2×10^6

A number is in **standard form** when:

$$7.2 \times 10^6$$

This part is written as a number between 1 and 10

This part is written as a power of 10

72×10^5 is *not* in standard form.

This part has *not* been written as a number between 1 and 10.

Make sure you know how to enter numbers in standard form on your calculator.

Exercise 8N

1 Express these numbers as powers of 10.
(a) 10×10
(b) $10 \times 10 \times 10$
(c) $10 \times 10 \times 10 \times 10 \times 10 \times 10$
(d) $10 \times 10 \times 10 \times 10 \times 10 \times 10 \times 10 \times 10 \times 10 \times 10$

For example, $10 \times 10 \times 10 \times 10 = 10^4$.

2 Write each number as a number of millions using decimals where necessary.
(a) 6 000 000 (b) 3 400 000 (c) 7 800 000
(d) 5 500 000 (e) 2 650 000 (f) 7 642 000

For example, 8 650 000 is 8.65 million.

3 Write these numbers in standard form.
(a) 3.1 million (b) 4.3 million (c) 0.5 million
(d) 2 400 000 (e) 7 800 000 (f) 8 600 000
(g) 4 000 000 (h) 9 000 000 (i) 400 000

4 Write these numbers as powers of 10.

(a) 1 thousand   (b) ten thousand

(c) ten   (d) 1 million

(e) 1 hundred thousand   (f) ten $\times$ ten $\times$ ten

> For example,
> $100 = 10 \times 10 = 10^2$.

5 Copy and complete this table.

	10^6	10^5	10^4	10^3	10^2	10^1	10^0	Standard form
(a)			7	4	0	0	0	7.4×10^4
(b)					2	6	0	
(c)		6	8	0	0	0	0	
(d)						4	5	
(e)	9	9	0	0	0	0	0	
(f)						6	2	
(g)							8	

> The first row is done for you.

6 Write these numbers in standard form.

(a) 16   (b) 4300   (c) 650 000

(d) 87 000 000   (e) 670   (f) 865

(g) 9 870 000   (h) 98 500   (i) 805 000 000 000

Changing from standard form to an ordinary number

Here is how to change 5.2×10^4 from standard form into an ordinary number:

$$
\begin{aligned}
5.2 \times 10^4 &= 5.2 \times 10 \times 10 \times 10 \times 10 \\
&= 52 \times 10 \times 10 \times 10 \\
&= 520 \times 10 \times 10 \\
&= 5200 \times 10 \\
&= 52\,000
\end{aligned}
$$

So $5.2 \times 10^4 = 52\,000$ as an ordinary number.

Exercise 8O

1 Change these numbers from standard form to ordinary numbers.

(a) 4.2×10^2 (b) 6.7×10^4 (c) 5.5×10^3

(d) 7.5×10^6 (e) 6.2×10^5 (f) 7.3×10^4

(g) 2.4×10^7 (h) 1.1×10^1 (i) 7.25×10^0

2 Evaluate these expressions, giving your answers in standard form.

(a) 25×36 (b) 640×15

(c) 45×900 (d) 25^4

Numbers less than 1

Very small numbers can also be written more conveniently in standard form.

You need to know how to write these as powers of 10:

$$0.1 = \frac{1}{10} \qquad = \qquad = 10^{-1}$$

$$0.01 = \frac{1}{100} = \frac{1}{10 \times 10} = 10^{-2}$$

$$0.001 = \frac{1}{1000} = \frac{1}{10 \times 10 \times 10} = 10^{-3}$$

$$0.0001 = \frac{1}{10\,000} = \frac{1}{10 \times 10 \times 10 \times 10} = 10^{-4}$$

Here is how to write 0.041 in standard form:

$$0.041 = \frac{4.1}{100} = 4.1 \times \frac{1}{100} = 4.1 \times \frac{1}{10^2} = 4.1 \times 10^{-2}$$

0.041 in standard form is 4.1×10^{-2}.

Here is how to change a number in standard form back to an ordinary decimal number:

$$2.4 \times 10^{-3} = \frac{2.4}{10^3} = \frac{2.4}{1000} = 0.0024$$

$$10^{-3} = \frac{1}{10^3}$$

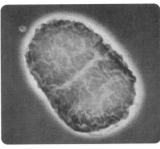

Biologists working with micro-organisms sometimes use standard form to record their sizes.
This bacterium is 0.000 21 cm long or 2.1×10^{-4} cm.
Magnification: ×14 400.

Start by writing 0.041 as a fraction with a numerator between 1 and 10.
Remember:
$$10^{-2} = \frac{1}{10^2}$$

Exercise 8P

1   Copy and complete this table.

	10^0	10^{-1}	10^{-2}	10^{-3}	10^{-4}	10^{-5}	Standard form
(a)	0	0	0	2	4		2.4×10^{-3}
(b)	0	2					
(c)	0	0	0	0	0	6	
(d)	0	1	5				
(e)	0	0	0	7			
(f)	0	0	0	0	4	5	
(g)	0	0	3	4	6		
(h)	0	0	0	1	2	5	

The first row is done for you.

2   Write these numbers in standard form.

(a)   0.002   (b)   0.15   (c)   0.0004

(d)   0.054   (e)   0.000 008   (f)   0.000 000 000 068

(g)   0.346   (h)   0.09   (i)   0.0056

3   Change these numbers back to ordinary decimal number form.

(a) 3.5×10^{-1} (b) 6.0×10^{-2} (c) 7.2×10^{-4}

(d) 2.2×10^{-3} (e) 1.35×10^{-5} (f) 5.33×10^{-6}

(g) 8.8×10^{-10} (h) 4.4×10^{-7} (i) 4.999×10^{-1}

8.12 Calculations in standard form

Example 17

Evaluate $(2.95 \times 10^5) \times (4.0 \times 10^3)$, giving your answer in standard form.

$$2.95 \times 4.0 \times 10^5 \times 10^3$$

Rearrange the expression with the powers of 10 on the right.

$$= 11.8 \times 10^8$$

Multiply the numbers on the left and the tens on the right.

11.8 is not between 1 and 10, so the answer is not in standard form yet.

11.8 is 1.18×10

so $(1.18 \times 10) \times 10^8 = 1.18 \times 10^9$

> Remember:
> The index rule for multiplying powers is:
> $$10^m \times 10^n = 10^{m+n}$$
> So
> $$10^5 \times 10^3 = 10^8$$

Worked examination question

(a)   The distance from the Earth to the Moon is approximately 250 000 miles. Express this number in standard form.

(b)   The distance from the Earth to the Sun is approximately 9.3×10^7 miles. Calculate the value of the expression

$$\frac{\text{distance from the Earth to the Moon}}{\text{distance from the Earth to the Sun}}$$

giving your answer in standard form.

(a) $250\,000 = 2.5 \times 10^5$

(b)   Substituting the distances in the expression gives $\dfrac{2.5 \times 10^5}{9.3 \times 10^7}$.

Here are two ways of evaluating this:

Method 1

$$\frac{2.5 \times 10^5}{9.3 \times 10^7}$$

$$= \frac{250\,000}{93\,000\,000} \quad \overset{\div\,10000}{=} \quad \frac{25}{9300}$$

$$\underset{\div\,10000}{}$$

$$= 0.002\,69 \text{ (to 3 s.f.)}$$

$$= 2.69 \times 10^{-3}$$

Method 2

$$\frac{2.5 \times 10^5}{9.3 \times 10^7}$$

$$= \frac{2.5 \times 1}{9.3 \times 10^2}$$

$$= 0.269 \times \frac{1}{10^2}$$

$$= 0.269 \times 10^{-2}$$

$$= 2.69 \times 10^{-3}$$

> Remember:
> The index rule for dividing powers is
> $$10^m \div 10^n = 10^{m-n}$$
> So
> $$10^5 \div 10^7 = 10^{(5-7)} = 10^{-2}$$

You can use standard form to help you make approximations and estimates.

Example 18

The distance from the Earth to the Sun is 93 000 000 miles. The Earth moves around the Sun in an approximately circular orbit. One complete orbit of the Sun takes 365 days.

Use standard form to help work out an estimate of the average speed of the Earth as it orbits the Sun.

The Earth orbits the Sun at a distance of approximately 90 000 000 or 9×10^7 miles.

The distance travelled by the Earth as it makes one orbit of the Sun is roughly equal to the circumference of a circle of radius 9×10^7 miles.

Using circumference $= 2\pi r$, the total distance travelled by the Earth in 365 days is

$$2 \times \pi \times 9 \times 10^7$$

> You should remember from previous work that
> $C = 2\pi r$

But π is about 3.1, so $2 \times \pi$ is just over 6.

$2 \times \pi \times 9$ is about 55.

Hence the distance travelled by the Earth in 365 days is approximately

$$55 \times 10^7 = 5.5 \times 10 \times 10^7$$
$$= 5.5 \times 10^8 \text{ miles}$$

365 days is 365×24 hours, or approximately

$$350 \times 20 = 7000$$
$$= 7 \times 10^3 \text{ hours}$$

So the Earth travels approximately 5.5×10^8 miles in approximately 7×10^3 hours.

Using the formula **average speed** $= \dfrac{\textbf{distance}}{\textbf{time}}$,

the speed of the Earth as it orbits the Sun is

$$\text{average speed} = \frac{5.5 \times 10^8}{7 \times 10^3}$$
$$= \frac{55}{7} \times \frac{10^7}{10^3}$$
$$= \frac{55}{7} \times 10^4$$

This sort of division should be done **mentally**.

> A good method for approximating $55 \div 7$ mentally is to think of the number closest to 55 in the 7 times table.
> $7 \times 8 = 56$.

$\dfrac{55}{7}$ is approximately 8

So the speed is approximately 8×10^4 miles per hour.

You could use different approximations to get your answer. For example, you could have approximated the distance from the Earth to the Sun as 100 000 000 miles, or 1×10^8 miles. This would give you a different answer.

In your exam you should make sure that you clearly state what approximations you are using.

Exercise 8Q

1 Evaluate these expressions, giving your answers in standard form.

   (a) $(6.4 \times 10^8) \times (1.5 \times 10^5)$

   (b) $(8.2 \times 10^7) \times (2.1 \times 10^{-4})$

   (c) $(5.25 \times 10^2) \times (1.6 \times 10^{-5})$

   (d) $(4.1 \times 10^{-6}) \times (3.6 \times 10^{-7})$

   (e) $\dfrac{6.4 \times 10^4}{2.2 \times 10^5}$

   (f) $\dfrac{5.6 \times 10^{-2}}{3.8 \times 10^7}$

   (g) $\dfrac{3.4 \times 10^{24}}{1.2 \times 10^{-6}}$

   (h) $\dfrac{8.8 \times 10^{25}}{8.8 \times 10^{22}}$

   (i) $(2.8 \times 10^3)^2$

   (j) $(2.2 \times 10^{-2})^3$

2 Evaluate these expressions, giving your answers in standard form.

   (a) $500 \times 600 \times 700$

   (b) 0.006×0.004

   (c) $\dfrac{0.08 \times 480}{180}$

   (d) $\dfrac{89\,000 \times 0.0086}{48 \times 0.25}$

   (e) $\dfrac{65 \times 120}{1500}$

   (f) $\dfrac{8.82 \times 5.007}{10\,000}$

   (g) $(12.8)^4$

   (h) $(36.4 \times 24.2)^{-3}$

3 The distance from the Earth to the Sun is approximately 93 000 000 miles. Light travels at a speed of approximately 300 000 kilometres per second.

   Use standard form to work out an estimate of the time it takes light to travel from the Sun to the Earth.

> 1 mile is approximately 1.6 kilometres.

4 The Earth's diameter is 1.27×10^4 km and the diameter of Mars is 6.79×10^3 km.

   (a) Which planet has the larger diameter?

   (b) What is the difference between their diameters?

   (c) What is the total if the two diameters are added?

   (d) How many times bigger is one planet than the other? (Hint: consider their volumes.)

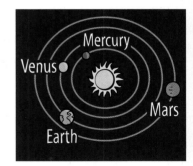

Mixed exercise 8

1 The auction price of a second hand car is 25% less than the retail price of the same car.

   (a) Calculate the auction price of a car which has a retail price of £4200.

   (b) Calculate the retail price of a car which has an auction price of £5400.

2   The price of a shirt has been reduced by $12\frac{1}{2}$% to £21 in a sale. What was the original selling price?

3   Asif puts £200 into a building society savings account. He leaves the money in the account for three years without withdrawing any money or adding any to his savings. The building society pays a compound interest rate of 4% per annum. How much money will Asif have in his account at the end of the three years?

4   On a test drive, a new car travels 307 miles on 37.7 litres of petrol. Work out the average number of miles the car travels on each litre of petrol. Give your answer correct to two significant figures.

5   In a competition, the total prize money is £2000.

    This is divided between the winners of the first, second and third prizes in the ratio 5 : 3 : 1 respectively.

    Calculate the value of the first prize.

6   The prices below include VAT at $17\frac{1}{2}$%.
    Calculate each price exclusive of VAT.

    (a)   £135 for a mountain bike

    (b)   £209 for a hammer drill

    (c)   £49 for a grill

7   A motorbike originally cost £2800. Its value has depreciated by 35%. What is the new value?

8   Dinosaurs roamed the Earth about 140 million years ago.

    (a)   Express 140 million in standard form.

    (b)   The average human life span is 72 years. How many average human life spans is 140 million years? Give your answer in standard form, correct to three significant figures.

9   In 2005, 623 118 students sat GCSE English. The entry fee for each GCSE student was £22.50. Use standard form to work out an estimate for the total entry fee in 2005 for all the GCSE English students.

10 The density of oak is $720\,\text{kg/m}^3$.
    Calculate the mass of a rectangular block of oak measuring 6 m by 0.7 m by 0.12 m.

11 Approximately $\frac{2}{3}$ of the Earth's surface is covered with water. The Earth can be assumed to be a sphere with a radius of 6350 km. Use standard form and the formula

    surface area of sphere = $4 \times \pi \times$ radius^2

    to work out an estimate of the area of the surface of the Earth covered by water.

Summary of key points

1 If an amount is increased by x% the new amount is $(100 + x)$% of the original amount. So the new amount = the old amount $\times (100 + x)$%.

2 If an amount is decreased by x% the new amount is $(100 - x)$% of the original amount. So the new amount = the old amount $\times (100 - x)$%.

3 The **percentage increase** is $\dfrac{\text{actual increase}}{\text{original price}} \times 100\%$

4 The **percentage decrease** is $\dfrac{\text{actual decrease}}{\text{original price}} \times 100\%$

5 **Compound interest** is interest paid on an amount and on the interest already earned.

6 To calculate compound interest, find the multiplier:

   Amount after n years = original amount $\times$ multiplier^n

7 **Average speed** $= \dfrac{\text{distance travelled}}{\text{time taken}}$ (typical units: km/h)

8 Volume flowing per second = cross-sectional area $\times$ speed (typical units: cm^3/s)

9 **Density** $= \dfrac{\text{mass}}{\text{volume}}$ (typical units: kg/m^3)

10 You can use **ratios** (such as $2:3$ and $5:4:7$) to show how things are divided or shared.

11 Large and small numbers can conveniently be represented in **standard form**. A number is in standard form when:

$$7.2 \times 10^6$$

This part is written as a number between 1 and 10

This part is written as a power of 10

⑨ Functions, lines, simultaneous equations and regions

In this chapter you will draw straight line graphs for functions, and use graphs to represent inequalities and simultaneous equations.

9.1 Functions

Here is a pattern sequence:

 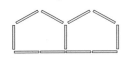 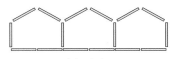 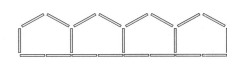

| 6 sticks | 11 sticks | 16 sticks | 21 sticks |

To work out the number of sticks in the pattern, the rule is:

'Multiply the pattern number by 5, then add 1'

> There is more on number sequences in Chapter 12.

You can show how the rule works by using a flowchart like this:

$$\text{Input} \rightarrow \boxed{\times 5} \rightarrow \boxed{+1} \rightarrow \text{Output}$$

$$
\begin{array}{cccc}
1 & \rightarrow & 5 & \rightarrow & 6 \\
2 & \rightarrow & 10 & \rightarrow & 11 \\
3 & \rightarrow & 15 & \rightarrow & 16 \\
4 & \rightarrow & 20 & \rightarrow & 21 \\
n & \rightarrow & 5n & \rightarrow & 5n + 1
\end{array}
$$

Instead of using the flow chart you can write $n \rightarrow 5n + 1$

You can say that n maps onto $5n + 1$

$n \rightarrow 5n + 1$ is called a function.

> You can let the letter n stand for any number. For this function, n must be a positive whole number.

A **function** is a rule which shows how one set of numbers relates to another set.

Exercise 9A

1 Work out the outputs for each function using the inputs 1, 2, 3 and 10.

> Remember the order of operations.

(a) $n \rightarrow 2n - 2$

(b) $n \rightarrow 3n + 2$

(c) $n \rightarrow 2(n + 3)$

(d) $n \rightarrow 3(n - 1)$

(e) $n \rightarrow \dfrac{n}{2} + 1$

(f) $n \rightarrow \dfrac{n + 3}{2}$

9.2 Graphs of linear functions with positive whole number inputs

Here is a simple function: $n \rightarrow 2n + 1$

Another way to look at a function is to let x
stand for the input and y stand for the output, so $x \rightarrow y$

But using $n \rightarrow 2n + 1$ we know that $x \rightarrow 2x + 1$

So y and $2x + 1$ must be worth the same and
you can write $y = 2x + 1$

> $y = 2x + 1$ is called an **equation**. The value of y is *always* equal to the value of $2x + 1$.

You now have two ways of showing the same rule or function:

$n \rightarrow 2n + 1$ or $y = 2x + 1$

$0 \rightarrow 1$
$1 \rightarrow 3$
$2 \rightarrow 5$
$3 \rightarrow 7$
$4 \rightarrow 9$
$5 \rightarrow 11$

x	0	1	2	3	4	5
y	1	3	5	7	9	11

This is called a **table of values**.

You can show the numbers in the table
as coordinates on a graph.

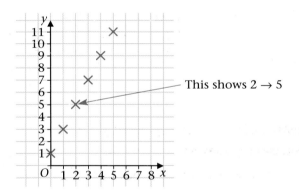

This shows $2 \rightarrow 5$

Exercise 9B

You will need graph paper for this exercise.

1   (a)   Write the function $n \rightarrow n + 2$ as an equation using x and y.
    (b)   Make a table of values for your equation using the values 0, 1, 2, 3, 4, 5 for x.
    (c)   Draw a graph to show the numbers from your table of values.

2   Do the same as in question **1** for each of these functions:
    (a) $n \rightarrow n + 4$ (b) $n \rightarrow 2n + 3$
    (c) $n \rightarrow 3n + 2$ (d) $n \rightarrow 2n$

3 **(a)** Complete the table of values for this sequence of matchstick patterns.

Number of squares x	1	2	3	4	5
Number of matchsticks y	4	7			

(b) Draw the graph for this table of values.

(c) Write down the equation of the graph passing through these points.

4 **(a)** Complete the table of values for this sequence of matchstick patterns.

Number of triangles x	1	2	3	4	5
Number of matchsticks y	3				

(b) Draw the graph for the table of values.

(c) Write down the equation of the graph passing through these points.

5 **(a)** Invent and draw a sequence of stick patterns for each of these graphs:

> Draw up a table of values for each graph.

(i)

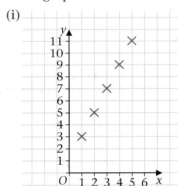

(ii)

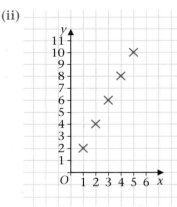

(b) Write an equation for each graph in part **(a)**.

9.3 Graphs of linear functions with continuous positive inputs

In questions **3** and **4** of Exercise 9B, x was the number of matchsticks. So x could take only whole number values. You cannot not join the points in those graphs because there are no values of x and y between those points.

If x can take any values, either whole numbers or decimals, we say x is **continuous**.

> The input values of x are **continuous** if x can take any value (whole numbers or decimals).

Look at the graph for $y = 2x + 1$.

x	0	1	2	3	4	5
y	1	3	5	7	9	11

The points lie on a straight line.

You can draw the straight line in this graph because x is continuous.

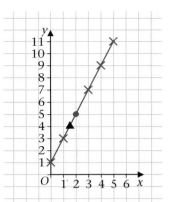

The x and y values for the point marked ● are 2 and 5.
So the point ● has coordinates (2, 5).

You say the coordinates (2, 5) 'satisfy' the equation $y = 2x + 1$.

Look at the point ▲ which has coordinates $(1\frac{1}{2}, 4)$.
Put $x = 1\frac{1}{2}$ in the equation $y = 2x + 1$.
You get $y = 2 \times 1\frac{1}{2} + 1 = 4$.
So when $x = 1\frac{1}{2}$, $y = 4$.

Do the same for other points on the straight line.

> $y = 2x + 1$ is the equation of a straight line. The coordinates of any point on the straight line 'satisfy' the equation.

> $x \rightarrow 2x + 1$ is called a **linear function** because its graph is a straight line.

You can draw graphs to solve real-life problems.

Example 1

Water flows steadily from a pipe into a storage tank. Every minute the depth of the water in the tank increases by 2 cm. When the water starts to flow the depth of water in the tank is already 3 cm.

(a) Draw a graph to show the relationship between the depth, y cm, of the water in the tank and the time, x minutes, after the water starts to flow.

(b) Use the graph to find the depth of water $3\frac{1}{2}$ minutes after the water starts to flow.

(a) Draw up a table of values for whole number values of x:

Time in minutes x	0	1	2	3	4
Depth in cm y	3	5	7	9	11

Draw the graph.

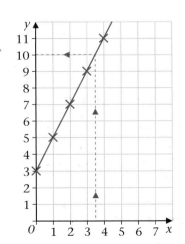

(b) Find $3\frac{1}{2}$ on the x-axis.
Go up to the line and across to find the corresponding value of y, which is 10.
The depth of the water $3\frac{1}{2}$ minutes after the water starts to flow is 10 cm.

Exercise 9C

You will need graph paper for these questions.

1 A narrow boat travels on a canal at a speed of 4 miles per hour.

(a) Complete this table of values:

Time travelled in hours x	0	1	2	3	4
Distance travelled in miles y	0	4			

(b) Draw the graph for the table of values.

(c) Use your graph to find how far the boat travels in $3\frac{1}{2}$ hours.

2 In Britain miles are used to measure large distances. In the rest of Europe kilometres are used. A good approximation to use to change miles into kilometres is 5 miles = 8 kilometres.

(a) Complete this table of values:

Distance in miles x	0	5	10	15	20	25
Distance in kilometres y	0	8				

(b) Draw the graph for the table of values.

(c) Use your graph to find how many kilometres are approximately the same as 13 miles.

Find 14 on the y-axis, go across to the line and down to the x-axis.

(d) How many miles are approximately the same as 14 kilometres?

3 Mortar is used to stick bricks together. To mix mortar you put 1 bucket of cement with 3 buckets of sand.

(a) Complete this table of values:

Buckets of cement x	0	1	2	3	4	5	6
Buckets of sand y	0	3					

I always use 3 buckets of sand to every bucket of cement.

(b) Draw the graph for the table of values.

(c) Use your graph to find how many buckets of cement you should put with 14 buckets of sand.

4 A cookery book gives this rule for cooking joints of lamb:

Cook for half an hour per 500 g plus half an hour.

(a) Complete this table of values:

Weight in kg x	$\frac{1}{2}$	1	$1\frac{1}{2}$	2	$2\frac{1}{2}$	3	$3\frac{1}{2}$
Cooking time in hours y	1	$1\frac{1}{2}$					

(b) Why isn't it sensible to start the values of x at 0? What is the biggest value you think it would be sensible to put for x?

(c) Draw the graph for the table of values.

(d) Use your graph to estimate how long it would take to cook a 2.3 kg joint of lamb.

9.4 Lines and parallel lines

The diagram shows the lines with equations
$y = 2x + 3$, $y = 2x + 1$ and $y = 2x - 1$.

In all three equations the number in front
of x (the **coefficient** of x) is 2, and the three
lines are all **parallel**.

The number term in each equation gives the
y-coordinate of the point where the line cuts
the y-axis.

For example, the line with equation
$y = 2x + 3$ cuts the y-axis at (0, 3).

This point is called its **intercept** on the y-axis.
So, the intercept on the y-axis of the line with
equation $y = 2x + 1$ is (0, 1) and the intercept on
the y-axis of the line with equation $y = 2x - 1$ is (0, −1).

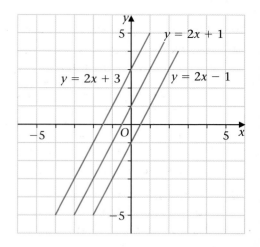

> An **intercept** is a point at which a line cuts the y-axis or
> the x-axis.

Example 2

(a) Draw the line with equation $y = 3x + 2$.

(b) Write its intercept on the y-axis.

(c) Write the equation of the line parallel to $y = 3x + 2$ whose
intercept is (0, −1).

(a) If you plot two points on the line, you can draw it,
but a third point is a useful check. All three points
should lie on the same straight line.

Here is a table of values:

x	−2	0	2
y	−4	2	8

When $x = -2$, $y = 3 \times -2 + 2 = -6 + 2 = -4$
When $x = 0$, $y = 3 \times 0 + 2 = 0 + 2 = 2$
When $x = 2$, $y = 3 \times 2 + 2 = 6 + 2 = 8$

(b) The intercept on the y-axis is (0, 2).

(c) The line parallel to $y = 3x + 2$ will have the same coefficient
of x, which is 3.

The equation is $y = 3x - 1$.

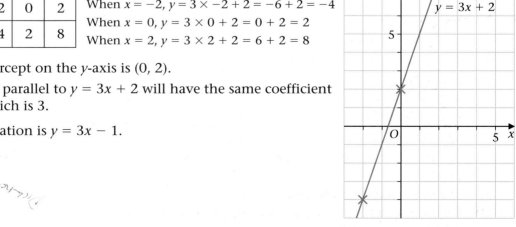

Exercise 9D

1   On separate diagrams, draw lines with these equations.

   (a) $y = x + 6$ (b) $y = 2x + 5$

   (c) $y = 3x + 1$ (d) $y = 3x - 4$

   (e) $y = -2x + 3$ (f) $y = \frac{1}{2}x - 2$

2   Write the equations of the two parallel lines in question **1**.

3   (a)   Without drawing the lines, write the equations of the two
            parallel lines in the following.

   (i) $y = 4x - 1$ (ii) $y = \frac{1}{4}x + 1$

   (iii) $y = x - 4$ (iv) $y = -4x + 1$

   (v) $y = \frac{1}{4}x - 1$ (vi) $y = -x + 4$

   (b)   Explain how you found your answer.

4   Write the intercepts on the y-axis of lines with these equations.

   (a) $y = 7x + 5$ (b) $y = 5x - 2$

   (c) $y = -3x - 5$ (d) $y = \frac{1}{4}x + 4$

5   A line is parallel to the line $y = 4x + 5$ and its intercept on the
   y-axis is (0, 3). Write the equation of the line.

6   A line is parallel to the line $y = 5x + 7$ and its intercept on the
   y-axis is (0, −3). Write the equation of the line.

7   A line is parallel to the line $y = 3x$ and its intercept on the
   y-axis is (0, 5). Write the equation of the line.

8   A line is parallel to the line $y = mx$ and its intercept on the
   y-axis is (0, c). Write the equation of the line.

9   The point with coordinates (2, 14) lies on the line with equation
   $y = 5x + c$. Find the value of c and write the line's intercept on
   the y-axis.

10   The point with coordinates (8, 1) lies on the line with equation
   $y = \frac{1}{2}x + c$. Find the value of c and write the line's intercept on
   the y-axis.

11   A line has equation $y = mx + 1$.

   (a)   Write the line's intercept on the y-axis.

   The point with coordinates (3, 16) lies on the line.

   (b)   Find the value of m.

12   A line is parallel to the line with equation $y = \frac{1}{3}x + 2$ and passes
   through the point with coordinates (12, 0).

   (a)   Find the equation of the line.

   (b)   Write its intercept on the y-axis.

13 (a) On the same diagram, draw lines with these equations.

    (i) $y = x + 4$ (ii) $y = 2x + 4$ (iii) $y = 3x + 4$

  (b) Write the intercept on the y-axis of each of the lines.

  (c) How does the coefficient of x in the equations affect the lines?

14 The diagram below shows the line $y = 3x - 4$.

Calculate the value of

(a) $\dfrac{AB}{BC}$ (b) $\dfrac{DE}{EF}$

This is the **gradient** of the line.

15 The diagram below shows the line $y = \frac{1}{2}x + 1$.

Calculate the value of

(a) $\dfrac{AB}{BC}$

(b) $\dfrac{DE}{EF}$

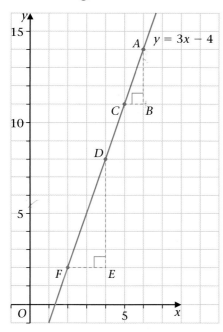

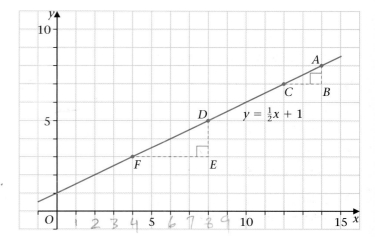

9.5 $y = mx + c$

Following on from Exercise 9D you should now see that

> **gradient** = $\dfrac{\text{change in } y\text{-direction}}{\text{change in } x\text{-direction}}$

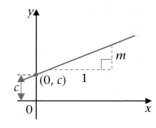

The straight line with equation $y = mx + c$ has a gradient of m and its intercept on the y-axis is $(0, c)$.

Example 3

Write down (a) the gradient and (b) the intercept on the y-axis of the line with equation $y = 7x - 3$.

(a) The gradient is 7.   (b) The intercept on the y-axis is $(0, -3)$.

Example 4

Write down (a) the gradient and (b) the intercept on the y-axis of the line with equation $y = -2x + 5$.

(a)   The gradient is -2.   (b)   The intercept on the y-axis is $(0, 5)$.

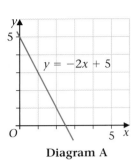

Diagram A shows the direction of the slope of a line with a **negative** gradient.

Diagram B shows lines with various gradients.

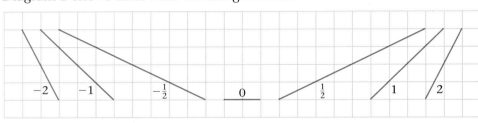

Diagram A

Diagram B

Lines with equation $y = mx + c$, with the same gradient (m), are parallel.

Example 5

Find (a) the gradient and (b) the intercept on the y-axis of the line with equation $2x - 3y = 18$.

Here the equation must be rearranged in the form $y = mx + c$

$$3y = 2x - 18$$
$$y = \tfrac{2}{3}x - 6$$

(a)   The gradient is $\tfrac{2}{3}$.   (b)   The intercept on the y-axis is $(0, -6)$.

> Rearrange
> $2x - 3y = 18$
> to $2x - 18 = 3y$
> So $3y = 2x - 18$
> Then divide by 3.

Example 6

The graph shows the cost, C pounds, of staying in a hotel for n nights.
C is given by the formula $C = an + b$.

Find the values of a and b.

$$\text{Gradient} = \frac{PQ}{QR} = \frac{40}{1}$$
$$= 40$$

so $a = 40$

Intercept on C-axis is $(0, 20)$

so $b = 20$

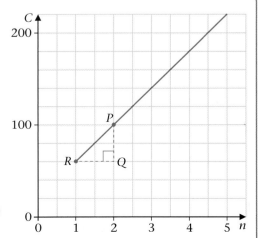

> Although n is not continuous you can draw the line to help work out the gradient.

> $PQ = £40$
> $QR = 1$ night

> If you put your ruler on the graph you will see that it would cross the C-axis at $(0, 20)$.

Exercise 9E

1 Write (i) the gradient and (ii) the intercept on the y-axis of lines with the equations

(a) $y = 5x + 4$ (b) $y = 2x - 7$

(c) $y = \frac{1}{4}x + 9$ (d) $y = -4x - 3$

(e) $y = 8 + 9x$ (f) $y = 7 - 2x$

2 Find (i) the gradient and (ii) the intercept on the y-axis of lines with the equations

(a) $2x + y = 3$ (b) $3x - y = 5$

(c) $x - 2y = 8$ (d) $x + 3y = 6$

(e) $3x - 5y = 10$ (f) $3x + 4y - 12 = 0$

3 The gradient of a line is 6 and its intercept on the y-axis is (0, 7). Write the equation of the line.

4 Find the equations of the lines shown on the diagram.

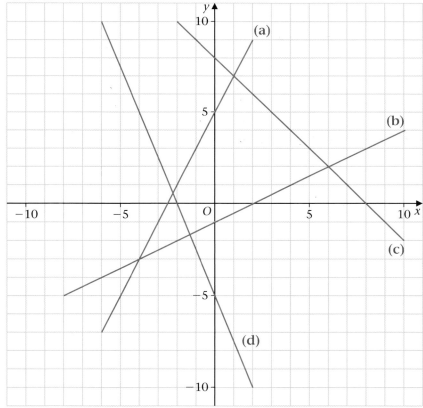

5 A line passes through the points with coordinates (0, 2) and (2, 8).

(a) Find the gradient of the line.

(b) Write the equation of the line.

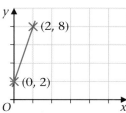

6 The gradient of a line is 4.
The point with coordinates (4, 23) lies on the line.
Find the equation of the line.

> It may help to draw a diagram for these questions.

7 The intercept on the y-axis of a line is (0, 11).
The point with coordinates $(-2, 17)$ lies on the line.
Find the equation of the line.

8 A line passes through the points with coordinates (2, 11) and (5, 23).

(a) Find the gradient of the line.

(b) Find the equation of the line.

9.6 Perpendicular lines

The diagram shows two pairs of perpendicular lines.

The gradient of line a is 2 and the gradient of line b is $-\frac{1}{2}$.

The gradient of line c is $-\frac{2}{3}$ and the gradient of line d is $1\frac{1}{2}$ or $\frac{3}{2}$.

In both cases, to obtain the second gradient, find the reciprocal of the first gradient and change its sign.

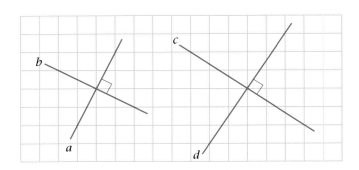

If a line has a gradient m, a line perpendicular to it has a gradient $-\dfrac{1}{m}$

> Reciprocal of
> $m = 1 \div m = \dfrac{1}{m}$

Notice that $2 \times -\frac{1}{2} = -1$ and $-\frac{2}{3} \times \frac{3}{2} = -1$.

> If two lines are perpendicular, the product of their gradients is -1.
> $$m \times \frac{1}{m} = -1$$

Example 7

A line is perpendicular to the line with equation $y = 4x + 7$ and its intercept on the y-axis is (0, 3). Find the equation of the line.

The line with equation $y = 4x + 7$ has a gradient of 4.

So a line perpendicular to it has a gradient of $-\frac{1}{4}$.

Its intercept on the y-axis is (0, 3).

The equation of the line is $y = -\frac{1}{4}x + 3$.

> $y = -\frac{1}{4}x + 3$ can also be written as $x + 4y = 12$.

Example 8

A line passes through the point with coordinates $(10, -6)$ and is perpendicular to the line with equation $5x - 4y = 20$.
Find the equation of the line.

Make y the subject of the equation $5x - 4y = 20$
$$y = \tfrac{5}{4}x - 5$$
The gradient of this line is $\tfrac{5}{4}$.

The gradient of the perpendicular is $-\tfrac{4}{5}$.

The equation of the perpendicular line is $y = -\tfrac{4}{5}x + c$ and it passes through $(10, -6)$ so $y = -6$ and $x = 10$ satisfy the equation
$$-6 = -\tfrac{4}{5}(10) + c$$
$$-6 = -8 + c$$
$$c = 2$$
The required equation is $y = -\tfrac{4}{5}x + 2$.

> $y = -\tfrac{4}{5}x + 2$ can also be written as $4x + 5y = 10$.

Exercise 9F

1   Write down the gradient of a line which is perpendicular to a line with gradient
   (a) 5   (b) 2   (c) 1
   (d) $\tfrac{3}{4}$ (e) $-2\tfrac{1}{2}$ (f) $-\tfrac{2}{7}$

2   Which of these lines is perpendicular to the line with equation $y = 5x - 3$? Give reasons for your answers.
   (a) $y = \tfrac{1}{5}x + 2$ (b) $y = -\tfrac{1}{5}x + 2$
   (c) $y = -5x - 2$ (d) $y = 5x + 2$

3   A line passes through the origin and is perpendicular to the line with equation $y = 3x - 2$. Find the equation of the line.

4   A line is perpendicular to the line with equation $y = x + 4$ and its intercept on the y-axis is $(0, 6)$. Find the equation of the line.

5   A line passes through the point with coordinates $(4, 1)$ and is perpendicular to the line with equation $y = -\tfrac{2}{5}x + 3$.
   Find the equation of the line.

6   Which of these lines is perpendicular to the line with equation $3x + 5y = 15$? Give reasons for your answer.
   (a) $3x - 5y = 30$ (b) $5x + 3y = 30$ (c) $5x - 3y = 30$

7   A line passes through the point with coordinates $(6, 1)$ and is perpendicular to the line with equation $2x - 3y = 6$.
   Find the equation of the line.

8   A line passes through the point with coordinates $(8, 1)$ and is perpendicular to the line with equation $4x + 3y = 12$.
   Find the equation of the line.

> Remember, if two lines are perpendicular, then they are at right angles to each other.

9.7 The mid-point of a line segment *AB*

Here is a straight line *AB*.
The straight line is called the line segment *AB*.

A has coordinates (1, 1), *B* has coordinates (3, 6).

M is halfway along the line segment *AB*.
M is called the **mid-point of the line segment *AB***.

You can find the coordinates of the mid-point.

Step 1 Add the *x*-coordinates and divide by 2
$$\frac{1 + 3}{2} = \frac{4}{2} = 2$$

Step 2 Add the *y*-coordinates and divide by 2
$$\frac{1 + 6}{2} = \frac{7}{2} = 3\tfrac{1}{2}$$

So the coordinates of the mid-point are $(2, 3\tfrac{1}{2})$.

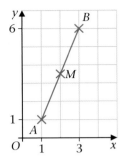

> *AB* is a segment of the line through *A* and *B*. This line continues infinitely in either direction.

> The **mid-point of the line segment *AB*** between $A(x_1, y_1)$ and $B(x_2, y_2)$ is $\left(\dfrac{x_1 + x_2}{2}, \dfrac{y_1 + y_2}{2}\right)$.

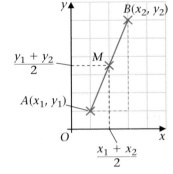

_____ **Example 9** _____

Work out the coordinates of the mid-point of the line segment *AB* where *A* is (2, 3) and *B* is (7, 11).

Mid-point *x*-coordinate is
$$\frac{2 + 7}{2} = \frac{9}{2} = 4\tfrac{1}{2}$$

Mid-point *y*-coordinate is
$$\frac{3 + 11}{2} = \frac{14}{2} = 7$$

So mid-point is $(4\tfrac{1}{2}, 7)$.

Exercise 9G

1   Work out the coordinates of the mid-point of the line segments:

    (a)   *A*(0, 0) and *B*(4, 5)   (b)   *C*(3, 1) and *D*(4, 9)

    (c)   *E*(5, 7) and *F*(1, 6)   (d)   *G*(1, 0) and *H*(5, 8)

    (e)   *I*(9, 10) and *J*(3, 7)

9.8 Simultaneous equations – graphical solutions

An *infinite number* of pairs of values of *x* and *y* will make the equation $2x + y = 6$ true.

$x = 0$ and $y = 6$, for example, satisfy this equation.

The coordinates of every point on the line represent a possible solution.

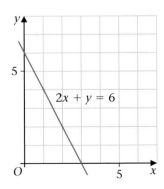

Similarly, an infinite number of pairs of values of x and y satisfy the equation $x + y = 5$.

The coordinates of the points on the line represent these solutions.

But there is only *one* pair of values which makes both $2x + y = 6$ and $x + y = 5$ true.

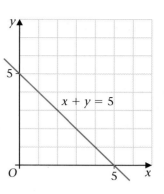

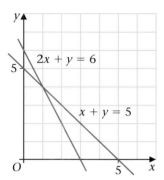

The diagram above shows a graphical solution to finding the pair.

The pair of values are the x and y coordinates of the point of intersection of the two lines:

$$x = 1 \quad \text{and} \quad y = 4$$

$2x + y = 6 \quad$ and $\quad x + y = 5 \quad$ are called **simultaneous equations** and their solution is $x = 1$, $y = 4$.

> Check by subtituting into the original equations:
> $2 \times 1 + 4 = 6$ ✓
> $1 + 4 = 5$ ✓

Example 10

Solve these simultaneous equations
$$x + 2y = 10$$
$$x - y = 1$$

Here are the tables of values:

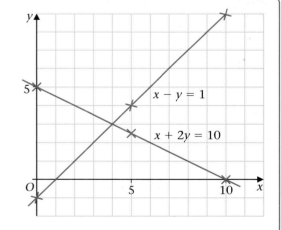

$x + 2y = 10$

x	0	5	10
y	5	2.5	0

$x - y = 1$

x	0	5	10
y	-1	4	9

Plot three points for each equation and draw straight lines.

The coordinates of the point of intersection of the lines gives the solution
$$x = 4 \quad \text{and} \quad y = 3$$

Check by substituting into the original equations:
$$4 + 2 \times 3 = 10 \quad ✓$$
$$4 - 3 = 1 \quad ✓$$

Simultaneous equations can be solved graphically, by drawing the graphs for the two equations and finding the coordinates of their point of intersection.

Exercise 9H

1 Write down four pairs of integer values of x and y which satisfy the equation $x + y = 10$.

2 Which of the following points lies on the line with equation $2x + 3y = 6$?

   (a) (2, 0)   (b) (3, 0)   (c) (6, −2)   (d) (−3, 4)

3 Write four pairs of integer values of x and y which satisfy the equation $2x − y = 8$.

4 Which two of the following equations are satisfied by $x = 3$ and $y = 5$?

   (a) $2x + 3y = 20$ (b) $3x − y = 4$

   (c) $5x − 2y = 5$ (d) $4x + 3y = 25$

5 Use the diagram to solve these simultaneous equations

$$x + 3y = 6$$
$$x + y = 4$$

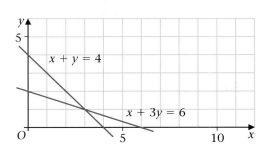

6 Draw appropriate straight lines to solve these simultaneous equations

   (a) $x + y = 3$ (b) $x + y = 6$
        $x − y = 1$ $x + 2y = 8$

   (c) $x + 2y = 6$ (d) $2x + 3y = 6$
        $x − y = 3$ $x + 2y = 5$

   (e) $x + y = 2$ (f) $3x − y = 9$
        $x − 3y = 6$ $x − 2y = −2$

> Draw a separate diagram for each part.

> Remember to draw the lines by plotting three points.

9.9 Simultaneous equations – algebraic solutions

Simultaneous equations can also be solved algebraically.

Example 11

Solve these simultaneous equations

$$2x + y = 12$$
$$x − y = 3$$

Here the two equations may be added together in order to eliminate y

$$
\begin{array}{r}
2x + y = 12 \\
\text{ADD} \quad \underline{x - y = 3} \\
3x = 15 \\
x = 5
\end{array}
$$

> $+y$ ADD $−y = 0$

To find y, substitute $x = 5$ into one of the equations.

$$5 − y = 3$$
$$y = 2$$

The solution is $x = 5$, $y = 2$.

When the coefficients of x or y in the two equations are equal, you may eliminate one of the unknowns by **subtracting** one equation from the other.

Example 12

Solve these simultaneous equations

$$4x + 3y = 5 \quad (1)$$
$$4x - 5y = 13 \quad (2)$$

> Call the equations (1) and (2).

In this case, adding equations (1) and (2) will not help.
Try subtracting:

$$
\begin{array}{r}
4x + 3y = 5 \quad (1)\\
\text{SUBTRACT} \quad 4x - 5y = 13 \quad (2)\\
\hline
8y = -8 \\
\hline
y = -1
\end{array}
$$

> $+ 3y - (-5y)$
> $= + 3y + 5y$
> $= 8y$

Substitute $y = -1$ into equation (1).

$$4x - 3 = 5$$
$$4x = 8$$
$$x = 2$$

The solution is $x = 2$, $y = -1$.

> Check:
> $4 \times 2 + 3 \times -1 = 5$ ✓
> $4 \times 2 - 5 \times -1 = 13$ ✓

It doesn't matter which of the two original equations you substitute into, but you should always remember to check that your values fit the other one as well.

Exercise 9I

Solve these simultaneous equations.

1 $x + 4y = 11$ **2** $3x + 2y = 14$ **3** $5x + 4y = 6$
     $x - y = 1$ $5x - 2y = 18$ $3x - 4y = 10$

4 $3x - 4y = 6$ **5** $4x + 2y = 5$ **6** $2x + 3y = 2$
     $5x - 4y = 2$ $8x - 2y = 1$ $8x + 3y = 17$

7 $x - 6y = 16$ **8** $4x - 2y = 9$ **9** $2x - 3y = 4$
     $2x + 6y = 5$ $4x - 5y = 18$ $2x + 3y = -8$

More algebraic solutions

When the equations do not have equal x or y coefficients, you need to multiply one or both of the equations before adding or subtracting.

Example 13

Solve these simultaneous equations.

$6x - 7y = 16$ (1)

$3x - 2y = 5$ (2)

Try multiplying equation (2) by 2. This gives:

$6x - 4y = 10$ (3)

> This gives the same coefficient of x in two equations.

Now subtract (1) from (3):

$6x - 4y = 10$ (3)

$6x - 7y = 16$ (1)

$(1) - (3)$: $3y = -6$

> $-4y - (-7y) = 3y$

$y = -2$

Substitute $y = -2$ into (2):

$3x + 4 = 5$

$3x\quad = 1$

$x\quad = \frac{1}{3}$

The solution is $x = \frac{1}{3}$, $y = -2$.

Example 14

Solve these simultaneous equations.

$2x + 3y = 12$ (1)

$3x - 4y = 1$ (2)

$(1) \times 4$: $8x + 12y = 48$ (3)

$(2) \times 3$: $9x - 12y = 3$ (4)

$(3) + (4)$: $17x = 51$

$x = 3$

> Multiply equation (1) by 4 and equation (2) by 3 to get $12y$ in each equation.

> Add, since the $12y$ terms have different signs.

Substitute $x = 3$ into (1):

$6 + 3y = 12$

$3y = 6$

$y = 2$

The solution is $x = 3$, $y = 2$.

Simultaneous equations can be solved algebraically by
- multiplying one or both equations by a number, if necessary
- adding or subtracting, and solving the resulting equation
- substituting this value back into the original equation.

Exercise 9J

Solve these simultaneous equations.

1 $8x + 3y = 35$
    $2x - 5y = 3$

2 $7x + 2y = 17$
    $5x + 6y = 3$

3 $5x - 4y = 5$
    $7x - 12y = 39$

4 $2x + 3y = 12$
     $3x - 4y = 1$

5 $6x - 5y = 23$
     $4x - 3y = 14$

6 $2x + 7y = 31$
     $5x - 3y = 16$

7 $3x - 8y = 11$
     $2x - 5y = 6$

8 $9x + 4y = 7$
     $8x - 5y = 49$

9 $6x + 7y = 22$
     $8x + 9y = 29$

10 $9x + 4y = 8$
      $6x + 5y = 3$

9.10 Solving problems using simultaneous equations

Example 15

The total cost of 4 small radiators and 3 large radiators is £159.
The total cost of 5 small radiators and 2 large radiators is £134.
Find the cost of one small radiator.

Let the cost of a small radiator be x pounds and the cost of a large radiator be y pounds.

Define any letters you introduce.

The cost, in pounds, of 4 small radiators is $4x$.
The cost, in pounds, of 3 large radiators is $3y$.
The total cost, in pounds, is therefore $4x + 3y$.

Express the information in two equations.

So $4x + 3y = 159$ (1)
Similarly $5x + 2y = 134$ (2)
(1) × 2: $8x + 6y = 318$ (3)
(2) × 3: $15x + 6y = 402$ (4)
(4) − (3): $7x = 84$
                 $x = 12$

Solve the equations.

One small radiator costs £12.

Remember to answer the original question.

Worked examination question

Mrs Rogers bought 3 blouses and 2 scarves. She paid £26. Miss Summers bought 4 blouses and 1 scarf. She paid £28. The cost of a blouse was x pounds. The cost of a scarf was y pounds.

(a) Use the information to write down two equations in x and y.

(b) Solve these equations to find the cost of one blouse.

(a) $3x + 2y = 26$ (1)
                    $4x + y = 28$ (2)

(b) (2) × 2: $8x + 2y = 56$ (3)
     (3) − (1): $5x = 30$
                         $x = 6$

One blouse costs £6.

Exercise 9K

1 The cost, C pounds, for a group of n students to visit
the aquarium is given by the formula $C = a + bn$.
It costs £25 for 5 students and £43 for 9 students.
Find the values of a and b.

2 For 6 hours of work at the normal rate of pay and one hour at the overtime rate, Mr Chung is paid £37. For 4 hours at normal rate and 3 hours at the overtime rate, he is paid £41. Find his normal rate of pay.

3 The sum of two numbers is 66. Their difference is 8. Find both numbers.

4 The total cost of a meal and a bottle of wine is £12.50. The meal costs £6.10 more than the bottle of wine. Find the cost of the meal.

> Work in pence.

5 Heather has 23 coins in her pocket. Some of them are 5p coins and the rest are 10p coins. The total value of the coins is £2.05. Find the number of 10p coins.

6 The cost, C pounds, of artificial Christmas trees is given by the formula $C = ah + b$, where h is the height in feet. A 3 foot tree costs £10 and a 6 foot tree costs £25. Find the values of a and b.

7 Cinema tickets for 2 adults and 3 children cost £12. The cost for 3 adults and 5 children is £19. Find the cost of an adult's ticket.

8 The total cost of 3 biros and 4 pencils is 98p. The total cost of 2 biros and 5 pencils is 91p. Find the cost of a pencil.

9 The total weight of 2 giraffes and 3 zebras is 2.5 tonnes. The total weight of 5 giraffes and 4 zebras is 5.2 tonnes. Find the weight, in kg, of a zebra.

10 The points with coordinates (3, 5) and (5, 12) lie on the line with equation $ax + by = 11$.
  (a) Find the values of a and b.
  (b) Find (i) the gradient of the line and
       (ii) its intercept on the y-axis.

> 1 tonne = 1000 kg

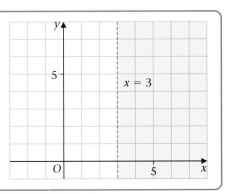

9.11 Regions

In Chapter 6 inequalities were represented on a number line.

An inequality may also be represented by a region on a graph.

Example 16

Sketch the region defined by the inequality $x > 3$.

First draw the line $x = 3$. The region for $x > 3$ is the **shaded part** to the right of the line $x = 3$.

Every point to the right of the line $x = 3$ satisfies this inequality.

The unshaded region is the unwanted part for $x > 3$. Because $x = 3$ is also in the unwanted region, the line is dotted.

Example 17

Sketch the region defined by the inequality $2x + y \geqslant 6$.

First draw the line $2x + y = 6$. If it is not obvious which side of the line satisfies the inequality, substitute the x and y coordinates of a point (often the origin) into the inequality.

In this case, letting $x = 0$ and $y = 0$ does not satisfy the inequality and so the origin lies in the unwanted (unshaded) region.

In this case, because the inequality is '**greater than or equal to**', the line is included in the region required. Hence it is drawn as a solid line.

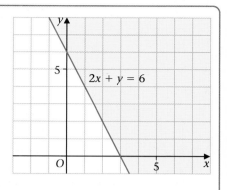

Example 18

Sketch the region defined by the inequality $y < 3x$.

First draw the line $y = 3x$. Test with a point not on the line, for example $(1, 0)$.

Letting $x = 1$ and $y = 0$ does satisfy the inequality and so $(1, 0)$ lies in the required (shaded) region.

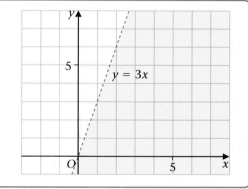

Example 19

Sketch the region which is defined by all of the inequalities

$$y < x + 1, \quad y < 7 - 2x \quad \text{and} \quad y > 2$$

Draw each of the lines

$$y = x + 1, \quad y = 7 - 2x \quad \text{and} \quad y = 2$$

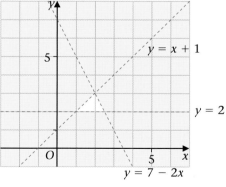

The region needed is   below $y = x + 1$
                            below $y = 7 - 2x$
and   above $y = 2$

So it is the unshaded region.

When you have more than one inequality, if you shade the parts that are **not needed**, then the blank region is the region required.

Note: the region does not include the boundary lines.

Exercise 9L

In questions **1–12**, draw diagrams and shade the regions which satisfy the inequalities.

1 $x < 4$	**2** $y \geq 2$	**3** $x \geq 0$
4 $-2 < y \leq 3$	**5** $x + y \leq 5$	**6** $x + 2y > 6$
7 $3x + 4y < 12$	**8** $2x - 5y < 10$	**9** $y \geq 2x$
10 $y \leq 3x - 2$	**11** $2x + 3y \geq 12$	**12** $3x - 2y > 6$

> Remember to draw dotted or solid lines depending on the inequality.

In questions **13–16**, write the inequalities which describe the **unshaded** regions.

13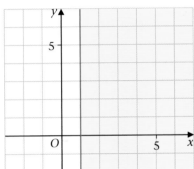

14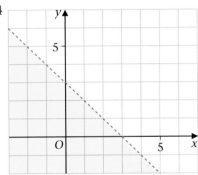

15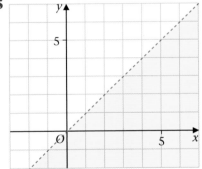

16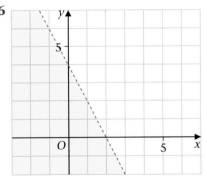

In questions **17–20**, draw diagrams and shade the regions which are not required, leaving unshaded regions which satisfy **all** the inequalities.

17 $x \geq 1, y \leq 4, y \geq x - 1$ **18** $x \geq 0, y \geq 0, x + y \leq 6$

19 $x < 5, y < \frac{1}{2}x, y \geq 0$ **20** $x \geq 0, y > 2x, 3x + 4y \leq 24$

21 Sketch the region defined by the inequalities
$$y > x + 1, \quad y > 7 - 2x \quad \text{and} \quad y < 5$$

22 Sketch the region defined by the inequalities
$$y \leq 2x + 1, \quad 2x + y \leq 9 \quad \text{and} \quad y \geq 3$$

23 Work out the three inequalities which define the shaded region in the diagram.

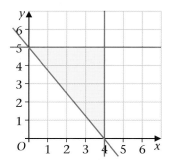

Note: the region should include the boundaries.

Mixed exercise 9

1

 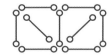

Pattern 1   Pattern 2   Pattern 3

Pattern number	1	2	3	4	5
Number of rods	5	9	13		

Copy and complete the table of values.
Draw the graph for the table of values.
Write down the equation of the graph.

2 On separate diagrams, draw lines with these equations
  (a) $y = 3x - 2$ (b) $y = \frac{1}{2}x + 1$ (c) $y = 5 - 2x$

3 For each of the lines in question **2**, write down the equation of a line that is parallel to it.

4 Work out the coordinates of the mid-point of the line segment AB, where A is the point $(-2, 5)$ and B is the point $(3, 5)$.

5 Find the equation of the line which is perpendicular to $y = 3x - 5$ and which passes through
  (a) $(0, 9)$ (b) $(3, 5)$

6 Find the gradient and the y-intercept for the lines with these equations
  (a) $y = 7x + 3.5$ (b) $3x + 2y = 6$
  (c) $3y - 6x = 15$ (d) $x + y + 2 = 0$

7 Solve these simultaneous equations algebraically
  (a) $3p + q = 11$ (b) $6x - 2y = 33$
       $p + q = 3$ $4x + 3y = 9$

8   *ABCD* is a rectangle. *A* is the point (0, 1).
C is the point (0, 6).

The equation of the straight line
through *A* and *B* is $y = 2x + 1$.

(a)   Find the equation of the
straight line through *D* and *C*.

(b)   Find the equation of the
straight line through *B* and *C*.

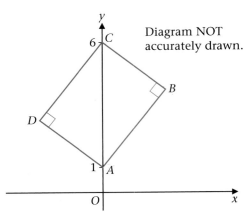

Diagram NOT
accurately drawn.

9   (a)   On a copy of the grid draw straight lines and use
shading to show the region **R** that satisfies
the inequalities

$$x \geqslant 2, \qquad y \geqslant x \quad \text{and} \quad x + y \leqslant 6$$

(b)   The point *P* with coordinates (x, y) lies inside
the region **R**. *x* and *y* are **integers**.

Write down all the possible points *P* could be.

10   The line with equation $6y + 5x = 15$ is drawn on
the graph below.

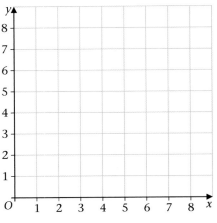

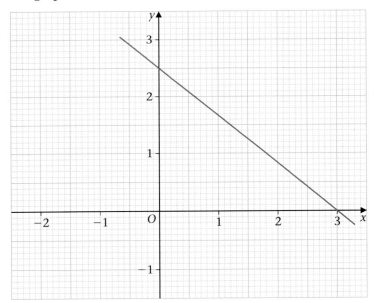

(a)   Rearrange the equation $6y + 5x = 15$ to make *y* the subject.

(b)   The point $(-21, k)$ lies on the line. Find the value of *k*.

(c)   (i)   On a copy of the grid, shade the region of points whose
coordinates satisfy these four inequalities:

$$y > 0, \qquad x > 0, \qquad 2x < 3, \qquad 6y + 5x < 15$$

Label this region **R**. *P* is a point in the region **R**. The coordinates of *P* are both integers.

(ii)   Write down the coordinates of *P*.

'Make *y* the subject'
means 'put in the
form $y = ...$'

Summary of key points

1 A **function** is a rule which shows how one set of numbers relates to another.

2 $y = 2x + 1$ is called an **equation**. The value of y is always equal to the value of $2x + 1$.

3 The input values of x are **continuous** if x can take any value (whole numbers or decimals).

4 $y = 2x + 1$ is the equation of a straight line. The coordinates of any point on the straight line 'satisfy' the equation.

5 $x \rightarrow 2x + 1$ is called a **linear function** because its graph is a straight line.

6 An **intercept** is a point at which a line cuts the y-axis or the x-axis.

7 $\textbf{Gradient} = \dfrac{\text{change in } y\text{-direction}}{\text{change in } x\text{-direction}}$

8 The straight line with equation $y = mx + c$ has a gradient of m and its intercept on the y-axis is $(0, c)$.

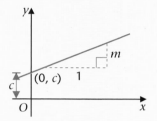

9 Lines with equation $y = mx + c$, with the same gradient (m), are parallel.

10 If a line has a gradient of m, a line perpendicular to it has a gradient of $-\dfrac{1}{m}$.

11 If two lines are perpendicular, the product of their gradients is -1.

$$m \times \frac{1}{m} = -1$$

12 The **mid-point of the line segment AB** between $A(x_1, y_1)$ and $B(x_2, y_2)$ is $\left(\dfrac{x_1 + x_2}{2}, \dfrac{y_1 + y_2}{2}\right)$.

13 **Simultaneous equations** can be solved
  - graphically, by drawing the graphs for the two equations and finding the coordinates of their point of intersection
  - algebraically, by multiplying one or both of the equations by a number, if necessary, and then adding or subtracting before dividing by the coefficient.

14 Regions on a graph can be used to represent inequalities.
For example, the shaded region represents the inequality $5x + 3y > 15$.

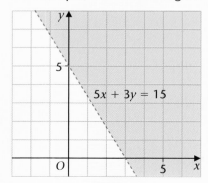

10 Presenting and analysing data 1

In this chapter you will calculate averages and measures of the spread of data and draw conclusions from them. The second part shows you how to draw and interpret time series and scatter diagrams.

> Some scientific calculators have statistical function keys to help you work out some calculations, for example the mean.

10.1 Averages for discrete data

The mean

> The **mean** of a set of data is the sum of the values divided by the number of values:
>
> $$\text{mean} = \frac{\text{sum of values}}{\text{number of values}}$$

A short way of writing the mean uses the Greek letter sigma Σ to represent the sum of a set of values:

> $$\bar{x} = \frac{\Sigma x}{n}$$
>
> where $\bar{x}$ is the mean of the values, Σx is the sum of the values and n is the number of values.

A male Emperor penguin goes without food for an average of 115 days each winter, while it looks after the egg or newly hatched chick and waits for the female to return.

The median

> The **median** is the middle value when the data is arranged in order of size.

If there is an even number of values in the data then the median is the mean of the middle two values.

> For a data set with n data values, the median is the $\dfrac{n+1}{2}$th value.

If there is an even number of data values in the set, $\dfrac{n+1}{2}$ will not

give a whole number. For 30 data values, for example, $\dfrac{n+1}{2} = 15\frac{1}{2}$.

The median is the mean of the data values either side, in this case the mean of the 15th and 16th values.

The mode

> The **mode** of a set of data is the value which occurs most often.

There can be more than one mode for a set of data.

Example 1

Here are the CPU speeds, in GHz, of eight computers in an office:

    1.5   1.9   2.4   1.7   2.7   1.7   1.6   2.1

Calculate

(a) the mean   (b) the median   (c) the mode.

(a) Calculate the **mean** by adding all the speeds and dividing by the number of computers

$$\text{mean} = \frac{1.5 + 1.9 + 2.4 + 1.7 + 2.7 + 1.7 + 1.6 + 2.1}{8} = \frac{15.6}{8} = 1.95$$

(b) Calculate the **median** by putting the speeds in order of increasing size and then calculating the mean of the two middle values:

    1.5   1.6   1.7   1.7   1.9   2.1   2.4   2.7

$$\text{median} = \frac{1.7 + 1.9}{2} = 1.8$$

(c) Find the **mode** by selecting the speed that appears most often:

    mode = 1.7

10.2 Stem and leaf diagrams

The ages of the 30 members of an aerobics class are:

19	22	31	17	8	12	23	47	53	47
19	46	38	59	47	52	21	58	54	26
32	47	55	62	64	36	37	43	15	51

These are presented as a stem and leaf diagram by using the first digit as stem, and the second digit as leaf:

```
19 is written 1 | 9          0 | 8
22 is written 2 | 2          1 | 9, 7
31 is written 3 | 1          2 | 2
                             3 | 1
17 is then    1 | 9, 7       4
                             5
and 8 is      0 | 8          6
```

So the complete diagram is:

```
0 | 8
1 | 9, 7, 2, 9, 5
2 | 2, 3, 1, 6
3 | 1, 8, 2, 6, 7
4 | 7, 7, 6, 7, 7, 3
5 | 3, 9, 2, 8, 4, 5, 1
6 | 2, 4
```

Now put the 'leaves' in order:

> You should include a key with a stem and leaf diagram.

```
0 | 8
1 | 2, 5, 7, 9, 9
2 | 1, 2, 3, 6
3 | 1, 2, 6, 7, 8
4 | 3, 6, 7, 7, 7, 7
5 | 1, 2, 3, 4, 5, 8, 9
6 | 2, 4
```

16th value ← 3
15th value → (1 row)
mode → 4 row

Key: 6 | 2 means 62 years

From this you can pick out the mode and identify the median.

Mode = 47

The median is the $\dfrac{30 + 1}{2} = 15\frac{1}{2}$th value = 40.5

> The average of the 15th and 16th values is
> $$\dfrac{38 + 43}{2} = 40.5$$

Exercise 10A

1   Here are the ages of a random sample of people on holiday in Spain in October.

```
53   55   45   65   68   60   57   75   62   58
57   74   60   55   49   54   64   82   70   58
64   73   67   62   56   66   83   60   77   63
```

Draw a stem and leaf diagram for this data (remember to use a key). Find the mode, median and mean.

2   Here are the masses (in kg) of 24 babies born in a hospital.

```
4.1   3.7   2.8   4.3   4.7   4.6   2.8   5.2   3.9   3.7   3.7   4.6
3.8   3.4   4.9   4.6   4.7   4.2   3.0   3.1   2.5   5.0   4.5   4.5
```

Draw a stem and leaf diagram for this data. Find the mean, median and mode.

3   The data shows an athlete's times in seconds, over a season, in the 400 m track event.

```
44.9   45.3   45.2   45.7   45.6   45.0   45.1   46.2   45.9
45.3   44.8   44.9   46.3   44.8   46.3   46.1   45.4   44.7
45.3   45.7   44.7   46.2   45.7   44.9   45.0   45.5   46.0
```

Use 44, 45 and 46 as the stem to produce a stem and leaf diagram. Work out the mode and median.

10.3 Using appropriate averages

The three different types of average are useful in different situations. The choice of which average you use or quote can be crucial; this is especially true when you do statistics-based coursework.

The **mean** is very useful when you need to quote a 'typical' value, provided that the data is quite closely grouped around the mean.

The mean can, however, be seriously influenced by extreme values. For example, consider the situation below.

Bob is the Director of a small company which employs 10 people, including himself. The salaries of the 10 people are:

Director	£80 000 per year
7 people earning	£20 000 per year
2 people earning	£12 000 per year.

The mean of these salaries is

$$\frac{80\,000 + 7 \times 20\,000 + 2 \times 12\,000}{10}$$

or $\dfrac{80\,000 + 140\,000 + 24\,000}{10} = £24\,400$

It would be silly to say that a 'typical' salary is £24 400 when in fact nobody at the company earns that amount and only one person earns above, and considerably above, that amount.

In a situation such as this, it would be far more sensible to use the **median** salary as the 'typical' value, i.e. £20 000.

The **mode** is useful in situations such as when you want to know:

- Which dress size is most common?
- Which brand of dog food is most popular?

The mode is an average which shops use when they calculate the stock they need.

> The **median** and **mode** are not influenced by extreme values, but the **mean** is influenced by extremes.

Example 2

There are 10 houses in Streetfield Close.

On Monday, the numbers of letters delivered to the houses are:

    0   2   5   3   34   4   0   1   0   2

Calculate the mean, mode and median of the number of letters.

Comment on your results.

$$\text{Mean} = \frac{0 + 2 + 5 + 3 + 34 + 4 + 0 + 1 + 0 + 2}{10}$$

$$= 5.1$$

$$\text{Mode} = 0$$

$$\text{Median} = 2$$

In this case the mean has been distorted by the large number of letters delivered to one of the houses. It is, therefore, not a good measure of a 'typical' number of letters delivered to any house in the close.

The mode (0) is also not a good measure of a 'typical' number of letters delivered to a house, since 7 out of the 10 houses do actually receive some letters.

The median (2) is perhaps the best measure of the 'typical' number of letters delivered to each house, since half of the houses received 2 or more letters and the other half received 2 or fewer letters.

Exercise 10B

1 Ten students submitted their Design portfolios which were marked out of 40. The marks they obtained were

37　34　34　34　29　27　27　10　4　28

(a) For these marks find
(i) the mode　(ii) the median　(iii) the mean.

(b) Comment on your results.

(c) An external moderator reduced all the marks by 3.
Find the mode, median and mean of the moderated marks.

2 The ages of the people in the Jones family are

Grandpa	79
John	42
Mary	44
Sarah	16
Peter	12
Lucy	3

(a) Calculate the mean age of the Jones family.

(b) State, with a reason, whether the mean age is a sensible measure of the age of a 'typical' member of the Jones family.

3 A newspaper headline reads:

'The average weekly wage in the UK is in excess of £250'

Explain why this headline could be misleading.

Activity – Citius, Altius, Fortius (Swifter, Higher, Stronger)

Use the internet to find out the total number of medals won by Great Britain in each of the Olympic Games since 1894.

Work out the mode, median and mean of the data.

Which is the best measure of average to use? Explain your answer.

10.4 The quartiles

The median divides a set of data arranged in order of size into two halves. The data can also be divided into four quarters.

When the data is arranged in ascending order of size:
- the **lower quartile** is the value one quarter of the way into the data
- the **upper quartile** is the value three quarters of the way into the data

For example, here are the results of asking 15 people how many cousins they have. The data is arranged in order of size.

Number of cousins   0 2 4 5 7 18 10 12 12 12 13 14 14 15 16

The lower quartile   The median is   The upper quartile
is the 4th value.   the 8th value.   is the 12th value.

So the lower quartile is 5 cousins, the median is 12 cousins, and the upper quartile is 14 cousins.

For a set of n data values arranged in order of size:

$$\text{lower quartile} = \left(\frac{n+1}{4}\right)\text{th value}$$

$$\text{median} = \left(\frac{n+1}{2}\right)\text{th value}$$

$$\text{upper quartile} = \frac{3(n+1)}{4}\text{th value}$$

Example 3

Shah records the numbers of hits on his web page for each of 11 consecutive days. Here are his results

     63 96 67 99 87 68 61 85 91 102 89

Work out (a) the lower quartile (b) the upper quartile.

(a) First arrange the numbers of hits in order of size

     61 63 67 68 85 87 89 91 96 99 102

The lower quartile is the $\left(\frac{11+1}{4}\right)$ = 3rd value, so it is 67.

(b) The upper quartile is the $\frac{3(11+1)}{4}$ = 9th value, so it is 96.

Example 4

The table shows the life expectancy of people in South America.

Country	Argentina	Bolivia	Brazil	Chile	Colombia	Ecuador
Life expectancy (years)	75.5	71.1	71.1	76.8	71.1	71.9

F. Guiana	Guyana	Surinam	Paraguay	Peru	Uruguay	Venezuela
76.7	63.1	69.2	74.4	70.9	75.9	73.8

Work out **(a)** the median   **(b)** the lower quartile   **(c)** the upper quartile of the data.

(a)   First arrange the values in order of size:

63.1   69.2   70.9   71.1   71.1   71.1   71.9   73.8   74.4   75.5   75.9   76.7   76.8

There are 13 values. The median is the $\dfrac{13 + 1}{2}$ = 7th value, so it is 71.9 years.

(b)   The lower quartile is the $\dfrac{13 + 1}{4} = 3\frac{1}{2}$th value.

You treat this as being the value halfway between the 3rd value (70.9) and the 4th value (71.1), so the lower quartile is 71 years.

(c)   The upper quartile is the $\dfrac{3(13 + 1)}{4} = 10\frac{1}{2}$th value.

You treat this as being the value halfway between the 10th value (75.5) and the 11th value (75.9), so the lower quartile is 75.7 years.

Exercise 10C

For each set of data below work out
(a) the median   (b) the lower quartile   (c) the upper quartile.

1   The numbers of seats won by the Labour party in General Elections from 1959 to 1998

258   317   363   287   301   319   269   209   229   271   418

2   The numbers of delegates for each country in the European parliament

18   25   6   24   14   6   14   78   99   24   13   78   9

13   6   5   27   54   24   14   7   54   19   78   24

3   The masses (in g) of a random sample of watch batteries

Mass of batteries

50.7	3, 5, 7
50.8	6, 9
50.9	0, 2, 3, 6
51.0	6
51.1	3, 6, 8, 9, 9
51.2	5
51.3	6, 8, 9

Key: 50.9 | 2 means 50.92 grams

10.5 Measures of spread

The range

The **range** is a measure of **spread**.

> The **range** of a set of data is the difference between the highest and lowest values
>
> range = highest value − lowest value

The range can be **greatly** affected by extreme (very large or very small) values.

The interquartile range

The quartiles are useful because together they can be used to give a better idea of the spread of a set of data than the range can.

> The **interquartile range** is the difference between the upper and lower quartiles:
>
> interquartile range = upper quartile − lower quartile

Unlike the range, the interquartile range is not distorted by extreme data values. It also tells you whether the data values are spread evenly throughout the range, or concentrated in the middle of the range.

For example, there are 12 houses in a close. The numbers of letters delivered to the houses one day are, in order:

   1   2   2   2   3   3   3   3   5   5   5   30

The range of these numbers is $30 - 1 = 29$.

The interquartile range is

   upper quartile − lower quartile = $5 - 2 = 3$

One house receives a lot of letters.

If the house that received 30 letters instead had received 8 letters then the range would have been $8 - 1 = 7$. This change in the extreme value from 30 to 8 makes no change to the interquartile range.

Because it is not affected by extreme values, the interquartile range is a good measure of the spread of the data as a whole, and particularly the spread either side of the median.

A large range tells you very little about how the data is spread.

A large interquartile range tells you that the middle 50% of the data is widely spread about the median.

A small interquartile range tells you that the middle 50% of the data is highly concentrated about the median.

10.6 Averages from frequency distributions

This section shows you how to find the mean, mode and median where the information is given in a frequency table.

A frequency table with the totals included is called a **frequency distribution**.

For a frequency distribution:

- the **mode** = the item of data which has the highest frequency

- the **median** = the middle value of the data = the $\dfrac{\Sigma f + 1}{2}$th value ← the sum of the frequencies

- the **mean**, $\bar{x} = \dfrac{\Sigma fx}{\Sigma f}$ ← the sum of all the ($f \times x$) values in the distribution
   ← the sum of the frequencies

Example 5

In a survey, the number of eggs in seagulls' nests in June was counted. The table shows the results.

Find

(a) the mode

(b) the median

(c) the mean number of eggs.

Number of eggs x	Frequency f
0	17
1	12
2	23
3	37
4	18
Total	107

(a) The mode is the value which occurs most often. In a frequency table this is the item of data which has the highest frequency. For this data the mode is 3 eggs. There are more nests with 3 eggs than with any other number of eggs.

(b) The median is the middle value of the data = the $\dfrac{\Sigma f + 1}{2}$th value.

There are 107 nests altogether. So the middle nest is the $\dfrac{107 + 1}{2} = 54$th nest.

Nests 1–17 have 0 eggs

Nests 18–29 have 1 egg

Nests 30–52 have 2 eggs

Nests 53–89 have 3 eggs

So nest 54 is a 3-egg nest. The median is 3 eggs.

(c) The mean is the total number of eggs divided by the total number of nests (the total frequency) $= \dfrac{\Sigma fx}{\Sigma f}$

Number of eggs x	Frequency f	Frequency × number of eggs fx
0	17	0
1	12	12
2	23	46
3	37	111
4	18	72
Totals	107	241

> Add a column to the table for fx.

> The number of eggs in 2-egg nests is $23 \times 2 = 46$

> The total number of eggs is the sum of all the fx values.

The mean $= \dfrac{\Sigma fx}{\Sigma f}$

$= \dfrac{241}{107} = 2.25$ eggs

Exercise 10D

1 For the data given, calculate the mean, mode and median number of cars per household. Use a table like this:

Cars per household x	Frequency f	fx
0	6	
1	20	
2	13	
3	4	
4	2	
Totals		

The frequency tables in questions **2–3** are written sideways in rows. You might find it easier to rewrite them in columns.

2 A factory takes random samples of 102 items and tests them for faults. Here are their results for last week

Number of faults per sample	0	1	2	3	4	5
Number of samples	76	13	6	3	3	1

Work out the mode, median and mean number of faults per sample.

> 'Per sample' tells you that you need to divide the total number of faults by the total number of samples. So the 'number of samples' is the frequency.

3 The table shows the number of goals Lucy scored in matches during the netball season.

Number of goals	0	1	2	3	4	5	6	7	8
Frequency	4	0	6	5	3	2	1	3	0

(a) How many goals scored is the mode?

(b) Work out the median number of goals scored

(c) Calculate the mean number of goals scored correct to 1 d.p.

10.7 Spread from frequency distributions

This section shows you how to work out the interquartile range from information given in a frequency distribution.

Example 6

The table shows the number of complaints per day to Ofcom about a popular TV soap.

Number of complaints per day	3	4	5	6	7	8	9	10	11	Total
Frequency	1	2	4	7	17	23	24	16	1	95

Work out **(a)** the range **(b)** the interquartile range.

(a) Range = highest value − lowest value = 11 − 3 = 8 complaints

(b) The lower quartile is the $\dfrac{95 + 1}{4}$ = 24th value.

Adding up the frequencies in order:

$$1 + 2 + 4 + 7 = 14$$

and $\quad 1 + 2 + 4 + 7 + 17 = 31$

As the 24th value is between the 14th and 31st values, it is one of the 17 values recorded as 7 complaints. So the lower quartile is 7 complaints.

Using the same method, the upper quartile is the $\dfrac{3(95 + 1)}{4}$ = 72nd value.

Adding up the frequencies in order:

$$1 + 2 + 4 + 7 + 17 + 23 = 54$$

and $\quad 1 + 2 + 4 + 7 + 17 + 23 + 24 = 78$

As the 72nd value is between the 54th and 78th values, it is one of the 24 values recorded as 9 complaints. So the upper quartile is 9 complaints.

The interquartile range is

upper quartile – lower quartile = 9 – 7 = 2 complaints

Exercise 10E

For each set of data below work out

(a)   the range   (b) the interquartile range.

1   The numbers of emails received in a day by 31 people in an office

Number of emails	0	1	2	3	4	5	6	7
Frequency (number of people)	4	11	8	6	1	0	0	1

2   The amounts received from the sale of £1 lottery tickets to 93 people

Amount (£)	1	2	3	4	5	6	7	8	9	10
Frequency	32	16	3	2	21	4	2	3	0	10

3   The sizes of shoes for hire at a bowling alley

Size	6	$6\frac{1}{2}$	7	$7\frac{1}{2}$	8	$8\frac{1}{2}$	9	$9\frac{1}{2}$	10	$10\frac{1}{2}$
Frequency	4	5	7	8	6	5	4	6	8	4

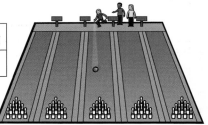

10.8 Averages from grouped data

Data in a frequency distribution may be grouped. You can **estimate** averages from grouped data.

The manager of a hotel checks the cost of telephone calls made by each guest by looking at the number of units used per call. She summarises the data about the phone calls in a frequency table, grouping it in class intervals.

Number of units	Frequency
1–5	73
6–10	161
11–15	294
16–20	186
21–25	65
over 25	11
Total	790

The $395\frac{1}{2}$th value is in this class interval.

The median: There were 790 phone calls so the position of the median data value is the $(790 + 1) \div 2 = 395\frac{1}{2}$th value.
It is in the class interval 11–15.
You cannot give an exact value for the median.

For grouped data, you can state the class interval that contains the median.

The modal class: The class interval 11–15 units per call has the highest frequency.

> For grouped data, the class interval with the highest frequency is called the **modal class**.

The modal class only makes sense as a measure of the average if the class intervals are the same.

The mean: Each class interval contains calls using different numbers of units, so you have not got the exact data you need to calculate the mean. But you can calculate an **estimate of the mean**. This is *not* a guess, but a calculation. Here is how to do it.

Assume each call in a class interval uses the number of units given by the middle value of the class interval the call is in.

> The middle value of the class interval 1–5 is 3. The middle value of class interval 6–10 is 8 and so on.

Number of units used per call	Frequency f	Middle value x	fx
1–5	73	3	219
6–10	161	8	1288
11–15	294	13	3822
16–20	186	18	3348
21–25	65	23	1495
over 25	11	28	308
Total, Σf	790	Total, Σfx	10 480

> The 161 calls did not all use 8 units each. The errors made by overestimating approximately balance those made by underestimating.

> A middle value must be chosen for calls over 25 units. This is a matter of judgement depending on the raw data. Here 28 is used.

Now you can calculate an estimate of the mean in a similar way to calculating the mean for ungrouped data (Section 10.6):

$$\text{estimate of mean} = \frac{\text{sum of (middle values} \times \text{frequencies)}}{\text{sum of frequencies}}$$

$$= \frac{\Sigma fx}{\Sigma f}$$

$$= \frac{10\,480}{790} = 13.3$$

> For grouped data you can **calculate an estimate of the mean** using the middle value of each class interval.

In the telephone calls example the data is discrete. You can use exactly the same method to find an estimate of the mean of continuous data. The method works even if the class intervals are not the same size and the middle values are not whole numbers.

> The estimate of the mean for grouped data is
> $$\bar{x} = \frac{\Sigma fx}{\Sigma f}$$
> where x is the mid-point of each class interval.

Example 7

A saleswoman kept a record of the mileage she travelled each day.
Calculate an estimate for the mean distance travelled from the
data in this table:

Mileage m (miles)	Frequency
$0 \leqslant m < 50$	2
$50 \leqslant m < 100$	8
$100 \leqslant m < 150$	17
$150 \leqslant m < 200$	21
$200 \leqslant m \leqslant 300$	2

This is 0 up to but not including 50.

First draw up a table of frequencies and mid-points.

Mileage	f	Mid-point x	fx
$0 \leqslant m < 50$	2	25	50
$50 \leqslant m < 100$	8	75	600
$100 \leqslant m < 150$	17	125	2125
$150 \leqslant m < 200$	21	175	3675
$200 \leqslant m \leqslant 300$	2	250	500
	$\Sigma f = 50$		$\Sigma fx = 6950$

The mid-point of the interval from 0 to 50 is $\frac{0+50}{2} = 25$

The estimate of the mean $\bar{x}$ is

$$\bar{x} = \frac{\Sigma fx}{\Sigma f}$$

$$= \frac{6950}{50} = 139 \text{ miles}$$

Exercise 10F

1 Hajra looked at a passage from a book. She recorded the number of words in each sentence in a frequency table using class intervals of 1–5, 6–10, 11–15, etc.

(a) Copy and complete the table.

(b) Work out an estimate for the mean number of words in a sentence.

Class interval	Frequency f	Middle value x	$f \times x$
1–5	16		48
6–10	28		
11–15	26	13	
16–20	14		
21–25	10		230
26–30	3		
31–35	1		
36–40	0		
41–45	2		86
Total, Σf		Total, Σfx	

2 The table shows how long couples had been married whose marriage ended in a divorce in 2005.

Length of marriage in completed years	0–2	3–4	5–9	10–14	15–19	20–24	25–29	30–40
Frequency as a %	9	14	27	18	13	10	5	4

Work out an estimate for the mean length of a marriage that ended in divorce in 2005.

3 In a darts match a record is kept of the scores for each throw of three darts.

Score	1–20	21–40	41–60	61–100	101–140	141–180
Frequency	3	17	25	56	8	3

(a) Write down the modal class.

(b) Write down the class interval in which the median lies.

(c) Work out an estimate for the mean score.

Activity – Maze

How long does it take you to get through this maze?

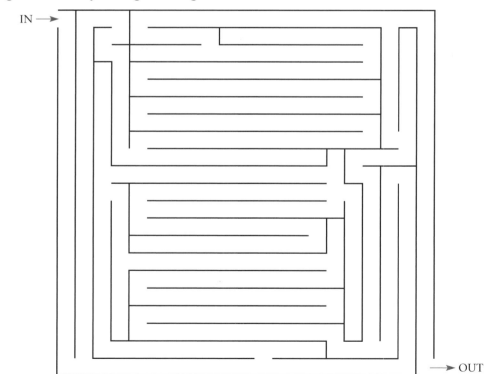

(a) Design your own maze and time how long it takes your friends to do it.

(b) Record the data in a frequency table using class intervals of equal width.

(c) Work out an estimate for the mean.

10.9 Drawing conclusions from data

In statistics it is easy to draw incorrect conclusions from data, especially if you do not have sufficient information.

For example, suppose that a group of girls and a group of boys took the same test marked out of 20. You are told that:

- the mean mark for the girls was 10
- the mean mark for the boys was 8

Can you now conclude that the girls have on the whole done better in the test than the boys?

It would be easy to conclude this because the mean mark for the girls is higher than for the boys. But a closer look shows that this conclusion need not be true.

Suppose only three girls and three boys took the test and that the marks were:

girls   9   10   11   boys   0   12   12

The mean for the girls is   The mean for the boys is

$$\frac{9 + 10 + 11}{3} = 10 \qquad \frac{0 + 12 + 12}{3} = 8$$

Although the mean for the girls is greater, it would not be valid to say that the girls did better than the boys. In fact two-thirds of the boys scored a mark higher than any of the girls.

To say whether the girls did better than the boys, you need further information. You need to know how the marks were distributed or spread out.

> To compare two sets of data you should use a measure of average and a measure of spread.

To help you see why, you can represent the scores in the test along an axis:

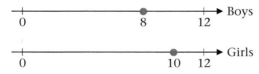

The red marks show the positions of the average marks for boys and girls. Here it looks as if the girls have done better than the boys.

But if we add brackets to show the measure of spread of the scores, the diagrams now look like this:

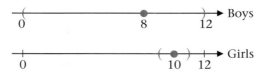

Boys: range = 12 − 0 = 12
Girls: range = 11 − 9 = 2

While the average score for the girls is greater than that for the boys, the **measures of spread** mean that many boys may have done better than any of the girls.

In the test results:

- if the mean mark for the girls had been greater than the mean mark for the boys

  and

- if the spread of marks for the girls had been about the same as the spread of marks for the boys

you would be justified in concluding that the girls did better in the test than the boys.

Example 8

A group of students took a science examination consisting of two papers, Paper 1 and Paper 2.

The table gives the median marks and the interquartile ranges of the marks on those papers.

Paper	Median (%)	Interquartile range (%)
1	63	12
2	51	38

Comment on these results.

Because the median mark on Paper 1 was greater than the median mark on Paper 2, it looks as if Paper 1 was easier than Paper 2 (or as if more students scored higher marks on Paper 1 than on Paper 2).

However, the interquartile range for Paper 2 was much greater than that for Paper 1.

So it could be that many students obtained a higher mark on Paper 2 than on Paper 1 – and so found Paper 2 to be the easier paper.

Exercise 10G

1 The mean weekly wage at Cromby and Sons is £300.

The mean weekly wage at Drapeway Ltd is £350.

Is it fair to say that the people at Drapeway Ltd are better paid than those at Cromby and Sons? Explain your answer.

2 The Year 11 students at Lucea High School sat a French exam consisting of two papers, Paper A and Paper B.

The table shows the median marks and the interquartile ranges of the marks on the two papers.

Paper	Median (%)	Interquartile range (%)
A	72	32
B	60	12

Comment on this data.

3 In the town of Reddshaw there are two golf clubs.

The table shows the median ages and the interquartile ranges of the ages of the club members.

Club	Median age	Interquartile range of ages
1	34	17
2	42	12

Explain whether or not it would be fair to say that the members of Club 2 are, in general, older than the members of Club 1.

10.10 Moving averages

This section shows you how to calculate moving averages and interpret time series.

> A graph showing how a given value changes over time is called a **time series graph**.

This graph shows the variation in the number of job vacancies in a district over three years.

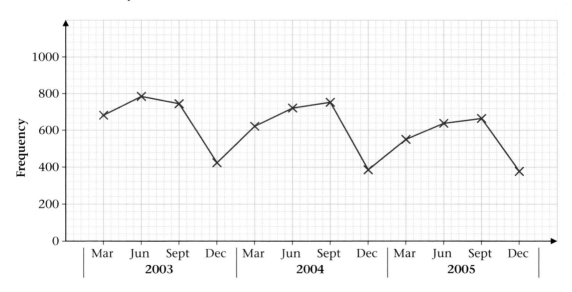

It is difficult to see from the graph whether the trend in the number of job vacancies has risen or fallen over the three years. There is too much variation within each year.

You can take account of this variation by working out a **moving average**. This takes the average of consecutive pieces of data.

Here is the table of results for the graph on the previous page:

Year	March	June	September	December
2003	682	785	742	423
2004	622	722	752	385
2005	550	639	661	376

There are four values for each year, so look at groups of four.

The first moving average, M_1, is the mean of the first four values:

$$M_1 = \frac{682 + 785 + 742 + 423}{4} = 658$$

> If you consider groups of four you are finding a **four-point moving average**.

The next moving average, M_2, is found by 'moving up one place'.
It is the mean of the values from June 2003 to March 2004:

$$M_2 = \frac{785 + 742 + 423 + 622}{4} = 643$$

The four-point moving averages for the whole data set are:

M_1	M_2	M_3	M_4	M_5	M_6	M_7	M_8	M_9
658	643	627	630	620	602	582	559	557

You can plot the four-point moving averages on the same graph as
your original data. Each average should be plotted at the mid-point of
the values from which it was generated.

You can now see that the number of job vacancies fell between the
beginning of 2003 and the end of 2005.

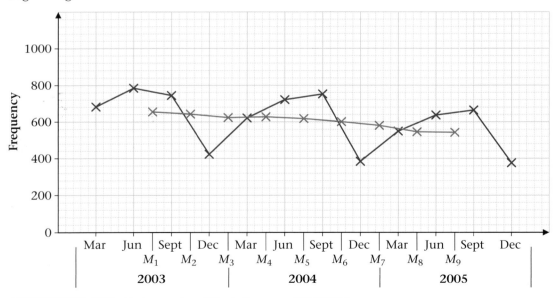

A moving average 'smooths out' the data to show the trend on a time
series graph.

Exercise 10H

1   Dr Singh keeps a record of the number of patients he treats for shingles. Here are his results for eight months:

Month	June	July	Aug	Sept	Oct	Nov	Dec	Jan
Number of patients	7	8	6	10	5	9	7	8

   (a)   Plot the time series graph.

   (b)   Calculate the three-point moving averages for the data and plot them on the same graph.

   (c)   Dr Singh says 'Shingles appears to be in decline'.
        Do you agree? Give a reason for your answer.

2   The table shows information about the quarterly gas bills for Alice's house, over a period of two years.

Year	Quarter			
	1	2	3	4
1	£196	£158	£80	£138
2	£216	£162	£96	£146

   (a)   Plot the data and the four-point moving averages on the same graph.

   (b)   Comment on your graph.

3   The table shows information about the numbers of passengers travelling by fast ferry to Jersey.

Year	2003				2004				2005			
Season	Sp	Su	Au	Wi	Sp	Su	Au	Wi	Sp	Su	Au	Wi
Number of passengers (thousands)	21	50	33	16	17	46	25	16	21	42	25	12

Draw suitable graphs and comment on the trend of the data.

> Use four-point moving averages.

10.11 Scatter diagrams

Statements such as 'Smoking can cause lung cancer' and 'Drink driving causes accidents' are often made in the media. Sometimes they are supported by data showing, for example, whether there is a relationship between a smoking habit and the chance of getting lung cancer.

This section shows you how to compare two sets of data to see whether there is a relationship between them.

For example, is there a relationship between the number of ice-creams sold at a kiosk and the average daytime temperature? James collected data to find out:

Average temperature (°C)	13	14	21	22	16	18	13	20	21	18	15	16	14
No. of ice-creams sold	5	9	51	48	20	30	15	32	42	37	23	25	14

He plotted each pair of values (13, 5), (14, 9), and so on, on a graph. This is called a **scatter diagram** or **scatter graph**.

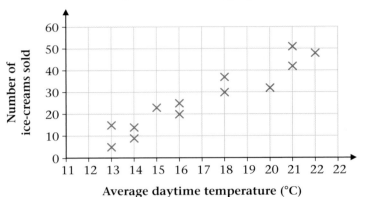

Average daytime temperature (°C)

There does appear to be a relationship between the average daytime temperature and the number of ice-creams sold: the hotter it is, the more ice-creams are sold.

Here is another example. On a journey Chandra noted down how many miles there were still to go. She did this every ten minutes:

Time (min)	10	20	30	40	50	60	70	80	90	100
Miles to go	72	60	54	42	40	35	24	18	10	0

Here is a scatter diagram showing this data:

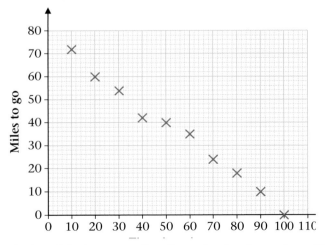

There is a relationship between the time she has been travelling and the number of miles still to go: the greater the travelling time, the fewer miles there are to go.

Sometimes there is no relationship between two sets of data.
Every Monday Trish recorded the temperature in °F and the rainfall in mm.
Her results were:

Temperature (°F)	74	70	63	68	65	64	60	51	54	56	50
Rainfall (mm)	1	0	2	7	5	1	8	2	11	4	6

A scatter graph of this data looks like this:

There does not appear to be any relationship.

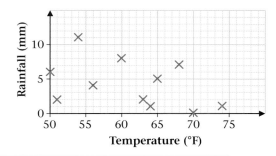

You can use a **scatter diagram** to show whether two sets of data are related.

Exercise 10I

1 The table shows the distance travelled and the depth of tread of a sample of motorcycle tyres.

Tyre	A	B	C	D	E	F	G	H	I
Distance travelled (1000 km)	15	9	2	23	30	45	19	6	36
Depth of tread (mm)	8.9	9.6	10.7	8.2	7.9	6.8	7.9	9.2	7.7

Draw a scatter diagram for the data and comment on any relationship between the depth of tread and the distance travelled.

2 The table below shows the number of hours of sunshine and the maximum temperature in ten British towns, on one particular day.

Maximum temperature (°C)	13	20	19	15	16	12	14	14	17
Number of hours of sunshine	11.7	15.2	15.4	13.2	11.8	9.8	10.2	12.4	13.7

(a) Plot this information as a scatter diagram.
(b) What does your diagram tell you about change in the maximum temperature as the number of hours of sunshine increases?

> **Activity – Height and arm-span**
> Is there a relationship between a person's height and their arm-span?
> (a) Collect data for your class.
> (b) Draw a scatter diagram to represent your data.
> (c) Describe the relationship.

10.12 Correlation and lines of best fit

Scatter diagrams can be used to show whether there is a relationship between two sets of data. Such a relationship is called a **correlation**.

> If the points on a scatter diagram are very nearly along a straight line there is a high **correlation** between the variables.

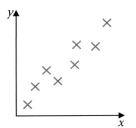

Positive correlation

When the values in one set increase as the values in the other set increase, the relationship between them is called **positive correlation**.

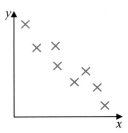

Negative correlation

When the values in one set increase as the values in the other set decrease the relationship between them is called **negative correlation**.

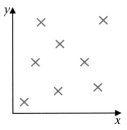

No correlation

When there is no linear relationship between the two sets of values, there is **no correlation**.

Line of best fit

When there is a relationship between two sets of data there is some correlation. You can show this on a scatter diagram by drawing the **line of best fit**. This is a line that shows the general trend of the relationship between the two sets of data. It may or may not pass through any of the data points themselves.

> A line that is drawn to pass as close as possible to all the plotted points on a scatter diagram is called the **line of best fit**.

This scatter diagram (from the data on page 204) shows the average daytime temperature plotted against the numbers of ice-creams sold:

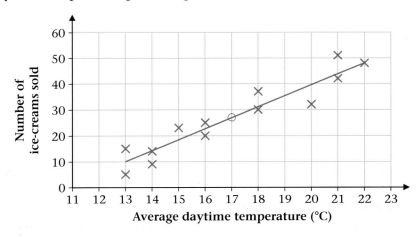

The line of best fit should really be drawn through the point with coordinates (mean average daytime temperature, mean number of ice-creams sold). Here this is (17, 27) and the point is circled on the graph. However, you would not be expected to do this in your exam. Drawing the line 'by eye' would be suggested.

The line of best fit has been drawn by eye. You should draw the line so that there are about the same number of plotted points on either side of it. Position the line so that the distances of the plotted points from it on each side balance each other out as much as possible.

The closer the points are to this line of best fit, the higher the correlation. The gradient of the line is not important, except that a vertical or horizontal line of best fit means that the variables are not connected. Here are some typical examples and their interpretation:

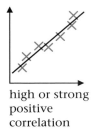
high or strong positive correlation

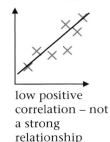
low positive correlation – not a strong relationship

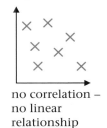
no correlation – no linear relationship

low negative correlation – not a strong relationship

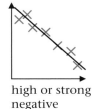
high or strong negative correlation

Example 9

The table shows the acidity of seven lakes near to an industrial plant and their distance from it.

Distance (km)	4	34	17	60	6	52	42
Acidity (pH)	3.0	4.4	3.3	7.0	3.2	6.8	5.2

(a) Draw a scatter graph to illustrate this data.

(b) Draw and label the line of best fit on your scatter graph.

(c) Use your line of best fit to predict the acidity of a lake at a distance of 25 km.

(d) For what range of values could you use your line of best fit to predict the acidity of a lake?

(a), (b)

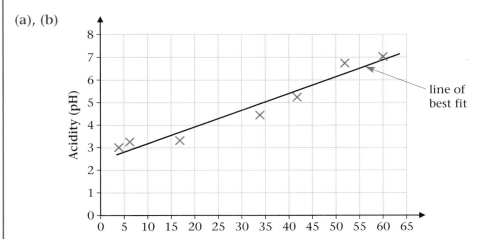

(c) Read up from 25 on the distance axis to the line of best fit, and across to the acidity axis. Using your line of best fit you can predict that the pH will be 4.3.

(d) Look at the range of the line of best fit. This is approximately from 4 km to 60 km.

Exercise 10J

1   The table shows the engine sizes of various cars and the distance they travel on one litre of petrol.

Engine size (litres)	0.9	1.2	2.0	1.0	1.9	1.3	1.5
Distance travelled (km)	13	10	8	12	7	11	9

   (a)   Draw a scatter diagram to represent this data.

   (b)   What type of correlation do you find?

   (c)   Draw a line of best fit and use it to estimate the distance travelled on one litre of petrol by a car with an engine size of 1.7 litres.

2   An anemometer is used to measure wind speed. The table gives the rotational speed of an anemometer for different wind speeds.

Wind speed (m/s)	1.0	1.2	1.4	1.6	1.8	2.0	2.2
Rotational speed (rev/min)	20	40	65	85	110	130	155

   (a)   Plot a scatter diagram and draw a line of best fit for the data.

   (b)   Comment on the relationship between the rotational speed of the anemometer and the wind speed.

   (c)   Use your graph to estimate the rotational speed of the anemometer when the wind speed is 1.5 m/s.

   (d)   Explain why it may not be appropriate to use your graph to estimate the rotational speed of the anemometer when the wind speed is 4 m/s.

> The reason is not that your graph is too small.

3   The scatter diagram shows the number of hours of sunshine and the maximum temperature in ten British towns on one day.

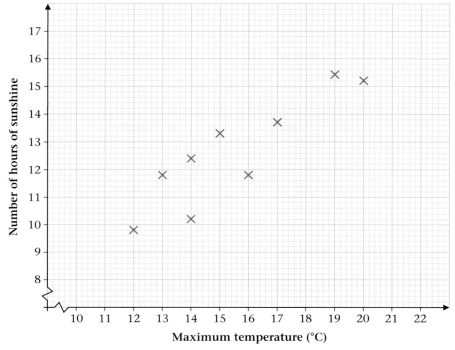

(a)   Copy the scatter diagram and draw a line of best fit on it.

(b)   Use your line of best fit to estimate
   (i)   the number of hours of sunshine when the maximum temperature was 18°C
   (ii)   the maximum temperature when the number of hours of sunshine recorded in the day was 12 hours.   [E]

Activity – Country area and population

Is there a correlation between the area of a country and the size of its population?

(a)   Use the internet to collect data for European countries.

(b)   Draw a scatter diagram to represent your data.

(c)   Comment on your findings.

Mixed exercise 10

1   In an experiment, 35 people were asked to guess the weight, in kg, of the lead singer in a rock group. Here are their guesses:

76	80	90	100	83	95	106	95	95
77	80	87	108	78	90	85	88	95
108	80	77	105	98	95	82	79	80
86	101	78	106	95	95	77	87	

> The total of these weights is 3137 kg

(a)   Draw a stem and leaf diagram for the data.

(b)   For this data work out
   (i)   the mode   (ii)   the median   (iii)   the mean
   (iv)   the range   (v)   the interquartile range.

2   The numbers 4, 8, 12, 17 and N are arranged in ascending order. The mean of the numbers is equal to the median. Find N.

3   A farmer wants to sell her sheep. Their weights, in kg, are given in the stem and leaf diagram.

Weight of sheep (kg)

15	0 2 3 3 7
16	0 1 3 9
17	5 7
18	7 8
19	5

> The total weight is 2340 kg

Key: 19 | 5 means 195 kg

(a)   Find the median weight and the mean weight.

(b)   In an advertisement the farmer claims that 'The average weight is 162 kg'.
   (i)   Which 'average' has the farmer used to describe her sheep?
   (ii)   Does this average describe the sheep fairly? Explain your answer.

4 The table gives information about the numbers of letters delivered one day to the houses in a street.

(a) How many houses are there in the street?

(b) Work out the range and the interquartile range for the data.

(c) Work out the mean number of letters delivered to each house in the street.

(d) Is this a sensible measure of average to describe the number of letters delivered to each house in the street?

Number of letters delivered	Number of houses (frequency)
0	3
1	14
2	2
3	7
4	0
5	3
6	1

5 Malvika is doing a geology project. She weighs 40 rock samples from her test site. Her results are summarised in the table.

Weight (grams)	$100 < w \leqslant 300$	$300 < w \leqslant 500$	$500 < w \leqslant 700$	$700 < w \leqslant 900$
Frequency	9	16	7	8

(a) Write down the modal class.

(b) Calculate an estimate of the total weight of the rock samples.

(c) Explain why this is an *estimate* of the total weight of the rock samples.

(d) Calculate an estimate for the mean.

6 The graph gives information about the quarterly gas bills of a householder.

(a) Using the information in the graph, calculate the four-point moving averages for the data.

(b) Comment on the trend of the gas bills.

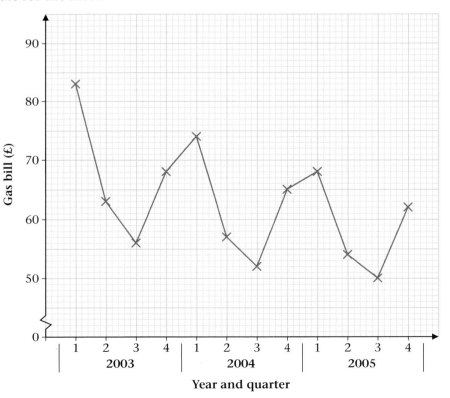

7 The table gives the numbers of text messages received by a student over a weekend.

	Morning	Afternoon	Evening
Friday			12
Saturday	15	9	17
Sunday	19	18	11
Monday	1		

(a) Calculate the values of a suitable moving average.

(b) Plot the data and the moving averages on the same graph.

(c) Comment on your graph.

8 The table shows information about the age and cranial capacity of eight fossils of a particular species of ape.

Fossil	A	B	C	D	E	F	G	H
Age (millions of years)	2.3	1.9	0.7	1	1.5	0.6	0.3	1.3
Cranial capacity (cm³)	37	81	173	170	88	158	205	128

Cranial capacity is a measure of the volume of the interior of the skull.

(a) Draw a scatter diagram to represent this data.

(b) Is there any evidence from the scatter diagram to suggest that there is a correlation between the age of a fossil and the cranial capacity for this species of ape? Explain your answer.

(c) Draw a line of best fit and use it to estimate the cranial capacity of a fossil ape that is 1.75 million years old.

9 The table gives information about eight types of aircraft.

Aircraft	Wingspan (x metres)	Maximum take-off weight (y tonnes)
B747-400	64.30	394.0
B747-200	38.00	108.8
DC10-30	50.40	263.5
LT-1011-200	47.35	219.0
CRJ-200 ER	21.21	23.1
VC10	44.55	114.5
A300 B4	44.84	165.0
HS Com 4	34.98	70.7

(a) Plot the data on a scatter diagram.

(b) Calculate the mean wingspan, $\bar{x}$ metres, and the mean maximum take-off weight, $\bar{y}$ tonnes, of these eight aircraft.

(c) On your scatter diagram
   (i) plot the point $(\bar{x}, \bar{y})$
   (ii) draw a line of best fit through the point $(\bar{x}, \bar{y})$.

(d) For what range of values of the wingspan is it suitable to use your line of best fit to predict the maximum take-off weight?

10 This back-to-back stem and leaf diagram gives information about the numbers of lessons needed by a group of students to pass their driving test:

Male students		Female students
8 6 3 2 2	1	3 6 7 9
7 6 4 3 3 2 1 0	2	3 3 4 5 5 6 7 8 9 9
6 6	3	1 2 5

Key for males:
6 | 3 means 36

Key for females:
3 | 1 means 31

(a) Copy and complete the following table.

	Median	Interquartile range
Male students		
Female students		

(b) Explain whether or not it would be fair to say that, in general, the male students passed their driving test after fewer lessons than the female students.

Summary of key points

1 The **mean** of a set of data is the sum of the values divided by the number of values:

$$\text{mean} = \frac{\text{sum of values}}{\text{number of values}} = \bar{x} = \frac{\Sigma x}{n}$$

where Σx is the sum of the values and n is the number of values.

2 The **median** is the middle value when the data is arranged in order of size.

3 For a data set with n data values, the median is the $\frac{n+1}{2}$th value.

4 The **mode** of a set of data is the value which occurs most often.

5 The **median** and **mode** are not influenced by extreme values, but the **mean** is influenced by extreme values.

6 When the data is arranged in ascending order of size:
   ● the **lower quartile** is the value one quarter of the way into the data
   ● the **upper quartile** is the value three quarters of the way into the data.

7 The **range** of a set of data is the difference between the highest and lowest values:
   range = highest value − lowest value

8 The **interquartile range** is the difference between the upper and lower quartiles:
   interquartile range = upper quartile − lower quartile

9 For a frequency distribution:
  - the **mode** = the item of data which has the highest frequency
  - the **median** = the middle value of the data = the $\dfrac{\Sigma f + 1}{2}$ th value
  - the **mean** = $\dfrac{\Sigma fx}{\Sigma f}$

  where Σf is the sum of the frequencies and Σfx is the sum of all the $(f \times x)$ values in the distribution.

10 For grouped data:
  - you can state the class interval that contains the median
  - the class interval with the highest frequency is called the **modal class**
  - you can **calculate an estimate of the mean** using the middle value of each class interval
  - the estimate of the mean for grouped data is

  $$\bar{x} = \dfrac{\Sigma fx}{\Sigma f},$$ where x is the mid-point of each class interval.

11 To compare two sets of data you should use a measure of average and a measure of spread.

12 A graph showing how a given value changes over time is called a **time series graph**.

13 A moving average 'smooths out' the data to show the trend on a time series graph.

14 You can use a **scatter diagram** to show whether two sets of data are related.

15 If the points on a scatter graph are very nearly along a straight line there is a high **correlation** between the variables.

16 A line that is drawn to pass as close as possible to all the plotted points on a scatter graph is called the **line of best fit**.

⑪ Estimation and approximation

In this chapter you will round values and measurements to different degrees of accuracy and use approximations to calculate estimates.

11.1 Rounding to a number of decimal places

Sometimes you will be asked to round (or correct) a decimal number to a given number of decimal places. For example:

Asafa Powell ran 100 m in 9.771 seconds. Round his time to 1 decimal place (1 d.p.).

9.771 has 3 decimal places (digits after the decimal point).

To round to 1 decimal place, count 1 place from the decimal point and look at the next digit. If it is 5 or more you need to round up.

Asafa Powell, 100 m world record holder, 14 June 2005.

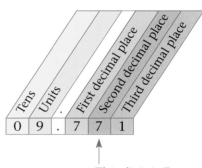

This digit is 7, so you need to round up

9.771 seconds to 1 decimal place (1 d.p.) is 9.8 seconds.

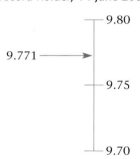

9.771 is closer to 9.8 than to 9.7

You can round (or correct) numbers to a given number of **decimal places (d.p.)**. The first decimal place is the first digit (zero or non-zero) after the decimal point.

Remember: In rounding, halves or 5s are usually rounded upwards.

Example 1

Round 46.382 to (a) 1 decimal place (b) 2 decimal places.

(a) Rounding to 1 d.p. (1 digit after the decimal point):

46.382
↑

The digit after the first decimal place is 8, so **round up**.

46.382 rounded to 1 d.p. is 46.4

(b) Rounding to 2 d.p.

46.382

The digit after the second decimal place is 2, so **round down**.

Notice that when you round down to 2 d.p. these two decimal places remain the same.

46.382 rounded to 2 d.p. is 46.38

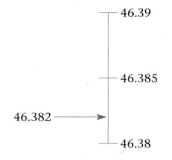

46.39

46.385

46.382 ⟶

46.38

46.382 is closer to 46.38 than to 46.39

Exercise 11A

1 Copy this interval diagram.

(a) Mark the position of 46.3826 on the diagram.

(b) Write 46.382 67 rounded to 3 d.p.

46.383

46.382

2 This table shows the times recorded for the first three athletes in an 800 m race:

Name	Time
R. Grey	1 min 45.4826 s
M. Hobson	1 min 45.4768 s
T. Knight	1 min 45.4817 s

(a) Write these times correct to 2 d.p.

(b) Write the names of the athletes in the order in which they finished the race.

3 Write these numbers correct to the number of decimal places given in brackets.

(a) 137.27 (1 d.p.)   (b) 0.673 81 (4 d.p.)

(c) 4.999 (1 d.p.)   (d) 8.999 (2 d.p.)

(e) 17.9939 (3 d.p.)   (f) 2.009 72 (4 d.p.)

4 Carry out the following calculations and give your answers correct to the number of decimal places given in brackets.

(a) 24.56×3.87 (3 d.p.)   (b) 3.764×2.593 (3 d.p.)

(c) 2.888×3.777 (2 d.p.)   (d) 13.799×12.752 (1 d.p.)

11.2 Rounding to a number of significant figures

You will often be asked to round answers to '2 significant figures' (2 s.f.) or '3 significant figures' (3 s.f.). 'Significant' means 'important'.

When you are estimating the number of people at an open-air concert you don't need to say that there were exactly 8732 people there. You can give the number to two significant figures:

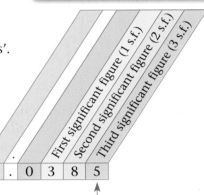

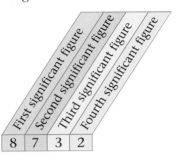

> Estimating the attendance as 8700 to 2 s.f. implies that the actual number is between 8650 and 8749 inclusive.

There were about 8700 people (to 2 s.f.).

This is called 'giving a value correct to two significant figures'.

You can round (or give) numbers to a given number of significant figures (s.f.). The first significant figure is the first non-zero digit in the number, counting from the left.

Rounding to a significant figure which is on the right of the decimal point is like the process used in rounding to decimal places. You look at the next digit after the significant figure:

0.0385 correct to 2 s.f. is 0.039

The digit after the second significant figure is 5.

Example 2

Round 642.803

(a) to 1 s.f.   (b) to 2 s.f.   (c) to 3 s.f.   (d) to 4 s.f.   (e) to 5 s.f.

You need the zeros to show the place value of the 6.

	1 s.f.	2 s.f.	3 s.f.	.	4 s.f.	5 s.f.	6 s.f.					.			
(a)	6	4	2	.	8	0	3	=	6	0	0				(to 1 s.f.)
(b)	6	4	2	.	8	0	3	=	6	4	0				(to 2 s.f.)
(c)	6	4	2	.	8	0	3	=	6	4	3				(to 3 s.f.)
(d)	6	4	2	.	8	0	3	=	6	4	2	.	8		(to 4 s.f.)
(e)	6	4	2	.	8	0	3	=	6	4	2	.	8	0	(to 5 s.f.)

You need the zero to show 5 significant figures.

Exercise 11B

1 Write these numbers correct to the number of significant figures given in brackets.

   (a) 3.168   (3 s.f.)   (b) 964.8   (3 s.f.)

   (c) 15.8   (2 s.f.)   (d) 9.9   (1 s.f.)

   (e) 55.7639 (5 s.f.)   (f) 55.898   (4 s.f.)

2 Using a calculator Samajit works out the answer to
113.12 × 1.7341 as 196.161 39

Write Samajit's answer correct to:

   (a) 3 decimal places   (b) 3 significant figures

   (c) 2 decimal places   (d) 2 significant figures

   (e) the nearest whole number   (f) the nearest 10

   (g) the nearest 100   (h) 7 significant figures.

3 43 672 spectators watched Draycott Town beat
Borrowash Rovers in the Cup Final.
Write the number of spectators correct to:

   (a) 4 s.f.   (b) 2 s.f.   (c) 1 s.f.

4 Work out the following calculations and give your answers correct to the number of significant figures given in brackets.

   (a) 54 × 36   (3 s.f.)   (b) 117 × 38   (3 s.f.)

   (c) 148 × 66 (2 s.f.)   (d) 235 × 364   (2 s.f.)

   (e) 65 × 23   (1 s.f.)   (f) 24 × 42 × 59 (1 s.f.)

5 Write these numbers correct to the number of significant figures given in brackets.

   (a) 0.0032   (1 s.f.)   (b) 0.000 876 (2 s.f.)

   (c) 0.008 97   (2 s.f.)   (d) 0.000 91   (1 s.f.)

   (e) 0.0099   (1 s.f.)   (f) 0.0004   (1 s.f.)

6 Carry out the following calculations and give your answers correct to the number of significant figures asked for.

   (a) 0.3 × 0.023 (correct to 1 s.f.)

   (b) 3.2 ÷ 48 (correct to 4 s.f.)

   (c) 0.007 × 0.000 41 (correct to 2 s.f.)

7 1 inch is approximately equal to 2.54 cm. Write

   (a) 1 cm in inches (correct to 2 s.f.)

   (b) 1 mm in inches (correct to 2 s.f.)

   (c) 1 metre in inches (correct to 3 s.f.)

   (d) 1 km in inches (correct to 1 s.f.)

11.3 Checking and estimating

Get into the habit of checking the answers to your calculations. Sometimes a wrong answer is obvious: 0.42 cm for the average height of a woman in a survey or 2000 km for the distance travelled by a car on a motorway in an afternoon are clearly incorrect.

Once you notice that an answer is obviously wrong you can repeat the calculation to try and find the correct answer.

Making an estimate

Another way of checking a calculation is to make an estimate.

> To estimate answers, round all the numbers to 1 significant figure and do the simpler calculation.

___

Example 3 ___

Estimate the answer to

(a) 5.12×2.79 (b) $19.67 \div 2.8$ (c) $\dfrac{2.75 \times 8.33}{5.23 + 2.74}$

(a) Round all the numbers to 1 significant figure.
An estimate for 5.12×2.79 is $5 \times 3 = 15$
(Actual answer: 14.2848)

(b) Round all the numbers to 1 significant figure:
$20 \div 3 = 6\frac{2}{3} = 7$ (to 1 significant figure).
(Actual answer: 7.025)

(c) Round all the numbers to 1 significant figure:
$\dfrac{3 \times 8}{5 + 3} = \dfrac{24}{8} = 3$ as an estimate.
(Actual answer: $22.9075 \div 7.97 = 2.874...$)

___

Worked examination question ___

$$\dfrac{10.25 + 29.75}{0.2 \times 45}$$

Show how you would estimate the answer to this expression without using a calculator. Write down your estimate.

Round all the numbers to 1 s.f.:

$$\dfrac{10 + 30}{0.2 \times 50} = \dfrac{40}{10} = 4 \text{ as an estimate.}$$

Exercise 11C

1   For each of parts (a) to (f):
   (i) Write down a calculation that could be used to estimate the answer.
   (ii) Work out an estimated answer.
   (iii) Use a calculator to work out the exact answer.
   (a) 8.66×9.56 (b) $3.75 \times (2.36 - 0.39)$
   (c) 37.12×4.33 (d) $(2095 \times 302) + 396$
   (e) $\dfrac{187.3 \times 75.4}{47.9}$ (f) $\dfrac{0.634 \times 0.0176}{0.0425}$

2   Trevor's calculator is faulty. It doesn't show the decimal point so Trevor has answered the following questions by writing down the digits from the display. By estimating, find the correct answers including the decimal points.

(a) $3.4 \times 2.5 = 85$	**(b)** $5.6 \times 7.23 = 40488$
(c) $2.42 \times 1.93 = 46706$	**(d)** $10.63 \times 12.64 = 1343632$
(e) $3.2 \times 5.1 \times 4.7 = 76704$	**(f)** $2.9 \times 0.32 \times 8.7 = 80736$

3   Estimate the answer to each calculation, giving your answers to the nearest whole number:
   (a) $36 \div 5$ (b) $47 \div 8$ (c) $103 \div 11$
   (d) $148 \div 49$ (e) $239 \div 31$ (f) $197 \div 19$
   (g) $281 \div 41$ (h) $796 \div 98$ (i) $352 \div 58$
   (j) $\dfrac{53 \times 3}{49}$ (k) $\dfrac{38 \times 5}{19}$ (l) $\dfrac{24 \times 11}{59}$
   (m) $\dfrac{41.2 \times 3.79}{3.17 \times 4}$ (n) $\dfrac{28.4 \times 70.17}{2.95 \times 4.21}$ (o) $\dfrac{61.8 \times 1.46}{39.2 - 18.8}$

11.4 Describing the accuracy of a measurement

Suppose you use a ruler marked in millimetres to measure a line as 237 mm, correct to the nearest millimetre.

The true length could be anywhere between 236.5 mm and 237.5 mm.

So the true length could be anywhere in a range between 0.5 mm below and 0.5 mm above the recorded value:

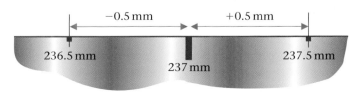

For any measurement you make:

> If you make a measurement correct to a given unit, the true value lies in a range that extends half a unit below and half a unit above the measurement.

Exercise 11D

1 These lengths are all measured to the nearest millimetre. Write down the smallest and largest values of the possible range of each length.

(a) 3.9 cm   (b) 17.620 m   (c) 53.0 cm

(d) 1.030 m   (e) 29 mm

2 Copy and complete this table showing the attendances to the nearest hundred at these schools.

School	Attendance to the nearest 100	Lowest value of range	Greatest value of range
Southpark	1400	1350	1450
Westown	400		
Eastgrove	800		
Northbury	1200		
Southone	2100		

The first one is done for you.

3 Kirsty ran a 1500 metre race. Her time to the nearest 0.01 second was 4 minutes 7.56 seconds. Write down the range of times between which her time could lie.

4 The distance between two motorway junctions is given as 17 miles to the nearest mile. Write down the range within which the true length could lie.

11.5 Upper and lower bounds

In the measuring example on page 219 the true length could be anywhere between 236.5 mm and 237.5 mm.

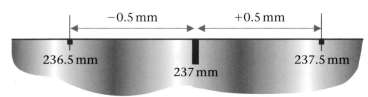

the smallest (or minimum) possible value is called the **lower bound**

the greatest (or maximum) possible value is called the **upper bound**

The **lower bound** and the **upper bound** are the minimum and maximum possible values of a measurement or calculation.

> For more on upper and lower bounds see Chapter 23.

Example 4

A rectangle has length 7.5 cm and width 4.3 cm, both measured to the nearest 0.1 cm.

Work out

(a) the lower bounds of the length and width

(b) the upper bounds of the length and width

(c) the maximum possible area (the upper bound of the area).

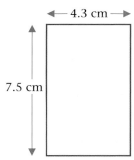

(a) The lower bound of the length is 7.45 cm.
The lower bound of the width is 4.25 cm.

(b) The upper bound of the length is 7.55 cm.
The upper bound of the width is 4.35 cm.

(c) The maximum possible area is:

$$\text{upper bound of length} \times \text{upper bound of width} = 7.55 \text{ cm} \times 4.35 \text{ cm}$$
$$= 32.8425 \text{ cm}^2$$

Exercise 11E

1 Each measurement is given correct to 3 decimal places. Write down
(i) the lower bound (ii) the upper bound.
(a) 0.234 km (b) 0.319 s (c) 0.545 kg
(d) 0.795 km (e) 9.417 s (f) 3.459 g

2 (a) Jafar has a piece of wood of length 30 cm, correct to the nearest centimetre. Write down the minimum possible length of the piece of wood.

(b) Fatima has a different piece of wood of length 18.4 cm, correct to the nearest millimetre. Write down the maximum and minimum lengths between which the length of her piece of wood must lie.

3 The length of each side of a square, correct to 2 significant figures, is 3.7 cm.

(a) Write down the least possible length of each side.

(b) Calculate the greatest and least possible perimeters of this square.

(c) When calculating the perimeter of the square, how many significant figures is it appropriate to give in the answer? Explain your answer.

(d) If this question had referred to a regular octagon, instead of a square, would your answer to part (c) have been the same? Explain your answer.   [E]

4 Stephanie ran 100 metres. The distance was correct to the nearest metre.

  (a) Write down the shortest distance Stephanie could have run.

  Stephanie's time for the run was 14.8 seconds. Her time was correct to the nearest tenth of a second.

  (b) Write down
    (i) her shortest possible time for the run
    (ii) her longest possible time for the run.

  (c) Calculate **(i)** the lower bound **(ii)** the upper bound for her average speed. Write down all the figures on your calculator display.

  (d) **(i)** Write down her average speed to an appropriate degree of accuracy.
    (ii) Explain how you arrived at your answer. [E]

5 The diagram represents a metal girder in the shape of a half cylinder. The radius of each semicircular end is 5.6 centimetres, correct to the nearest tenth of a centimetre.

  (a) Write down the upper and lower bounds of the radius.

  The length of the girder is 1.20 metres, correct to the nearest centimetre.

  (b) Write down the upper and lower bounds of the length.

  (c) Calculate the upper and lower bounds of the volume of the girder. Give your answer in cubic centimetres (cm^3).

1.20 m

5.6 cm

Diagram NOT accurately drawn

> For the formula for volume of a cylinder see Section 22.4.

6 Brazil has an area of 8 500 000 km^2 correct to the nearest 100 000 km^2.

  (a) Write down the limits between which the area of Brazil must lie.

  The population density of a country is the average number of people per km^2 of the country.

  Brazil has a population of 184 million correct to the nearest million.

  (b) Calculate the maximum and minimum values of the population density of Brazil. [E]

7 The formula $S = \dfrac{F}{A}$ is used in engineering.

  $F = 810$, correct to 2 significant figures.
  $A = 2.93$, correct to 3 significant figures.

  (a) For the value of F, write down
    (i) the upper bound
    (ii) the lower bound.

(b) For the value of A, write down

(i) the upper bound   (ii) the lower bound.

(c) Calculate **(i)** the upper bound and **(ii)** the lower bound for the value of S for these values of F and A. Write down all the figures on your calculator display.

(d) Write down this value of S correct to an appropriate number of significant figures.   [E]

Mixed exercise 11

1 Round these numbers to the number of decimal places given in brackets.

(a) 9.4915   (1 d.p.)   (b) 0.0455   (3 d.p.)

(c) 7.699   (2 d.p.)   (d) 7.4317   (2 d.p.)

2 Write these numbers to the number of significant figures given in brackets.

(a) 423.58   (3 s.f.)   (b) 0.214 73   (4 s.f.)

(c) 10 769   (2 s.f.)   (d) 90 873   (1 s.f.)

3 Write down a suitable estimation to check each answer:

(a) $56.25 \div 1.25 = 45$ (b) $3.256 \times 1.5 = 4.884$

(c) $109.46 - 83.27 = 26.19$ (d) $56.21 + 19.78 = 75.99$

(e) $\dfrac{20.47 + 3.82}{2.67 + 5.05} = 3.146$ (f) $\dfrac{8.37 \times 3.81}{4.47 + 4.74} = 3.463$

4 A distance of 15.2 km has been rounded to 1 decimal place. Write down

(a) the largest value of the possible range

(b) the smallest value of the possible range.

5 To the nearest centimetre, $x = 4$ cm and $y = 6$ cm.

(a) Calculate the upper bound for the value of xy.

(b) Calculate the lower bound for the value of $\dfrac{x}{y}$.

Give your answer correct to 3 significant figures.   [E]

6 The diagram represents two metal spheres of different sizes.
The radius of the smaller sphere is r cm.
The radius of the larger sphere is R cm.
$r = 1.7$ correct to 1 decimal place.
$R = 31.0$ correct to 3 significant figures.

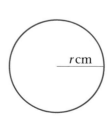
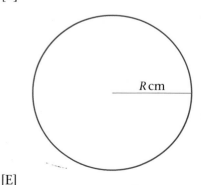

(a) Write down the upper and lower bounds of r and R.

(b) Find the smallest possible value of $R - r$.   [E]

Summary of key points

1 You can round (or give) numbers to a given number of **decimal places (d.p.)**.
The first decimal place is the first digit (zero or non-zero) after the decimal point.

2 You can round (or give) numbers to a given number of significant figures (s.f.).
The first significant figure is the first non-zero digit in the number, counting from the left.

3 To estimate answers, round all the numbers to 1 significant figure and do the simpler calculation.

4 If you make a measurement correct to a given unit, the true value lies in a range that extends half a unit below and half a unit above the measurement.

5 The **lower bound** and the **upper bound** are the minimum and maximum possible values of a measurement or calculation.

12 Sequences and formulae

In this chapter you will find formulae to describe sequences and use formulae from mathematics and other subjects.

12.1 Finding terms of a sequence

These diagrams show the triangular numbers:

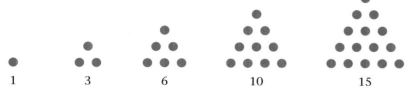

These numbers are a **sequence** – a succession of numbers formed according to a rule.

Each number in a sequence is called a **term**. 6 is the third term in the sequence of triangular numbers.

Sometimes you will be given an algebraic formula for the rule for the nth term of a sequence. You can use this to find any term in the sequence. The nth term is often written as u_n.

> The nth term is sometimes called the **general term**.

When you know a formula for the nth term of a sequence, u_n, you can calculate any term in the sequence by substituting a value for n in the formula. n must be a positive integer ($n = 1, 2, 3 \ldots$).

Example 1

The nth term of a sequence is given by $u_n = 2n - 1$.   Find the first two terms of the sequence.

Substituting $n = 1$, the first term $u_1 = (2 \times 1) - 1 = 1$
Substituting $n = 2$, the second term $u_2 = (2 \times 2) - 1 = 3$

The first two terms are 1 and 3.

Example 2

For a given sequence $u_n = 10 - 3n$.   Find u_2.

Substituting $n = 2$, the second term $u_2 = 10 - (3 \times 2) = 4$

Example 3

For a given sequence $u_n = 2n^2 + 3$.   Find the fifth term of the sequence.

Substituting $n = 5$, the fifth term $u_5 = (2 \times 5^2) + 3 = 53$

Exercise 12A

Find the first five terms of the sequences in questions **1–14**.

1 $u_n = 2n$ **2** $u_n = n + 2$ **3** $u_n = 3n + 7$

4 $u_n = 15 - 2n$ **5** $u_n = n^2 + 5$ **6** $u_n = 3n^2 - 1$

7 $u_n = (n + 1)^2$ **8** $u_n = n^3$ **9** $u_n = 2^n$

10 $u_n = (-2)^n$ **11** $u_n = (\frac{1}{2})^n$ **12** $u_n = \frac{1}{2}n(n + 1)$

13 $u_n = \dfrac{1}{n}$ **14** $u_n = \dfrac{n}{n + 1}$

15 Find the sixth term of the sequence with $u_n = 4n - 7$.

Finding which term has a given value

The number 105 is a term in the sequence given by the formula
$u_n = 7(n + 4)$.

Which term in the sequence has this value?

You can find out by finding the value of n when $u_n = 105$.
Substituting $u_n = 105$ in the formula gives the equation
$$7(n + 4) = 105$$
$$7n + 28 = 105$$
$$7n = 77$$
$$n = 11$$

> For more on
> manipulating formulae
> see Section 12.4.

So 105 is the 11th term of the sequence.

___ **Example 4** ___

A sequence is given by the formula $u_n = 2n^2 + 5$.
For which value of n does $u_n = 103$?

Substituting $u_n = 103$ in the formula gives the equation:
$$2n^2 + 5 = 103$$
$$2n^2 = 98$$
$$n^2 = 49$$
$$\text{so} \quad n = 7$$

> Notice that $n = -7$ is
> also a solution of $n^2 = 49$,
> but for a sequence n must
> be a positive integer.
> There is no -7th term in
> a sequence.

Exercise 12B

Find the value of n for which u_n has the given value.

1 $u_n = 6n + 7,\ \ u_n = 55$ **2** $u_n = 7(n + 4),\ \ u_n = 112$

3 $u_n = 7 - 3n,\ \ u_n = -44$ **4** $u_n = \dfrac{2n + 3}{3n + 1},\ \ u_n = \dfrac{17}{22}$

5 $u_n = n^2 + 5,\ \ u_n = 149$ **6** $u_n = 3n^2 - 8,\ \ u_n = 235$

7 $u_n = (n - 8)^2,\ \ u_n = 169$ **8** $u_n = 2n^2 + 12,\ \ u_n = 140$

9 $u_n = 3^n,\ \ \ \ \ \ \ \ u_n = 2187$ **10** $u_n = n(n - 1),\ \ u_n = 380$

12.2 Finding the formula for a sequence

Arithmetic sequences

The first six terms of a sequence are

   9,   14,   19,   24,   29,   34

By looking for a pattern in the sequence you can find a formula for the nth term u_n. This is sometimes called the **general term**.

A good way to start is to find the differences between successive terms:

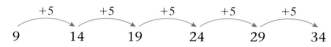

> A sequence in which the differences between successive terms are equal is called an **arithmetic sequence**.

The terms go up in 5s, so the nth term will include n lots of 5, or $5n$.

For the first term $n = 1$, so $5n = 5$. But the first term is 9, which is 4 more than $5n$. This suggests a formula of the form

   $u_n = 5n + 4$

Trying a few values of n gives

   $u_1 = (5 \times 1) + 4 = 9$ ✓
   $u_2 = (5 \times 2) + 4 = 14$ ✓
   $u_3 = (5 \times 3) + 4 = 19$ ✓
   $u_4 = (5 \times 4) + 4 = 24$ ✓

So the formula $u_n = 5n + 4$ does describe the sequence.

> The **general term** for an arithmetic sequence is of the form $u_n = an + b$. The value of a is the difference between successive terms in the sequence.

Example 5

The first five terms of a sequence are

   5,   3,   1,   −1,   −3

Find a formula for the nth number in the sequence, u_n.

As each term is 2 less than the previous one, the formula for u_n will include $-2n$. But $u_n = -2n$ would give the sequence

   -2, -4, -6, -8,   ...

For the sequence to start at 5, add 7 to each term, giving

   5,   3,   1,   −1,   −3,   ...

The formula that describes the sequence is $u_n = -2n + 7$

Exercise 12C

Your answers to Exercise 12A may be helpful here.

Find formulae for u_n to describe each of these sequences.

Your answers to Exercise 12A may be helpful here.

1   3, 4, 5, 6, 7, ...   2   5, 7, 9, 11, 13, ...

3   7, 11, 15, 19, 23, ...   4   9, 8, 7, 6, 5, ...

5   12, 10, 8, 6, 4, ...   6   7, 4, 1, −2, −5, ...

12.3 More about formulae

As well as describing sequences, formulae can also be used to describe other relationships between quantities.

> An algebraic formula can be used to describe a relationship between two sets of values.

For example, the formula $C = \pi d$ describes the relationship between the diameter of a circle d and its circumference C.

Notice the difference between a formula and an equation:

<table>
<tr>
<td>

$C = \pi \times d$ is a **formula**.

If you substitute **any** value for d the formula will give you a corresponding value for C.

Similarly if you substitute any value for C it will give you a corresponding value for d.

</td>
<td>

$C = \pi \times 2.5$ is an **equation**.

The equation is only true for **one value** of C ($C = 7.85$ to 2 d.p.).

</td>
</tr>
</table>

You need to be able to evaluate formulae by substituting given values into them, including negative numbers and fractions.

Example 6

The value of s can be found from the formula $s = pt^2 + q$.
Calculate the value of s when

(a) $p = 1$, $q = 2$ and $t = \frac{1}{2}$ (b) $p = 5$, $q = -1$ and $t = -2$

(a) $s = 1 \times (\frac{1}{2})^2 + 2$
      $= \frac{1}{4} + 2$
      $= 2\frac{1}{4}$

(b) $s = 5 \times (-2)^2 - 1$
      $= 5 \times 4 - 1$
      $= 20 - 1$
      $= 19$

Worked examination question 1

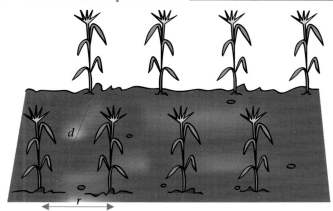

The diagram above is taken from a book about growing maize. The distance between the rows of plants is d metres. The spacing between the plants in each row is r metres.

The number, P, of plants per hectare is given by the formula

$$P = \frac{10\,000}{dr}$$

Given that $d = 0.8$ and $r = 0.45$, calculate the value of P. Give your answer correct to 2 significant figures.

$$P = \frac{10\,000}{0.8 \times 0.45}$$

$$= 27\,777.777$$
$$= 28\,000 \text{ (correct to 2 s.f.)}$$

Exercise 12D

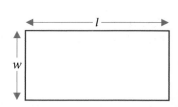

1   $y = 2x - 5$ Calculate the value of y when $x = -3$

2   $A = lw$ Calculate the value of A when $l = 8\frac{1}{2}$ and $w = 6\frac{1}{4}$

3   $P = 2(l + w)$ Calculate the value of P when $l = 8.3$ and $w = 6.8$

4   $C = \pi d$ Calculate the value of C correct to 3 s.f. when $d = 8.37$

5   $y = 3x^2 + 2$ Calculate the value of y when $x = -1$

6   $A = \pi r^2$ Calculate the value of A correct to 3 s.f. when $r = 3.9$

7   $A = 2\pi r(r + h)$ Calculate the value of A correct to 3 s.f. when $r = 3.7$ and $h = 5.9$

8   $s = \frac{1}{2}gt^2$ Calculate the value of s when $g = -10$ and $t = 3$

9   $x = L(1 + at)$ Calculate the value of x when $L = 100$, $a = 0.000\,019$ and $t = 80$

10 $v = u + at$ Calculate the value of v when $u = 5$, $a = -8$ and $t = 6$

11 $V = \frac{1}{3}\pi r^2 h$ Calculate the value of V correct to 3 s.f. when $r = 2.6$ and $h = 6.7$

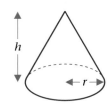

12 $I = \dfrac{PRT}{100}$ Calculate the value of I when $P = 850$, $R = 5\frac{3}{4}$ and $T = 4$

13 $s = ut + \frac{1}{2}at^2$ Calculate the value of s when $u = -3$, $t = 5$ and $a = -4$

14 $D = v + \dfrac{v^2}{20}$ Calculate the value of D correct to 2 s.f. when $v = 31.6$

15 $c = \sqrt{a^2 + b^2}$ Calculate the value of c correct to 3 s.f. when $a = 3.7$ and $b = 8.3$

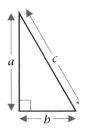

16 $f = \dfrac{uv}{u + v}$ Calculate the value of f when $u = 3.6$ and $v = -5.4$

17 $r = \sqrt{\dfrac{A}{\pi}}$ Calculate the value of r correct to 3 s.f. when $A = 37.2$

18 $r = \sqrt{\dfrac{A}{4\pi}}$ Calculate the value of r correct to 3 s.f. when $A = 21.3$

19 $T = 2\pi\sqrt{\dfrac{L}{g}}$ Calculate the value of T correct to 3 s.f. when $L = 23.4$ and $g = 9.8$

20 $T = \dfrac{2Mmg}{M + m}$ Calculate the value of T correct to 2 s.f. when $M = 3.8$, $m = 1.7$ and $g = 9.8$

12.4 Manipulating formulae

Sometimes when you substitute values into a formula you get an equation to solve.

Example 7

$C = \pi d$ Calculate the value of d when $C = 23.2$, giving your answer correct to 3 s.f.

$$23.2 = \pi d$$

Dividing each side by π gives

$$d = \frac{23.2}{\pi} = 7.38 \text{ (to 3 s.f.)}$$

> Remember:
> Whatever you do to one side of an equation, you must also do to the other side.

Example 8

$v = u + at$ Calculate the value of t when $v = 39$, $u = 4$ and $a = 3$

$39 = 4 + 3t$

$35 = 3t$ ———————————— Subtract 4 from both sides.

$t = 11\frac{2}{3}$ ———————————— Divide both sides by 3.

Worked examination question 2

The air temperature T °C outside an aircraft flying at a height of h feet is given by the formula

$$T = 26 - \frac{h}{500}$$

The air temperature outside an aircraft is -52 °C. Calculate the height (h) of the aircraft.

$$-52 = 26 - \frac{h}{500}$$

$$\frac{h}{500} - 52 = 26$$ ——————Add $\frac{h}{500}$ to both sides.

$$\frac{h}{500} = 78$$ ——————Add 52 to both sides.

$$h = 39\,000$$ ——————Multiply both sides by 500.

The height of the aircraft is $39\,000$ feet.

Exercise 12E

1 $C = \pi d$ Calculate the value of d correct to 2 s.f. when $C = 56.8$

2 $A = lw$ Calculate the value of l correct to 2 d.p. when $A = 32.94$ and $w = 6.1$

3 $C = 2\pi rh$ Calculate the value of h correct to 2 s.f. when $C = 800$ and $r = 5.9$

4 $y = 5x + 7$ Calculate the value of x when $y = 11$

5 $v = u + at$ Calculate the value of a when $v = 17$, $u = -3$ and $t = 5$

6 $I = \dfrac{PRT}{100}$ Calculate the value of R when $I = 45$, $P = 250$ and $T = 4$

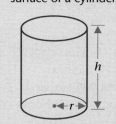

$C = 2\pi rh$ is the formula for the area of the curved surface of a cylinder.

7 $y = 8 + \frac{1}{4}x$ Calculate the value of x when $y = 6$

8 $y = 9 - 4x$ Calculate the value of x when $y = 14$

9 $y = 5 - \frac{1}{2}x$ Calculate the value of x when $y = 2$

10 $P = 2(l + w)$ Calculate the value of w when $P = 42$ and $l = 13\frac{1}{2}$

11 $S = \frac{1}{2}n(a + d)$ Calculate the value of a when $S = 72$, $d = 3$ and $n = 18$

12 $A = 2\pi r(r + h)$ Calculate the value of h correct to 3 s.f. when $A = 540$ and $r = 4.9$

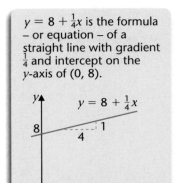

$y = 8 + \frac{1}{4}x$ is the formula – or equation – of a straight line with gradient $\frac{1}{4}$ and intercept on the y-axis of $(0, 8)$.

13 $y = 2x^2$ Calculate the two possible values of x when $y = 98$

14 $A = \pi r^2$ Calculate the positive value of r correct to 3 s.f. when $A = 50$

15 $s = ut + \frac{1}{2}at^2$ Calculate the value of u when $s = 121$, $t = 5$ and $a = 8$

16 $c = \sqrt{a^2 + b^2}$ Calculate the positive value of a when $c = 29.9$ and $b = 11.5$

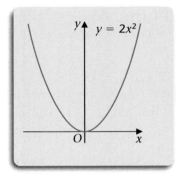

17 $r = \sqrt{\dfrac{A}{\pi}}$ Calculate the value of A correct to 3 s.f. when $r = 8$

18 $V = \frac{1}{3}\pi r^2 h$ Calculate the value of h correct to 3 s.f. when $V = 100$ and $r = 2.5$

19 $y = \dfrac{20}{x + 1}$ Calculate the value of x when $y = 8$

20 $f = \dfrac{uv}{u + v}$ Calculate the value of v when $f = 12$ and $u = -8$

12.5 Changing the subject of a formula

x is the **subject** of a formula when it appears on its own on one side of the formula and does not appear on the other side.

Sometimes you will need to rearrange a formula so that a letter such as x is on one side of the equals sign with the rest of the formula on the other side. This is called **changing the subject** of the formula. You can use the same methods as you use to rearrange equations.

Example 9

Rearrange $ax + b = c$ to make x the subject of the formula.

$\qquad ax = c - b$ ——— Subtract b from both sides.

$\qquad x = \dfrac{c - b}{a}$ ——— Divide both sides by a.

Example 10

Make x the subject of the formula $p - 5x = q$.

$\qquad p = 5x + q$ —— Add $5x$ to both sides.

$\qquad p - q = 5x$ ——— Subtract q from both sides.

$\qquad \dfrac{p - q}{5} = x$ ——— Divide both sides by 5.

Exercise 12F

In questions **1–14** make x the subject of the formula.

1 $x + p = q$ **2** $x + 7 = t$ **3** $x - a = b$

4 $5x = w$ **5** $nx = p$ **6** $\frac{1}{3}x = f$

7 $\dfrac{x}{m} = d$ **8** $\dfrac{ax}{b} = c$ **9** $nx - p = q$

10 $5 - ax = b$ **11** $c - dx = h$ **12** $p - 2x = 3x + q$

13 $a - 4x = b - 7x$ **14** $\dfrac{x}{m} + n = p$

Worked examination question 3

The formula below gives the Body Mass Index, I, of a person who is h metres tall and weighs W kg.

$$I = \frac{W}{h^2}$$

Make h the subject of the formula.

$\qquad Ih^2 = W$ ——— Multiply both sides by h^2.

$\qquad h^2 = \dfrac{W}{I}$ ——— Divide both sides by I.

$\qquad h = \sqrt{\dfrac{W}{I}}$ ——— Take the square root of both sides.

h is now on its own on one side – it is the subject of the formula.

Worked examination question 4

The approximate range, R miles, of a radar mounted x feet above sea level is given by the formula

$$R = \sqrt{2x}$$

Make x the subject of the formula.

$R^2 = 2x$ ——— Square both sides.

$\frac{1}{2}R^2 = x$ ——— Divide both sides by 2.

x is on its own on one side. The formula can be written as $x = \frac{1}{2}R^2$.

Exercise 12G

In questions **1–30** make the letter in square brackets the subject of the formula.

1 $V = IR$ [I]   **2** $A = lw$ [l]

3 $C = \pi d$ [d]   **4** $V = lwh$ [w]

5 $v = u + at$ [u]   **6** $v = u + at$ [t]

7 $C = 2\pi rh$ [h]   **8** $I = \dfrac{PRT}{100}$ [R]

9 $P = 2(l + w)$ [l]   **10** $A = 2\pi r(r + h)$ [h]

11 $\dfrac{y}{x} = 3$ [x]   **12** $\dfrac{PV}{T} = k$ [T]

13 $y = 5x^2$ [x]   **14** $E = kx^2$ [x]

15 $A = \pi r^2$ [r]   **16** $S = \frac{1}{2}n(a + d)$ [n]

17 $S = \frac{1}{2}n(a + d)$ [a]   **18** $y = kx^2$ [x]

19 $V = \frac{1}{3}\pi r^2 h$ [h]   **20** $V = \frac{1}{3}\pi r^2 h$ [r]

21 $c = \sqrt{a^2 + b^2}$ [a]   **22** $a = 2\sqrt{b^2 - 2}$ [b]

23 $t = \sqrt{\dfrac{2s}{g}}$ [s]   **24** $T = 2\pi\sqrt{\dfrac{L}{g}}$ [L]

25 $x^2 + y^2 = 1$ [y]   **26** $v^2 = u^2 + 2as$ [u]

27 $v = w\sqrt{a^2 - x^2}$ [x]   **28** $y = 3\sqrt{1 - x^2}$ [x]

29 $y = \dfrac{5 + x}{1 - x}$ [x]   **30** $f = \dfrac{uv}{u + v}$ [u]

12.6 Substituting one formula in another

You can sometimes eliminate one of the letters from a formula by substituting an expression from a second formula.

Example 11

Two formulae used when designing electrical circuits are $P = IV$ and $V = IR$.

Find a formula for P in terms of I and R.

In $P = IV$, replace V by IR:

$$P = I \times IR$$
$$P = I^2R$$

Example 12

The length of a rectangle is l and its area is A.
Find a formula for its perimeter, P, in terms of l and A.

Let w represent the width of the rectangle.

$$P = 2l + 2w \qquad (1)$$
$$A = lw \qquad (2)$$

Make w the subject of formula (2):

$$w = \frac{A}{l}$$

In formula (1), replace w by $\frac{A}{l}$:

$$P = 2l + \frac{2A}{l}$$

> Formula (1) gives P in terms of l and w. To find P in terms of l and A, you need to replace the term $2w$ with a term in l and A.

Exercise 12H

1 $P = IV$ and $V = IR$.
   Find a formula for P in terms of V and R.

2 $s = vt$ and $v = \frac{1}{2}gt$.
   Find a formula for s in terms of g and t.

3 $A = \pi r^2$ and $r = \dfrac{d}{2}$
   Find a formula for A in terms of d.

4 $a = \dfrac{v^2}{r}$ and $v = rw$
   Find a formula for a in terms of r and w.

5 The area of a square is A and its perimeter is P.

(a) Find a formula for A in terms of P.

(b) Find a formula for P in terms of A.

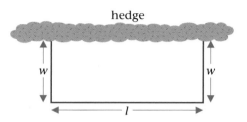

Let x represent the length of the square's sides and express A and P in terms of x.

6 $x = at^2$ and $y = 2at$.
Find a formula for x in terms of a and y.

7 $A = 2\pi r^2 + 2\pi rh$ and $V = \pi r^2 h$.
Find a formula for A in terms of r and V.

8 A farmer encloses sheep in a rectangular pen using fencing for three sides and a long hedge for the other side. The length of the pen is l and its width is w. The total length of fencing used is F and the area enclosed is A. Find a formula for F in terms of A and l.

Mixed exercise 12

1 These are the first five numbers of a simple sequence:

$$3, \quad 12, \quad 27, \quad 48, \quad 75$$

(a) Write down the next two numbers of the sequence.

(b) Write down, in terms of n, an expression for the nth term of this sequence. [E]

2 Write down the next two numbers for each sequence.

(a) 7, 11, 15, 19, 23, ...   (b) 30, 27, 24, 21, 18, ...

(c) 1, 2, 4, 8, 16, ...   (d) 3, 6, 10, 15, 21, ...

3 Make t the subject of $v = 7t + 30$ [E]

4 Make g the subject of $T = 2\pi\sqrt{\dfrac{a}{g}}$

5 Make r the subject of $A = 4\pi r^2$

6 $y = 4t + 3$ and $x = 2t - 1$.
Find a formula for y in terms of x.

7 Make v the subject of $f = \dfrac{uv}{u + v}$

8 Make u the subject of the formula $D = ut + kt^2$ [E]

9 A formula used in science is $s = \frac{1}{2}gt^2$

Tracy has to work out the value of s when $g = 9.81$ and $t = 23.67$

Before using her calculator to work out the value of s, Tracy decides to obtain an estimate for s.

(a) (i) Write down values for g and t which Tracy could use in her calculation for an estimate of s.

(ii) Calculate the estimate of s that these values would give.

(b) Use your calculator to work out the actual value of s, giving your answer correct to two decimal places.

10 The formula for the volume V of a cylinder with circular base of radius r and vertical height h is

   $V = \pi r^2 h$

The radius of the base and the vertical height of a cylinder are measured to the nearest millimetre and quoted as $r = 7.4\,\text{cm}$ and $h = 10.8\,\text{cm}$, whilst $\pi = 3.1416$, correct to 4 decimal places.

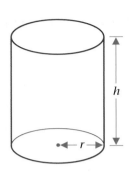

   (a) Without using a calculator and showing all your working, give sensible approximations for π, r and h and use these to find an estimate for the volume of the cylinder.

   (b) Use your calculator and the quoted values of π, r and h to calculate the volume of the cylinder, giving your answer in cubic centimetres, correct to three significant figures.

   (c) Using $\pi = 3.1416$, find the maximum possible volume of the cylinder, giving your answer correct to two decimal places.

11 In Florida the speed limit for motor vehicles is 55 miles per hour. The 'on the spot' fine for motorists caught exceeding this limit is 4 dollars for each mile per hour over 55 miles per hour.

   (a) Write down a formula for f, the 'on the spot' fine in dollars, for a motorist caught travelling at v miles per hour. You may assume that $v > 55$.

   (b) Make v the subject of the formula.   [E]

Summary of key points

1 When you know a formula for the nth term of a sequence, u_n, you can calculate any term in the sequence by substituting a value for n in the formula. n must be a positive integer ($n = 1, 2, 3 \ldots$).

2 The **general term** for an arithmetic sequence is of the form $u_n = an + b$. The value of a is the difference between successive terms in the sequence.

3 An algebraic formula can be used to describe a relationship between two sets of values.

4 x is the **subject** of a formula when it appears on its own on one side of the formula and does not appear on the other side.

13 Measure and mensuration

This chapter is about measurement of lengths, areas and volumes – mainly of geometrical shapes.

In the GCSE exam you will need to be able to apply a number of formulae. *Some* of them will appear on the formula sheet, but you need to learn the formulae and understand how they are derived.

In this chapter you will also consider how the accuracy of measurement affects calculation of areas.

13.1 Continuous and discrete data

When someone says their age is 15 years, this is really an approximation. The age quoted is the number of complete years. You can quote the age more exactly as, say, 15 years and 200 days. But this is still only an approximation, to the nearest day. The age could be quoted even more accurately as, say 15 years, 200 days, 7 hours, 35 minutes and 10 seconds, but this is still only an approximation, now given to the nearest second.

Time passes in a **continuous** way, so if someone quotes his or her age, then they are a little older at the end of their statement than they were at the beginning. No matter how accurate the quote, it is never more than an approximation.

A similar argument is true for a measurement such as length. We might say that the length of a metal rod is 72 cm, but we are dependent on the accuracy of the measuring instrument.

For some data, these arguments do not apply. You might, for instance, be asked to quote the number of GCSE exams you are taking. This will be a whole number and you can quote it exactly. It is an example of **discrete** data.

The figure does not have to be a whole number. For instance, 10 students could take 76 GCSE exams between them. Then the mean number of GCSEs taken per person is $76 \div 10 = 7.6$; this is still an exact number.

13.2 The measurement of continuous data

You will often see examples of continuous data. For instance, a road sign might read:

The distance is quoted to the nearest mile. The actual distance could be anywhere between 75.5 and 76.5 miles.

London 76

Example 1

The dimensions of a rectangular lawn are given as

width = 5 m   length = 7 m

Each measurement is given correct to the nearest metre.

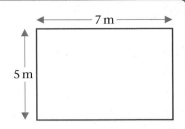

(a)   Write the longest and shortest possible values for the width and the length of the rectangle.

(b)   Calculate the largest and smallest possible values for the area of the lawn.

(a)   For the width, the measurement of 5 m means that the true width can be anywhere between 4.5 m and 5.5 m.
      The shortest possible width is 4.5 m and the longest possible width is 5.5 m.

      For the length, the measurement of 7 m means that the true length can be anywhere between 6.5 m and 7.5 m.
      The shortest possible length is 6.5 m and the longest possible length is 7.5 m.

(b)   The area of a rectangle = width × length.
      To find the smallest area we combine the shortest width with the shortest length. This gives a smallest possible area of 4.5 × 6.5 = 29.25 m².
      The largest area is found by combining the longest width with the longest length. This gives a largest possible area of 5.5 × 7.5 = 41.25 m².

Exercise 13A

1   A piece of carpet is rectangular. Its dimensions are quoted to the nearest 10 cm as

    width = 3.4 m   length = 4.6 m

    (a)   Write the shortest and longest possible widths for the carpet.

    (b)   Write the shortest and longest possible lengths for the carpet.

    (c)   Calculate the smallest and largest possible areas for the carpet.

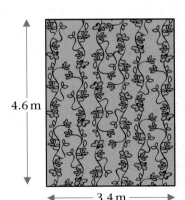

2   A car travels at a constant speed of 70 km per hour for a time of 2 hours and 45 minutes. Its speed is quoted to the nearest km/h and the time is quoted to the nearest minute. Calculate

    (a)   the maximum possible distance that the car could have travelled

    (b)   the minimum distance the car could have travelled.

13.3 Perimeter and area of triangles and quadrilaterals

You will need to know and apply the formulae for the perimeters and areas of many different shapes.

Rectangle

Perimeter $= 2(a + b)$

Area $= a \times b = ab$

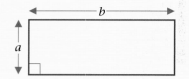

Triangle

Perimeter $= a + b + c$

Area $= \frac{1}{2} \times$ base $\times$ height

$\qquad = \frac{1}{2}bh$

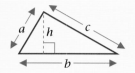

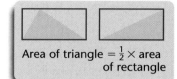

Area of triangle $= \frac{1}{2} \times$ area of rectangle

Parallelogram

Perimeter $= 2(a + b)$

Area $= bh$

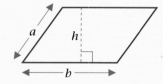

Trapezium

Perimeter $=$ sum of the lengths of the four sides

Area $= \frac{1}{2}(a + b) \times h \quad$ or $\quad \frac{1}{2}(a + b)h$

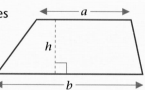

Example 2

Work out the area of triangle XYZ.

If you take YZ as the base, the height is XW.

The area of triangle XYZ is $\frac{1}{2} \times$ base $\times$ height $= \frac{1}{2} \times 7 \times 5 = \dfrac{35}{2}$

The area of triangle XYZ is $17.5\,\text{cm}^2$.

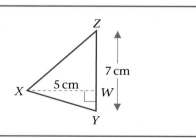

Example 3

Use the formula to calculate the area of the trapezium.

Area of trapezium $= \frac{1}{2}(a + b)h = \frac{1}{2}(3 + 10) \times 4$

$\qquad\qquad\qquad = \frac{1}{2} \times 13 \times 4 = 13 \times 2$

$\qquad\qquad\qquad = 26\,\text{cm}^2$

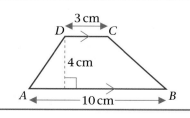

Example 4

Find the shaded area.

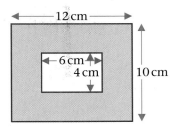

It is easiest to find the area of the larger rectangle and subtract the area of the smaller rectangle.

Large rectangle area: $12 \times 10 = 120 \text{ cm}^2$
Small rectangle area: $6 \times 4 = 24 \text{ cm}^2$
Area of shaded region: $120 - 24 = 96 \text{ cm}^2$

Example 5

$ABCD$ is a trapezium with AB parallel to DC.
The lengths of AB and CD are in the ratio $1:2$.
The perpendicular distance between AB and $CD = 12$ cm.
The area of $ABCD = 72 \text{ cm}^2$.

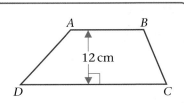

Calculate the length of AB.

The statement that the lengths of AB and CD are in the ratio $1:2$
means that $AB = \frac{1}{2}CD$.

Let the length of $AB = x$ cm, so the length of $CD = 2x$ cm.

$$\text{Area of } ABCD = \frac{1}{2}(AB + CD) \times h = \frac{1}{2}(x + 2x) \times 12$$
$$= \frac{1}{2}(3x) \times 12 = 18x$$

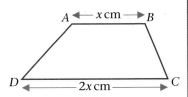

But the area $= 72 \text{ cm}^2$

so $\quad 18x = 72 \quad x = \dfrac{72}{18} = 4 \quad AB = 4 \text{ cm}$

Exercise 13B

1 Work out the area and perimeters of these shapes:

(a)
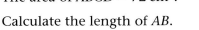
5 cm
10.5 cm

(b)
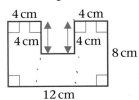

2 (a) Find, in its simplest form, an expression for the perimeter
of this rectangle in terms of x.

(b) Given that the perimeter is 50 cm, calculate the value of x.

(c) Calculate the area of the rectangle.

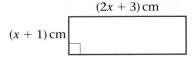

3 The area of a square in cm^2 is numerically equal to the perimeter of the square in cm.

    (a) Calculate the length of a side of the square.

    (b) Calculate the area of the square.

4 A farmer uses exactly 1000 metres of fencing to fence off a square field. Calculate the area of the field.

5 Calculate the area of each of these triangles:

    (a)
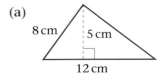

    (b)
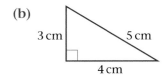

    (c)
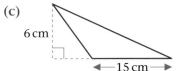

6 **(a)** Write down an expression for the perimeter of the triangle in terms of x.

    (b) If the perimeter of the triangle is 29, calculate the value of x.

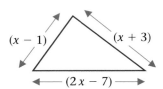

7 Calculate the area of each parallelogram.

    (a)
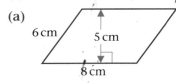

    (b)
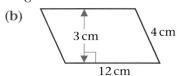

    (c)
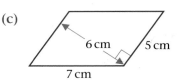

8 $ABCD$ is a parallelogram.

    $BC = AD = 14$ cm $AB = DC = 8$ cm

    The perpendicular distance between AD and BC is 5 cm.

    (a) Calculate the area of the parallelogram.

    (b) Calculate the perpendicular distance between AB and DC.

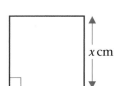

9 The diagram shows a parallelogram and a square. These two shapes have equal areas.

    Calculate the value of x (the side of the square).

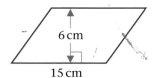

10 Find the shaded area in each diagram.

    (a)
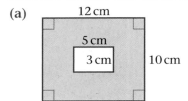

    (b)
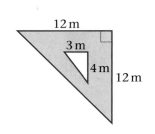

    (c)

11 PQRS is a trapezium with sides PQ and SR parallel. The perpendicular distance between PQ and SR = 8 cm. PQ = 6.4 cm and SR = 8.2 cm. Calculate the area of PQRS.

12 ABCD is a trapezium with sides AB and DC parallel. The perpendicular distance between AB and DC = 18 cm. The lengths of AB and DC are in the ratio

$$AB : DC = 2 : 3$$

The area of ABCD = 54 cm². Calculate the length of AB.

13.4 Circumference and area of circles

Diameter = 2 × radius   or $d = 2r$

Circumference = $2\pi r = \pi d$

Area = $\pi r^2 = \dfrac{\pi d^2}{4}$

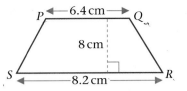

Worked examination question 1

The diagram shows one face of a sheet of metal in the form of a rectangle with a semicircle cut out from one end.

Calculate the area of the face of the sheet of metal.

Radius of semicircle = $\frac{1}{2} \times 0.6 = 0.3$ m

    Shaded area = area of rectangle − area of semicircle

            = $0.6 \times 0.8 - \frac{1}{2}\pi\,(0.3)^2$

            = $0.48 - \frac{1}{2}\pi \times 0.09$

            = $0.48 - 0.141$

            = 0.339 m² (correct to 3 d.p.)

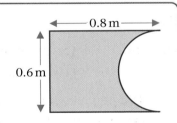

Exercise 13C

1 Calculate the circumference and area of each circle.

(a)

5 cm

(b)

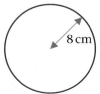

8 cm

(c)

12 cm

2 Mrs de Silva lays out a circular flower bed of diameter 5 metres.

    (a) Calculate the circumference of the flower bed.

    (b) Calculate the area of the flower bed.

3 The circumference of a circle in centimetres is numerically equal to the area of the circle in square centimetres. Show that the radius of the circle must be 2 cm.

4 The circumference of a circle is 15 cm.

    (a) Calculate the radius of the circle.

    (b) Calculate the area of the circle.

5 The area of a circle is 114 cm². Calculate the circumference of the circle.

6 Calculate the area and perimeter of the semicircle

    (a) giving your answer correct to 2 d.p.

    (b) leaving your answer in terms of π.

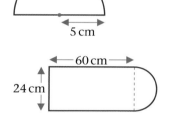

7 The diagram represents a sheet of metal. It consists of a rectangle of length 60 cm and width 24 cm, and a semicircle. Calculate

    (a) the perimeter of the sheet of metal

    (b) the area of the sheet of metal.

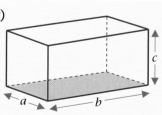

13.5 Volume and surface area of 3-D shapes

Cuboid

Total length around the edges = $4(a + b + c)$

Surface area = $2(ab + ac + bc)$

Volume = abc

In the special case when the cuboid is a cube, then

    $a = b = c$

and

        total distance around the edges = $12a$

              surface area = $6a^2$

                  volume = a^3

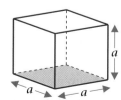

Prisms

A cuboid is a special case of a prism. It is a rectangular prism.

For a general prism:

Surface area = 2 × area of base + total area of vertical faces

Volume = area of base × vertical height
= area of base × h

Example 6

Calculate the height of a prism which has a base area of 25 cm² and a volume of 205 cm³.

Volume = area of base × h

$205 = 25 \times h$

$h = \dfrac{205}{25}$

So $\qquad h = 8.2$ cm

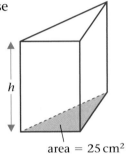

area = 25 cm²

Worked examination question 2 (Part question)

ABCDEF is a triangle-based prism.

The angle $A\hat{B}C = 90°$.

$AB = x$ cm, $BC = (x + 3)$ cm and $CD = 8$ cm.

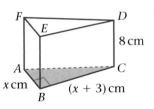

8 cm

x cm $(x + 3)$ cm

The volume of the prism = 40 cm³.

Show that $x^2 + 3x - 10 = 0$

Volume of the prism = area of base × height

Area of base = $\frac{1}{2} \times AB \times BC$

$= \frac{1}{2} x \times (x + 3) = \frac{1}{2} x(x + 3)$

So the volume = $\frac{1}{2} x(x + 3) \times 8 = 4x(x + 3)$

But the volume = 40, so

$4x(x + 3) = 40$

$x(x + 3) = 10$ (dividing through by 4)

$x^2 + 3x = 10$

So, rearranging, or taking 10 from each side:

$x^2 + 3x - 10 = 0$ (what you were asked to show)

Note: a full question such as this in the exam would almost certainly go on to ask you to solve the equation

$x^2 + 3x - 10 = 0$

This is a quadratic equation. Solving quadratic equations is covered in Chapters 17 and 19.

Worked examination question 3

The diagram shows a solid door that is made from a rectangle with a semicircular top.

(a) Work out the area of the cross-section of the door.

(b) The door is made of metal and has a constant thickness of 5 cm.
Work out the volume of the door.

(c) The density of the metal is 7.8 grams per cm³.
Work out the mass of the door.

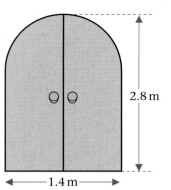

(a) The radius of the semicircle is half of the width of the door, so

radius of semicircle = $\frac{1}{2} \times 1.4 = 0.7$ metres

area of the semicircle = $\dfrac{\pi r^2}{2} = \dfrac{\pi \times 0.7^2}{2} = 0.769\,690\,2$ m²

area of rectangular part = $1.4 \times 2.1 = 2.94$ m²

So the total area of the cross-section of the door is

$2.94 + 0.769\,690\,2 = 3.709\,690\,2 = 3.71$ m²
(to 2 decimal places)

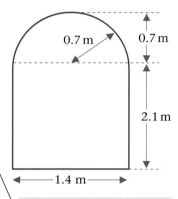

To avoid rounding errors you should not round answers that precede the final answer.

(b) The volume of the door can be found using the formula for the volume of a prism:

volume = area of cross-section × depth (thickness)

The area of the door's cross-section is 3.709 690 2 m².

The thickness of the door is 5 cm.

Before applying any formula you need to make sure your units are consistent. It is meaningless to multiply 3.709 690 2 m² by 5 cm. You need to convert centimetres into metres:

5 cm = 0.05 m

Volume of door = 3.709 690 2 × 0.05
              = 0.19 m³ (correct to 2 decimal places)

(c) The relationship between mass, volume and density is

density = $\dfrac{\text{mass}}{\text{volume}}$ or   mass = density × volume

Again you need to be consistent with the units.
The density is given as 7.8 grams per cm³.
The calculated volume is 0.19 m³.

$$1\,m = 100\,cm$$
so $\quad 1\,m^3 = 100 \times 100 \times 100 = 1\,000\,000\,cm^3$

Hence the volume of the door $= 0.19 \times 1\,000\,000$
$$= 190\,000\,cm^3$$

The mass of the door $=$ density $\times$ volume
$$mass = 7.8 \times 190\,000$$
$$= 1\,500\,000 \text{ grams}$$
(correct to 2 significant figures)

It is sensible to give the most appropriate units.
You can change grams to kilograms by dividing by 1000.

So the mass of the door is $\dfrac{1\,500\,000}{1000} = 1500\,kg$

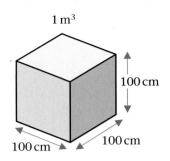

1 m³

100 cm

100 cm   100 cm

**Choosing the right
degree of accuracy**
The density of the metal
was given in the question
to 2 significant figures.
You should give your
answer to the same
degree of accuracy.

Exercise 13D

1 Calculate the volume of a cube of side length
  (a) 5 cm
  (b) 12 cm
  (c) 3.8 cm

2 A cube has a volume of 1000 cm³. Calculate
  (a) the length of a side of the cube
  (b) the surface area of the cube.

3 Calculate the volume of a cuboid with sides of length
  (a) 5 cm, 6 cm and 6 cm
  (b) 4.5 cm, 9.2 cm and 11.6 cm

4 The volume of the cuboid *ABCDEFGH* is 384 cm³.
  The edges *AB*, *BC* and *AF* are in the ratio
  $$AB : BC : AF = 1 : 2 : 3$$
  Calculate
  (a) the length of *AB*
  (b) the surface area of the cuboid.

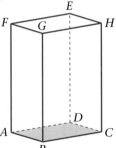

5 The diagrams show a prism and its base, which is a trapezium.

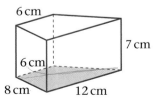
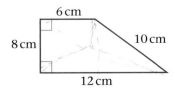

  (a) Calculate the volume of the prism.
  (b) Calculate the surface area of the prism.

6 *ABCDEF* is a triangle-based prism.
Angle *ABC* = 90°, *AB* = 5 cm, *BC* = 8 cm and *CD* = 12 cm.
Calculate the volume of *ABCDEF*.

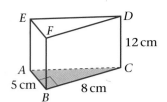

7 *ABCDEF* is a wedge of volume 450 cm³.
Angle *ABC* = 90°, *AB* = 5 cm and *CD* = 15 cm.
Calculate the length of *BC*.

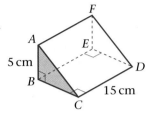

8 This is the base of a prism of vertical height 24 cm.

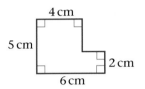 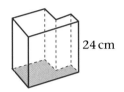

Calculate the volume of the prism.

13.6 1-D, 2-D or 3-D?

The number line goes in one direction
(either horizontally or vertically).

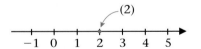

> The number line is **1-dimensional** or **1-D**.
> You can describe position on the number line using
> one number or coordinate, for example (2).

Coordinate grids and flat shapes go in two directions.

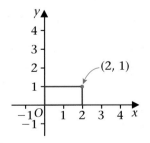

> Flat shapes are **2-dimensional** or **2-D**.
> You can describe position on a flat shape using two
> numbers or coordinates, for example (2, 1).

Solid shapes go in three directions.

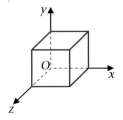

> Solid shapes are **3-dimensional** or **3-D**.
> You can describe position in a solid shape
> using three numbers or coordinates,
> for example (4, 1, 2).

This diagram shows a 3-D grid. The *x*-, *y*- and *z*-axes are all at right angles to each other. The lengths of the edges of the cuboid are 2, 2 and 1 units.

To get to point *P* from *O*, you go 3 units along the *x*-axis, then 1 unit parallel to the *y*-axis, then 2 units parallel to the *z*-axis.

The coordinates of point *P* are (3, 1, 2).

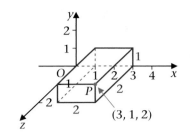

Example 7

The diagram represents a cuboid on a 3-D grid.
OR = 2 units, *OP* = 4 units and *OS* = 3 units.
Find the coordinates of
(a) *S* (b) *P* (c) *R* (d) *V* (e) *U* (f) *O*
(g) Find the mid-point of *PV*.

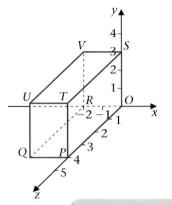

(a) To get to *S* from *O* you go 0 along the *x*-axis,
3 units up the *y*-axis and 0 units parallel
to the *z*-axis.
So *S* = (0, 3, 0)
(b) Similarly *P* = (0, 0, 4)
(c) *R* = (−2, 0, 0)
(d) *V* = (−2, 3, 0)
(e) To get to *U* from *O* you go −2 units along the *x*-axis,
3 units parallel to the *y*-axis and 4 units parallel
to the *z*-axis.
So *U* = (−2, 3, 4)
(f) *O* = (0, 0, 0)
(g) Mid-point of $PV = \left(\dfrac{0 + -2}{2}, \dfrac{0 + 3}{2}, \dfrac{4 + 0}{2}\right)$

$= (-1, 1\frac{1}{2}, 2)$

> Remember to give the coordinates in the order (*x*, *y*, *z*).

> *A* (*x*, *y*) *B* (*p*, *q*)
> Mid-point of *AB*
> $= \dfrac{x + p}{2}, \dfrac{y + q}{2}$

Exercise 13E

1 Say whether each shape is 1-D, 2-D or 3-D.
(a) pentagon (b) pyramid (c) rectangle (d) cone

2 Write down all the 3-dimensional coordinates from this list:
(4)   (2, 1, 3)   (6, 5)   (4, 3, 1)   (6)
(10, 7)   (5, 2, 12)   (8, 2)   (27, 3, 11)   (245)

3 Write down the coordinates of
each vertex of this cuboid:

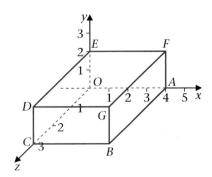

4 (a) Write down the coordinates of each vertex of this cuboid.
   (b) Write down the mid-points of
      (i) *DG*
      (ii) *EB*

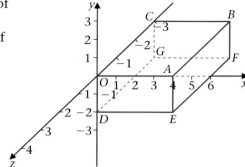

13.7 Dimension theory

It can be easy to forget the formulae for perimeters, areas and volumes or to confuse them.

For instance, the formulae for the circumference and the area of a circle are

$$C = 2\pi r \qquad A = \pi r^2$$

which are somewhat similar. You could easily confuse one with the other.

In the formula for the circumference, r represents the radius of the circle. It has units of length. You say that r has **dimension** = 1

In the formula for the circumference, π is just a number (about equal to 3.14).
It is dimensionless. You say it has dimension = 0.
The 2 in the formula is also just a number.
So the 2 is also dimensionless (dimension = 0).

The formula for the area of a circle is:

$$A = \pi r^2$$

Here π is again dimensionless.

The r^2 part is length × length (it has units of area).
So r^2 has dimension = 2.

___ **Example 8** _____

Check the dimensions of $V = abc$ for the volume of a cuboid.

a, b and c each have dimension = 1

So abc has dimension = 3, which is correct for a volume.

> An incorrect formula may be dimensionally correct. Dimension theory only helps you rule out *some* incorrect formulae!

All expressions for length or distance have dimension = 1.
All expressions for area have dimension = 2.
All expressions for volume have dimension = 3.
Numbers (like π and 2) are dimensionless.

This is called **dimension theory**. It gives a simple method for checking that you may have the correct formula for a given situation, since a correct formula will have the same dimensions on both sides.

Example 9

Yvonne is using a formula sheet to work out the volume and the area of the curved surface of a cylinder. Unfortunately a blot on the sheet covers up part of the information.

Her sheet looks like this:

$$= \pi r^2 h$$
$$= 2\pi rh$$

Yvonne knows that one of these gives the volume and the other the area of the curved surface.

How could she tell which formula is which?

Volume has dimension = 3.
In the expression $\pi r^2 h$, π is a number, r is a length and h is a length.
π is dimensionless. r^2 has dimension = 2. h has dimension = 1.
So $\pi r^2 h$ has dimension = 3.
So this is a formula for a volume.

In the expression $2\pi rh$,
2 and π are dimensionless whilst r and h each have dimension = 1.
So $2\pi rh$ has dimension = 2 and is an expression for an area.

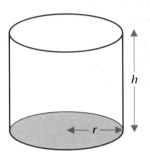

You should be able to use **dimension theory** to confirm whether formulae are for length, area or volume:

- length has dimension = 1
- area has dimension = 2
- volume has dimension = 3

Exercise 13F

1 Here are two expressions related to a sphere of radius r:

$$4\pi r^2 \quad \text{and} \quad \frac{4\pi r^3}{3}$$

One of these gives the volume of a sphere and the other the surface area of the sphere. State, with an explanation, which formula gives the surface area of a sphere.

2   The volume of a solid is given by $V = \lambda a^n b^m$, where λ is dimensionless, a and b have dimensions of length and n and m are non-negative integers. State the possible values of n and m.

3   The table below shows some expressions.
The letters x, y and z represent lengths.

Copy the table. Place a tick in the appropriate column for each expression to show whether the expression can be used to represent a length, an area, a volume or none of these.

Expression	Length	Area	Volume	None of these
$x + y + z$				
xyz				
$xy + yz + xz$				

4   Here are three expressions.

Expression	Length	Area	Volume	None of these
$\pi a^2 b$				
$\pi b^2 + 2h$				
$2ah$				

a, b and h are lengths. π and 2 are numbers and have no dimensions.

Copy the table and place a tick in the correct column to show whether each expression can be used for length, area, volume or none of these.

Mixed exercise 13

1   The area of a pane of glass is $12\,000 \text{ cm}^2$.
Work out the area of this pane of glass in m^2.

2   The volume of a large packing case is 3.4 m^3.
Work out the volume of this packing case in cm^3.

3   Here is a cuboid. The rectangular base has width 4 m and length 5 m. The height is 300 cm. The dimensions are all quoted correct to the nearest metre.

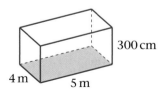

   (a)   Calculate the maximum possible volume of the cuboid.

   (b)   Calculate the maximum possible total surface area.

4   The diagram represents the plan of a sports field. The field is in the form of a rectangle with semicircular pieces at each end.

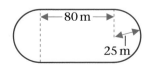

Calculate

   (a)   the perimeter of the field   (b)   the area of the field.

5 Calculate the area of the trapezium *ABCD*.

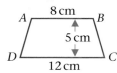

6 The circumference of a circle is 44 cm.
Calculate the area of the circle.

7 Calculate the volume and surface
area of the wedge *ABCDEF*
in which

Angle $A\hat{B}C = 90°$

$AB = 5$ cm, $AC = 13$ cm,
$BC = 12$ cm   and $CD = 20$ cm

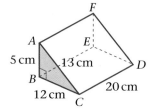

8 The cuboid has a square base of side length *x* cm.
The height of the cuboid is *h* cm.
The volume of the cuboid is 50 cm³.

(a) Show that $x = \sqrt{\dfrac{50}{h}}$.

(b) Calculate the value of *x* when *h* = 4.

(c) Calculate the value of *h* when *x* = 5.

9 (a) List the coordinates of all the
vertices of the cuboid.

(b) Write down the coordinates
of the mid-point of

(i) *AD*
(ii) *FC*

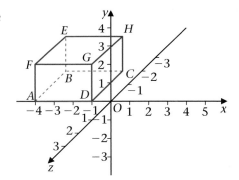

10 In the expression $\mu a^2 h$, μ is dimensionless and *a* and *h* have
dimensions of length.
Explain whether the expression represents a length, an area or
a volume.

11 An estate agent quotes the dimensions of a rectangular floor of a
room as 4.3 metres by 3.7 metres, both correct to 10 cm.
(a) Write down the maximum and minimum possible values
for each of these dimensions.
(b) Calculate the maximum and minimum possible values for
the area of the floor, giving your answers correct to three
significant figures in each case.

Summary of key points

1   Perimeter of a rectangle $= 2(a + b)$
Area of a rectangle $= a \times b = ab$

2   Perimeter of a triangle $= a + b + c$
Area of a triangle $= \frac{1}{2} \times$ base $\times$ height
$= \frac{1}{2}bh$

3   Perimeter of a parallelogram $= 2(a + b)$
Area of a parallelogram $= bh$

4   Perimeter of a trapezium = sum of the lengths of the four sides
Area of a trapezium $= \frac{1}{2}(a + b)h$

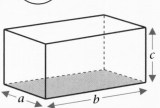

5   Diameter of a circle $= d = 2 \times$ radius $= 2r$
Circumference $= 2\pi r = \pi d$
Area of a circle $= \pi r^2 = \dfrac{\pi d^2}{4}$

6   Total length around the edges of a cuboid $= 4(a + b + c)$
Surface area of a cuboid $= 2(ab + ac + bc)$
Volume of a cuboid $= abc$

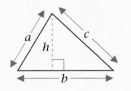

7   Surface area of a prism $= 2 \times$ area of base + total area of vertical faces
Volume of a prism = area of base $\times h$

8   You should be able to use **dimension theory** to confirm whether formulae are for length, area or volume:
- length has dimension $= 1$
- area has dimension $= 2$
- volume has dimension $= 3$

14 Simplifying algebraic expressions

In this chapter you will extend your skills in simplifying algebraic expressions, including dealing with indices and algebraic fractions.

14.1 Index notation

$x \times x$ is usually written as x^2 and is read as 'x squared'.

$x \times x \times x$ is usually written as x^3 and is read as 'x cubed'.

When x is multiplied by itself any number of times, you say it is raised to a 'power' or 'index'. For example, $x \times x \times x \times x \times x$ is usually written as x^5 and is read as 'x to the power of five' or 'x to the fifth'.

> In the expression x^n the number x is called the **base** and the number n is called the **index** or power.

x^1, read as 'x to the power of one', is usually written just as x.

> $x^1 = x$

The mathematical plural of index is **indices**.

14.2 Multiplying expressions involving indices

Multiplying $x^m \times x^n$

In number:

$$4^2 \times 4^3 = 4^{2+3} = 4^5$$

For more on this, see Chapter 1.

In algebra:

$$x^2 \times x^3 = (x \times x) \times (x \times x \times x)$$
$$= x^5$$

So $\quad x^2 \times x^3 = x^{2+3} = x^5$

For more on this, see Chapter 2.

In general, the rule for multiplying algebraic expressions involving powers is

> $x^m \times x^n = x^{m+n}$

You may find it easier to remember this rule as: 'To multiply numbers which have the same base, add the indices'.

Example 1

Simplify $2x^3 \times 5x^4$.

$$2x^3 \times 5x^4 = 2 \times x^3 \times 5 \times x^4$$
$$= 2 \times 5 \times x^3 \times x^4$$
$$= 10 \times x^{3+4}$$
$$= 10x^7$$

> A common error is to forget that the numbers must have the same base. For example:
> $$2^3 \times 3^2 \neq 6^5$$
> ($\neq$ means 'is **not equal** to').

The rule for multiplying algebraic expressions involving powers can also be used with more complicated expressions like this:

Example 2

Simplify $4a^2b^3 \times 3ab^2 \times 2b$.

$$4a^2b^3 \times 3ab^2 \times 2b = 4 \times a^2 \times b^3 \times 3 \times a^1 \times b^2 \times 2 \times b^1$$
$$= 4 \times 3 \times 2 \times a^2 \times a^1 \times b^3 \times b^2 \times b^1$$
$$= 24 \times a^{2+1} \times b^{3+2+1}$$
$$= 24a^3b^6$$

Multiplying out $(x^m)^n$

$(x^2)^3$ means $(x^2) \times (x^2) \times (x^2)$ or $x^2 \times x^2 \times x^2$

so $(x^2)^3 = x^{2+2+2} = x^6 = x^{2 \times 3}$

Similarly $(x^3)^4$ means $x^3 \times x^3 \times x^3 \times x^3 = x^{3+3+3+3} = x^{12} = x^{3 \times 4}$

In general, the rule for multiplying algebraic expressions involving powers is

$$(x^m)^n = x^{m \times n} = x^{mn}$$

Example 3

Simplify   (a) $(x^2)^6$ (b) $(xy^2)^3$ (c) $(2x^3)^5$

(a) $(x^2)^6 = x^{2 \times 6} = x^{12}$

(b) $(xy^2)^3 = (x)^3 \times (y^2)^3 = x^3 \times y^{2 \times 3} = x^3 \times y^6 = x^3y^6$

(c) $(2x^3)^5 = (2)^5 \times (x^3)^5 = 32 \times x^{3 \times 5} = 32 \times x^{15} = 32x^{15}$

Exercise 14A

> Simplify $x + 2x + 3x$ first.

Simplify these expressions.

1	$x^3 \times x^5$	**2**	$y^4 \times y^4$	**3**	$6x \times x^6$	**4**	$a^3 \times 5a$
5	$2a^2 \times 3a \times 4a^3$	**6**	$2y^3 \times 4y^2 \times 6$	**7**	$2x^2 \times 5x^3y^2$	**8**	$3a^2b^3 \times 2a^3 \times 4b^2$
9	$ab \times abc^2$	**10**	$2a^2b \times 3b^2c^3 \times 3c^2$	**11**	$2x^2 \times 3y^3$	**12**	$(x^2)^4$
13	$(3a^2)^2$	**14**	$(2x^2)^3$	**15**	$2(a^4)^5$	**16**	$2(xy^2)^3 \times x$
17	$(3x^4)^2 \times 7$	**18**	$7(a^2)^3 \times 2a^3b^2 \times (ab)^4$	**19**	$(x + 2x + 3x)^2$	**20**	$(x \times x + 2x^2)^3$

14.3 Dividing expressions involving indices

Chapter 1 shows you how to divide numbers in index form. For example

$$5^6 \div 5^4 = 5^2$$

In the same way:

$$x^6 \div x^4 = \frac{\cancel{x} \times \cancel{x} \times \cancel{x} \times \cancel{x} \times x \times x}{\cancel{x} \times \cancel{x} \times \cancel{x} \times \cancel{x}}$$

$$= x^2$$

In general: $x^m \div x^n = x^{m-n}$

You may find it easier to remember this rule as 'To divide numbers which have the same base, subtract the indices'.

Example 4

Simplify   (a) $x^6 \div x^2$ (b) $8a^5 \div 4a^3$ (c) $15x^4y^3 \div 3xy^2$

(a) $x^6 \div x^2 = x^{6-2} = x^4$ (b) $8a^5 \div 4a^3 = \frac{8}{4}a^{5-3} = 2a^2$

(c) $15x^4y^3 \div 3xy^2 = 15x^4y^3 \div 3x^1y^2 = \frac{15}{3}x^{4-1}y^{3-2} = 5x^3y^1 = 5x^3y$

Exercise 14B

Simplify these expressions.

1 $x^7 \div x^4$ **2** $6a^5 \div 2a^2$ **3** $18x^4y^3 \div 3x^3$ **4** $x^2 \div x^4$

5 $a^7 \div a$ **6** $6a^5 \div 4xa^2$ **7** $12x^4 \div 3x^4$ **8** $x^4 \div x^2$

9 $15q^5 \div 3q^3$ **10** $6r^5 \div 3r$ **11** $8e^8 \div 4e^4$ **12** $12p^8 \div 4p^3$

14.4 Zero, negative and fractional indices

The zero index

Here is how to find the value of an expression to the power zero.
2^0 can be written as:

$$2^3 \div 2^3 = 2^{3-3} = 2^0$$

But $2^3 \div 2^3 = \dfrac{2^3}{2^3} = 1$

so $2^0 = 1$

Putting $n = m$ in the rule for dividing algebraic expressions involving powers gives

$$x^m \div x^m = x^{m-m}$$

So $\dfrac{x^m}{x^m} = x^0$

$$1 = x^0$$

$x^0 = 1$ for all non-zero values of x.

gives 'Error' on most calculators.

What does your calculator display if you input

(a) a small negative number instead of the first 0

(b) a small positive decimal number instead of the second 0?

Negative indices

You can also find the value of an expression to a negative power.
For example, 6^{-2} can be written as

$$6^3 \div 6^5 = 6^{3-5} = 6^{-2}$$

So $\dfrac{\cancel{6} \times \cancel{6} \times \cancel{6}}{\cancel{6} \times \cancel{6} \times \cancel{6} \times 6 \times 6} = 6^{-2}$

or $\dfrac{1}{6^2} = 6^{-2}$

Putting $m = 0$ in the rule for dividing algebraic expressions involving powers gives

$$x^0 \div x^n = x^{0-n}$$

So $\dfrac{x^0}{x^n} = x^{-n}$

which gives $\dfrac{1}{x^n} = x^{-n}$

$$x^{-n} = \dfrac{1}{x^n}$$

> **Remember:**
> $$x^0 = 1$$
> so $\dfrac{x^0}{x^n} = \dfrac{1}{x^n}$

> The reciprocal of
> $$x^n = 1 \div x^n = \dfrac{1}{x^n}$$

This result can be written in words as: x^{-n} means the **reciprocal** of x^n.

Fractional indices

To see what a fractional power means, compare these expressions involving square roots.

$$\sqrt{25} \times \sqrt{25} = 25$$

$$\sqrt{x} \times \sqrt{x} = x = x^1$$

and $x^{\frac{1}{2}} \times x^{\frac{1}{2}} = x^{\frac{1}{2} + \frac{1}{2}} = x^1$

So $x^{\frac{1}{2}} = \sqrt{x}$

Similarly

$$\sqrt[3]{x} \times \sqrt[3]{x} \times \sqrt[3]{x} = x$$

and $x^{\frac{1}{3}} \times x^{\frac{1}{3}} \times x^{\frac{1}{3}} = x$

So $x^{\frac{1}{3}} = \sqrt[3]{x}$

> **Remember:**
> $\sqrt[3]{}$ means 'cube root'.

In general:

$$x^{\frac{1}{n}} = \sqrt[n]{x}$$

Example 5

Find the value of $8^{-\frac{1}{3}}$

$$8^{-\frac{1}{3}} = \dfrac{1}{8^{\frac{1}{3}}} = \dfrac{1}{\sqrt[3]{8}} = \dfrac{1}{2}$$

Exercise 14C

Find the value of these expressions.

1 $25^{\frac{1}{2}}$ **2** $16^{\frac{1}{4}}$ **3** $64^{\frac{1}{3}}$ **4** $49^{-\frac{1}{2}}$ **5** $81^{-\frac{1}{4}}$

6 $216^{\frac{1}{3}}$ **7** $125^{-\frac{1}{3}}$ **8** $100^{-\frac{1}{2}}$ **9** $(\frac{1}{4})^{\frac{1}{2}}$ **10** $(0.064)^{\frac{1}{3}}$

14.5 Combining indices

You can use the rule for multiplying algebraic expressions involving indices to show that

$$(\sqrt[n]{x})^m = (x^{\frac{1}{n}})^m = x^{\frac{1}{n} \times m} = x^{\frac{m}{n}}$$

and that $\quad \sqrt[n]{x^m} = (x^m)^{\frac{1}{n}} = x^{m \times \frac{1}{n}} = x^{\frac{m}{n}}$

$$\frac{1}{n} \times m = \frac{m}{n}$$

$$x^{\frac{m}{n}} = (\sqrt[n]{x})^m \quad \text{or} \quad x^{\frac{m}{n}} = \sqrt[n]{x^m}$$

Example 6

Find the value of $81^{\frac{3}{4}}$

Here $m = 3$ and $n = 4$. Substituting in the rule above gives

$$81^{\frac{3}{4}} = (\sqrt[4]{81})^3 = (3)^3 = 27$$

Example 7

Simplify   (a) $(25x^2)^{\frac{1}{2}}$ (b) $(8x^6)^{-\frac{1}{3}}$

(a) $(25x^2)^{\frac{1}{2}} = (25)^{\frac{1}{2}}(x^2)^{\frac{1}{2}}$

$\qquad = \sqrt{25}(x^{2 \times \frac{1}{2}})$

$\qquad = 5x^1$

$\qquad = 5x$

(b) $(8x^6)^{-\frac{1}{3}} = (8)^{-\frac{1}{3}}(x^6)^{-\frac{1}{3}}$

$\qquad = \frac{1}{\sqrt[3]{8}}(x^{6 \times (-\frac{1}{3})})$

$\qquad = \frac{1}{2}x^{-2} \quad \text{or} \quad \frac{1}{2x^2}$

Exercise 14D

1 Find the value of

(a) $8^{\frac{1}{3}}$ (b) $1000^{-\frac{2}{3}}$ (c) $32^{\frac{2}{5}}$ (d) $(0.25)^{-\frac{5}{2}}$ (e) $27^{\frac{4}{3}}$

2 Find the value of

(a) $25^{\frac{3}{2}}$ (b) $16^{\frac{5}{4}}$ (c) $4^{\frac{5}{2}}$ (d) $8^{-\frac{2}{3}}$

(e) $(0.125)^{-\frac{5}{3}}$ (f) $125^{\frac{2}{3}}$ (g) $(\frac{1}{8})^{\frac{2}{3}}$ (h) $64^{-\frac{5}{6}}$

(i) $1296^{\frac{1}{4}}$ (j) $(\frac{27}{8})^{\frac{2}{3}}$ (k) $(\frac{343}{512})^{-\frac{2}{3}}$ (l) $25^{-\frac{3}{2}}$

3 Simplify

(a) $(3x^0)^2$

(b) $a^{-2} \div \dfrac{1}{a^4}$

(c) $(\sqrt{xy})^3 \times y$

(d) $(9x^2)^{\frac{1}{2}}$

(e) $(8x^{-3})^{\frac{1}{3}}$

(f) $(25a^{-4})^{-\frac{1}{2}}$

(g) $(81x^2)^{\frac{3}{2}}$

(h) $32\sqrt{x} \times (16x^2)^{-\frac{1}{4}}$

(i) $4^{\frac{3}{2}} \times (9y^4)^{\frac{1}{2}}$

14.6 Equations involving indices

Example 8

Find the value of k in

(a) $x^k = \dfrac{1}{x^{-2}}$

(b) $x^k = \sqrt{x} \div \dfrac{1}{x^3}$

(c) $x^k = \dfrac{x \times \sqrt[3]{x^4}}{(\sqrt[4]{x})^2}$

(d) $4^{k+1} = 8$

(a) $\quad x^k = \dfrac{1}{x^{-2}}$

$\quad\quad = (x^{-2})^{-1}$

Using the rule for multiplying out $(x^m)^n$ gives

$\quad\quad x^k = x^2$

So $\quad k = 2$ —————— Comparing indices.

(b) $\quad x^k = \sqrt{x} \div \dfrac{1}{x^3}$

$\quad\quad = x^{\frac{1}{2}} \div x^{-3}$

$\quad\quad = x^{\frac{1}{2} - (-3)}$

$\quad\quad = x^{\frac{1}{2} + 3} = x^{3\frac{1}{2}}$

So $\quad k = 3\frac{1}{2}$

(c) $\quad x^k = \dfrac{x \times \sqrt[3]{x^4}}{(\sqrt[4]{x})^2}$

$\quad\quad = \dfrac{x^1 \times x^{\frac{4}{3}}}{x^{\frac{2}{4}}}$

$\quad\quad = \dfrac{x^{1 + \frac{4}{3}}}{x^{\frac{1}{2}}}$

$\quad\quad = \dfrac{x^{\frac{7}{3}}}{x^{\frac{1}{2}}}$

$\quad\quad = x^{\frac{7}{3} - \frac{1}{2}}$

$\quad\quad = x^{\frac{11}{6}}$

So $\quad k = \frac{11}{6}$

(d) $\quad\quad 4^{k+1} = 8$

$\quad\quad = 4 \times 2$

$\quad\quad = 4 \times \sqrt{4}$

$\quad\quad = 4^1 \times 4^{\frac{1}{2}}$

$\quad\quad = 4^{1\frac{1}{2}}$

So $\quad k + 1 = 1\frac{1}{2}$

$\quad\quad k = \frac{1}{2}$

Alternative method:
Because 4 and 8 are multiples of 2, you can also answer part **(d)** like this:

$4^{k+1} = 8$

$(2^2)^{k+1} = (2)^3$

$2^{2k+2} = 2^3$

$2k + 2 = 3$

$2k = 1$

$k = \frac{1}{2}$

Exercise 14E

1 Find the value of k in

(a) $x^k = \sqrt{x}$

(b) $y^k = 1$

(c) $a^k = 1 \div a^3$

(d) $x^k = \sqrt[3]{x^2}$

(e) $y^k = y^2\sqrt{y^3}$

(f) $a^k = \sqrt{a} \div \dfrac{1}{a^4}$

(g) $x^{k+1} = (x^{-2})^{-3}$

(h) $2^k = 16$

(i) $3^k = 27$

(j) $25^k = 5$

(k) $9^k = 27$

(l) $4^{\frac{k}{2}} = 8$

14.7 Algebraic fractions

$\dfrac{9x^4}{6x}$ and $\dfrac{3(x + 8)}{3x}$ are examples of **algebraic fractions**.

Fractions can be simplified by cancelling *only* if the numerator and the denominator have a common factor.

Example 9

Simplify:

(a) $\dfrac{9x^4}{6x}$

(b) $\dfrac{3(x + 8)}{3x}$

(a) $\dfrac{9x^4}{6x} = \dfrac{3 \times \cancel{3} \times \cancel{x} \times x^3}{2 \times \cancel{3} \times \cancel{x}} = \dfrac{3x^3}{2}$

(b) $\dfrac{3(x + 8)}{3x} = \dfrac{\cancel{3} \times (x + 8)}{\cancel{3} \times x} = \dfrac{x + 8}{x}$

> Check this answer using the rule for dividing indices.

> You can't cancel the x's here as x isn't a common factor of x and $(x + 8)$.

Multiplying and dividing algebraic fractions

When you multiply algebraic fractions, you should cancel any factor which is common to both numerator and denominator. When no further cancelling is possible you can then multiply the remaining numerators and denominators.

Example 10

Simplify $\dfrac{x^4}{y} \times \dfrac{xy}{z} \times \dfrac{z}{x^3}$

$\dfrac{x^4}{y} \times \dfrac{xy}{z} \times \dfrac{z}{x^3} = \dfrac{x \times \cancel{x^3}}{\cancel{y}} \times \dfrac{x \times \cancel{y}}{\cancel{z}} \times \dfrac{\cancel{z}}{\cancel{x^3}}$ ——————— Cancel by x^3, then by y, then by z.

$= \dfrac{x}{1} \times \dfrac{x}{1} \times \dfrac{1}{1} = \dfrac{x \times x \times 1}{1 \times 1 \times 1} = \dfrac{x^2}{1} = x^2$

> Dividing by an algebraic fraction is equivalent to multiplying by its reciprocal.
> For example, to divide by $\dfrac{a}{b}$ you can multiply by $\dfrac{b}{a}$.

To see why this works, notice that

$$1 = \dfrac{a}{b} \times \dfrac{b}{a}$$

Dividing both sides by $\dfrac{a}{b}$ gives

$$1 \div \dfrac{a}{b} = 1 \times \dfrac{b}{a}$$

So dividing 1 by $\dfrac{a}{b}$ is the same as multiplying 1 by the reciprocal of $\dfrac{a}{b}$.

> The reciprocal of $\dfrac{a}{b}$ is $\dfrac{b}{a}$.

Example 11

Simplify $3x \div \dfrac{a}{b}$

$$3x \div \frac{a}{b} = \frac{3x}{1} \times \frac{b}{a} = \frac{3xb}{a}$$

Example 12

Simplify $\dfrac{x^2(x+2)}{y^2} \div \dfrac{x(x+4)}{y^3}$

$$\frac{x^2(x+2)}{y^2} \div \frac{x(x+4)}{y^3} = \frac{x^2(x+2)}{y^2} \times \frac{y^3}{x(x+4)}$$ ——— Multiply by the reciprocal.

$$= \frac{\cancel{x} \times x \times (x+2)}{\cancel{y^2}} \times \frac{\cancel{y^2} \times y}{\cancel{x} \times (x+4)}$$ ——— Cancel by x and then by y^2.

$$= \frac{x(x+2)}{1} \times \frac{y}{x+4} = \frac{xy(x+2)}{x+4}$$ ——— No further cancelling is possible here.

Exercise 14F

Simplify these expressions.

1 $\dfrac{xy}{4y}$

2 $\dfrac{abc}{acd}$

3 $\dfrac{24x^2}{6x}$

4 $\dfrac{8st}{6sr}$

5 $\dfrac{12a^2}{b^3} \times \dfrac{b^4}{4a}$

6 $\dfrac{4(x-2)}{2(x-2)}$

7 $\dfrac{2(x+2)}{4x}$

8 $\dfrac{9(x+4)}{3(x+4)(x-4)}$

9 $\dfrac{y^3(y-2)}{x^2} \div \dfrac{y(y-6)}{x^5}$

14.8 Adding and subtracting algebraic fractions

To add or subtract algebraic fractions you can use the same methods as for ordinary numerical fractions.

Both fractions must have the same denominator. If they do not, find their lowest common multiple (LCM) and change each fraction to an equivalent fraction with this denominator.

> For more on this, see Chapter 4.

Here is a numerical example as a reminder.

To add $\frac{1}{3}$ and $\frac{1}{5}$ find the lowest common multiple of 3 and 5. The LCM is 15.

Change each fraction to an equivalent fraction with this denominator:

$$\frac{1}{3} \xlongequal{\times 5} \frac{5}{15} \qquad \frac{1}{5} \xlongequal{\times 3} \frac{3}{15}$$

So $\frac{1}{3} + \frac{1}{5}$ is equivalent to

$$\frac{5}{15} + \frac{3}{15} = \frac{8}{15}$$

In general, to add two fractions $\dfrac{1}{n} + \dfrac{1}{m}$ you can change them to equivalent fractions with denominator $m \times n$.

$$\frac{1}{n} + \frac{1}{m} = \frac{m+n}{mn}$$

The following examples show you how to add two algebraic fractions, but a similar process can be used to subtract or add three or more fractions.

Example 13

Write as a single fraction in its lowest terms:

(a) $\dfrac{2}{3} + \dfrac{1}{4}$
(b) $\dfrac{2}{y} + \dfrac{1}{z}$
(c) $\dfrac{2}{(x+1)} + \dfrac{1}{(x+2)}$

(a) $\dfrac{2}{3} + \dfrac{1}{4}$

$= \dfrac{2 \times 4}{3 \times 4} + \dfrac{1 \times 3}{4 \times 3}$

$= \dfrac{8}{12} + \dfrac{3}{12}$

$= \dfrac{8+3}{12}$

$= \dfrac{11}{12}$

(b) $\dfrac{2}{y} + \dfrac{1}{z}$

$= \dfrac{2 \times z}{y \times z} + \dfrac{1 \times y}{z \times y}$

$= \dfrac{2z}{yz} + \dfrac{y}{zy}$

$= \dfrac{2z + y}{yz}$

(c) $\dfrac{2}{(x+1)} + \dfrac{1}{(x+2)}$

$= \dfrac{2(x+2)}{(x+1)(x+2)} + \dfrac{1(x+1)}{(x+2)(x+1)}$

$= \dfrac{2x+4}{(x+1)(x+2)} + \dfrac{x+1}{(x+2)(x+1)}$

$= \dfrac{2x+4+x+1}{(x+1)(x+2)}$

$= \dfrac{3x+5}{(x+1)(x+2)}$

When the denominators have no common factor, the LCM is the product of all the denominators.

The next example shows you how to find the LCM when the denominators have a common factor, by writing the denominators in terms of their prime factors.

For more on prime factors see Section 1.1.

Example 14

Write as a single fraction:

(a) $\dfrac{5}{12} - \dfrac{2}{9}$

(b) $\dfrac{5}{x(x+1)} - \dfrac{2}{x^2}$

In both examples, the denominators have a common factor.

To find the LCM first split each denominator into its prime factors.

Then write the denominator of the first fraction and multiply it by the 'remaining' factors of the other denominators.

(a) $\dfrac{5}{12} - \dfrac{2}{9}$

(b) $\dfrac{5}{x(x+1)} - \dfrac{2}{x^2}$

> $12 = 2 \times 2 \times 3$ and $9 = 3 \times 3$
> The LCM is $2 \times 2 \times 3 \times 3 = 36$

> $x(x+1) = x \times (x+1)$ and $x^2 = x \times x$
> The LCM is $x \times (x+1) \times x = x^2(x+1)$

$= \dfrac{5 \times 3}{12 \times 3} - \dfrac{2 \times 4}{9 \times 4}$

$= \dfrac{5 \times x}{x(x+1) \times x} - \dfrac{2(x+1)}{x^2(x+1)}$

$= \dfrac{15}{36} - \dfrac{8}{36}$

$= \dfrac{5x}{x^2(x+1)} - \dfrac{2(x+1)}{x^2(x+1)}$

$= \dfrac{15 - 8}{36}$

$= \dfrac{5x - 2(x+1)}{x^2(x+1)}$

$= \dfrac{7}{36}$

$= \dfrac{5x - 2x - 2}{x^2(x+1)}$

$= \dfrac{3x - 2}{x^2(x+1)}$

Checking your solutions

In the GCSE exam you should check your algebra solutions by choosing suitable numbers for the unknown (usually x). When checking work on algebraic fractions, do not choose a value of x which makes a denominator zero.

For example, in part **(b)** of Example 14 you could choose $x = 1$ to check.

This makes the expression in the question $\dfrac{5}{2} - 2 = \dfrac{1}{2}$ which checks

with the expression in the answer $\dfrac{3 - 2}{1^2(1 + 1)} = \dfrac{1}{2}$.

Exercise 14G

1 Write each expression as a single fraction in its lowest terms.

(a) $\dfrac{x}{5} + \dfrac{2x}{5}$

(b) $\dfrac{1}{x} + \dfrac{1}{y}$

(c) $\dfrac{2}{x} + \dfrac{3}{5x}$

(d) $\dfrac{2x}{3} + \dfrac{x}{4}$

(e) $\dfrac{a-1}{3} + \dfrac{a+2}{4}$

(f) $\dfrac{x+3}{2} - \dfrac{2x-1}{5}$

(g) $\dfrac{2}{x+1} - \dfrac{1}{x+2}$

(h) $\dfrac{1}{x} + \dfrac{1}{2-x}$

(i) $\dfrac{1}{2} + \dfrac{2}{y-5}$

2 Find the LCM for the denominators

(a) 6 and 9

(b) x^2 and $x(x-1)$

(c) $4(x-1)(x+1)$ and $6(x+1)$

(d) $x^3(1-x)$ and $x^2(1+x)$

3 Write each expression as a single fraction in its lowest terms.

(a) $\dfrac{a}{6} + \dfrac{a}{9}$

(b) $\dfrac{2}{x^2} - \dfrac{1}{x(x-1)}$

(c) $\dfrac{x}{4(x-1)(x+1)} - \dfrac{1}{6(x+1)}$

(d) $\dfrac{2}{x^3(1-x)} - \dfrac{1}{x^2(1+x)}$

4 Winston was asked to write
$$\dfrac{6}{(x-2)} - \dfrac{4}{(x+2)}$$
as a single fraction in its lowest terms. In his solution he has made two mistakes.

Find where these errors occur and write a correct solution.

$$\dfrac{6}{(x-2)} - \dfrac{4}{(x+2)} \qquad LCM = (x-2)(x+2)$$

$$= \dfrac{6(x+2)}{(x-2)(x+2)} - \dfrac{4(x-2)}{(x+2)(x-2)} = \dfrac{6(x+2)-4(x-2)}{(x-2)(x+2)}$$

$$= \dfrac{6x+12-4x-8}{(x-2)(x+2)} = \dfrac{2x+4}{(x-2)(x+2)} = \dfrac{2(x+2)}{(x-2)(x+2)}$$

$$= \dfrac{\overset{1}{2}}{x-\underset{1}{2}} = \dfrac{1}{x-1}$$

14.9 Factorising

Chapter 2 introduces factorisation as the opposite of expanding brackets. For example, here is how to factorise $2a^2 + 6a$:

$$2a^2 + 6a = 2 \times a \times a + 2 \times 3 \times a$$

The highest common factor (HCF) of the two terms is $2 \times a$, so $2a$ will go outside the brackets:

$$2a^2 + 6a = 2a(a+3)$$

This section shows you how to factorise more complicated algebraic expressions. In each case the first step is to 'take out' the HCF.

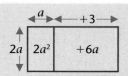

Another way of thinking about this is to see the factors as the lengths of the sides of rectangles. Products such as $2a^2$ are the areas of rectangles.

For a reminder on how to find the HCF see Section 1.1.

Example 15

Factorise the expression $14ax^3 + 7x^2$.

$14ax^3 + 7x^2$ ———————— Look for common factors of the terms $14ax^3$ and $7x^2$.

$= 7x^2(2ax + 1)$ ———————— $7x^2$ is a common factor so it can be taken outside the bracket.

Example 16

Factorise the expression $cd^2 + 5c$.

$cd^2 + 5c$ ———————— Look for common factors of the terms cd^2 and $5c$.

$= c(d^2 + 5)$ ———————— c is a common factor so it can be taken outside the bracket.

Exercise 14H

Factorise each of these expressions.

1 (a) $8x^2 + 4x$ (b) $6p^2 + 3p$ (c) $6x^2 - 3x$

   (d) $3b^2 - 9b$ (e) $12a + 3a^2$ (f) $15c - 10c^2$

   (g) $21x^4 + 14x^3$ (h) $16y^3 - 12y^2$ (i) $6d^4 - 4d^2$

2 (a) $ax^2 + ax$ (b) $pr^2 - pr$ (c) $ab^2 - ab$

   (d) $qr^2 - q^2$ (e) $a^2x + ax^2$ (f) $b^2y - by^2$

   (g) $6a^3 - 9a^2$ (h) $8x^3 - 4x^4$ (i) $18x^3 + 12x^5$

3 (a) $12a^2b + 18ab^2$ (b) $4x^2y - 2xy^2$

   (c) $4a^2b + 8ab^2 + 12ab$ (d) $4x^2y + 6xy^2 - 2xy$

   (e) $12ax^2 + 6a^2x - 3ax$ (f) $a^2bc + ab^2c + abc^2$

Don't forget to check your solutions.

14.10 Factorising by pairing

An expression such as $ab + cd + bc + ad$ can be factorised by pairing terms that have common factors:

$ab + cd + bc + ad = ab + bc + ad + cd$ ———————— Put terms in pairs that have common factors.

$= b(a + c) + d(a + c)$ ———————— Bracket term must be a factor of both pairs.

$= (a + c)(b + d)$ ———————— $(a + c)$ is a facor of both pairs.

                   $(a + c)(b + d)$ is the same as $(b + d)(a + c)$.

Example 17

Factorise $5a(x - y) + (y - x)$ completely.

$5a(x - y) + (y - x) = 5a(x - y) - 1(x - y)$ ———————— $(y - x) = -1(x - y)$

$= (x - y)(5a - 1)$ ———————— $(x - y)$ is a common factor.

Example 18

Factorise $2ab - 12cd + 6ac - 4bd$ completely.

$2ab - 12cd + 6ac - 4bd$

$\quad = 2[ab - 6cd + 3ac - 2bd]$ ———— 2 is the HCF.

$\quad = 2[ab + 3ac - 2bd - 6cd]$ ———— Group in pairs that have a common factor.

$\quad = 2[a(b + 3c) - 2d(b + 3c)]$ ———— Factorise in pairs to get $(b + 3c)$ as a common factor.

$\quad = 2(b + 3c)(a - 2d)$

> Before factorising the second pair, leave a space then write in the bracket term like this:
>
> $2[a(b + 3c) \; ? \; ? \; (b + 3c)]$
>
> That should help you find the correct common factor and correct sign.

Exercise 14I

Factorise each expression completely.

1 $x(p + q) + y(p + q)$

2 $3(x + y) - z(x + y)$

3 $5(x - y) + x(y - x)$

4 $10(a - b) - 2x(b - a)$

5 $xy + 2x + zy + 2z$

6 $ay + 2b + 2ab + y$

7 $x - b - x^2 + bx$

8 $x^2 + yz - xy - zx$

9 $2ab - 10 + 2a - 10b$

10 $3a - 3ab + 6bc - 6c$

14.11 Factorising quadratic expressions

$2x^2 - 5x + 4$, $3x^2 - 27$, $y^2 + 20y$ and t^2 are all examples of quadratic expressions.

> An expression of the form $ax^2 + bx + c$, with $a \neq 0$, is called a **quadratic expression** in x.

The expression $3y^2 - 20y$ is a quadratic expression in y.

> In the expression $ax^2 + bx + c$
> - the coefficient of x^2 is a
> - the coefficient of x is b
> - the constant term is c.

> c is a number.

Example 19

Write the quadratic expression in x in which the coefficient of x^2 is 1, the coefficient of x is -2 and the constant term is -3.

$\quad 1x^2 + (-2)x + (-3) = x^2 - 2x - 3$

To factorise quadratic expressions of the form $ax^2 + bx$, take out the HCF.

For example $3x^2 + 18x = 3x(x + 6)$.

Chapter 2 shows you how to expand expressions such as $(x - 3)(2x + 5)$ to get $2x^2 - x - 15$. You can reverse this process to factorise a quadratic expression.

Factorising quadratic expressions of the form $ax^2 + bx + c$

Example 20

Factorise $x^2 - 8x + 15$.

Step 1   Check the equation is in the form $ax^2 + bx + c$.

Step 2   Multiply the coefficient of x^2 by the constant term:

$$a \times c = 15$$

Step 3   Find the two factors of this product which add to give the coefficient of x.

Here the factors are -3 and -5.

> Factors of 15 are:
> 1 and 15   ✗
> 3 and 5   ✗
> −1 and −15   ✗
> −3 and −5   ✓

Step 4   Rewrite the bx term using these two factors:

$$-3x - 5x$$

Step 5   Proceed as for factorising by pairing:

$$x^2 - 3x - 5x + 15 = x(x - 3) - 5(x - 3) = (x - 5)(x - 3)$$

To factorise $ax^2 + bx + c$, start by looking for two numbers whose product is ac and whose sum is b.

Example 21

Factorise $6x^2 - 45 - 3x$.

Step 1   Take out any HCF and apply the remaining steps to the other factor

$3[2x^2 - 15 - x]$

> The other factor is $2x^2 - 15 - x$

Step 2   Rearrange into the form $ax^2 + bx + c$

$3[2x^2 - x - 15]$

Step 3   Multiply the coefficient of x^2 by the **constant term**. Find the two factors of this product which add to give the **coefficient of x**

$2 \times (-15) = -30$

> The other factor is $2x^2 - 15 - x$
> Try:
> −30 and 1 ✗
> −15 and 2 ✗
> −10 and 3 ✗
> −6 and 5 ✓

Step 4   Rewrite the bx term using these two factors:

$3[2x^2 - 6x + 5x - 15]$

Step 5   Proceed as for factorising by pairing:

$3[2x(x - 3) + 5(x - 3)]$
$= 3[(x - 3)(2x + 5)]$

Example 22

Factorise $14x - 6 - 4x^2$.

Step 1 $14x - 6 - 4x^2 = 2[7x - 3 - 2x^2]$ ————— 2 is the HCF.

Step 2 $\qquad\qquad = 2[-2x^2 + 7x - 3]$ ————— Rearrange

Step 3 $(-2) \times (-3) = (+6)$

You need factors of $+6$ which add to $+7$.

$\qquad (+2) + (+3) = 5$ ✘
$\qquad (+6) + (+1) = +7$ ✔

$14x - 6 - 4x^2 = 2[-2x^2 + 7x - 3]$

Step 4 $\qquad\qquad = 2[-2x^2 + 6x + 1x - 3]$

Step 5 $\qquad\qquad = 2[-2x(x - 3) + 1(x - 3)]$

$\qquad\qquad = 2[(x - 3)(-2x + 1)]$ •

> This can also be expressed as:
>
> $2[(x - 3)(1 - 2x)]$ or
> $-2(x - 3)(2x - 1)$

Example 23

Find two numbers whose product P is -60 and whose sum S is -7.

The product is negative, so the two numbers must have different signs (and therefore one number must be odd and the other even).

The greater number must have the negative sign because the sum is negative.

Try: $+1, -60$ ✘ $+3, -20$ ✘
$\qquad +4, -15$ ✘ $+5, -12$ ✔

Exercise 14J

1   Factorise

(a) $x^2 + 3x$

(b) $8x + 24$

(c) $x^2 + 3x + 8x + 24$

(d) $x^2 + 6x$

(e) $2x + 12$

(f) $x^2 + 6x + 2x + 12$

(g) $x^2 - 5x$

(h) $8x - 40$

(i) $x^2 - 5x + 8x - 40$

(j) $x^2 + 5x + 3x + 15$

(k) $x^2 + 4x - 8x - 32$

(l) $x^2 - 6x - 2x + 12$

(m) $x^2 + 4x - 3x - 12$

(n) $x^2 + 15 + 3x + 5x$

(o) $x^2 + 63 - 7x - 9x$

> Your first step must be to look for the HCF.

2   Write the quadratic expression in x in which the coefficient of x^2 is 1, the coefficient of x is -5 and the constant term is -6.

3 In each of the following, find two numbers whose product is P and whose sum is S.

    (a) $P = 6$, $S = -5$ (b) $P = 6$, $S = 7$

    (c) $P = 12$, $S = -8$ (d) $P = 60$, $S = 19$

    (e) $P = -24$, $S = -2$ (f) $P = -24$, $S = 10$

    (g) $P = -48$, $S = -13$ (h) $P = -48$, $S = 8$

4 Factorise

    (a) $x^2 - 5x + 4$ (b) $x^2 + 7x + 10$

    (c) $x^2 - 2x - 15$ (d) $5x + 6 + x^2$

    (e) $10 - 11x + x^2$ (f) $y^2 - y - 12$

5 Factorise each expression completely.

    (a) $2x^2 - 5x + 3$ (b) $6x^2 + 42x + 60$

    (c) $2y^2 - y - 10$ (d) $6a^2 + 10a + 4$

    (e) $4x^2 - 5x - 6$ (f) $75 + 35x - 20x^2$

    (g) $5x^2 + 3 - 16x$ (h) $x^4 + 5x^2 + 6$ •————— | Let $x^2 = y$. |

    (i) $2x^4 + 14x^2 + 24$

6 Factorise $x^{2n} + 9x^n + 8$. •————— | Let $x^n = y$. |

7 Factorise $x^{4n+1} + 6x^{2n+1} + 8x$ completely. •—————

| Question 7 is harder than the others. Only attempt it if you are confident with factorising expressions. |

After further practice, you may be able to factorise some of the quadratics more directly. Diagrams can help, as shown in the next example.

Example 24

Factorise $2x^2 + 7x - 15$.

You need to find factors of -30 whose sum is $+7$.

These are $+10$ and -3

$2x^2 + 7x - 15 = (2x\ \underline{?}\ \ \underline{?}\ )(x\ \underline{?}\ \ \underline{?}\ )$ •

$= (2x\ ?\ 3)(x\ ?\ 5)$ •

$= (2x - 3)(x + 5)$ •

	$2x$	
x	$2x^2$	

	$2x$	-3
x	$2x^2$	$-3x$
5	$+10x$	

	$2x$	-3
x	$2x^2$	$-3x$
5	$+10x$	-15

14.12 The difference of two squares

The expression $x^2 - y^2$ is called a difference of two squares.
There is an easy way to factorise an expression like this.

Here is a square of side x with a square of side y cut out of the top left
corner. The part A which is left has an area $x^2 - y^2$.

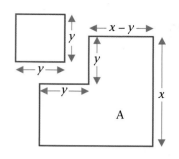

Imagine cutting along the dotted line and moving the shaded piece B
around to the bottom end of C.

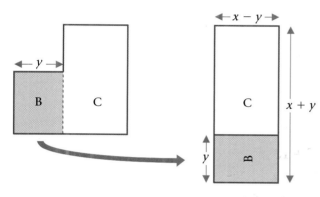

From the diagram you can see that

$$\text{area A} = \text{area B} + \text{area C}$$
$$= \text{area (B + C)}$$

and $x^2 - y^2 = (x - y)(x + y)$

So to factorise the difference of two squares you can use

$$x^2 - y^2 = (x - y)(x + y)$$

___ **Example 25** ___

Factorise completely:
(a) $x^2 - 16$ (b) $3x^2 - 27$ (c) $18x^2 - 2(x + 5)^2$

(a) $x^2 - 16$
$\quad = x^2 - 4^2$
$\quad = (x - 4)(x + 4)$

(b) $3x^2 - 27$
$\quad = 3[x^2 - 9]$
$\quad = 3[x^2 - 3^2]$
$\quad = 3(x - 3)(x + 3)$

(c) $18x^2 - 2(x + 5)^2$
$\quad = 2[9x^2 - (x + 5)^2]$
$\quad = 2[(3x)^2 - (x + 5)^2]$
$\quad = 2[3x - (x + 5)][3x + (x + 5)]$
$\quad = 2(2x - 5)(4x + 5)$

Two other results which you should memorise are:

$$x^2 + 2ax + a^2 = (x + a)^2$$

$$x^2 - 2ax + a^2 = (x - a)^2$$

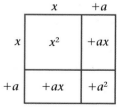

So, for example, putting $a = 6$ in $x^2 + 2ax + a^2 = (x + a)^2$ you get

$$x^2 + 12x + 36 = (x + 6)^2$$

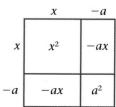

Exercise 14K

1 Factorise each expression completely.

(a) $x^2 - 1$ (b) $x^2 - 25$ (c) $4x^2 - 9$

(d) $9x^2 - (x-3)^2$ (e) $2p^2 - 98$ (f) $2x^2 - 8y^2$

(g) $x^2 + 4x + 4$ (h) $x^2 - 6x + 9$ (i) $2x^2 + 28x + 98$

2 Factorise each expression completely.

(a) $x^2 - 4$ (b) $x^2 - 3x$ (c) $x^2 - 3x - 4$

(d) $4x^2 - 36$ (e) $4x^2 - 70x$ (f) $4x^2 - 70x - 36$

(g) $x^2 + 2x + 1$ (h) $x^2 + 2x$ (i) $2x - 28 + 8x^2$

3 Factorise these expressions in as straightforward a way as you can. Copy (a)–(f) which have been partly done for you and complete the blanks. Check your factors by mentally expanding your answers.

> For a reminder on expanding expressions involving brackets see Section 2.7.

(a) $x^2 - 5x + 4$ •————— $x^2 + 5x + 4 + (x - 4)(x \underline{\hspace{1cm}})$

(b) $x^2 + 9x - 36$ •————— $x^2 + 9x - 36 = (x + 12)(\underline{\hspace{1cm}})$

(c) $x^2 + 4x - 32$ •————— $x^2 + 4x - 32 = (x + 8)(\underline{\hspace{1cm}})$

(d) $2x^2 - 7x + 5$ •————— $2x^2 - 7x + 5 = (x - 1)(\underline{\hspace{1cm}})$

(e) $4x^2 - 3x - 7$ •————— $4x^2 - 3x - 7 = (\underline{\hspace{1cm}} - 7)(\underline{\hspace{1cm}} + 1)$

(f) $5x^2 - 7x - 6$ •————— $5x^2 - 7x - 6 = (5x \underline{\hspace{1cm}})(\underline{\hspace{1cm}})$

(g) $x^2 - 4x + 3$ (h) $y^2 + 7y + 10$ (i) $x^2 + 2x - 15$

(j) $x^2 - 4x - 21$ (k) $y^2 + 8y + 16$ (l) $4x^2 - 20x + 25$

(m) $5x^2 - 13x + 6$ (n) $4x^2 + x - 5$ (o) $5x^2 + 7x - 6$

14.13 Further fractional algebraic expressions

This section shows you how to combine the skills developed in the previous two sections to simplify expressions which contain more difficult algebraic fractions.

The steps to be taken to do this are:

Step 1 Factorise all expressions completely.

Step 2 Cancel fully.

Step 3 Complete the fraction calculation.

Example 26

Simplify $\dfrac{2x - 4}{x^2 - 5x + 6} \div \dfrac{3x^2 + 9x}{x^2 - 9}$

$$\dfrac{2x - 4}{x^2 - 5x + 6} \div \dfrac{3x^2 + 9x}{x^2 - 9} = \dfrac{2(x - 2)}{(x - 3)(x - 2)} \div \dfrac{3x(x + 3)}{(x - 3)(x + 3)} \quad \text{—— Step 1}$$

$$= \dfrac{2}{(x - 3)} \div \dfrac{3x}{(x - 3)} \quad \text{—— Step 2}$$

$$= \dfrac{2}{(x - 3)} \times \dfrac{(x - 3)}{3x} \quad \text{—— Multiply by the reciprocal.}$$

$$= \dfrac{2}{(x - 3)} \times \dfrac{(x - 3)}{3x} \quad \text{—— Cancel by } (x - 3).$$

$$= \dfrac{2}{3x}$$

Example 27

Write $\dfrac{6x - 30}{2x^2 - 50} + \dfrac{21}{x^2 + 3x - 10}$ as a single fraction.

$$\dfrac{6x - 30}{2x^2 - 50} + \dfrac{21}{x^2 + 3x - 10} = \dfrac{{}^3 6(x - 5)}{2(x - 5)(x + 5)} + \dfrac{21}{(x + 5)(x - 2)} \quad \text{—— Factorise fully.}$$

> Take out the HCF, giving
> $2x^2 - 50 = 2(x^2 - 25)$
> Then use the difference of two squares for
> $x^2 - 25$

$$= \dfrac{3}{(x + 5)} + \dfrac{21}{(x + 5)(x - 2)} \quad \text{—— Cancel by } 2(x - 5).$$

$$= \dfrac{3(x - 2)}{(x + 5)(x - 2)} + \dfrac{21}{(x + 5)(x - 2)} \quad \text{—— LCM is } (x + 5)(x - 2).$$

$$= \dfrac{3(x - 2) + 21}{(x + 5)(x - 2)}$$

$$= \dfrac{3x + 15}{(x + 5)(x - 2)}$$

$$= \dfrac{3(x + 5)}{(x + 5)(x - 2)} \quad \text{—— Factorise } 3x + 15.$$

$$= \dfrac{3}{x - 2}$$

Exercise 14L

Simplify each expression as fully as possible.

1 $\dfrac{x^2 - 4}{2x - 4}$

2 $\dfrac{x^2 - 5x + 4}{x^2 - 16}$

3 $\dfrac{4x + 8}{4x^2 - 8x} \div \dfrac{x^2 - 4}{x^2 + 2x}$

4 $\dfrac{x^2 + 11x + 18}{x^2 + 3x + 2}$

5 $\dfrac{x^2 + 6x + 8}{x^2 + 4x}$

6 $\dfrac{1}{x + 1} - \dfrac{x - 2}{x^2 + 3x + 2}$

7 $\dfrac{x^2 - 5x + 4}{x^2 - 16} + \dfrac{x^2 + 11x + 18}{x^2 + 6x + 8}$

Mixed exercise 14

1 Simplify (a) $\dfrac{x^6}{x^2}$ (b) $(y^4)^3$ (c) $(m^{-4})^{-2}$ (d) $2t^2 \times 3r^3t^4$ [E]

2 Factorise each expression fully.

     (a) $2p^2 - 4pq$ (b) $(p+q)^2 + 5(p+q)$ (c) $x^2 + 7x + 6$

3 Factorise (a) $3x^2 + 6xy$ (b) $2(x-5)^2 + 3(x-5)$

4 Simplify (a) $\dfrac{3(y+2)}{(y+2)^2}$ (b) $\dfrac{3x-3}{x^2-1}$

5 Simplify $\dfrac{4x^2 - 9}{2x^2 - 5x + 3}$

6 Simplify $\dfrac{2}{x+1} + \dfrac{3}{x-1} - \dfrac{5}{x^2-1}$

7 Simplify (a) $p^2 \times p^7$ (b) $x^8 \div x^3$ (c) $\dfrac{y^4 \times y^3}{y^5}$ [E]

8 Evaluate (a) 3^{-2} (b) $36^{\frac{1}{2}}$ (c) $27^{\frac{2}{3}}$ (d) $\left(\dfrac{16}{81}\right)^{-\frac{3}{4}}$ [E]

9 (a) Factorise $9x^2 - 6x + 1$

     (b) Simplify $\dfrac{6x^2 + 7x - 3}{9x^2 - 6x + 1}$ [E]

10 Simplify $\dfrac{5}{x-1} - \dfrac{5}{x^2 - 3x + 2}$

11 Simplify $\dfrac{4x^2 - 25}{4x^2 + 8x - 5}$

12 For the identity $(px + 3)^2 - 2x^2 + qx = 7x^2 + 12x + r$ find the values of p, q and r.

13 p is a prime number not equal to 7.

     (a) Write down the HCF of $49p$ and $7p^2$.

     (b) x and y are different prime numbers

         (i) Write down the HCF of the two expressions x^2y and xy^2.

         (ii) Write down the LCM of the two expressions x^2y and xy^2.

                                                    [E]

14 Find the value of k in

     (a) $4^k = 64$ (b) $x^k = x^2 \div \dfrac{1}{x^5}$ (c) $\dfrac{1}{9} = 3^k$

Summary of key points

1 In the expression x^n, the number x is called the **base** and the number n is called the **index** or power.

2 $x^1 = x$

3 $x^m \times x^n = x^{m+n}$

4 $(x^m)^n = x^{m \times n} = x^{mn}$

5 $x^m \div x^n = x^{m-n}$

6 $x^0 = 1$ for all non-zero values of x.

7 $x^{-n} = \dfrac{1}{x^n}$

8 $x^{\frac{1}{n}} = \sqrt[n]{x}$

9 $x^{\frac{m}{n}} = (\sqrt[n]{x})^m$ or $x^{\frac{m}{n}} = \sqrt[n]{x^m}$

10 Dividing by an algebraic fraction is equivalent to multiplying by its reciprocal.

For example, to divide by $\dfrac{a}{b}$ you can multiply by $\dfrac{b}{a}$.

11 $\dfrac{1}{n} + \dfrac{1}{m} = \dfrac{m+n}{mn}$

12 An expression of the form $ax^2 + bx + c$, with $a \neq 0$, is called a **quadratic expression** in x.

13 In the expression $ax^2 + bx + c$
  - the coefficient of x^2 is a
  - the coefficient of x is b
  - the constant term is c.

14 To factorise $ax^2 + bx$, take out the highest common factor (HCF).

15 To factorise $ax^2 + bx + c$, start by looking for two numbers whose product is ac and whose sum is b.

16 $x^2 - y^2 = (x - y)(x + y)$ is called the **difference of two squares**.

17 $x^2 + 2ax + a^2 = (x + a)^2$

18 $x^2 - 2ax + a^2 = (x - a)^2$

15 Pythagoras' theorem

Pythagoras was a Greek mathematician and philosopher who lived during the sixth century BC. The theorem that carries his name, **Pythagoras' theorem**, is perhaps the best known theorem in the whole of mathematics. It is also extremely useful, very important and a regular feature in GCSE exam papers.

In this chapter you will use Pythagoras' theorem to find the lengths of sides in right-angled triangles.

15.1 Pythagoras' theorem

Pythagoras' theorem states that:

> For any **right-angled triangle**, the area of the square formed on the hypotenuse is equal to the sum of the areas of the squares formed on the other two sides.

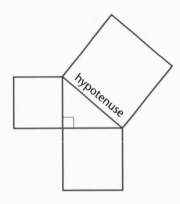

In symbols:

$$c^2 = a^2 + b^2$$

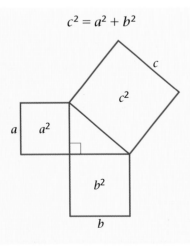

Confirming the theorem

The area of the small
square is 9 square units.

The area of the middle
square is 16 square units.

The area of the large
square is 25 square units.

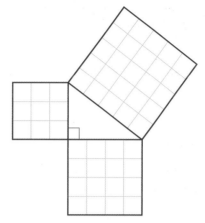

Area of small square + area of middle square = area of large square

   9 + 16 = 25

Exercise 15A

1 On squared paper draw a right-angled triangle PQR with the
90° angle at Q. Make $PQ = 4.5$ units and $QR = 6$ units.

Draw squares on the three sides of the triangle.

Confirm, by counting squares, that the area of the square
drawn on the hypotenuse PR, that is the largest square, is

   $4.5^2 + 6^2 = 20.25 + 36 = 56.25$ squares

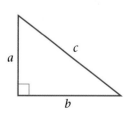

2 (a) Draw squares of side 5 cm, 12 cm and 13 cm on squared paper
and cut them out.

(b) Draw triangle PQR on squared paper, with $PQ = 5$ cm,
$QR = 12$ cm and a right angle at Q.

(c) Use your cut-out squares to confirm that

$$\text{area of square on } PQ + \text{area of square on } QR = \text{area of square on } PR$$

15.2 Finding lengths using Pythagoras' theorem

Pythagoras' theorem is used to find the length of one side of a
right-angled triangle, when the other two lengths are known, using
the formula $c^2 = a^2 + b^2$.

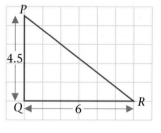

___ **Example 1** _____

Calculate the length of the side marked x.

   $x^2 = 24^2 + 7^2$
   $= 576 + 49$
   $= 625$
   $x = \sqrt{625}$
   $= 25$ cm

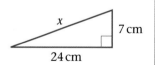

Example 2

In triangle LMN, angle $N = 90°$, $LN = 13$ cm and $MN = 7$ cm.

Calculate LM correct to one decimal place.

First make a sketch.

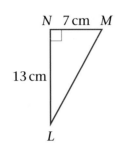

$$c^2 = a^2 + b^2$$
$$LM^2 = NM^2 + NL^2$$
$$= 7^2 + 13^2$$
$$= 49 + 169$$
$$= 218$$

So $LM = \sqrt{218}$
$$= 14.8 \text{ cm (to 1 d.p.)}$$

> The side opposite the right angle is the **hypotenuse**.

$\sqrt{218}$ is not a square number, so $\sqrt{218}$ cannot be worked out exactly as an ordinary number. It is a non-terminating decimal number.

However $\sqrt{218}$ is an exact value. A number written like this (the square root of an integer) is called a **surd**. A surd is always the positive square root.

> For more on surds see Chapter 23.

If you were not allowed to use a calculator, you could leave your answer in surd notation as $\sqrt{218}$ cm.

Exercise 15B

1 Calculate the lengths marked with letters in these triangles:

(a)

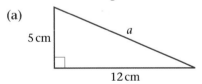

(b)

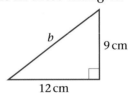

(c)

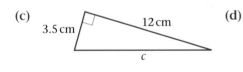

(d)

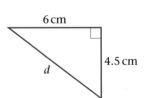

2 Give your answers correct to one decimal place.

(a) Find YZ.

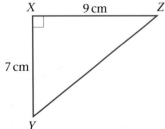

(b) Find SU.

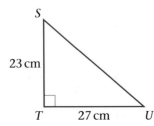

(c) Find *JL*.

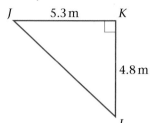

(d) Find *AC*.

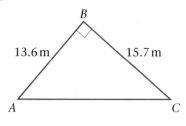

3 A rectangle measures 10 cm by 24 cm.
Calculate the length of a diagonal.

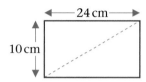

4 The diagonal of a square has a length of 20 cm.
Calculate the length of a side of the square.

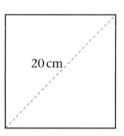

15.3 Using Pythagoras' theorem to find one of the shorter sides of a triangle

You can use the formula $a^2 + b^2 = c^2$ to calculate the longest side c of a right-angled triangle. If you need to find one of the shorter sides (a, for example) it is easier to change the formula so that a^2 is on its own on one side.

Pythagoras' theorem states that

$$a^2 + b^2 = c^2$$

Subtract b^2 from both sides:

$$a^2 + b^2 - b^2 = c^2 - b^2$$
$$a^2 = c^2 - b^2$$

___ **Example 3** ___

In triangle *DEF*, angle *F* is a right angle, *DF* = 5 cm and *DE* = 7 cm. Calculate the length of *EF*.

$$a^2 = c^2 - b^2$$
$$EF^2 = 7^2 - 5^2$$
$$= 49 - 25$$
$$= 24$$

So $EF = \sqrt{24}$

$$= 4.898... = 4.9 \text{ cm (correct to 1 d.p.)}$$

> Always check that the hypotenuse is still the longest side. If not, look for your mistake.

Exercise 15C

1 Calculate the lengths marked with letters in these triangles.

(a)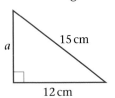

15 cm
a
12 cm

(b)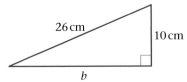

26 cm
10 cm
b

(c)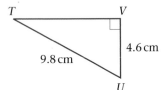

c
18 cm
19.5 cm

(d)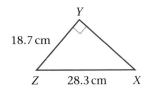

36 cm
d
37.5 cm

2 Give your answers correct to one decimal place.

(a) Find *OP*.

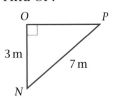

O   P
3 m
7 m
N

(b) Find *RS*.

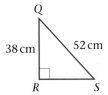

Q
38 cm   52 cm
R   S

(c) Find *TV*.

T   V
4.6 cm
9.8 cm
U

(d) Find *XY*.

Y
18.7 cm
Z   28.3 cm   X

3 An isosceles triangle *XYZ* has *XY* = 26 cm, *ZX* = 26 cm and *ZY* = 18 cm.

Calculate the height of triangle *XYZ*.

4

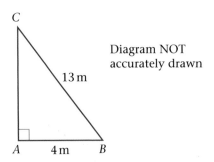

C
Diagram NOT accurately drawn
13 m
A   4 m   B

ABC is a right-angled triangle. *AB* is of length 4 m and *BC* is of length 13 m.

Calculate the length of *AC*.   [E]

5 A ladder extends to 10 metres in length and is placed 2 metres from a vertical wall on horizontal ground. How high up the wall will the ladder reach, to the nearest centimetre?

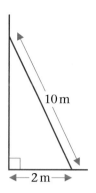

6 A yacht leaves a port, P, and travels 43 km due South to a buoy, B.
At B the yacht turns due East and travels to a lighthouse, L.
At L, the yacht turns again and travels in a straight line back to P.
The straight line distance from L to P is 54 km.
Calculate the total distance travelled by the yacht.

15.4 Pythagoras' theorem applied twice

Sometimes you need to find the length of one side before you can find the length of another side. You may need to apply Pythagoras' theorem twice.

Example 4

Calculate the lengths marked x and y.

First $\quad x^2 = 3^2 + 4^2$
$\qquad \quad = 9 + 16$
$\qquad \; x = \sqrt{25}$
$\qquad \quad = 5$ cm

Then $\quad y^2 + x^2 = 13^2$
$\qquad \quad y^2 + 5^2 = 13^2$
$\qquad \quad y^2 + 25 = 169$
$\qquad \qquad y^2 = 169 - 25$
$\qquad \qquad \; y = \sqrt{144}$
$\qquad \qquad \quad = 12$ cm

Exercise 15D

1 Calculate each of the lengths marked with a letter.

(a)

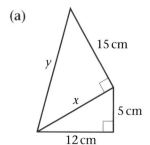

(b)

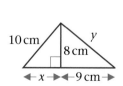

(c)

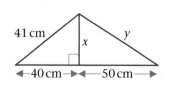

15.5 The distance between two points

Example 5

A is the point with coordinates (2, 1).
B is the point with coordinates (8, 9).

Find the straight line distance from *A* to *B*.

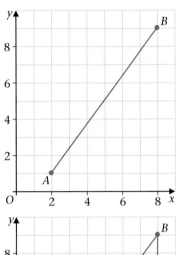

This can be done using Pythagoras' theorem.
First we make a triangle *ABM*, where the angle at
M is 90°, *M* is horizontally across from *A* and
vertically below *B*.

$AM = 6$ units $BM = 8$ units

Using Pythagoras' theorem: $AB^2 = AM^2 + BM^2$
$$= 6^2 + 8^2$$
$$= 36 + 64$$
$$= 100$$
$$AB = 10 \text{ units}$$

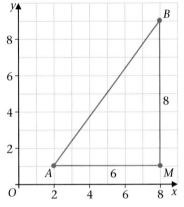

The distance between two points on a
coordinate grid is
$$d = \sqrt{(x_2 - x_1)^2 + (y_2 - y_1)^2}$$

Exercise 15E

1   Work out the distance between the
points *P* and *Q*.

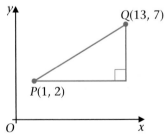

2   Calculate the distance between the points in each pair.

   (a)   (1, 2) and (5, 5)   (b)   (3, 5) and (5, 9)

   (c)   (–1, 4) and (7, 12)   (d)   (–2, 6) and (9, –4)

   (e)   (8, –3) and (12, –6)   (f)   (–2, –3) and (–7, 8)

> You may find it helpful to
> make a sketch first.

3 Calculate the distance PQ.

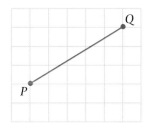

> Each square of the grid is 1 unit by 1 unit.

15.6 Non right-angled triangles

When all angles are acute

The labels L, M and S are used to indicate the lengths of the longest side, the middle side and the shortest side, respectively.

The area of the largest square is 9 square units.
The area of the middle square is 8 square units.
The area of the smallest square is 5 square units.

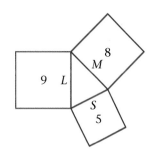

> In a triangle with sides of length L (longest), M (middle) and S (shortest), if $L^2 < M^2 + S^2$ then the angles are all acute.

When one angle is obtuse

In this diagram one angle is obtuse, and L, the longest side, is opposite this angle.

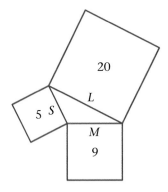

The area of the largest square is 20 square units.
The area of the middle square is 9 square units.
The area of the smallest square is 5 square units.

> In a triangle with sides of length L (longest), M (middle) and S (shortest), if $L^2 > M^2 + S^2$ then one angle is obtuse.

Example 6

The lengths of the sides of a triangle are 3 cm, 5 cm and 6 cm.

Work out whether the angles of this triangle are all acute, whether one is obtuse or whether one is a right angle.

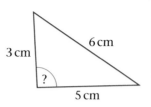

The longest side, $L = 6$ cm

The middle side, $M = 5$ cm

The shortest side, $S = 3$ cm

$$L^2 = 6^2 = 36 \qquad M^2 = 5^2 = 25 \qquad S^2 = 3^2 = 9$$

$$M^2 + S^2 = 25 + 9 = 34$$

so $\quad L^2 > M^2 + S^2$

So one of the angles is obtuse.

Exercise 15F

1 You will need some squared paper. On it draw

   (a) a triangle where all angles are acute

   (b) a triangle where one angle is obtuse.

   Using the symbols L, M and S for the longest, middle and shortest sides respectively, convince yourself that in case (a) $L^2 < M^2 + S^2$ and in case (b) $L^2 > M^2 + S^2$.

2 The numbers in the brackets are the lengths in centimetres of the sides of triangles.

   In each case, work out whether the angles of the triangle are all acute, whether one angle is obtuse or whether the triangle is right-angled.

> When
> $L^2 = M^2 + S^2$
> the triangle is **right-angled**.

   (a) (4, 7, 9)   (b) (5, 11, 12)   (c) (3, 14, 15)

   (d) (9, 41, 40)   (e) (17, 10, 13)   (f) (8, 8, 12)

15.7 Pythagorean triples

Any set of three numbers, usually positive whole numbers (a, b, c), which satisfy (fit into) the relationship

$\quad c^2 = a^2 + b^2$ (assuming c to be the largest)

is known as a **Pythagorean triple**.

The two most famous Pythagorean triples are

$\quad$ (3, 4, 5)   and   (5, 12, 13)

They are worth remembering.

Exercise 15G

1   Check that (3, 4, 5) and (5, 12, 13) are Pythagorean triples.

2   Which of the following are Pythagorean triples?
    Show all your working.

   (a) (6, 8, 10)   (b) (7, 24, 25)   (c) (9, 40, 41)

   (d) (8, 23, 30)   (e) (11, 60, 61)   (f) (1, 1, 2)

15.8 Pythagoras' theorem in three dimensions

You can use Pythagoras' theorem in three dimensions to find the length of the longest diagonal of a cuboid.

> The longest diagonal is often called a **space diagonal**.

Example 7

A cuboid measures 4 cm by 5 cm by 8 cm.
Work out the length of the longest diagonal of this cuboid.

One of the longest diagonals is *HB*.

By Pythagoras' theorem:

$$HB^2 = HD^2 + DB^2$$

$$HD = 8$$

> The four diagonals *HB*, *GA*, *EC* and *FD* are all equal. You can find the length of any one.

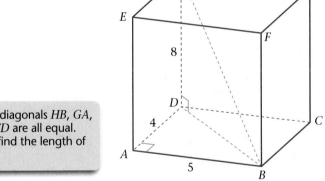

To find the length of *DB*, look at the base *ABCD*.

Using Pythagoras' theorem: $DB^2 = AB^2 + AD^2$

$$= 5^2 + 4^2$$

$$= 25 + 16$$

$$DB^2 = 41$$

HB can now be found: $HB^2 = HD^2 + DB^2$

$$= 8^2 + 41$$

$$= 64 + 41$$

$$= 105$$

$$= \sqrt{105}$$

$$= 10.25 \text{ cm (to 2 d.p.)}$$

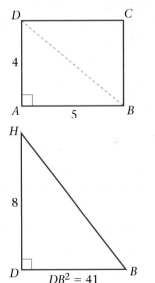

The length of the longest diagonal of a cuboid with dimensions a, b, c is

$$d = \sqrt{a^2 + b^2 + c^2}$$

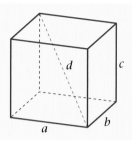

Exercise 15H

1 Calculate the length of the longest diagonal of a cuboid which has sides of length 5 cm, 8 cm and 12 cm.

2 Calculate the length of the longest diagonal of a cuboid which has sides of length 6.2 mm, 8.3 mm and 15.1 mm.

3 *VABC* is a triangle-based pyramid.

The base *ABC* is a triangle, right-angled at *A*. The vertex *V* is vertically above *A*. $AB = 5$ cm, $AC = 12$ cm and $VA = 17$ cm.

(a) Calculate the lengths of   (i) *BC*   (ii) *VB*   (iii) *VC*.

(b) Work out whether the angle *BVC* is acute, obtuse or a right angle.

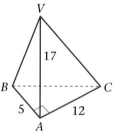

15.9 The equation of a circle

___ **Example 8** ___

Find the equation of the circle with centre at the origin (0, 0) and radius 3 units.

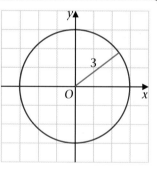

Mark any general point, *P*, on the circle, with coordinates (x, y).

Using Pythagoras' theorem:

$$x^2 + y^2 = 3^2 \ (= 9)$$

The result will be true for *every* point on the circle.

The *equation* of this circle is

$$x^2 + y^2 = 3^2 \quad \text{or} \quad x^2 + y^2 = 9$$

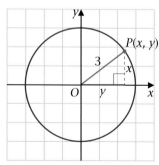

The equation of any circle, centre the origin and radius r units is

$$x^2 + y^2 = r^2$$

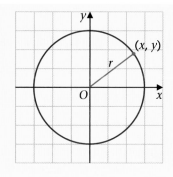

Exercise 15I

1 Write down the equation of each circle.

(a)
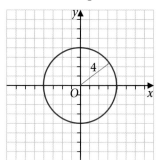

(b)
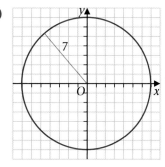

2 Find the equation of the circle
   (a) radius 5 units, centre the origin
   (b) radius 12 units, centre the origin
   (c) radius 15 units, centre the origin.

3 A circle has its centre at the origin and passes through the point (5, 12).
   Find the equation of this circle.

4 A circle of radius 5 units is drawn with its centre at the origin.
   Find the coordinates of 8 points which lie on the circle and have integer values for their coordinates.

Mixed exercise 15

1 Calculate the length PQ.

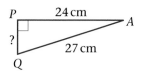

2 The diagram shows a ladder of length 8 metres resting between horizontal ground and a vertical wall.

The distance from the base of the wall to the bottom of the ladder is 5 metres.

Calculate the height of the top of the ladder above the ground.

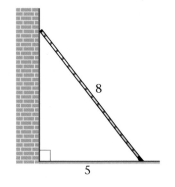

3 A yacht leaves a harbour, *H*, and travels due North for an unknown distance until it reaches a marker buoy, *B*.

At *B* the yacht turns due East and travels for 30 km to a lighthouse, *L*.

At *L* the yacht turns again and travels in a straight line back to *H*. The distance from *L* to *H* is recorded as 43 km.

Calculate the distance from *H* to *B*.

4 Calculate the lengths marked *x* and *y*.

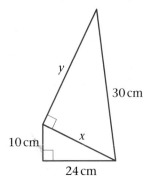

5 Points *A* and *B* have coordinates $(-3, 7)$ and $(5, -4)$, respectively. Calculate the distance *AB*.

6 *ABC* is a triangle.
   AB = 7 cm   *AC* = 8 cm   and   *BC* = 11 cm
   Work out whether the angles of *ABC* are all acute, whether one is obtuse or whether the triangle is right-angled.

7 Showing all your working, explain which of these sets are Pythagorean triples:

   (a)   (4, 6, 8)

   (b)   (10, 24, 26)

   (c)   (1, 2, 3)

8 A cuboid measures 8 cm by 12 cm by 15 cm.
   Calculate the length of the longest diagonal of the cuboid.

9 *ABCDEF* is a prism.

The horizontal base *ABCD* is a rectangle
with *AB* = 12 cm and *BC* = 20 cm.
The vertical face *ADEF* is also a rectangle
with *AF* = 5 cm.
Angles *FAB* and *EDC* are right angles.

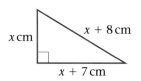

Calculate the lengths

(a)   *FB*   (b)   *BD*   (c)   *FC*.

10 Calculate the value of *x*.

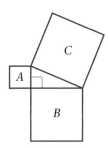

11 Find the equation of the circle with radius 4 units, centre the
origin.

12 A circle has its centre at the origin and passes through the point
(8, 15). Write down the equation of this circle.

13 **A special challenge**
Pythagoras' theorem states that for squares
A, *B* and *C*,

   area of *C* = area of *A* + area of *B*

But do the shapes drawn on the sides of the
right-angled triangle have to be squares?
Justify your answer.

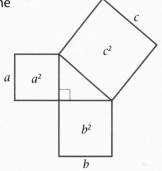

> This challenge question is
> outside the GCSE
> specification for
> Pythagoras' theorem.
> You can try it if you have
> completed all the other
> work on Pythagoras.

Summary of key points

1 For any **right-angled triangle**, the area of the square formed on the
hypotenuse is equal to the sum of the areas of the squares formed
on the other two sides:

   $c^2 = a^2 + b^2$

2 The distance between two points on a coordinate grid is

   $d = \sqrt{(x_2 - x_1)^2 + (y_2 - y_1)^2}$

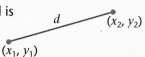

3 In a triangle with sides of length *L* (longest), *M* (middle) and *S* (shortest):
* if $L^2 < M^2 + S^2$ the angles are acute
* if $L^2 > M^2 + S^2$ the angle is obtuse
* if $L^2 = M^2 + S^2$ one angle is a right angle.

4 Some well-known Pythagorean triples are

(3, 4, 5)   (5, 12, 13)   (6, 8, 10)   (7, 24, 25)

5 The length of the longest diagonal of a cuboid with dimensions a, b, c is

$$d = \sqrt{a^2 + b^2 + c^2}$$

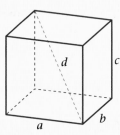

6 The equation of any circle, centre the origin and radius r units is

$$x^2 + y^2 = r^2$$

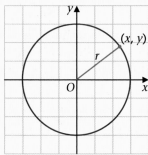

16 Basic trigonometry

This chapter introduces the trigonometric functions sine, cosine and tangent, often shortened to sin, cos and tan.

Trigonometry (or 'trig' for short) looks at how the sides and angles of a triangle are related to each other.

Trigonometry was first used in astronomy and navigation. How to tell the time from the position of the Sun was just one of the many practical problems that contributed to its development. Sundials work because shadows move and change length as the height of the Sun changes in the sky. But without trigonometry, sundials wouldn't have been possible.

Today, trigonometry is used by engineers, surveyors, architects and others who need to work out relationships connecting distances and angles. Trigonometry is always tested in GCSE exams.

16.1 Right-angled triangles and the trigonometric ratios

In any right-angled triangle you can name the sides in relation to the angles:

- side a is **opposite** to angle x
- side b is **adjacent** to angle x
- side c is the **hypotenuse** (opposite the right angle).

Sin x

Consider a right-angled triangle OAM with hypotenuse 1 unit in length.

When you apply the function **sine** to the angle x in the triangle you get the length of the side opposite x, AM.

The vertical height of the triangle OAM is $\sin x$.

Using similar triangles, when the length of the hypotenuse is doubled, the vertical height is also doubled.

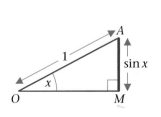
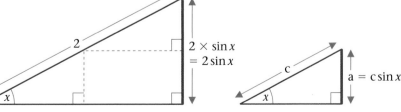
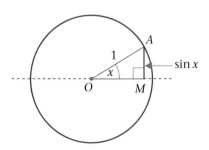

In the general case, when the length of the hypotenuse is c units and the vertical height is a units, then the length of a is given by

$a = c \sin x$

Making $\sin x$ the subject of the formula gives:

$\sin x = \dfrac{a}{c}$

$\sin x = \dfrac{\text{length of side opposite } x}{\text{length of hypotenuse}}$ or $\sin x = \dfrac{\text{opposite}}{\text{hypotenuse}}$

$\sin x = \dfrac{\text{opp}}{\text{hyp}}$

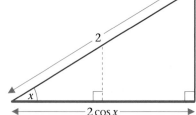

Cos x

Consider again the triangle OAM

When you apply the function **cosine** to the angle x in the triangle you get the length of the side adjacent to x, OM.

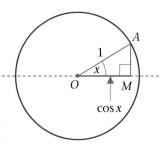

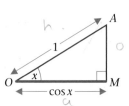 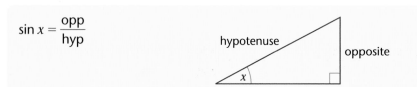 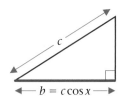

Making $\cos x$ the subject of the formula $b = c \cos x$ gives

$\cos x = \dfrac{b}{c}$

$\cos x = \dfrac{\text{length of side adjacent to } x}{\text{length of hypotenuse}}$ or $\cos x = \dfrac{\text{adjacent}}{\text{hypotenuse}}$

$\cos x = \dfrac{\text{adj}}{\text{hyp}}$

Tan x

The tangent ratio is a little different.

When you apply the function **tangent** to the angle x in the triangle OAM you get the length of the tangent, AT.

Imagine the triangle OAT turned over, so OA is the base.

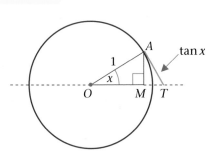

Consider this sequence of diagrams:

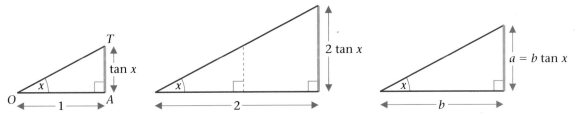

Making tan x the subject of the formula $a = b \tan x$ gives

$$\tan x = \frac{a}{b}$$

$$\tan x = \frac{\text{length of side opposite angle } x}{\text{length of side adjacent to angle } x} \quad \text{or} \quad \tan x = \frac{\text{opposite}}{\text{adjacent}}$$

$$\tan x = \frac{\text{opp}}{\text{adj}}$$

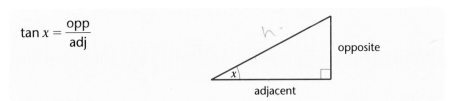

You need to understand how to use the three trigonometric ratios for your GCSE exam.

Example 1

Write down which trigonometric ratio is needed to calculate the angle θ in each of these triangles:

(a)

(b)

(c)

(a) The given sides are opposite to angle θ and hypotenuse so **sine** is needed.

(b) The given sides are opposite and adjacent to angle θ so **tangent** is needed.

(c) The given sides are adjacent to angle θ and hypotenuse so **cosine** is needed.

Example 2

Write down which trigonometric ratio is needed to calculate the side AB.

Side BC is adjacent to the given angle. Side AB is the hypotenuse.

$$\cos \theta = \frac{\text{adj}}{\text{hyp}}, \text{ so the ratio needed is } \textbf{cosine}.$$

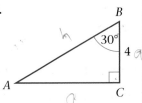

Exercise 16A

Write down which trigonometric ratio is needed to calculate the side or angle marked x in each of these triangles.

1 (a)
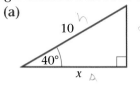

(b)
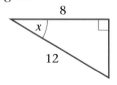

(c)
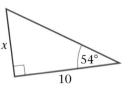

(d)
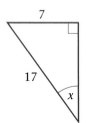

(e)
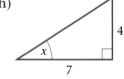

(f)
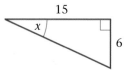

(g)
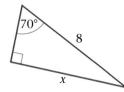

(h)

(i)
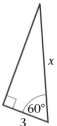

2 Shaun stands 25 m away from his house, which is built on flat ground. He uses a clinometer to measure the angle between the ground and the top of his house. The angle is 20°.

Estimate the height of Shaun's house.

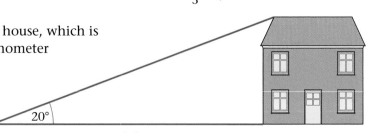

16.2 Using a calculator

You can use a calculator to find the values of sin x, cos x and tan x for different values of x.

Example 3

Find cos 43°.

The calculator display is 0.731 353 701

so   cos 43° = 0.731 353 701

In most cases there is no need to give results to more than four decimal places,

so   cos 43° = 0.7314

> Check that you know how to use the trig functions on your calculator.

When given the value of any trigonometric function, you can use the inverse function to find the value of the angle.

The inverses of sin, cos and tan are $\sin^{-1}$, $\cos^{-1}$ and $\tan^{-1}$. If you are given sin x, for example, use $\sin^{-1}(\sin x)$ to find x.

Example 4

Given that tan x = 1.2411, find the value of x.

$$x = \tan^{-1} 1.2411$$

The calculator display is 51.140 326 82.
So, correct to two decimal places:

$$x = 51.14°$$

> Your calculator may have INV, $\sin^{-1}$ or arc buttons for finding inverse trig functions.

Exercise 16B

1 Find these:

(a) sin 47°   (b) cos 58°

(c) tan 21°   (d) tan 83°

(e) cos 25°   (f) sin 60°

(g) tan 106°   (h) cos 93°

(i) sin 132°

Give your answers correct to 4 d.p.

2 Find each angle x (correct to 0.1°) when

(a) sin x = 0.3524   (b) cos x = 0.1364

(c) tan x = 1.4142   (d) tan x = 0.4365

(e) cos x = 0.9854   (f) sin x = 0.8856

16.3 Using trigonometric ratios to find angles

Example 5

Calculate the size of the angle at A.

Since 4 is adjacent to A and 5 is the hypotenuse we use the cosine ratio:

$$\cos A = \frac{\text{adj}}{\text{hyp}} = \frac{4}{5} = 0.8$$

$$A = \cos^{-1} 0.8$$

So   A = 36.87° (correct to 2 d.p.)

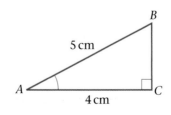

Exercise 16C

In this exercise all of the lengths are in cm. Calculate each of the angles marked with a letter, correct to 0.1°.

1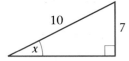

2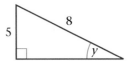

3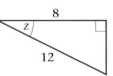

4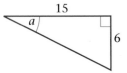

5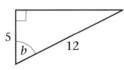

6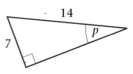

7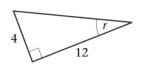

8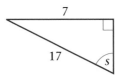

9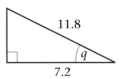

16.4 Using trigonometric ratios to find the lengths of sides

___ **Example 6** ___

Calculate the length of the side marked y.

For the given angle, y is opposite and 12 is adjacent.
So we use the tangent ratio:

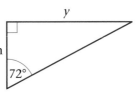

$$\tan 72° = \frac{\text{opp}}{\text{adj}} = \frac{y}{12}$$

So
$$y = 12 \times \tan 72°$$
$$= 12 \times 3.0777$$
$$= 36.93 \text{ cm (correct to 2 d.p.)}$$

___ **Worked examination question 1** ___

The distance from Alton, A, to Burton, B, is 10 km. The bearing of B from A is 030°. A transmitter, T, is due North of A and due West of B.

(a) (i) Calculate the distance of T from B.
    (ii) Calculate the distance of T from A.
        Give your answer to the nearest kilometre.

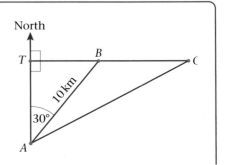

Clowne, C, is 8 km due east of Burton.

(b) Use your answers to **(a)** to calculate angle $T\hat{A}C$.

Hence give the three-figure bearing of C from A, correct to the nearest degree.

(c) A helicopter flies at the same height from A to B to C. If the average speed for the journey is 160 km/h, calculate the time taken. Give your answer correct to the nearest minute.

(a) **(i)** TB is opposite 30° and the AB is hypotenuse so we use the sine ratio:

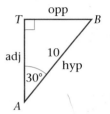

$$\frac{TB}{10} = \sin 30°$$

$$TB = 10 \times \sin 30°$$

So $TB = 5$ km

(ii) TA is adjacent to 30° and AB is the hypotenuse, so we use the cosine ratio:

$$\frac{TA}{10} = \cos 30°$$

$$TA = 10 \times \cos 30°$$

$$= 8.66 \text{ km}$$

So $TA = 9$ km, correct to the nearest km.

> Alternatively, using Pythagoras' theorem:
> $TA^2 + TB^2 = AB^2$
> $TA^2 + 25 = 100$
> $TA^2 = 100 - 25 = 75$
> $= \sqrt{75}$
> $TA = 8.66$ km

(b) Consider the diagram:

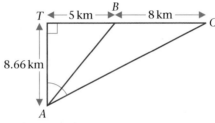

Then, for angle $T\hat{A}C$

$$TC = 13 = \text{opp} \quad \text{and} \quad TA = 8.66 = \text{adj}$$

So use the tangent ratio:

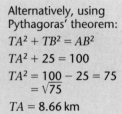

$$\tan T\hat{A}C = \frac{\text{opp}}{\text{adj}} = \frac{13}{8.66} = 1.501\,15$$

$$T\hat{A}C = \tan^{-1} 1.501\,15$$

$$= 56.33°$$

So, correct to the nearest degree, the bearing of C from A is 056°.

(c) The total distance from A to B to C is

$$10 + 8 = 18 \text{ km}$$

Distance = speed $\times$ time

So time = $\dfrac{\text{distance}}{\text{speed}}$

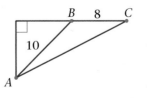

> This is not trigonometry but is an example of how trigonometry can be combined with another topic in exam questions.

$$= \frac{18}{160} = 0.1125 \text{ hours}$$

and 0.1125 hours = $0.1125 \times 60 = 6.75$ minutes

So time = 7 minutes, correct to the nearest minute.

Exercise 16D

1 All lengths are in centimetres. Calculate each length marked with a letter, correct to 2 d.p.

(a)
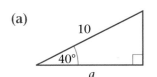

(b)
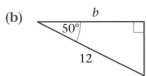

(c)
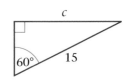

(d)
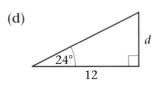

(e)
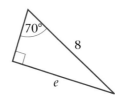

(f)
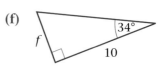

(g)
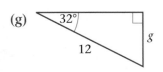

(h)
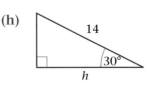

(i)
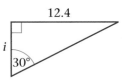

2 A lighthouse, L, is 34 km due North of a helicopter base, B.
A helicopter, H, is 21 km due West of L.
Calculate the bearing of H from B.

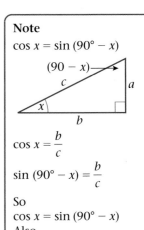

16.5 More about trigonometric functions

You have seen how useful the trig functions are for calculating lengths and angles in right-angled triangles. You have worked with angles between 0 and 90°, but you can apply sin, cos and tan to angles outside this range.

The three trig functions are shown on this unit circle, where $0 \leqslant x < 90°$.

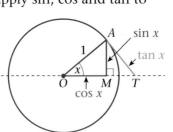

> **Note**
> $\cos x = \sin (90° - x)$
>
> $\cos x = \dfrac{b}{c}$
> $\sin (90° - x) = \dfrac{b}{c}$
> So
> $\cos x = \sin (90° - x)$
> Also
> $\sin x = \cos (90° - x)$

From the diagram you can look at sine, cosine and tangent in all 4 quandrants. That is:

1st quadrant: angles between 0° and 90°
2nd quadrant: angles between 90° and 180°
3rd quadrant: angles between 180° and 270°
4th quadrant: angles between 270° and 360°

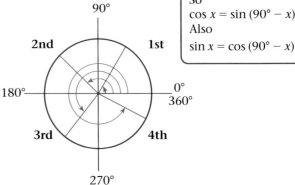

You can even consider other angles outside the range 0° to 360°, such as −50° and 410°.

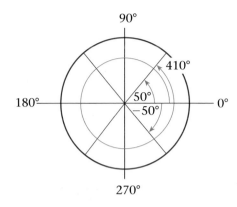

Consider the sine of these angles:

Correct to 4 d.p., your calculator gives:

$$\sin(-50°) = -0.7660$$
$$\sin 130° = 0.7660$$
$$\sin 410° = 0.7660$$

Compare these with:

$$\sin 50° = 0.7660$$

You can see that:

$$\sin(-50°) = -\sin 50°$$
$$\sin 130° = \sin 50°$$
$$\sin 410° = \sin 50°$$

Here are some useful results. You can explain them by looking at the trigonometric ratios in the unit circle.

$\sin(-x) = -\sin x$	$\cos(-x) = \cos x$	$\tan(-x) = -\tan x$
$\sin(180° - x) = \sin x$	$\cos(180° - x) = -\cos x$	$\tan(180° - x) = -\tan x$
$\sin(360° + x) = \sin x$	$\cos(360° + x) = \cos x$	$\tan(360° + x) = \tan x$

Exercise 16E

Write an equivalent trig ratio for each of the following. The first is done for you.

1 (a) $\sin(-60°) = -\sin 60°$ (b) $\sin 410°$ (c) $\sin 50°$

2 (a) $\cos 110°$ (b) $\cos(-85°)$ (c) $\cos 405°$

3 (a) $\tan(-45°)$ (b) $\tan 130°$ (c) $\tan 390°$

16.6 The graph of sin x

You can draw the graph of $y = \sin x$ for values of x from $0°$ to $360°$.

First make a table of values:

x	$\sin x$
$0°$	0
$30°$	0.5
$45°$	0.7071
$60°$	0.8660
$90°$	1
$120°$	0.8660
$135°$	0.7071
$150°$	0.5
$180°$	0
$210°$	-0.5
$225°$	-0.7071
$240°$	-0.8660
$270°$	-1
$300°$	-0.8660
$315°$	-0.7071
$330°$	-0.5
$360°$	0

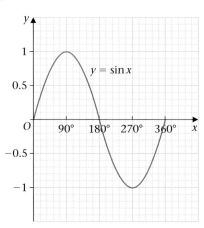

Graphical calculators are really helpful for plotting graphs of trig functions.

Key values of sin x
From this diagram:

hypotenuse $= \sqrt{2}$
(Pythagoras)

so $\sin 45° = \dfrac{1}{\sqrt{2}} = 0.707$

From this diagram:

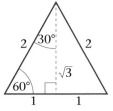

height $= \sqrt{3}$ (Pythagoras)

so $\sin 60° = \dfrac{\sqrt{3}}{2} = 0.866$

and $\sin 30° = \dfrac{1}{2} = 0.5$

The extended graph of $y = \sin x$ is

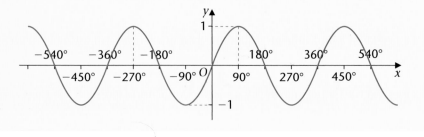

$y = 0$, i.e. $\sin x = 0$ when $x = 0, 180°, -180°, \ldots$

The important features of the graph are:

- it repeats itself every $360°$; we say that it has a **period** of $360°$
- it has maximum values (the largest values of y) of 1, which occur when $x = 90°, 450°, -270°, \ldots$
- it has minimum values (the smallest values of y) of -1, which occur when $x = -90°, 270°, -450°, \ldots$

$\sin(-x) = -\sin x \qquad \sin(180° - x) = \sin x \qquad \sin(180° + x) = -\sin x \qquad \sin(360° + x) = \sin x$

16.7 The graph of cos x

You can draw the graph of $y = \cos x$ for values of x from $0°$ to $360°$.

x	$\cos x$
$0°$	1
$30°$	0.8660
$45°$	0.7071
$60°$	0.5
$90°$	0
$120°$	-0.5
$135°$	-0.7071
$150°$	-0.8660
$180°$	-1
$210°$	-0.8660
$225°$	-0.7071
$240°$	-0.5
$270°$	0
$300°$	0.5
$315°$	0.7071
$330°$	0.8660
$360°$	1

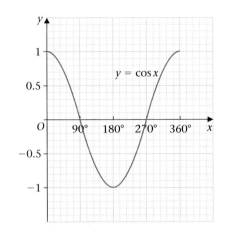

You can confirm this on a graphical calculator.

Key values of cos x
From this diagram:

$$\cos 45° = \frac{1}{\sqrt{2}} = 0.707$$

From this diagram:

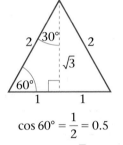

$$\cos 60° = \frac{1}{2} = 0.5$$

$$\cos 30° = \frac{\sqrt{3}}{2} = 0.866$$

The extended graph of $y = \cos x$ is

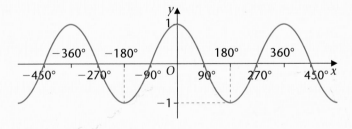

$y = 0$, i.e. $\cos x = 0$ when $x = 90°, 270°, -90°, \ldots$

The important features of this graph are:

- it repeats itself every $360°$, so it has a period of $360°$
- it has a maximum value of 1, which occurs when $x = 0°, 360°, -360°, \ldots$
- it has a minimum value of -1, which occurs when $x = 180°, -180°, 540°, \ldots$

The graph confirms
$\cos(90° - x) = \sin x$
$\cos x = \sin(90° - x)$

$$\cos(-x) = \cos x \qquad \cos(180° - x) = -\cos x \qquad \cos(180° + x) = -\cos x \qquad \cos(360° + x) = \cos x$$

16.8 The graph of tan x

You can draw the graph of $y = \tan x$ for values of x from $0°$ to $90°$.

Choosing some key values:

x	tan x
0°	0
30°	0.5774
45°	1
60°	1.7320
90°	infinite

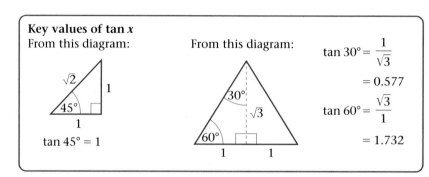

Key values of tan x
From this diagram: From this diagram:

$\tan 30° = \dfrac{1}{\sqrt{3}}$

$= 0.577$

$\tan 60° = \dfrac{\sqrt{3}}{1}$

$= 1.732$

$\tan 45° = 1$

The fact that tan 90° is infinite causes a problem.

In the range from 0° to 90°, the graph of $y = \tan x$ is like this:

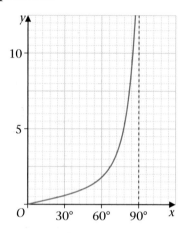

Angles between 90° and 270°

Taking an angle just greater than 90°, say 95°, you can get an idea of what happens to the tangent graph when x is just over 90° and look at some important values for x.

tan 95° is negative and large

x	tan x
95°	−11.43
135°	−1
180°	0
225°	1
270°	infinite
290°	−2.75
315°	−1
360°	0

In the range from 90° to 270° the graph of $y = \tan x$ is like this:

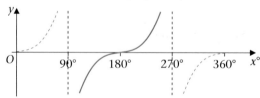

The extended graph of $y = \tan x$ is

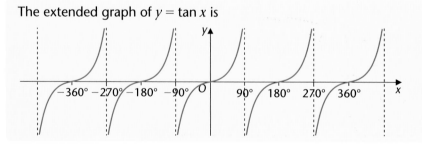

The important features of this graph are:
- it repeats itself every 180°, so it has a period of 180°
- it is infinite at $x = \pm 90°, \pm 270°, \ldots$

$\tan(-x) = -\tan x \qquad \tan(180° - x) = -\tan x \qquad \tan(180° + x) = \tan x \qquad \tan(360° + x) = \tan x$

Worked examination question 2

(a) Draw the graph of $y = 3 \sin 2x$ for values of x from $-180°$ to 180°.

(b) State the period of the graph.

(c) State the maximum and minimum values of $3 \sin 2x$ and the values of x at which these occur.

(a) Start with a table of values or use a graphical calculator.

x	$2x$	$3 \sin 2x$
$-180°$	$-360°$	0
$-135°$	$-270°$	3
$-90°$	$-180°$	0
$-45°$	$-90°$	-3
0°	0°	0
45°	90°	3
90°	180°	0
135°	270°	-3
180°	360°	0

For simplicity intervals of 45° have been chosen.

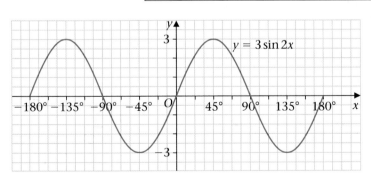

$y = 3 \sin 2x$

(b) From the drawing you can see that the graph repeats itself every 180°, so the period of the graph is 180°.

(c) The maximum value of $3 \sin 2x$ is 3.
This occurs when $x = -135°$ and 45°.

The minimum value of $3 \sin 2x$ is -3.
This occurs when $x = -45°$ and 135°.

Exercise 16F

1 Using a table of values or a graphical calculator, show that for the range 270° to 360° the graph of $y = \tan x$ is like this:

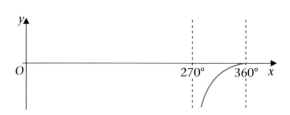

2 Using a table of values or a graphical calculator, show that the graph of $y = \tan x$ for the range −90° to 0° is like this:

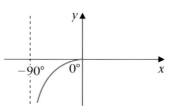

3 (a) For values of x from −180° to 180°, plot the graph of $y = 5 \cos 3x$.

   (b) State the period of the graph.

   (c) Find the maximum and minimum values of $5 \cos 3x$ and state the values of x at which these occur.

4 (a) Plot the graph of $y = 2 \tan x$ for values of x from 0° to 360°.

   (b) Find the period of the graph.

5 For values of x from 0° to 360°, plot the graphs of

   (a) $y = \sin 4x$ (b) $y = \sin \frac{1}{2}x$

   Comment on the periods of each of the two graphs.

6 Investigate graphs in the form $y = A \sin Bx$, where A and B are positive integers. Comment on the period of each graph and the maximum and minimum values in connection to A and B.

7 Investigate graphs in the form $y = A$ (trig function) Bx, where A and B can also take fractional values. Comment on your findings.

16.9 Solving a trigonometric equation

____ **Example 7** _____

Solve the equation $2 \sin x = 1$

giving all values of x in the range −360° to 360°.

$$2 \sin x = 1$$

so $\sin x = \frac{1}{2}$

From the calculator:

$$x = 30°$$

This is the first solution and you can use a sketch of the graph of sin x to find any others.

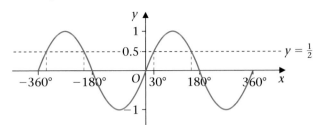

The sketch shows that there are three other solutions to the equation in the range $-360°$ to $360°$.

By examining the symmetries of the graph, you can see that the four solutions are

$x = 30°,\ 180° - 30°,\ -180° - 30°,\ -360° + 30°$

So the full set of solutions is

$x = 30°,\ 150°,\ -210°,\ -330°$

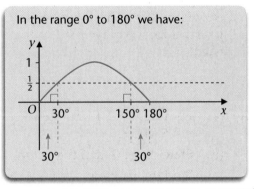

In the range 0° to 180° we have:

A trigonometric equation may have several solutions in any given range.

Exercise 16G

1   For values of x in the range $-360°$ to $360°$, solve the equation

   $2 \cos x = 1$

2   (a)   Show that one solution of the equation

   $5 \sin x = 2$

   is $x = 23.6°$ correct to 1 d.p.

   (b)   Hence find all solutions to the equation

   $5 \sin x = 2$

   which are in the range 0° to 720°, giving your answers correct to 1 d.p.

3   Find all solutions to the equation

   $\tan x = 2$

   which are in the range $-180°$ to $180°$.

4   (a)   Sketch the graph of

   $y = 3 \cos x$

   for values of x from 0° to 360°.

   (b)   Hence, or otherwise, obtain all the solutions to the equation

   $3 \cos x = 1$

   which are between 0° and 360°.

5 (a) Show that the equation
$$5 \sin 3x = 4$$
has a solution $x = 17.71°$ correct to 2 decimal places.

(b) Hence or otherwise, obtain all solutions to the equation
$$5 \sin 3x = 4$$
which are in the range $-180°$ to $180°$.

6 Suggest suitable equations for the following graphs.
Check your answers with a graphical calculator or computer.

(a)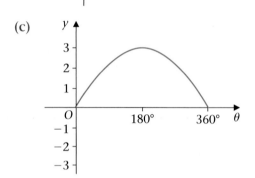

(b)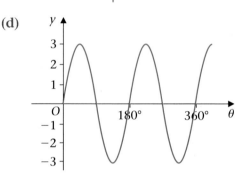

(c)

(d)

Mixed exercise 16

1 *ABC* is a right-angled triangle.
AB is of length 4 m and *BC* is of length 13 m.

(a) Calculate the length of *AC*.

(b) Calculate the size of angle $A\hat{B}C$.

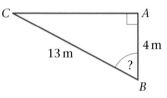

2 The diagram represents the frame, *PQRS*, of a roof.

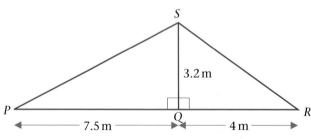

$PQ = 7.5$ m $QR = 4$ m $SQ = 3.2$ m

(a) Calculate the length of *PS*.

(b) Calculate the size of the angle $S\hat{R}Q$.

3 The diagram shows a ladder *LD* of length 12 m resting against a vertical wall. The ladder makes an angle of 40° with the horizontal.

Calculate the distance *BD* from the base of the wall to the top of the ladder.

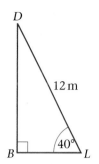

4 A lighthouse, *L*, is 43 km due North of a harbour, *H*. A marker buoy, *B*, is 17 km due East of *L*.

Calculate the bearing of *B* from *H*.

> You may find it helpful to draw a sketch.

5 Explain why it would be impossible to draw the triangle *ABC*.

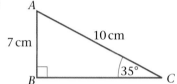

6 Coming in to land, a small aeroplane starts its descent at a vertical height of *h* metres above the horizontal land.

The aeroplane descends along a straight line at a constant angle of depression of 12°.

From starting its descent to touching down, the aeroplane travels through a distance of 6000 m.

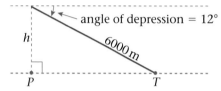

(a) Calculate the vertical height, *h*, at which the aeroplane starts its descent.

At the start of its descent, the aeroplane is vertically above a point *P* on the ground. It touches down at a point *T*.

(b) Calculate the distance *PT*.

7 This is the cross-section, *ABCD*, of a valley.

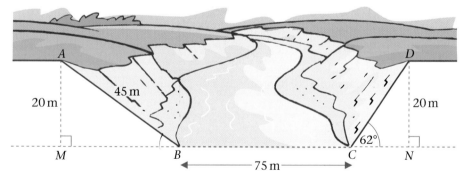

The vertical heights *AM* and *DN* above the horizontal base *BC* of the valley are both equal to 20 metres.

     AB = 45 m   angle *NCD* = 62°   *BC* = 75 m

Calculate

(a) the size of the angle *MBA*

(b) the distance from *A* to *D* across the top of the valley.

8 (a) Sketch the graph of $y = 2 \cos x$ for values of x from $0°$ to $360°$.

   (b) Hence find all solutions of the equation

   $$2 \cos x = -1$$

   which are in the range $0°$ to $360°$.

9 The height of the tide, h metres, in a river estuary is modelled by the function

   $$h = 3 \sin 20t + 10$$

   where t is the number of hours after midnight.

   (a) What is the greatest depth of the water?

   (b For how long is the water over 8 metres in depth for the first time, starting from midnight?

10 Find all the solutions between $0°$ and $360°$ for the following functions:

   (a) $3 \sin \theta = 2$ (b) $5 \cos \theta + 2 = 0$ (c) $3 \cos \theta - 1 = 0$

Summary of key points

1 The three basic **trigonometric functions** are $\sin x$, $\cos x$ and $\tan x$.

2 $\sin x = \dfrac{\text{opp}}{\text{hyp}}$

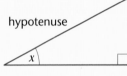

3 $\cos x = \dfrac{\text{adj}}{\text{hyp}}$

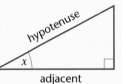

4 $\tan x = \dfrac{\text{opp}}{\text{adj}}$

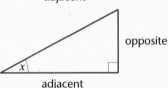

5 The extended graph of $y = \sin x$ is

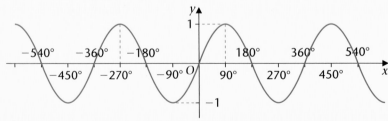

6 $\sin(-x) = -\sin x$ $\sin(180° - x) = \sin x$ $\sin(180° + x) = -\sin x$ $\sin(360° + x) = \sin x$

7 The extended graph of $y = \cos x$ is

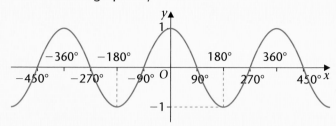

8 $\cos(-x) = \cos x$ $\qquad \cos(180° - x) = -\cos x$ $\qquad \cos(180° + x) = -\cos x$ $\qquad \cos(360° + x) = \cos x$

9 The extended graph of $y = \tan x$ is

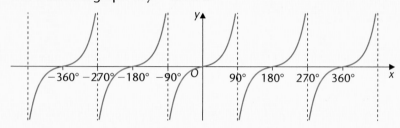

10 $\tan(-x) = -\tan x$ $\qquad \tan(180° - x) = -\tan x$ $\qquad \tan(180° + x) = \tan x$ $\qquad \tan(360° + x) = \tan x$

11 A trigonometric equation may have several solutions in any given range.

17 Graphs and equations

In this chapter you will draw graphs of equations involving powers of x, and for real-life situations. You will also use trial and improvement methods to solve equations involving powers.

17.1 Graphs of quadratic functions

A **quadratic function** is one in which the highest power of x is x^2.
For example $x^2 - 7$, $2x^2 - 3x + 2$ and $3x^2 + 4x$ are quadratic functions.

The graph of a quadratic function is a U-shaped curve called a **parabola**.

The simplest parabola has the equation $y = x^2$.

Here is a **table of values** for $y = x^2$:

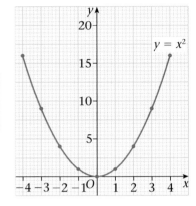

When $x = -4$:
$y = (-4)^2$
$\quad = -4 \times -4$
$\quad = 16$

x	-4	-3	-2	-1	0	1	2	3	4
y	16	9	4	1	0	1	4	9	16

You need to draw a smooth curve through the set of points.
Here is the parabola. Its line of symmetry is the y-axis.

Example 1

(a) Draw the graph of $y = 2x^2 - 4x - 3$, taking values of x from -2 to 4.
(b) Draw in the line of symmetry.
(c) Label the minimum point M with a cross.
(d) Write down the minimum value of y.

(a) Here is a table of values:

x	-2	-1	0	1	2	3	4
$2x^2$	8	2	0	2	8	18	32
$-4x$	$+8$	$+4$	0	-4	-8	-12	-16
-3	-3	-3	-3	-3	-3	-3	-3
y	13	3	-3	-5	-3	3	13

Here is the graph:

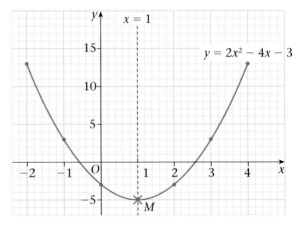

(b) The line of symmetry is $x = 1$. (c) See graph. (d) $y = -5$

Exercise 17A

(a) Draw graphs with the following equations, taking values of x from -4 to 4.

(b) In each case draw the line of symmetry and write down its equation.

(c) Label each minimum point M with a cross.

(d) Write down the minimum or maximum value of y.

1 $y = x^2 + 5$ **2** $y = x^2 - 10$

3 $y = 3x^2$ **4** $y = \frac{1}{2}x^2$

5 $y = -x^2$ **6** $y = -2x^2$

7 $y = x^2 + 2x$ **8** $y = x^2 + 3x$

9 $y = (x + 1)^2$ **10** $y = (x - 2)^2$

11 $y = x^2 - 4x - 1$ **12** $y = x^2 - 3x - 5$

13 $y = x^2 - x + 4$ **14** $y = 2x^2 + 3x - 5$

15 $y = 3x^2 - 4x + 2$

17.2 Graphs of cubic functions

> A **cubic function** is one in which the highest power of x is x^3.
> For example $x^3 - 3x + 2$, $5x^3$ and $2x^3 + 5x^2 + 4x - 7$ are all cubic functions.

The simplest cubic function has the equation $y = x^3$.

x	-3	-2	-1	0	1	2	3
y	-27	-8	-1	0	1	8	27

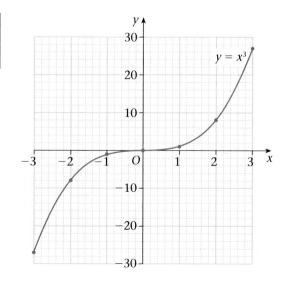

When $x = -3$:

$$y = (-3)^3$$
$$= (-3) \times (-3) \times (-3)$$
$$= -27$$

A tracing of this graph will fit on top of itself if you rotate it through $180°$ about the origin $(0, 0)$.
The graph has **rotational symmetry** of order 2 about the origin.

All graphs of cubic functions have this S-shape.

Example 2

Draw the graph of $y = x^3 - 2x^2 - 4x$, taking values of x from -2 to 4.

Here is a table of values for
$y = x^3 - 2x^2 - 4x$:

x	-2	-1	0	1	2	3	4
y	-8	1	0	-5	-8	-3	16

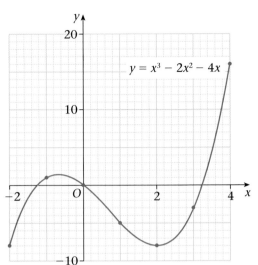

When $x = -2$:
$y = (-2)^3 - 2 \times (-2)^2 - (4 \times -2)$
$\quad = -8 - 8 + 8$
$\quad = -8$

When $x = 3$:
$y = 3^3 - 2 \times 3^2 - (4 \times 3)$
$\quad = 27 - 18 - 12$
$\quad = -3$

Exercise 17B

Draw graphs with the following equations, taking values of
x from -3 to 3.

1 $y = x^3 + 5$

2 $y = x^3 - 10$

3 $y = 2x^3$

4 $y = \frac{1}{2}x^3$

5 $y = -x^3$

6 $y = -2x^3$

7 $y = (x - 2)^3$

8 $y = (x + 1)^3$

9 $y = x^3 - 2x^2$

10 $y = x^3 + x^2$

11 $y = x^3 + 5x$

12 $y = x^3 - 5x$

13 $y = x^3 + x^2 - 8x$

14 $y = x^3 + 3x - 2$

15 $y = x^3 - 2x^2 + 2$

17.3 Graphs of reciprocal functions

To find the **reciprocal** of a number or expression divide it into 1.

For example:

- the reciprocal of 2 is $1 \div 2 = \frac{1}{2}$

- the reciprocal of x is $1 \div x = \dfrac{1}{x}$

Example 3

(a) Make a table of values for $y = \dfrac{1}{x}$ and draw the graph.

(b) Is y undefined for any value of x?

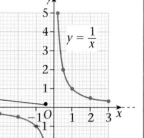

When x is very small and +ve y is $\dfrac{1}{\text{very small}}$ and +ve so y is very big and +ve.

(a) Here is a table of values for $y = \dfrac{1}{x}$:

x	-3	-2	-1	-0.5	-0.2	0.2	0.5	1	2	3
y	-0.3	-0.5	-1	-2	-5	5	2	1	0.5	0.3

When $x = -3$:
$$y = \frac{1}{-3} = -\frac{1}{3}$$
$$= -0.3 \text{ to 1 d.p.}$$

There is a **discontinuity** (break) in the graph at $x = 0$.

(b) y is undefined when $x = 0$ as division by 0 is undefined.

The lines $y = x$ and $y = -x$ are lines of symmetry of the graph $y = \dfrac{1}{x}$.

The graph approaches the axes but does not touch them. The axes are **asymptotes**.

A line which a graph approaches without touching is called an **asymptote**.

Example 4

(a) Make a table of values for $y = 3 - \dfrac{2}{x}$ and draw the graph.

(b) Draw in the asymptotes on the graph.

(c) Write down their equations

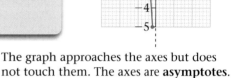

Asymptotes are usually drawn with dotted lines.

(a)

x	-3	-2	-1	-0.5	-0.2	0.2	0.5	1	2	3
y	3.7	4	5	7	13	-7	-1	1	2	2.3

When $x = -3$:
$$y = 3 - \frac{2}{-3}$$
$$= 3 + \frac{2}{3}$$
$$= 3.7 \text{ to 1 d.p.}$$

When $x = -0.5$:
$$y = 3 - \frac{2}{-0.5}$$
$$= 3 + 4$$
$$= 7$$

(b) See graph.

(c) The asymptotes are $x = 0$ and $y = 3$.

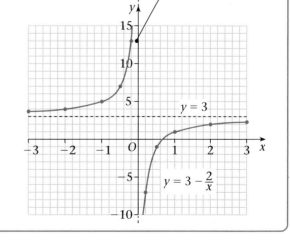

Exercise 17C

Draw graphs for the following equations. Use the same x-values as in Example 4. Draw in the asymptotes and write down their equations.

It will help if you make a table of values first.

1 $y = \dfrac{2}{x}$

2 $y = \dfrac{-1}{x}$

3 $y = \dfrac{-3}{x}$

4 $y = 4 + \dfrac{1}{x}$

5 $y = 5 - \dfrac{4}{x}$

6 $y = \dfrac{4}{x} - 3$

17.4 Graphs of the form $y = ax^3 + bx^2 + cx + d + \dfrac{e}{x}$

Sometimes you will need to plot graphs of equations that have a combination of cubic, quadratic, linear and reciprocal terms.

For example: $y = ax^3 + bx^2 + cx + d + \dfrac{e}{x}$

Here a, b, c, d, e are numbers (also called coefficients).

At GCSE level, if you are asked to plot graphs of this form, at least two of the coefficients will be zero.

___ Example 5 _____

Make a table of values for $y = x + \dfrac{1}{x}$ and draw the graph.

Here is a table of values:

You could check this using a graphical calculator, computer or graph plotter.

x	0.1	0.2	0.5	1	2	3	4	5
y	10.1	5.2	2.5	2	2.5	3.3	4.3	5.2

when $x = 0.1$:
$y = 0.1 + \frac{1}{0.1}$
$= 0.1 + 10$
$= 10.1$

when $x = 0.2$:
$y = 0.2 + \frac{1}{0.2}$
$= 0.2 + 5$
$= 5.2$

when $x = 2$:
$y = 2 + \frac{1}{2}$
$= 2.5$

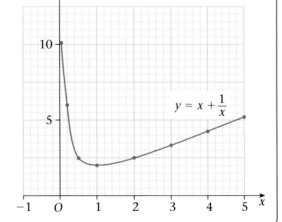

$y = x + \dfrac{1}{x}$

Example 6

Make a table of values for $y = x^2 + 3 + \dfrac{2}{x}$ and draw the graph.

Here is a table of values:

x	-3	-2	-1	-0.5	-0.2	0.2	0.5	1	2	3
y	11.3	6	2	-0.8	-7.0	13.0	7.3	6	8	12.7

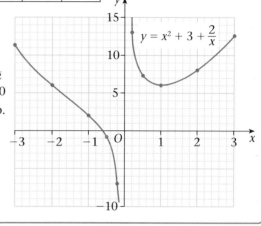

When $x = -3$:
$$y = (-3)^2 + 3 + \tfrac{2}{-3}$$
$$= 9 + 3 - \tfrac{2}{3}$$
$$= 11\tfrac{1}{3}$$
$$= 11.3 \text{ to 1 d.p.}$$

When $x = 0.2$:
$$y = 0.2^2 + 3 + \tfrac{2}{0.2}$$
$$= 0.04 + 3 + 10$$
$$= 13.0 \text{ to 1 d.p.}$$

Exercise 17D

Draw graphs with the following equations. Use the values of x given
in brackets to make a table of values.

1 $y = x^2 - 4x + 1$ (integers from -1 to 4)   2 $y = x^3 + 2x^2 - 3$ (integers from -3 to 2)

3 $y = x^3 - x^2 - 2x$ (integers from -2 to 3)   4 $y = x^3 - 5x + 2$ (integers from -3 to 3)

5 $y = 1 + 2x - x^2$ (integers from -2 to 4)   6 $y = 2x + \dfrac{1}{x}$ (values as in Example 6)

7 $y = x^2 - \dfrac{1}{x}$ (values as in Example 6)   8 $y = x + 2 + \dfrac{2}{x}$ (values as in Example 5)

9 $y = x^3 - \dfrac{1}{x}$ (0.2, 0.5, 1, 2, 3)   10 $y = x^2 - x + \dfrac{2}{x}$ (values as in Example 6)

17.5 Using graphs to solve equations

The solutions to the quadratic equation $x^2 - 4 = 0$
are $x = 2$ and $x = -2$. Notice that these are the values
of x where the graph of $y = x^2 - 4$ cuts the x-axis.

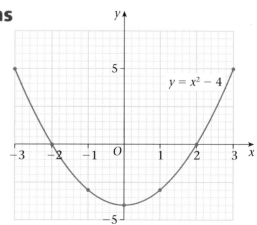

> The **solutions** of a quadratic equation are the
> values of x where the graph cuts the x-axis.

A similar method can be used to find approximate
solutions to higher order equations. There is more
about this in Chapter 19.

Worked examination question 1

(a) Make a table of values for $y = x^2 - 2x - 2$.

(b) Plot the points represented by the values in your table on a grid and join them with a smooth curve.

(c) Use your graph to solve the equation $x^2 - 2x - 2 = 0$. Give your answers correct to 1 d.p.

(a) Here is a table of values:

x	-2	-1	0	1	2	3	4
x^2	4	1	0	1	4	9	16
$-2x$	4	2	0	-2	-4	-6	-8
-2	-2	-2	-2	-2	-2	-2	-2
y	6	1	-2	-3	-2	1	6

(b) Here is the graph:

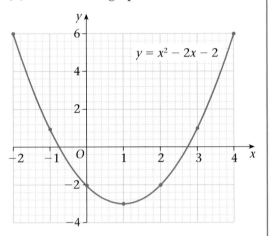

(c) The solutions to the equation $x^2 - 2x - 2 = 0$ are the values of x where the graph cuts the x-axis.

These are $x = -0.7$ and $x = 2.7$

Exercise 17E

Solve the following equations correct to 1 d.p. by drawing appropriate graphs. Use the values of x given in brackets to make a table of values.

1 $x^2 - 7x + 8 = 0$ (integers from 0 to 8)

2 $x^2 - x - 3 = 0$ (integers from -3 to 4)

3 $2x^2 - 3x - 7 = 0$ (integers from -3 to 4)

4 $5x^2 - 8x + 2 = 0$ (integers from -1 to 3)

5 $x^3 - 4x + 1 = 0$ (integers from -3 to 3) •————— There are 3 solutions.

6 $x^3 + x^2 - 3x = 0$ (integers from -3 to 2)

7 $x^3 - 2x^2 - 1 = 0$ (integers from -2 to 3) •————— There is only 1 solution.

8 $x^3 - 4x^2 + 4x = 0$ (integers from -1 to 4)

9 $x - \dfrac{5}{x} = 0$ (0.2, 0.5, 1, 2, 3, 4)

10 $x^2 - 3x + \dfrac{1}{x} = 0$ (0.2, 0.5, 1, 2, 3, 4)

17.6 Trial and improvement methods for solving equations

You can use trial and improvement methods to solve an equation to any degree of accuracy.

You would normally use a calculator for your calculations.

___ **Example 7** ___

Find the positive solution to $x^3 - 3x - 1 = 0$.
Give your answer correct to 1 d.p.

x	0	1	2
$x^3 - 3x - 1$	-1	-3	1

There is a solution to $x^3 - 3x - 1 = 0$ between $x = 1$ and $x = 2$.

It is not obvious whether the solution is nearer to 1 or to 2.

You could try 1.1, 1.2, ... until you find the first value that gives $x^3 - 3x - 1 > 0$.
It often saves time to try the middle value of an interval to establish in which half of the interval the solution lies:

x	1.5
$x^3 - 3x - 1$	-2.125 — too small

So there is a solution to $x^3 - 3x - 1 = 0$ between $x = 1.5$ and $x = 2$

x	1.6	1.7	1.8	1.9
$x^3 - 3x - 1$	-1.704	-1.187	-0.568	0.159

too small   too large

So the solution is between $x = 1.8$ and $x = 1.9$

0.159 is nearer to 0 than -0.568 is. This suggests that the solution is nearer to 1.9 than it is to 1.8, but you must confirm this by trying $x = 1.85$:

> You must always go to the next decimal place to confirm the solution, so here try 1.85.

x	1.85
$x^3 - 3x - 1$	$-0.218\,38$

There is a solution to $x^3 - 3x - 1 = 0$ between $x = 1.85$ and $x = 1.9$

So the solution is $x = 1.9$ (to 1 d.p.)

You can use a **trial and improvement method** to solve an equation. Try a value in the equation and use your result to improve your estimate. Repeat, getting closer and closer to the correct value.

Worked examination question 2

Using the method of trial and improvement, or otherwise, find the positive solution of $x^3 + x = 187$.
Give your answer correct to 1 d.p.

x	$x^3 + x$	
5	130	too small
6	222	too large
5.5	171.875	too small
5.6	181.216	too small
5.7	190.893	
5.65	186.012	

The soltion lies between 5.65 and 5.7

$x = 5.7$ (to 1 d.p.)

Exercise 17F

Find the positive solutions to these equations by trial and improvement.
Give your answers to the level of accuracy stated.

1 $x^3 - x - 1 = 0$ (1 d.p.) **2** $x^3 - x - 7 = 0$ (1 d.p.)

3 $x^3 - 4x - 1 = 0$ (1 d.p.) **4** $x^3 + 2x - 20 = 0$ (2 d.p.)

5 $x^3 + x = 300$ (1 d.p.) **6** $x^3 + x^2 = 700$ (2 d.p.)

7 $x^3 + 5x - 200 = 0$ (2 s.f.) **8** $x^3 - 4x = 300$ (3 s.f.)

9 $x^3 + 2x^2 = 120$ (2 d.p.) **10** $x^2 - \dfrac{2}{x} = 5$ (2 s.f.)

17.7 Solving problems by trial and improvement

Look at this problem:

Find the radius of a circle whose area is 100 cm².

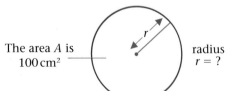

The area A is 100 cm²

radius $r = ?$

You can find the radius by guessing it and trying it in the formula

$A = \pi r^2$

If at first you don't succeed, try, try and try again!

Sometimes you can solve problems like this using trial and improvement.

Example 8

Use a trial and improvement method to find the radius r of a circle whose area is 100 cm².
Your answer should be accurate to 2 d.p.

r	πr^2	
5	78.539...	too small
7	153.938...	too big
6	113.097...	too big
5.5	95.033...	too small
5.75	103.868...	too big
5.65	100.287...	too big
5.64	99.932...	too small
5.645	100.110...	too big

Method:

Guess a value for r.
Calculate πr^2.
Compare your answer with 100.

You can now see that the radius r lies between 5.64 cm and 5.645 cm. Therefore $r = 5.64$ cm to 2 d.p.

Exercise 17G

1 Use a trial and improvement method to find the length l of the side of a cube whose volume V is 100 cm³.

The formula is $V = l^3$

Your answer should be accurate to 2 d.p.

2 Use a trial and improvement method to find the length l of the side of a cube whose total surface area A is 50 cm².

The formula is $A = 6l^2$

Your answer should be accurate to 2 d.p.

3 Convert a temperature of 22 °C to Fahrenheit (°F) by trial and improvement.

Your answer should be accurate to 1 d.p.

The formula to convert from °F to °C is $(F-32) \times 5 \div 9$

22°C

18°C

17.8 Graphs that describe real-life situations

Graphs can be used to describe a wide variety of real-life situations.
You may have to interpret or sketch graphs of this type.

Two important real-life graphs are distance–time graphs and speed–time graphs.

Distance–time graphs

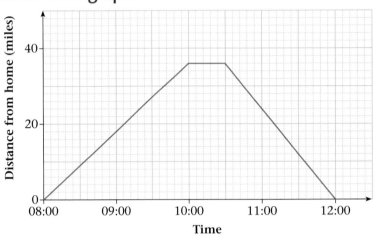

The diagram shows the distance–time graph for a cycle ride. Between
08:00 and 10:00 the cyclist travels 36 miles. This part of the graph
is a straight line because she travels at a constant speed.

Her constant speed is 36 miles ÷ 2 hours = 18 mph, which is the
gradient of the line.

Between 10:00 and 10:30 (the horizontal part of the graph) the
cyclist is stationary, having a rest perhaps.

Between 10:30 and 12:00 she travels back home at a constant
speed of 36 miles ÷ 1.5 hours = 24 mph.

> On a **distance–time graph**, the gradient gives the speed.

Speed–time graphs

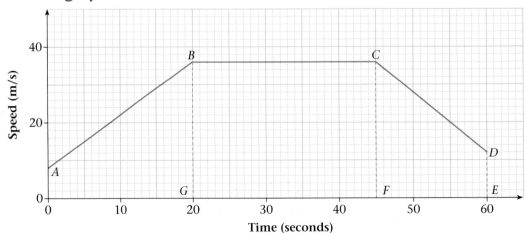

The diagram shows the speed–time graph for one minute of a car's journey.

In the first 20 seconds, its speed increases from 8 m/s to 36 m/s.
The car's speed increases steadily, so AB is a sraight line.
The gradient of the line AB gives the rate of increase of speed:

$$\frac{36 - 8}{20} = 1.4 \text{ m/s per second}$$

Rate of increase of speed is called **acceleration** and the units in this case are m/s per second or m/s^2.

In the first 20 seconds the car has a constant acceleration of 1.4 m/s^2.

Between $t = 20$ and $t = 45$, the car travels at a constant speed of 36 m/s.
Between $t = 45$ and $t = 60$, its speed decreases steadily.
The gradient of the line CD gives the constant acceleration as -1.6 m/s^2.

Negative acceleration is called **deceleration** and you could say that the deceleration between $t = 45$ and $t = 60$ is 1.6 m/s^2.

When the car is travelling at a constant speed of 36 m/s for 25 seconds, between $t = 20$ and $t = 45$, it travels a distance of $36 \times 25 = 900$ m. The area of rectangle $BCFG$ represents this distance.

Similarly, the area of trapezium $ABGO$ represents the distance travelled between $t = 0$ and $t = 20$:

$$\tfrac{1}{2}(8 + 36) \times 20 = 440 \text{ m}$$

The area of trapezium $CDEF$ represents the distance travelled between $t = 45$ and $t = 60$:

$$\tfrac{1}{2}(36 + 12) \times 15 = 360 \text{ m}$$

So you can work out the total distance travelled by the car in the whole minute:

$$900 + 440 + 360 = 1700 \text{ m}$$

> On a **speed–time graph**, the gradient gives the acceleration.
> When the graph is a curve, draw a tangent and find its gradient.

Sometimes 'velocity' is used instead of 'speed'.

> The area under a speed–time graph gives the distance travelled.

Drawing graphs to show water levels

The vases **A**, **B**, **C** and **D** have circular cross-sections and they contain water.

They all start off with the same depth of water. The water is pumped out of the vases at the same steady rate.

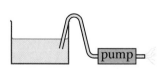

A   **B**   **C**   **D**

The graph shows the relationship between the water level in each vase and the volume of water pumped out of it.

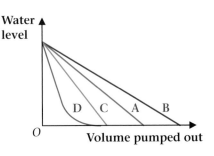

- Vase **B** has a bigger area of cross-section than **A**. When the same volume has flowed out of each vase the water level in **B** remains higher than in **A**. So the graph for **B** is less steep than the graph for **A**.
- The water level in vase **C** drops even more quickly than in **A**, so the graph for **C** is steeper.
- In vase **D** the level drops more quickly and steadily at first, then it gradually drops more slowly. The graph is straight to begin with (steeper than for **C**), but then it is curved to show the changing speed.

Worked examination question 3

A DJ can control the sound level of the records he plays at a club.

The sketch graph is a graph of sound level against time while one record was being played.

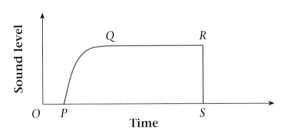

(a) Describe how the sound level changed between *P* and *Q* on the graph while the record was being played.

(b) Give one possible reason for the third part, *RS*, of the sketch graph.

(a) The sound level increases quickly at first and then more slowly. It then levels out.

(b) The sound level suddenly drops to zero. Possible reasons for this are the end of the record or a power cut.

Exercise 17H

1 Sharon and Tracey cycle from Sinton to Coseley and back again. The graph shows their journeys.

(a) Who sets off first?

(b) Describe what happens at *A*.

(c) Describe what happens at *B*.

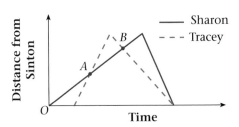

2 Heather walks home from school at a steady speed. Halfway home, she has a rest before continuing her journey at the same speed. Sketch a distance–time graph to show her journey.

3 The diagram shows a distance–time graph for a coach trip from Birmingham to Bristol and back again.

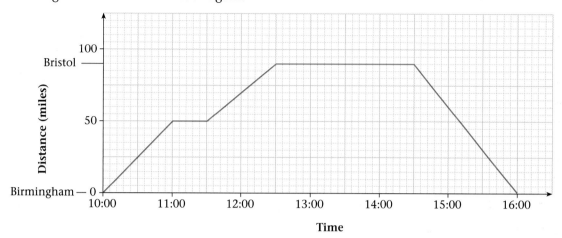

(a) At what time did the coach first stop?

(b) At what time did the coach reach Bristol?

(c) For how long did the coach stay in Bristol?

(d) Calculate the speed of the coach on the return journey.

4 A car in traffic travelled for 20 seconds at 15 m/s, stopped for 15 seconds and then travelled for 25 seconds at 10 m/s. Draw a distance–time graph for this 60-second period.

5 The diagram shows a speed–time graph for 10 seconds of a train's journey

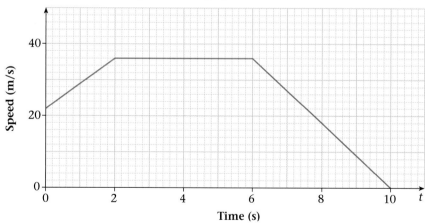

(a) Find the acceleration between $t = 0$ and $t = 2$.

(b) Describe the train's motion between $t = 2$ and $t = 6$.

(c) Find the acceleration between $t = 6$ and $t = 10$.

(d) Find the distance travelled by the train in the 10 seconds.

6 A car accelerates from rest at 5 m/s² for 6 seconds, travels with constant speed for 10 seconds and decelerates at 3 m/s² for 4 seconds.

(a) Draw a speed–time graph for this 20-second period.

(b) Find the total distance travelled by the car in the 20 seconds.

7 Here are some graphs:

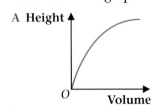

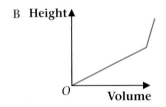

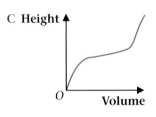

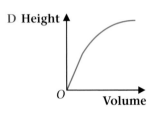

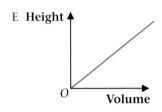

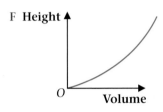

Coloured liquid is poured into the containers below. For each one write down the letter of the graph which best illustrates the relationship between the height of the liquid and the volume in the container.

(a)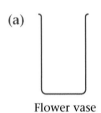

Flower vase

(b)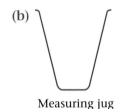

Measuring jug

(c)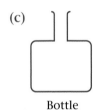

Bottle

(d)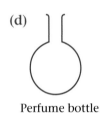

Perfume bottle

8 Water is poured into each of these containers. Sketch a graph to show the relationships between the water level and the volume of water in each container.

(a)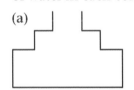

(b)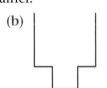

(c)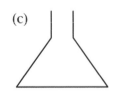

(d)

9 Nigel drives 100 miles. Sketch a graph to show the relationship between his average speed and the time he takes.

10 Sketch a graph to show how the value of a new car changes over a number of years.

11 (a) One evening Mr Fish marks a set of maths exam papers. From Graph A describe the three parts of his evening.

(b) Graph B shows the number of customers in a Post Office during one day.
   (i) What was the busiest part of the day?
   (ii) What happened in the time between P and Q?

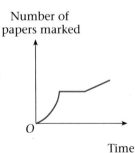

Graph A

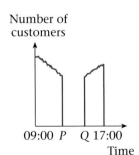

Graph B

12 The diagrams show the shapes of five graphs, A, B, C, D and E.

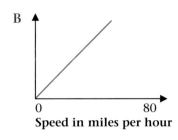

A

0   80
Speed in miles per hour

B

0   80
Speed in miles per hour

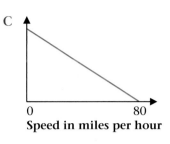

C

0   80
Speed in miles per hour

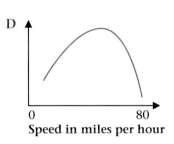

D

0   80
Speed in miles per hour

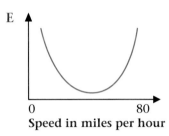

E

0   80
Speed in miles per hour

The vertical axes have not been labelled.
On one of the graphs, the missing label is 'Speed in km per hour'.

(a) Write down the letter of this graph.

On one of the graphs the missing label is 'Petrol consumption
in miles per gallon'. It shows that the car travels furthest on
1 gallon of petrol when it is travelling at 56 miles per hour.

(b) Write down the letter of this graph.   [E]

Mixed exercise 17

 1 The equation $x^3 + 4x = 100$ has one solution which is a positive number.
Use the method of trial and improvement to find this solution.

Give your answer to 1 decimal place.
You must show all your working.   [E]

 2 The equation $x^3 - 2x = 67$ has a solution between 4 and 5.
Use the method of trial and improvement to find this solution.

Give your answer to 1 decimal place.
You must show all your working.   [E]

3 Draw the graph of $y = 2 - \dfrac{3}{x}$ for values of x from -3 to $+3$.

Draw in the asymptotes and label them with their equations.

4 Plot the graph of $x^3 - 3x^2 + 14$ for values of x from -2 to $+3$.
Use your graph to find the solutions of

(a) $x^3 - 3x^2 + 14 = 0$

(b) $x^3 - 3x^2 + 4 = 0$

Draw $y = 3x$.

(c) $x^3 - 3x^2 + 14 = 3x$

(d) $2x^3 - 6x^2 = 9x - 12$

Divide by 2, then add 14
to both sides.

5 **(a)** Copy and complete the table of values for $y = x^2 - 6x + 10$.

x	0	1	2	3	4	5	6
y	10						10

(b) Hence draw the graph of $y = x^2 - 6x + 10$.

(c) Use your graph to find an estimate for the minimum value of y.

(d) Use a graphical method to find estimates of the solutions to the equation $x^2 - 3x + 1 = 2x - 4$.

6 Jim invests a sum of money for 30 years at 4% per annum compound interest.

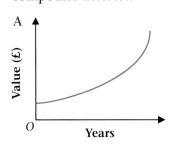

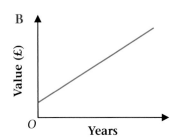

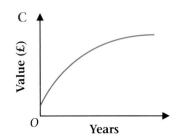

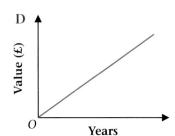

Write down the letter of the graph which best shows how the value of Jim's investment changes over the 30 years.

7 **(a)** Copy and complete the table of values for $y = x^3 - 3x^2 + 2$.

x	-2	-1	0	1	2	3	4
y		-2					

(b) Hence draw the graph of $y = x^3 - 3x^2 + 2$.

(c) Use your graph to find estimates, correct to 1 decimal place where appropriate, for the solutions of

    (i) $x^3 - 3x^2 + 2 = 0$

    (ii) $x^3 - 3x^2 - 4 = 0$ [E]

8   Here is a velocity–time graph for the first 6 seconds of the movement of an object:

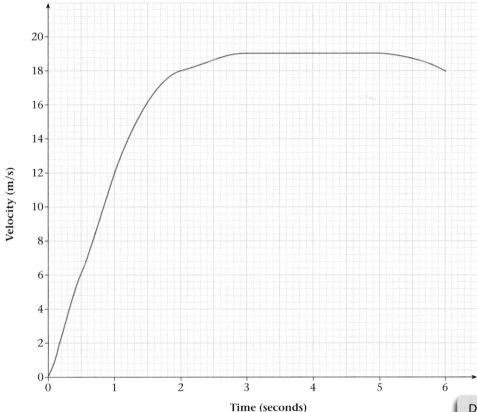

Velocity (m/s) / Time (seconds)

(a)   Calculate an estimate for the object's acceleration at $1\frac{1}{2}$ seconds. Give your answer in m/s².

> Draw a tangent (a straight line that touches the curve) at $(1\frac{1}{2}, 16)$.

(b)   Calculate an estimate for the distance the object travelled in the first 6 seconds.   [E]

Summary of key points

1   A **quadratic function** is one in which the highest power of x is x^2. For example $x^2 - 7$.

2   The graph of a quadratic function is a U-shaped curve called a **parabola**.

3   A **cubic function** is one in which the highest power of x is x^3. For example $x^3 - 3x + 2$.

4   To find the **reciprocal** of a number or expression divide it into 1.
    For example, the reciprocal of x is $1 \div x = \dfrac{1}{x}$.

5   A line which a graph approaches without touching is called an **asymptote**.

6   The **solutions** of a quadratic equation are the values of x where the graph cuts the x-axis.

7   You can use a **trial and improvement method** to solve an equation. Try a value in the equation and use your result to improve your estimate. Repeat, getting closer and closer to the correct value.

8   On a **distance–time graph**, the gradient gives the speed.

9   On a **speed–time graph**, the gradient gives the acceleration.
    When the graph is a curve, draw a tangent and find its gradient.

10   The area under a speed–time graph gives the distance travelled.

18 Proportion

The GCSE exam regularly contains questions involving proportion. In this chapter you will solve a variety of problems using proportion.

18.1 Direct proportion

Joan volunteers to take part in a charity fund raising walk. She is given 50 pence for each kilometre she walks.

The further she walks the more money she raises. If she doubles the distance she walks then she doubles the money she raises. But if she only walks half the distance then she raises only half the money.

This is an example of direct proportionality. The money collected, y, is **directly proportional** to the distance walked, x.

The symbol $\propto$ means 'is proportional to'.

$y \propto x$ means 'y is **directly proportional** to x'.

A simple 'rule of thumb' method to check whether one quantity is directly proportional to another is to try these two tests:

- If one quantity is zero, is the other also zero?
- If one quantity doubles, does the other quantity also double?

Exercise 18A

1 Which of the following could be examples of direct proportionality?
   (a) profit made by selling goods
   (b) final examination results and effort put into work
   (c) area of a square and the length of one side
   (d) the area of a rectangle with one side of constant length and the length of the other side
   (e) time and distance travelled at a constant speed
   (f) the height and weight of a student

2 Given that $w \propto t$ and $w = 8$ when $t = 6$, find
   (a) t when w doubles in value   (b) w when t halves in value.

3 Given that $a \propto b$ and $a = 10$ when $b = 8$, find
   (a) b when a increases to 30   (b) a when b decreases to 2.

4 Given that $d \propto s$ and $d = 36$ when $s = 16$, find
   (a) s if d decreases by 9   (b) d if s increases to 24.

18.2 Graphs that show direct proportion

In a science experiment the time it takes a small pump to fill a water cylinder is recorded. The height of the water in the cylinder is measured at regular time intervals.

Here are the results:

Height of water (cm)	2.1	4.2	6.3	8.4
Time taken (s)	5	10	15	20

The graph on the right shows the result of plotting height against time and joining the points.

A straight line connects the points. This indicates that the two quantities, height and time taken to reach that height, are in direct proportion.

> When a graph of two quantities is a straight line through the origin, one quantity is **directly proportional** to the other.

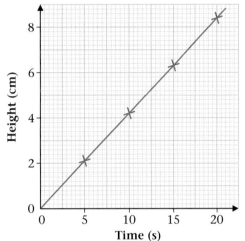

This means that if one quantity changes then the other changes in the same ratio.

Notice that this is consistent with the 'rule of thumb' in the previous section:

- If one quantity is zero, the other is also zero.
- If one quantity doubles, the other quantity also doubles.

You can use a graph to help find the rule connecting quantities that are in direct proportion to one another.

Example 1

In which of these graphs is s directly proportional to r?

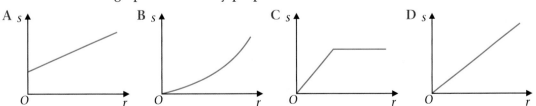

Graph D is a straight line through the origin, showing that s is directly proportional to r.

Example 2

The length of a shadow cast by a tree at midday is directly proportional to the height of the tree. At midday a tree 8 metres tall casts a shadow 10 metres long.

(a) Find a relationship connecting the height of the tree and the length of its shadow at midday.

(b) The shadow of another tree at midday is 8 metres long. How tall is the tree?

(c) Calculate the length of the shadow at midday of a tree 10 metres tall.

(a) The height and the shadow length are directly proportional to one another. The relationship can be shown on a straight line graph like this:

The equation of a straight line is

$$y = mx + c$$

For direct proportion the straight line passes through the origin so

$$c = 0 \quad \text{and} \quad y = mx$$

The gradient m of the line can be used to find the rule connecting h and s. Replacing y with h, and x with s, gives

$$h = ms$$

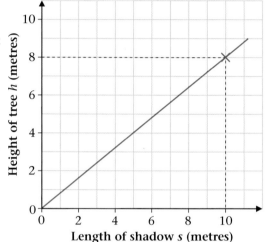

From the graph, the gradient $m = \frac{8}{10} = 0.8$.

So the relationship connecting the height of the tree and the length of the shadow is

$$h = 0.8s$$

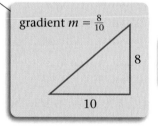

gradient $m = \frac{8}{10}$

For more on gradients see Chapter 9.

(b) The relationship $h = 0.8s$ can be used to find the height h of a tree whose shadow is $s = 8$ metres long:

$$h = 0.8 \times 8 = 6.4 \text{ metres}$$

The height of the tree is 6.4 metres when the shadow is 8 metres long.

(c) The relationship can also be used to find the length of shadow for a tree of height $h = 10$ metres:

$$h = 0.8s$$

So $\quad 10 = 0.8s$

$$\frac{10}{0.8} = s \quad\text{——— Divide both sides by 0.8}$$

$$12.5 = s$$

For a tree height of 10 metres, the shadow is 12.5 metres long.

Exercise 18B

1   At 10 am a tree 10 metres high casts a shadow 8 metres long.
The length of the shadow (*l* metres) is directly proportional to
the height (*h* metres) of the tree.
   (a)   Sketch a graph to show this information.
   (b)   Find a rule connecting *h* and *l*.
   (c)   Find *h* when *l* = 12.5.
   (d)   Find *l* when *h* = 15.

2   In which of these graphs is *x* directly proportional to *y*?

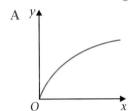   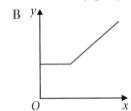   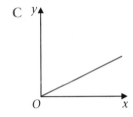

3   The area (*A* cm²) of a shape is directly proportional to the length
(*l* cm) of one of its sides. When *l* = 14, *A* = 42.
   (a)   Sketch a graph to show this information, with area on the
vertical axis.
   (b)   Work out a rule connecting *A* and *l*.
   (c)   What is the area of a similar shape with a side length of 32.4 cm?
   (d)   Find *l* when *A* = 24.6.

4   In the recipe for Yorkshire puddings, 2 eggs (*e*) are required to
make 24 small Yorkshire puddings (*Y*).
   (a)   Sketch the graph of *Y* = 12*e* for *e* = 0 up to 10.
   (b)   Explain why the equation *Y* = 12*e* represents the number of
eggs to make a certain number of small Yorkshire puddings.
   (c)   How many eggs would be needed to make 84 small Yorkshire
puddings?

18.3 Using ratios to find proportionality rules

You can use ratios to help find the rule connecting quantities that are
in direct proportion to one another.

Example 3

The mass of a silver trophy is directly proportional to its height.

A trophy of height 8.2 cm has a mass of 1.148 kg.
   (a)   Find a rule connecting the mass *m* and height *h* of a trophy.
   (b)   Find the mass of a similar trophy with a height of 15.5 cm.
Give your answer correct to 2 d.p.

(a)   Instead of drawing a graph you can present this information in a table:

Mass m (kg)	1.148	?
Height h (cm)	8.2	15.5

The mass to height ratio is

$$1.148 : 8.2 \quad \text{or} \quad \frac{1.148}{8.2} = 0.14$$

The rule connecting the mass m and height h is

$$m = 0.14h$$

(b)   Using the rule, when $h = 15.5$ then $m = 0.14 \times 15.5 = 2.17$

The mass of the second trophy is 2.17 kg.

Exercise 18C

1   These tables each show two variables which are directly proportional to each other.

For each of these tables find
  (i)   a rule connecting the two variables
  (ii) the missing values correct to 1 d.p.

(a)
w	8.4	?
h	8	12

(b)
h	12.0	16.8
s	20.5	?

(c)
p	7.8	?
l	6.8	27.6

(d)
a	110.4	10.4
p	?	23.4

2   The voltage V across a resistor (in volts) is directly proportional to the current I flowing through it (in amps).

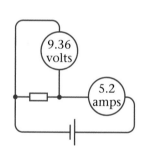

   (a)   Write down the readings of the two meters.
   (b)   Find a rule connecting the voltage and current.
   (c)   Calculate the current when the voltage increases to 11.52 volts.
   (d)   Calculate the voltage when the current is 2.46 amps.

3   Kirstine obtains these readings during an experiment, where V is proportional to I:

Voltage, V	2.8	4.6	8.4	12.6	22.4	28
Current, I	1	2	3	4.5	8	10

Unfortunately there is an error in her table.

   (a)   Plot the points and draw a graph of voltage V against current I.
   (b)   Which reading is wrong, and what should it be?
   (c)   Find the rule connecting V and I.
   (d)   Find V when $I = 3.5$
   (e)   Calculate I when $V = 20.5$, giving your answer correct to 3 d.p.

4   Given that $r \propto t$ and $r = 2.4$ when $t = 12.8$, find correct to
    2 decimal places
    (a) r when $t = 8.9$ (b) t when $r = 3.1$

5   The cost of a bottle of white correction fluid is directly
    proportional to the amount of fluid in the bottle. A bottle
    containing 20 ml of fluid costs 72p. Find
    (a)   a rule connecting the cost and the amount
    (b)   the cost of a bottle containing 100 ml of fluid.

6   On a car journey the distance travelled, d, is directly proportional
    to the time taken, t. In 2 hours a car travels 76 miles. Find
    (a)   a rule connecting d and t
    (b)   the distance travelled in 5 hours
    (c)   the time taken to travel a distance of 418 miles.

18.4 Writing proportionality formulae

When y is directly proportional to x, you can write a **proportionality
statement** and a **formula** connecting y and x:

• $y \propto x$ is the proportionality statement
• $y = kx$ is the proportionality formula, where k is the constant of
  proportionality.

Example 4

The extension E of a spring is directly proportional to the force
F pulling the spring. The extension is 8 cm when a force of
30 N is pulling it. Calculate the extension when the force is 9 N.

As $E \propto F$ the proportionality formula can be written as:

> Force is measured in
> newtons (N).

   $E = kF$

To find the value of the constant of proportionality k,
substitute $E = 8$ and $F = 30$ in the formula:

   $8 = k \times 30$

   $\dfrac{8}{30} = k$ ——— Divide both sides by 30

So the proportionality formula is

   $E = \dfrac{8}{30} F$

When the force F is 9 N, the extension E is

   $E = \dfrac{8}{30} \times 9 = 2.4 \text{ cm}$

To find the value of k, the constant of proportionality, substitute known
values of y and x into $y = kx$.

Exercise 18D

1 y is proportional to x so that $y = kx$.
Given that $y = 148$ when $x = 12$, find the value of k.
Calculate the value of y when $x = 7$.

2 p is proportional to q so that $p = kq$.
Given that $p = 34$ when $q = 51$, find the value of k.
Calculate the value of p when $q = 7$.

3 z varies in direct proportion to x.
Write a formula for z in terms of x.
Given that $z = 3\frac{3}{5}$ when $x = 6$, find the value of z when $x = 14$.

4 z varies in direct proportion to w.
Write a formula for z in terms of w.
Given that $z = 315$ when $w = 7$, find the value of z when $w = 105$.

5 y is proportional to x. Given that $y = 12$ when $x = 4$, calculate the value of
  (a) y when $x = 5$ (b) y when $x = 13$ (c) x when $y = 10$.

6 F varies directly as E. Given that $F = 300$ when $E = 120$, calculate the value of
  (a) F when $E = 90$ (b) F when $E = 500$ (c) E when $F = 180$.

7 Given that $y \propto p$, calculate the values missing from this table:

p	1		10
y		4	28

8 The volume V of liquid in a tube is proportional to the height h of the tube. When the height of the liquid in the tube is 7 cm, the volume of liquid is 10 cm³. Calculate the volume of liquid when the height of liquid is 4 cm.

9 The distance travelled by the tip of the second hand on a clock is proportional to the time elapsed. Calculate the increase in the distance travelled by the tip of the second hand when the time elapsed increases by 40%, giving your answer as a percentage.

10 The height h of an elephant is directly proportional to the diameter d of its footprint. A baby elephant has a height of 150 cm and the diameter of its footprint is 20 cm.
  (a) Find an equation connecting h and d.
  (b) Calculate the height of an elephant with a footprint of diameter 25 cm.
  (c) Calculate the diameter of the footprint of an elephant whose height is 3.2 m.

18.5 Square and cubic proportionality

Sets of Russian dolls are made to fit inside each other. They are exactly the same shape as each other and are mathematically **similar**.

One of the dolls is 8 cm high and has a surface area of 115.2 cm².

The surface area S of each doll is proportional to the square of its height h. This can be written as the proportionality statement

$S \propto h^2$

The proportionality formula can be written as

$S = kh^2$

> Reminder:
> When two shapes are **similar**, corresponding lengths on the two shapes are in proportion.

The constant of proportionality k can be found by substituting information about the 8 cm high doll into the formula:

$$115.2 = k \times 8^2$$
$$64k = 115.2$$
$$k = \frac{115.2}{64}$$
$$k = 1.8$$

So the proportionality formula is

$S = 1.8h^2$

The formula can be used to find the surface area of a doll of height 10 cm:

$$S = 1.8 \times 10^2$$
$$= 1800 \text{ cm}^2$$

In this example one quantity (the surface area) is proportional to the **square** of the other quantity (the height).

> When y is directly proportional to the **square** of x:
> * $y \propto x^2$ is the proportionality statement
> * $y = kx^2$ is the proportionality formula, where k is the constant of proportionality.

Example 5

The volume of a Russian doll is proportional to the cube of its height. The volume of the 8 cm high doll is 64 cm³.
Find the volume of the 10 cm high doll.

The proportionality statement is

$V \propto h^3$

The proportionality formula can be written as

$V = kh^3$

The value of the constant k can be found using the information about the 8 cm high doll:

$$64 = k \times 8^3$$

$$64 = 512k$$

$$k = \frac{64}{512} = \frac{1}{8}$$

$$k = 0.125$$

So the proportionality formula is

$$V = 0.125 \times h^3$$

The formula can be used to find the volume of the 10 cm high doll:

$$V = 0.125 \times 10^3$$

$$= 125 \text{ cm}^3$$

In this example one variable is proportional to the **cube** of the other.

When y is directly proportional to the **cube** of x:
- $y \propto x^3$ is the proportionality statement
- $y = kx^3$ is the proportionality formula, where k is the constant of proportionality.

Worked examination question 1

The distance D moved by an object is proportional to the square of the time t for which it is moving.

(a) Express D in terms of t and a constant of proportionality k.

(b) When $t = 10$ seconds, $D = 500$ metres.
Calculate:
  (i) the value of D when $t = 4$ seconds
  (ii) the value of t when $D = 720$ metres.

(a) $D \propto t^2$

   So $D = kt^2$

(b) (i) Substitute $t = 10$ seconds and $D = 500$ metres into the formula $D = kt^2$:

$$500 = k \times 10^2$$

$$500 = 100k$$

$$5 = k \quad\text{——— Divide both sides by 100.}$$

So the formula is

$$D = 5t^2$$

When $t = 4$ seconds:

$$D = 5 \times 4^2$$
$$= 5 \times 16$$
$$= 80 \text{ metres}$$

When $t = 4$ seconds, $D = 80$ metres.

(ii) When $D = 720$ metres:

$$720 = 5t^2$$
$$\frac{720}{5} = t^2 \quad\text{———— Divide by 5.}$$
$$144 = t^2$$
$$\sqrt{144} = t \quad\text{———— Take the square root.}$$
$$t = 12 \text{ seconds}$$

When $D = 720$ metres, $t = 12$ seconds.

Exercise 18E

1   y is directly proportional to the square of x so that $y = kx^2$.
    Given that $y = 20$ when $x = 2$, calculate the value of k.
    Calculate the value of y when $x = 4$.

2   z varies in direct proportion to the square of x so that $z = kx^2$.
    Given that $z = 100$ when $x = 4$, calculate the value of

    (a)   the proportionality constant k

    (b) z when $x = 5$

    (c) x when $z = 10$.

3   s varies in direct proportion to the cube of t, so that $s = kt^3$.
    Given that $s = 54$ when $t = 3$, calculate the value of

    (a)   the proportionality constant k

    (b) s when $t = 4$

    (c) t when $s = 128$.

4   l is directly proportional to the cube of m.

    (a)   Write a formula for l in terms of m.

    (b)   Given that the value of l when $m = 0.4$ is 3.2,
          (i)   calculate the value of l when $m = 0.8$
          (ii) calculate the value of m when $l = 43.2$

5   The resistance R to the motion of a train varies directly as
    the square of the speed v of the train. The resistance to
    motion is 100 000 N when the speed is 20 metres per second.
    Calculate the resistance to motion when the speed of the
    train is 10 metres per second.

6 A stone is dropped down a well a distance d metres. The value of d is directly proportional to the square of the time of travel t seconds.

When $t = 2$, $d = 20$. Calculate the value of d when

(a) $t = 3$

(b) $t = 4.5$

7 The variables p and q are related so that p is directly proportional to the square of q. Complete this table for values of p and q:

q	0.5	2	
p		12	27

8 z varies in direct proportion to the cube of t. When $t = 5$ the value of z is 0.25.

Calculate the value of z when $t = 1$.

9 The length of a pendulum l is directly proportional to the square of the period T of the pendulum.
Given that a pendulum which has a period of 3 seconds is 2.25 metres long, calculate the length of a pendulum which has a period of

(a) 2 seconds

(b) 2.5 seconds.

> The period of a pendulum is the time it takes to swing away from its starting position and back again.

10 A varies in direct proportion to the square of l. When the value of l is multiplied by 2, by what amount is A multiplied?

11 The variable z is directly proportional to the cube of the variable w. Complete the table for these variables:

w	2	1	
z	16		6.75

12 A set of models of a road bridge is made.
The surface area of the road is directly proportional to the square of the height h of the model.
The height of Model A is 50% greater than the height of Model B. By what percentage is the area of road surface in Model A greater than the area of road surface in Model B?

13 The pressure on a diver under water is directly proportional to the square of her depth below the surface of the water.
The diver has reached a depth of 10 metres.
How much further must she descend for the pressure to double?

18.6 Inverse proportion

When one quantity increases while the other decreases proportionally, the quantities are in **inverse proportion** to one another. (When one quantity decreases, the other will increase proportionally.)

$y \propto \dfrac{1}{x}$ means 'y is **inversely proportional** to x'.

When y is inversely proportional to x:

- $y \propto \dfrac{1}{x}$ is the proportionality statement
- $y = k \times \dfrac{1}{x}$ or $y = \dfrac{k}{x}$ are ways of writing the proportionality formula,
  where k is the constant of inverse proportionality.

Another way of describing such a relationship is to say that y is directly proportional to $\dfrac{1}{x}$.

Example 6

In a physics experiment the pressure P and volume V of a quantity of gas are measured at constant temperature.

The pressure is inversely proportional to the volume. When the pressure is 2 bar, the volume of the gas is 150 cm³.

Calculate the pressure of the gas when the volume is decreased to 100 cm³.

P is inversely proportional to V so

$$P \propto \frac{1}{V}$$

The formula is

$$P = \frac{k}{V}$$

The constant k is found by substituting the pressure $P = 2$ and the volume $V = 150$:

$$2 = \frac{k}{150}$$

so $k = 300$

This gives the formula

$$P = \frac{300}{V}$$

When $V = 100$:

$$P = \frac{300}{100}$$

$$= 3 \text{ bar}$$

> The bar is a unit of pressure.
> 1 bar $= 10^5$ pascals

Example 7

In Newtonian physics the force of attraction F between two bodies is inversely proportional to the **square** of the distance r between their centres.

When the distance between their centres is 1 unit, the force of attraction is 16 units.

Calculate the force of attraction when the distance between their centres is 2 units.

F is inversely proportional to r^2, so the proportionality statement is

$$F \propto \frac{1}{r^2}$$

The planet Jupiter has over 30 moons, four of which are visible with binoculars. The moon shown here is called Ganymede.

The formula is

$$F = \frac{k}{r^2}$$

where k is constant for these two bodies only. (The force between the bodies also depends on their masses.)

The constant k is found by using the fact that when $r = 1$ unit, the force $F = 16$ units:

$$16 = \frac{k}{1^2}$$

so $16 = k$

The formula is

$$F = \frac{16}{r^2}$$

When the distance between the centres of the two bodies is 2 units the force is

$$F = \frac{16}{2^2}$$

$$= 4 \text{ units}$$

> In Newtonian physics the Moon orbits the Earth because there is a force of attraction across space given by the formula
>
> $$F = \frac{G M_E M_M}{r^2}$$
>
> In Einsteinian physics each mass is in free fall through space–time, which is curved by their presence. This curvature gives rise to the orbit.

Worked examination question 2

The resistance R of 1 metre of cable of a certain material is inversely proportional to the square of the radius r of the cable.

When the radius is 5 mm, the resistance of 1 metre of the cable is 0.06 ohms.

(a)   Find a formula connecting R and r.

(b)   Calculate the value of R when r is 4 mm.

(a) $R \propto \dfrac{1}{r^2}$

So $R = \dfrac{k}{r^2}$

where k is a constant.

Substitute $R = 0.06$ and $r = 5$ into the formula:

$$0.06 = \dfrac{k}{5^2}$$

so $k = 1.5$

So the formula is

$$R = \dfrac{1.5}{r^2}$$

(b) When $r = 4$:

$$R = \dfrac{1.5}{4^2}$$

$= 0.094$ ohms (correct to 2 significant figures).

Exercise 18F

1 y is inversely proportional to x so that $y = \dfrac{k}{x}$

When $x = 6$ the value of y is 12.

   (a)   Find the value of k.

   (b)   Calculate the value of y when $x = 10$.

2 z is inversely proportional to t so that $z = \dfrac{k}{t}$

When $t = 0.3$ the value of z is 16.

   (a)   Find the value of k.

   (b)   Calculate the value of z when $t = 0.5$

3 z is inversely proportional to the square of w so that $z = \dfrac{k}{w^2}$

When $w = 4$, $z = 32$.

   (a)   Find the value of k.

   (b)   Calculate the value of z when $w = 2$.

4 p is inversely proportional to the square of q so that $p = \dfrac{k}{q^2}$

When $q = 0.5$, $p = 10$.

   (a)   Find the value of k.

   (b)   Calculate the value of p when $q = 0.25$

5 An essay typed at 50 characters to the line is 372 lines long. The number of lines in the essay is inversely proportional to the number of characters to the line. Calculate the number of lines that the essay will have when it is typed with

    (a)   80 characters to the line

    (b)   64 characters to the line.

6 In a mathematics investigation, students draw a series of rectangles which all have the same area. The length l of each rectangle is inversely proportional to its width w.

    (a)   Write the proportionality statement and the formula for this situation.

    (b)   What does the constant of proportionality represent?

7 The light intensity I at a distance d from a light source varies inversely as the square of the distance from the source. At a distance 1 cm from the light source the light intensity is 64 units. Calculate the light intensity at a distance 4 cm from the light source.

8 The frequency f of sound varies inversely as the wavelength w. The frequency of middle C is 256 hertz and the wavelength of this note is 129 cm.

    (a)   Find the equation connecting frequency f and the wavelength w.

    (b)   Calculate the frequency of the note with a wavelength of 86 cm.

    (c)   Calculate the wavelength of a note whose frequency is 344 Hz.

> The unit of frequency is the hertz (Hz) (cycles per second).

Mixed exercise 18

1 An electrical heater uses 16 units of electricity in 5 hours. The amount of electricity used is directly proportional to the time. How much electricity will the heater use in 3 hours?

2 A builder makes concrete paving slabs. All the slabs are the same thickness. The cross-sections of all the slabs are equilateral triangles. The mass of concrete is proportional to the square of the length of one of the sides. A slab with side 60 cm has mass 50 kg. Calculate the mass of a slab with side 20 cm.

3 A spaceship covers a distance of 18 000 kilometres in 3 hours 20 minutes. The distance travelled is directly proportional to the time taken. Calculate the distance that the spaceship will travel in 5 hours 20 minutes.

4   A hang-glider pilot, at a height of h metres above the sea, can see up to a distance of s kilometres.
It is known that h is proportional to the square of s.

   (a)   Given that $h = 140$ when $s = 16$, find the formula for h in terms of s.

   (b)   Calculate the height of the hang-glider when the pilot can just see a lighthouse which is 24 kilometres away.

5   The variables y and x are related by

$$y \propto \frac{1}{x}$$

When $x = 3$, $y = 30$. Calculate the value of y when $x = 5$.

6   The height h reached by a ball thrown up into the air varies in direct proportion to the square of the speed v at which the ball is thrown. A ball thrown at a speed of 20 metres per second reaches a height of 20 metres.
Calculate the height reached by a ball thrown at a speed of

   (a)   25 metres per second

   (b)   30 metres per second.

7   The time taken for a journey on a motorway is inversely proportional to the average speed for the journey. The journey takes 1 hour 30 minutes when the average speed is 54 miles per hour.
Calculate the time taken, in hours and minutes, for this journey when the average speed is 45 miles per hour.

8   The energy stored in a battery is proportional to the square of the diameter of the battery, for batteries of the same height. One battery has a diameter of 2.5 cm and stores 1.6 units of energy. Another battery of the same height has a diameter of 1.5 cm.
Calculate the energy stored in the second battery.

9   In a factory, chemical reactions are carried out in spherical containers.

The time, T minutes, that the chemical reaction takes is directly proportional to the square of the radius, R cm, of the spherical container.
When $R = 120$, $T = 32$. Find the value of T when $R = 150$.

10   The shutter speed, S, of a camera varies inversely as the square of the aperture setting, f. When $f = 8$, $S = 125$.

   (a)   Find a formula for S in terms of f.

   (b)   Hence, or otherwise, calculate the value of S when $f = 4$.

11 The graphs of y against x represent four different types of proportionality.

A
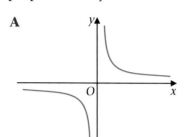

B
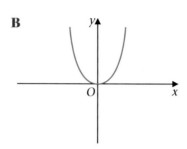

C
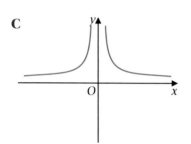

D
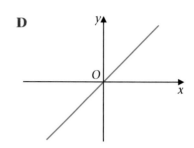

Write down the letter of the graph which represents each of these statements:

(a) y is directly proportional to x

(b) y is inversely proportional to x

(c) y is proportional to the square of x

(d) y is inversely proportional to the square of x.

Summary of key points

1 The symbol $\propto$ means '**is proportional to**'.

2 $y \propto x$ means 'y is **directly proportional** to x'.

3 When a graph of two quantities is a straight line through the origin, one quantity is **directly proportional** to the other.

4 When y is directly proportional to x, you can write a **proportionality statement** and a **formula** connecting y and x:

- $y \propto x$ is the proportionality statement
- $y = kx$ is the proportionality formula, where k is the constant of proportionality.

5 To find the value of k, the constant of proportionality, substitute known values of y and x into $y = kx$.

6 When y is directly proportional to the **square** of x:

- $y \propto x^2$ is the proportionality statement
- $y = kx^2$ is the proportionality formula, where k is the constant of proportionality.

7 When y is directly proportional to the **cube** of x:

- $y \propto x^3$ is the proportionality statement
- $y = kx^3$ is the proportionality formula, where k is the constant of proportionality.

8 $y \propto \dfrac{1}{x}$ means 'y is **inversely proportional** to x'.

9 When y is inversely proportional to x:

- $y \propto \dfrac{1}{x}$ is the proportionality statement

- $y = k \times \dfrac{1}{x}$ or $y = \dfrac{k}{x}$ are ways of writing the proportionality formula, where k is the constant of inverse proportionality.

⑲ Quadratic equations

In this chapter you will solve quadratic equations and equations involving algebraic fractions.

Equations of the form $ax^2 + bx + c = 0$, where $a \neq 0$ are called quadratic equations. Graphs of quadratic equations were introduced in Chapter 17.

Quadratic equations can be used to represent a wide variety of situations, for example the diving shown on the right.

> The quadratic equation $ax^2 + bx + c = 0$, with $a \neq 0$, has two solutions (or roots), which may be equal.

When solving quadratic equations it is helpful to remember that multiplying two numbers to get the answer 0 is only possible if one of the numbers is itself 0. For example:

- if $7 \times y = 0$ then y must be 0
- if $x \times 3 = 0$ then x must be 0.

> In general, if $xy = 0$ then either $x = 0$ or $y = 0$ (or both).

The solution of $7t^2 - t - 4 = 0$ gives the time in seconds to dive into the pool.

There are three algebraic methods for solving quadratic equations: by factorising, by completing the square, and by using a formula.

In each method the first step is to rearrange the quadratic equation into the form $ax^2 + bx + c = 0$.

19.1 Solving quadratic equations by factorising

Example 1

Solve the equation $x^2 = 8x$.

$$x^2 = 8x$$

$$x^2 - 8x = 0 \quad \underline{\quad} \text{ Rearrange into the form } ax^2 + bx + c = 0$$

$$x(x - 8) = 0 \quad \underline{\quad} \text{ Factorise}$$

So either $\quad x = 0 \quad$ or $\quad x - 8 = 0$

So $x = 0$, $x = 8$ are the two solutions or roots of the equation $x^2 = 8x$.

A common error is to divide both sides by x. This loses the $x = 0$ solution. Never divide by any term which can take value 0.

Example 2

Solve the equation $(3x + 2)(2x - 1) = 3$.

$$(3x + 2)(2x - 1) = 3$$
$$6x^2 + x - 2 = 3 \quad \text{—— Expand the brackets}$$
$$6x^2 + x - 5 = 0 \quad \text{—— Rearrange into the form } ax^2 + bx + c = 0$$
$$(6x - 5)(x + 1) = 0 \quad \text{—— Factorise}$$

So either $\qquad 6x - 5 = 0 \quad$ or $\quad x + 1 = 0$
$$6x = 5 \quad \text{or} \quad x = -1$$

The two solutions are $\quad x = \frac{5}{6} \quad$ and $\quad x = -1$

> For a reminder on how to factorise quadratic expressions see Chapter 14.

Example 3

Solve the equation $y^2 - 5y + 18 = 2 + 3y$.

$$y^2 - 5y + 18 = 2 + 3y$$
$$y^2 - 8y + 16 = 0 \quad \text{—— Rearrange into the form } ay^2 + by + c = 0$$
$$(y - 4)(y - 4) = 0 \quad \text{—— Factorise}$$

So either $\qquad y - 4 = 0 \quad$ or $\quad y - 4 = 0$
$$y = 4 \quad \text{or} \quad y = 4$$

$y = 4$ is the only solution. In this example the two roots of the quadratic equation are equal. This parabola touches the y-axis (horizontal axis) only once.

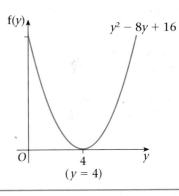

Example 4

Solve the equation $2a^2 - 162 = 0$.

$$2a^2 - 162 = 0$$
$$2(a^2 - 81) = 0 \quad \text{—— Take out the common factor 2}$$
$$2(a - 9)(a + 9) = 0 \quad \text{—— Using the difference of two squares}$$

For this equation to be 0, either
$$a - 9 = 0 \quad \text{or} \quad a + 9 = 0$$
$$a = 9 \quad \text{or} \quad a = -9$$

Write these roots as $a = \pm 9$.

> This is read as 'a equals positive or negative 9'.

Solving equations of the type $y^2 = k$

You can solve equations of the type $y^2 = k$ by taking the square root of both sides, but you must remember to write the $\pm$ sign.

If $y^2 = k$ then $y = \pm\sqrt{k}$

Example 5

Solve the equation $(3x - 1)^2 = 64$.

$(3x - 1)^2 = 64$

$(3x - 1) = \pm 8$ —— Take the square root of both sides

$3x = 1 \pm 8$ —— Add 1 to both sides

So either $\quad 3x = 1 + 8 \quad$ or $\quad 3x = 1 - 8$

$\qquad 3x = 9 \qquad$ or $\quad 3x = -7$

The solutions are $\quad x = 3$ and $x = -\frac{7}{3}$

Exercise 19A

1 Solve these equations.

(a) $(x - 1)(x + 2) = 0$ 　　　(b) $x(x - 3) = 0$

(c) $(x - 4)(2x + 1) = 0$ 　　　(d) $2x(2x - 3) = 0$

(e) $(2x - 1)(3x - 4) = 0$ 　　　(f) $2(x - 5)(x - 2) = 0$

2 Solve these quadratic equations in y.

(a) $y^2 - 3y = 0$ 　　　(b) $y^2 - 16 = 0$

(c) $y^2 - 3y - 4 = 0$ 　　　(d) $y^2 + 11y + 28 = 0$

(e) $2y^2 + 5y - 3 = 0$ 　　　(f) $12y^2 - 7y - 12 = 0$

(g) $4y^2 - 11y + 6 = 0$ 　　　(h) $2y^2 - 18y = 0$

(i) $6y^2 - 11y - 7 = 0$

3 Solve these equations.

(a) $y^2 = 25$ 　　　(b) $3x^2 = x$

(c) $3y^2 + 5y = 2$ 　　　(d) $2x^2 = 8$

(e) $5x = 2x^2$ 　　　(f) $y^2 + 2y = 35$

(g) $(a - 2)(a + 1) = 10$ 　　　(h) $(2x - 1)^2 = 3x^2 - 2$

(i) $6a^2 = 7 - 11a$ 　　　(j) $(3x - 2)^2 = x^2$

4 Explain what is wrong with Julie's method to solve the equation $6y^2 = 12y$. Write out a correct solution.

$6y^2 = 12y$

$\dfrac{6y^2}{6y} = \dfrac{12y}{6y}$

$y = 2$

19.2 Completing the square

To help solve quadratic equations you can complete the square.

Here is the diagram for $x^2 + 10x$:

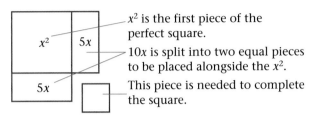

x^2 is the first piece of the perfect square.

10x is split into two equal pieces to be placed alongside the x^2.

This piece is needed to complete the square.

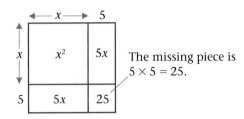

The missing piece is $5 \times 5 = 25$.

You can see that $x^2 + 10x + 5^2 = (x + 5)^2$, which is a perfect square.

So to complete the square for $x^2 + 10x$, you need to subtract 5^2 from the perfect square $(x + 5)^2$.

So $x^2 + 10x = (x + 5)^2 - 5^2$.

In general:

Completing the square: $x^2 + bx = \left(x + \dfrac{b}{2}\right)^2 - \left(\dfrac{b}{2}\right)^2$

You can use this formula when the coefficient of x^2 is 1.

The next example shows you what to do if the coefficient of x^2 is not 1.

Example 6

Write $2x^2 - 12x$ in the form $p(x + q)^2 + r$, where p, q and r are constants to be determined.

$2x^2 - 12x = 2[x^2 - 6x]$ ——————— Take out the coefficient of x^2 as a common factor

$2[x^2 - 6x] = 2[(x - 3)^2 - 9]$ ——————— Complete the square for $x^2 - 6x$ by using $b = -6$ in the formula

$\qquad\qquad = 2(x - 3)^2 - 18$

So $\qquad p = 2,\ q = -3$ and $r = -18$.

It is possible to write all quadratic expressions in the form $p(x + q)^2 + r$.

Exercise 19B

1 Write the following in the form $(x + q)^2 + r$.

(a) $x^2 + 4x$ (b) $x^2 - 14x$ (c) $x^2 + 3x$ (d) $x^2 + x$

(e) $x^2 - x$ (f) $x^2 - 4x$ (g) $x^2 + 7x$ (h) $x^2 - 10x$

2 Write the following in the form $p(x + q)^2 + r$.

(a) $2x^2 + 16x$ (b) $3x^2 - 12x$ (c) $2x^2 + x$ (d) $5x^2 - 15x$

(e) $2x^2 + 2x$ (f) $4x^2 - 8x$ (g) $3x^2 - 15x$ (h) $7x^2 - 28x$

19.3 Solving quadratic equations by completing the square

At first glance it *looks* as though $x^2 + 10x + 18 = 0$ cannot easily be solved by factorising. However, this quadratic can be solved by completing the square. The next example shows you how.

> You would need to find two numbers with a product of 18 and a sum of 10.

Example 7

Solve the equation $x^2 + 10x + 18 = 0$. Leave your answer in **surd** form.

$$x^2 + 10x + 18 = 0 \quad\text{------- Check that the coefficient of } x^2 \text{ is 1}$$
$$x^2 + 10x = -18 \quad\text{------- Subtract 18 to get constant term on RHS}$$
$$(x + 5)^2 - 25 = -18 \quad\text{------- Complete the square for } x^2 + 10x$$
$$(x + 5)^2 = 7 \quad\text{------- Add 25 to both sides}$$
$$x + 5 = \pm\sqrt{7} \quad\text{------- Square root both sides}$$
$$x = -5 \pm\sqrt{7} \quad\text{--- Subtract 5 from both sides}$$

> **Surds** are numbers that are left in square root form. They cannot be written as a fraction. $\sqrt{2}$, $\sqrt{3}$ and $\sqrt{7}$ are examples of surds.

The solutions (roots) of $x^2 + 10x + 18 = 0$ are $x = -5 + \sqrt{7}$ and $x = -5 - \sqrt{7}$.

Example 8

Solve the equation $2y^2 - 3y - 8 = 0$. Give your answers correct to 2 d.p.

$$2y^2 - 3y - 8 = 0$$
$$2\left(y^2 - \tfrac{3}{2}y - 4\right) = 0 \quad\text{------- Take out the coefficient of } y^2$$
$$y^2 - \tfrac{3}{2}y - 4 = 0 \quad\text{------- Divide by 2}$$
$$y^2 - \tfrac{3}{2}y = 4 \quad\text{------- Add 4 to get constant term on RHS}$$
$$\left(y - \tfrac{3}{4}\right)^2 - \left(\tfrac{3}{4}\right)^2 = 4 \quad\text{------- Complete the square}$$
$$\left(y - \tfrac{3}{4}\right)^2 = 4 + \left(\tfrac{3}{4}\right)^2$$
$$y - \tfrac{3}{4} = \pm\sqrt{\tfrac{73}{16}} \quad\text{------- Take the square root of both sides}$$
$$y = \tfrac{3}{4} \pm \sqrt{\tfrac{73}{16}} \quad\text{------- Add } \tfrac{3}{4} \text{ to both sides}$$
$$= 0.75 \pm 2.136\ldots$$
$$y = 0.75 + 2.136\ldots \quad\text{or}\quad y = 0.75 - 2.136\ldots$$

So $\qquad\qquad y = 2.89 \quad\text{or}\quad y = -1.39 \quad$ correct to 2 d.p.

Exercise 19C

1 Solve these equations. Leave your answers in surd form.
  (a) $x^2 + 10x + 3 = 0$ (b) $x^2 - 8x = 2$ (c) $2x^2 + 18x + 6 = 0$
  (d) $3x^2 - 6x + 1 = 0$ (e) $2y^2 - 3y - 4 = 0$ (f) $2y^2 = 4y + 7$
  (g) $3x^2 = -5x + 3$ (h) $2x^2 - 7x = 2$ (i) $3x^2 = 8x + 1$

2 Solve these equations. Give your answers correct to 2 d.p.
  (a) $x^2 - 4x = 3$ (b) $x^2 - 6x + 2 = 0$ (c) $2y^2 + 4y = 7$
  (d) $2x^2 - 3x - 3 = 0$ (e) $3y^2 - 6y = 1$ (f) $4y^2 - y = 8$
  (g) $x^2 - 4x - 3 = 0$ (h) $5x^2 + 3x - 1 = 0$ (i) $3y^2 - 5y = 3$

19.4 Solving quadratic equations by using a formula

The steps shown in the previous section can be applied to the general quadratic equation $ax^2 + bx + c = 0$. This gives a formula which can be used to solve **all quadratic equations**. This section shows a proof of the formula and how to apply it.

Here are two quadratic equations – one with number coefficients, the other a 'general' form of a quadratic with letter coefficients. The same steps are followed to solve both quadratics. Work your way down to the solutions, comparing the effect of the steps at each stage.

$2x^2 - 6x + 1 = 0$		$ax^2 + bx + c = 0$, where $a \neq 0$
$2[x^2 - 3x + \frac{1}{2}] = 0$	Take out the coefficient of x^2	$a\left[x^2 + \frac{b}{a}x + \frac{c}{a}\right] = 0$
$x^2 - 3x + \frac{1}{2} = 0$	Divide by the coefficient of x^2	$x^2 + \frac{b}{a}x + \frac{c}{a} = 0$
$x^2 - 3x = -\frac{1}{2}$	Rearrange so constant term is on RHS	$x^2 + \frac{b}{a}x = -\frac{c}{a}$
$[x - \frac{3}{2}]^2 - (-\frac{3}{2})^2 = -\frac{1}{2}$	Complete the square	$\left[x + \frac{b}{2a}\right]^2 - \left(\frac{b}{2a}\right)^2 = -\frac{c}{a}$
$[x - \frac{3}{2}]^2 = (-\frac{3}{2})^2 - \frac{1}{2}$	Rearrange so constant term is on RHS	$\left[x + \frac{b}{2a}\right]^2 = \left(\frac{b}{2a}\right)^2 - \frac{c}{a}$
$= \frac{9}{4} - \frac{1}{2}$	Simplify RHS to a single fraction	$= \frac{b^2}{4a^2} - \frac{c}{a}$
$= \frac{9-2}{4} = \frac{7}{4}$	For a reminder on how to simplify algebraic fractions see Chapter 14.	$= \frac{b^2 - 4ac}{4a^2}$
$x - \frac{3}{2} = \pm\sqrt{\frac{7}{4}}$	Square root both sides, remembering the $\pm$	$x + \frac{b}{2a} = \pm\sqrt{\frac{b^2 - 4ac}{4a^2}}$
$= \pm\frac{\sqrt{7}}{\sqrt{4}}$	Note that $\sqrt{\frac{p}{q}} = \frac{\sqrt{p}}{\sqrt{q}}$ There is more on manipulating surds in Section 23.3.	$= \pm\frac{\sqrt{b^2 - 4ac}}{\sqrt{4a^2}}$
$x = \frac{3}{2} \pm \frac{\sqrt{7}}{2}$	Get x on its own	$x = -\frac{b}{2a} \pm \frac{\sqrt{b^2 - 4ac}}{2a}$
$= \frac{3 \pm \sqrt{7}}{2}$	Write as a single fraction	$= \frac{-b \pm \sqrt{b^2 - 4ac}}{2a}$

The roots of the quadratic equation $ax^2 + bx + c = 0$, where $a \neq 0$, are given by the formula

$$x = \frac{-b \pm \sqrt{b^2 - 4ac}}{2a}$$

This formula will be on your GCSE exam formulae sheet. You do not have to remember it but you *must* know how to use it.

Example 9

Solve the equation $x(x + 3) = 2$. Give your answers correct to 2 d.p.

$$x(x + 3) = 2$$
$$x^2 + 3x = 2$$
$$1x^2 + 3x - 2 = 0$$

> A common error is to think that x^2 is $0x^2$ and put $a = 0$. To avoid this error write x^2 as $1x^2$.

Comparing with $ax^2 + bx + c = 0$ gives $a = 1$, $b = 3$, $c = -2$.

Using the formula

$$x = \frac{-b \pm \sqrt{b^2 - 4ac}}{2a}$$

gives the solutions

$$x = \frac{-3 \pm \sqrt{3^2 - 4(1)(-2)}}{2(1)}$$

$$= \frac{-3 \pm \sqrt{9 + 8}}{2}$$

$$= \frac{-3 \pm \sqrt{17}}{2}$$

So $\quad x = \dfrac{-3 + 4.123...}{2} \quad$ or $\quad x = \dfrac{-3 - 4.123...}{2}$

$$x = 0.5615... \quad \text{or} \quad x = -3.5615...$$

The two solutions (roots) are $x = 0.56$ and $x = -3.56$ correct to 2 d.p.

Example 10

Solve the equation $2y^2 - y - 4 = 0$. Leave your answers in surd form.

$2y^2 - 1y - 4 = 0$ is a quadratic in y. Comparing with $ay^2 + by + c = 0$ gives $a = 2$, $b = -1$, $c = -4$.

Using the formula

$$y = \frac{-b \pm \sqrt{b^2 - 4ac}}{2a}$$

gives the solutions

$$y = \frac{-(-1) \pm \sqrt{(-1)^2 - 4(2)(-4)}}{2(2)}$$

$$= \frac{1 \pm \sqrt{1 + 32}}{4}$$

$$= \frac{1 \pm \sqrt{33}}{4}$$

In surd form, the roots are $y = \dfrac{1 + \sqrt{33}}{4} \quad$ and $\quad y = \dfrac{1 - \sqrt{33}}{4}$

The quadratic $x^2 + 10x + 18 = 0$ was introduced in Section 19.3. We said that at first glance it could not easily be solved by factorising because two numbers with product 18 and sum 10 would need to be found. In fact two such numbers can be found. The next example shows you how.

Example 11

Find two numbers whose sum is 10 and whose product is 18.

Half the sum is 5.

Let one number be $5 - x$. Then the other must be $5 + x$ (so that they add up to 10).

The product of the numbers is 18 so

$$(5 - x)(5 + x) = 18$$
$$25 - x^2 = 18 \quad\text{——— Difference of two squares}$$

This gives $x^2 = 7$ so $x = \sqrt{7}$ or $x = -\sqrt{7}$

The two numbers are $5 - \sqrt{7}$ and $5 + \sqrt{7}$.

Exercise 19D

In this exercise use the formula $x = \dfrac{-b \pm \sqrt{b^2 - 4ac}}{2a}$

1 Write down the values of $b^2 - 4ac$ in these equations:

(a) $x^2 + 3x + 1 = 0$ (b) $x^2 - 2x - 1 = 0$

(c) $2x^2 + 6x - 1 = 0$ (d) $8x^2 - 9 = 0$

(e) $4 - 3x - 2x^2 = 0$ (f) $2x^2 = 2x + 3$

2 Solve the equations in question **1**.
Give your answers correct to 2 d.p.

3 Solve the following equations.
Give your answers correct to 2 d.p.

(a) $x^2 - 4x + 1 = 0$ (b) $x^2 - 5x + 1 = 0$

(c) $4x^2 + 9x + 1 = 0$ (d) $4x^2 - 2x = 3$

(e) $1 = x^2 - 8x + 2$ (f) $(x + 4)^2 = 2(x + 7)$

4 Find two numbers whose sum is 18 and whose product is 21.

5 Find two numbers whose difference is 6 and whose product is 15.

19.5 Equations involving algebraic fractions

You simplified algebraic fractions in Chapter 14. This section shows you how to apply this skill to algebraic equations. For example:

if $\dfrac{x}{4} = \dfrac{3}{4}$ then $x = 3$.

In general, if $\dfrac{p}{r} = \dfrac{q}{r}$ then $p = q$.

To solve equations involving algebraic fractions:

Step 1 Write both sides of the equation with the same denominator.

Step 2 Use the numerators to write an equation.

Step 3 Solve the resulting equation.

Example 12

Solve the equation $\dfrac{2}{y+1} - \dfrac{3}{2y+3} = \dfrac{1}{2}$

Give your answer correct to 3 d.p.

$$\dfrac{2}{y+1} - \dfrac{3}{2y+3} = \dfrac{1}{2}$$

The LCM of the three denominators is $2(y+1)(2y+3)$.

Write each fraction with the LCM as denominator:

$$\dfrac{2 \times 2(2y+3)}{2(y+1)(2y+3)} - \dfrac{3 \times 2(y+1)}{2(y+1)(2y+3)} = \dfrac{(y+1)(2y+3)}{2(y+1)(2y+3)}$$

If $\dfrac{p}{r} = \dfrac{q}{r}$, then $p = q$, so ignore the denominators.

$$4(2y+3) - 6(y+1) = (y+1)(2y+3)$$
$$8y + 12 - 6y - 6 = 2y^2 + 5y + 3$$
$$2y + 6 = 2y^2 + 5y + 3$$
$$0 = 2y^2 + 3y - 3$$

Comparing this equation with the general quadratic $0 = ay^2 + by + c$ gives $a = 2$, $b = 3$, $c = -3$.

Using the formula:

$$y = \dfrac{-3 \pm \sqrt{3^2 - 4(2)(-3)}}{2(2)}$$

$$= \dfrac{-3 \pm \sqrt{33}}{4}$$

$$= \dfrac{-3 \pm 5.744\ldots}{4}$$

So $y = 0.686$ or $y = -2.186$ correct to 3 d.p.

> Remember: when the denominators have no common factor, the LCM is the product of the denominators.

Exercise 19E

Solve the following equations. Give fractional answers exactly. In all other cases give your answers correct to 2 d.p.

1 $\dfrac{1}{x} + \dfrac{1}{x+1} = \dfrac{7}{12}$

2 $2x + \dfrac{1}{x} = 5$

3 $\dfrac{5}{2x+1} + \dfrac{6}{x+1} = 3$

4 $\dfrac{1}{x-1} - \dfrac{1}{x} = 8$

5 $\dfrac{1}{x-3} - \dfrac{3}{x+2} = \dfrac{1}{2}$

6 $\dfrac{2}{y+1} + \dfrac{3}{2y+3} = 1$

19.6 Problems leading to quadratic equations

The equation for question **1** in Exercise 19E would be the correct one to use to solve the following problem:

Find consecutive integers whose reciprocals add up to $\frac{7}{12}$.

When answering questions in the GCSE exam, candidates often find it harder to obtain the quadratic equation to represent a problem than to solve it. You will find these steps useful.

Finding the equation to represent a problem

Step 1 Where relevant, draw a diagram and put all the information on it.

Step 2 Use x to represent the unknown which you have been asked to find.

Step 3 Use other letters to identify any other relevant unknowns.

Step 4 Look for information given in the question which links these letters to x and write them down.

Step 5 Try simple numbers for the unknowns and see if this helps you to find a method.

Step 6 Make sure that the units on both sides of your equation are the same.

When applied to problem solving, one of the roots of the quadratic equation is often not a solution to the problem and must be abandoned (with an explanation).

Example 13

In a right-angled triangle, the hypotenuse is 6 cm longer than the shortest side. The third side is 2 cm shorter than the hypotenuse. Find a quadratic equation in the form $ax^2 + bx + c = 0$ which when solved leads to the length of the shortest side.

Let the shortest side be x cm.

Mark the other sides y cm and z cm like this:

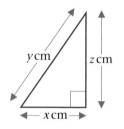

The hypotenuse is 6 cm longer than x
so $y = x + 6$.

The third side is 2 cm shorter than y
so $z = y - 2$ or $z = x + 6 - 2 = x + 4$.

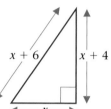

> The units of length are the same.

Using Pythagoras' theorem:

$$x^2 + (x + 4)^2 = (x + 6)^2$$

$$x^2 + x^2 + 8x + 16 = x^2 + 12x + 36$$

$$x^2 - 4x - 20 = 0$$

> Test:
> If $x = 1$ then
> $y = 1 + 6 = 7$ and
> $z = 1 + 4 = 5$.
>
> This confirms the relative size of each side.

Solving this quadratic equation gives $x = 6.899$ or -2.899.
As length is positive the only answer is 6.9 cm to 2 s.f.

Example 14

Lisa cycled from Norton to Sufton, a distance of 34 km, and then cycled the same distance back.

Lisa's average speed on the outward journey was x km/h.

On the return journey the wind was behind her so she cycled 2 km/h faster and completed the return journey 16 minutes quicker than the outward journey.

(a)　Write down, in terms of x, the time (in hours) taken for

　　(i)　the outward journey

　　(ii)　the return journey.

(b)　Show that $x^2 + 2x - 255 = 0$.

(c)　Calculate Lisa's average speed on the return journey.

(a)　(i)　Using time taken $= \dfrac{\text{distance}}{\text{average speed}}$

　　　the time for the outward journey is $\dfrac{34}{x}$ hours.

　　(ii)　Let Lisa's average speed on the return journey be y km/h.
　　　y is 2 more than x so $y = x + 2$.

　　　The time for the return journey is $\dfrac{34}{y} = \dfrac{34}{x + 2}$ hours.

(b) The time taken for the outward journey is 16 minutes more than for the return journey.

Time for outward journey − time for return journey = 16 minutes.

$$\frac{34}{x} - \frac{34}{x+2} = \frac{4}{15}$$

> The units must be the same.
> 16 min = $\frac{16}{60} = \frac{4}{15}$ hours

$$\frac{34 \times 15(x+2) - 34 \times 15(x)}{15x(x+2)} = \frac{4x(x+2)}{15x(x+2)}$$

$$510x + 1020 - 510x = 4x^2 + 8x$$

$$0 = 4x^2 + 8x - 1020$$

$$0 = 4(x^2 + 2x - 255)$$

So $x^2 + 2x - 255 = 0$.

(c) You need to find the value of y, which equals $x + 2$.

From part (b): $x^2 + 2x - 255 = 0$

$$(x + 17)(x - 15) = 0$$

> You can solve this equation by using the formula, but factorising is quicker if you can spot it.

So $\qquad x + 17 = 0 \quad$ or $\quad x - 15 = 0$

$$x = -17 \quad \text{or} \quad x = 15$$

Lisa's speed cannot be negative, so $x = 15$.

Lisa's average speed on the return journey is $x + 2 = 15 + 2 = 17$ km/h

Exercise 19F

Use quadratic equations to solve these problems.

1 Find two consecutive integers whose reciprocals add up to $\frac{9}{20}$.

2 The hypotenuse of a right-angled triangle is $2x$ cm.
The lengths of the other two sides are $(x + 1)$ cm and $(x + 3)$ cm.
Calculate the value of x.

3 The sum of the square of a number and 5 times the number itself is 24. Find the two possible values of the number.

4 The length of a rectangular piece of carpet is 4 m longer than its width. The area of the carpet is 16 m². Calculate the width of the carpet. Give your answer correct to the nearest centimetre.

5 One week a syndicate of x people won £400 in the National Lottery.

(a) Write down, in terms of x, how much each person should receive, if each gets an equal share.

We have won £400 in the lottery.

(b) If the prize had been won the previous week when there were $(x + 2)$ people in the syndicate, each person would have received £10 less.

Show that $x^2 + 2x - 80 = 0$.

(c) Calculate the amount of money won by each person in the syndicate.

6 A right-angled triangle is cut from the corner of a rectangular piece of card 15 cm by 8 cm.

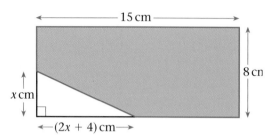

The base of the triangle is $(2x + 4)$ cm and its height is x cm.

The remaining piece of card (**green**) has an area of 89 cm². Calculate the value of x, giving your answer correct to 2 s.f.

7 Joe is training for a long-distance cycle race. One day he cycles for x hours in the morning and travels a distance of 84 km.

    (a) Write down, in terms of x, Joe's average speed in km/h.

    (b) In the afternoon he cycles for 1 hour more to travel the same distance and his average speed is 2 km/h slower than in the morning.

        Show that $x^2 + x - 42 = 0$.

    (c) Calculate the number of hours Joe cycles in the afternoon.

19.7 Solving linear and quadratic equations simultaneously

In Chapter 9 you found the coordinates of the point of intersection of two straight lines by solving the equations of the lines simultaneously.

In this section you will find the coordinates of the points of intersection of a straight line and a quadratic curve.

Let us consider the curve $y = x^2$ and three different lines.

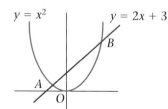 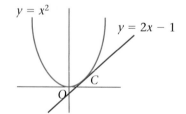 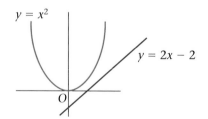

Diagram 1 The line $y = 2x + 3$ cuts the curve at two points, A and B.

Diagram 2 The line $y = 2x - 1$ just touches at the curve at the point C.

Diagram 3 The line $y = 2x - 2$ misses the curve completely.

The following three examples correspond to the three cases above.

Example 15

Find the coordinates of the points of intersection A and B of the line $y = 2x + 3$ and the curve $y = x^2$.

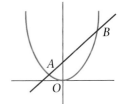

The points A and B both lie on the curve and on the line, so their coordinates must satisfy both equations:

$$y = x^2 \qquad (1)$$
$$\text{and} \quad y = 2x + 3 \qquad (2)$$

Eliminating y by substituting (1) into (2) gives

$$x^2 = 2x + 3$$
$$x^2 - 2x - 3 = 0 \text{ ————— Rearrange}$$
$$(x + 1)(x - 3) = 0 \text{ ————— Factorise}$$

So $\qquad x + 1 = 0 \quad \text{or} \quad x - 3 = 0$

which gives $\qquad x = -1 \quad \text{or} \quad x = 3$

> The linear equation is the equation of the straight line.

Substitute these values of x in the linear equation (2):

When $x = -1$, $\quad y = 2(-1) + 3 = 1$

When $x = 3$, $\qquad y = 2(3) + 3 = 9$

> From the sketch the x-coordinate of A is negative.

So the line $y = 2x + 3$ intersects the curve $y = x^2$ at the points $A(-1, 1)$ and $B(3, 9)$.

Example 16

(a) Solve the simultaneous equations $y = x^2$ and $y = 2x - 1$.

(b) Interpret your solution to part (a) geometrically.

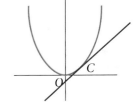

(a) $\qquad y = x^2 \qquad (1)$

$\qquad \text{and} \quad y = 2x - 1 \qquad (2)$

Eliminating y by substituting (1) into (2) gives:

$$x^2 = 2x - 1$$
$$x^2 - 2x + 1 = 0 \text{ ————— Rearrange}$$
$$(x - 1)(x - 1) = 0 \text{ ————— Factorise}$$

So $\qquad x - 1 = 0$ (repeated root)

which gives the single value $\qquad x = 1$

Substitute this value of x in the linear equation (2):

When $x = 1$, $y = 2(1) - 1 = 1$

(b) The line $y = 2x - 1$ meets the curve $y = x^2$ in just one point, $(1, 1)$.

The line $y = 2x - 1$ is a tangent to the curve $y = x^2$ at the point $(1, 1)$.

Example 17

(a) Show that the x-coordinate of any point of intersection of the line $y = 2x - 2$ and the curve $y = x^2$ would need to satisfy the equation $(x - 1)^2 = -1$.

(b) Interpret the result in part (a) geometrically.

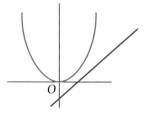

(a)
$$y = x^2 \qquad (1)$$
$$y = 2x - 2 \qquad (2)$$

Eliminating y by substituting (1) into (2) gives:
$$x^2 = 2x - 2$$
$$x^2 - 2x + 2 = 0 \quad \text{———— Rearrange}$$
$$x^2 - 2x + 1 = -2 + 1 \quad \text{——— Complete the square}$$
$$(x - 1)^2 = -1 \quad \text{——— Factorise}$$

The x-coordinate would need to satisfy this equation.

(b) For real numbers, $(x - 1)^2$ is always ≥ 0. So there are no real number solutions to the equation, and the line $y = 2x - 2$ never intersects the curve $y = x^2$.

> You can use **complex numbers** to solve this type of equation. You can learn about complex numbers at A-level.

Solving a linear equation ($y = px + q$) and a quadratic equation ($y = ax^2 + bx + c$) simultaneously:
• Find y in terms of x from the linear equation (or x in terms of y).
• Substitute for y (or x) in the quadratic equation.
• Solve the resulting quadratic equation for x (or y).
• Substitute the values of x (or y) into the linear equation to find y (or x).
If the roots of the quadratic equation are equal, the line will be a tangent to the curve.

Exercise 19G

1   Solve these pairs of simultaneous equations.
(a) $y = x^2$ and $y = 16$
(b) $y = x^2$ and $y = 12x - 36$
(c) $y = x^2$ and $y = 2x + 35$
(d) $y = x^2$ and $7x + 18$
(e) $y = 2x^2$ and $y = x + 3$
(f) $y = 3x^2$ and $y = 7x + 6$
(g) $y = 2 - x^2$ and $y = x - 4$
(h) $y = 12x^2 - 5$ and $y = 7 - 7x$

2   Find the coordinates of the points of intersection of these lines and curves.
(a) $y = 3 - 2x$ and $y = x^2$
(b) $y = 3x + 10$ and $y = 4x^2$
(c) $y = x + 5$ and $y = x^2 - 2x - 5$
(d) $y = 3x + 7$ and $y = 7 - 6x^2$

3   In each part determine the number of points of intersection of the line and the curve, and find the coordinates of any points of intersection.
(a) $y = x^2 + 3$ and $y = 2$
(b) $y = x^2$ and $y = 10x - 25$
(c) $y = 2x^2$ and $y = 5x + 7$
(d) $y = x^2 + 3x + 6$ and $y = 1 - x$
(e) $y = 5x^2$ and $y = 14x + 3$
(f) $y = 4x^2$ and $y = 10x + 6$
(g) $y = 3x - x^2$ and $y = 4 - x$
(h) $y = 12x^2 - 5x$ and $y = 6x + 5$

19.8 The intersection of a line and a circle

In Chapter 15 you saw that the equation of a circle with centre $(0, 0)$ and radius r is $x^2 + y^2 = r^2$.

Example 18

(a) Show that the x-coordinate of each point of intersection of the line $y = 2x + 2$ and the circle $x^2 + y^2 = 8$ must satisfy the equation $5x^2 + 8x - 4 = 0$.

(b) Hence find the coordinates of the points where the line $y = 2x + 2$ cuts the circle $x^2 + y^2 = 8$.

(a) The points of intersection lie on both the circle and the line, so their coordinates must satisfy both equations:

$$x^2 + y^2 = 8 \qquad (1)$$

$$\text{and} \qquad y = 2x + 2 \qquad (2)$$

Eliminating y by substituting (2) into (1) gives

$$x^2 + (2x + 2)^2 = 8$$

$$x^2 + 4x^2 + 8x + 4 = 8 \quad\text{———— Expand the brackets}$$

$$5x^2 + 8x - 4 = 0 \quad\text{———— Rearrange as required}$$

(b) Factorise: $(5x - 2)(x + 2) = 0$

So $\qquad\qquad\quad 5x - 2 = 0 \quad\text{or}\quad x + 2 = 0$

which gives $\qquad\quad x = \frac{2}{5} \quad\text{or}\qquad x = -2$

Substitute these values of x in the linear equation (2):

When $x = \frac{2}{5}$, $y = 2\left(\frac{2}{5}\right) + 2 = 2\frac{4}{5}$

When $x = -2$, $y = 2(-2) + 2 = -2$

The line $y = 2x + 2$ cuts the circle $x^2 + y^2 = 8$ at the points $\left(\frac{2}{5}, 2\frac{4}{5}\right)$ and $(-2, -2)$.

Example 19

(a) Solve the simultaneous equations $4y + 3x = 25$ and $x^2 + y^2 = 25$.

(b) Interpret your solution to part (a) geometrically.

(a) $\qquad\qquad 4y + 3x = 25 \qquad (1)$

$\qquad\qquad\quad x^2 + y^2 = 25 \qquad (2)$

From (1): $\qquad y = \dfrac{25 - 3x}{4} \qquad (3)$

Eliminating y by substituting (3) into (2) gives

$$x^2 + \frac{(25 - 3x)^2}{16} = 25$$

$$16x^2 + (25 - 3x)^2 = 400 \quad\text{———— Multiply each term by 16}$$

$$16x^2 + 625 - 150x + 9x^2 = 400 \quad\text{———— Expand the brackets}$$

$$25x^2 - 150x + 225 = 0 \quad\text{———— Rearrange}$$

$$25(x^2 - 6x + 9) = 0 \quad\text{———— Factorise}$$

$$25(x - 3)^2 = 0$$

So $\qquad x - 3 = 0$ (repeated root)

which gives $\qquad x = 3$

Substitute in (3):   When $x = 3$, $y = \dfrac{25 - 9}{4} = 4$

$x = 3$, $y = 4$ is the only solution of the simultaneous equations.

(b) The line $4y + 3x = 25$ is a tangent to the circle $x^2 + y^2 = 25$ at the point (3, 4).

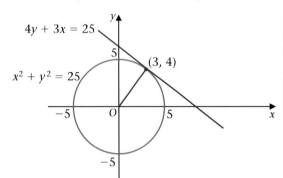

$4y + 3x = 25$

$x^2 + y^2 = 25$

(3, 4)

The gradient of the radius from (0, 0) to (3, 4) is $\frac{4}{3}$.

The gradient of the line $4y + 3x = 25$, or $y = -\frac{3}{4}x + \frac{25}{4}$, is $-\frac{3}{4}$.

From Section 9.6, since $\frac{4}{3} \times -\frac{3}{4} = -1$, the radius and the line $4y + 3x = 25$ are perpendicular. A radius is always perpendicular to a tangent.

Exercise 19H

1   In each part solve the simultaneous equations and interpret your solution geometrically.

(a) $y = 3$ and $x^2 + y^2 = 25$

(b) $y = x + 5$ and $x^2 + y^2 = 25$

(c) $x^2 + y^2 = 25$ and $y = 3x + 13$

(d) $x^2 + y^2 = 25$ and $y + x + 1 = 0$

(e) $x^2 + y^2 = 50$ and $y = x - 10$

(f) $x^2 + y^2 = 50$ and $y + 8x = 1$

(g) $y = 3x - 10$ and $x^2 + y^2 = 10$

(h) $y + 2x = 5$ and $x^2 + y^2 = 5$

(i) $x^2 + y^2 = 2$ and $y + 7x + 10 = 0$

(j) $3y + 4x = 6$ and $x^2 + y^2 = 4$

2   The line $y = x - 3$ intersects the circle $x^2 + y^2 = 25$ at the points A and B.

(a) Show that the x-coordinates of A and B satisfy the equation $x^2 - 3x - 8 = 0$.

(b) Hence show that the coordinates of A and B are

$$\left(\frac{3 + \sqrt{41}}{2}, \frac{-3 + \sqrt{41}}{2} \right) \text{ and } \left(\frac{3 - \sqrt{41}}{2}, \frac{-3 - \sqrt{41}}{2} \right)$$

3   (a) Show that the line $y = x + 4$ is a tangent to the circle $x^2 + y^2 = 8$.

(b) (i)   Find the equation of the tangent to the circle $x^2 + y^2 = 8$ that is parallel to $y = x + 4$ and write down the coordinates of the point where it touches the circle.

(ii) Find the distance between these two parallel tangents.

(c) Give a general result for the distance between two parallel tangents to any circle of the form $x^2 + y^2 = r^2$.

19.9 Solving equations graphically

Finding graphical solutions to quadratic equations was introduced in Chapter 17.

Unless graphs have been drawn for you, you could use a lot of valuable time in an exam drawing a graph accurate enough to solve an equation. Also, graphical solutions are less accurate. So you should only use a graphical approach to solve a quadratic equation if a question asks you to.

Although most cubic and higher power equations have to be solved graphically rather than algebraically at GCSE level, you should still consider an algebraic approach if x is a common factor.

Example 20

Solve the equation $x^3 - 4x^2 + 4x = 0$.

$$x^3 - 4x^2 + 4x = 0$$
$$x(x^2 - 4x + 4) = 0$$
$$x(x - 2)(x - 2) = 0$$

This leads to the solutions $x = 0$ or $x = 2$ (repeated).

Notice that the 'double solution' at $x = 2$ indicates that the graph of $y = x^3 - 4x^2 + 4x$ just touches the x-axis at $x = 2$. The x-axis is a tangent to this curve at $x = 2$.

Here is a graph of $y = x^3 - 4x^2 + 4x$:

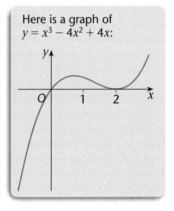

Using a given graph to solve equations

Sometimes you will need to rearrange the equation to be solved to make use of a given graph.

Example 21

The graph of $y = 2 + x - x^2$, for $-3 \leqslant x \leqslant 3$, is shown.

(a) By drawing lines on the graph, solve the following equations, giving your answers to (i) correct to 1 d.p.

    (i) $1 + x - x^2 = 0$

    (ii) $2 - x - x^2 = 0$

(b) Find the equation of the line you would draw on the graph to solve the equation $2x^2 - x = 6$.

(c) Explain how the graph can be used to show that the equation $x^2 - x + 3 = 0$ has no real solutions.

(d) By drawing a suitable curve on the graph, solve the equation $x^3 - x^2 - 2x + 6 = 0$.

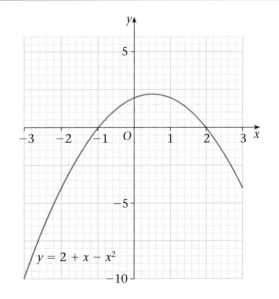

(a) (i) To solve $1 + x - x^2 = 0$ using the graph:

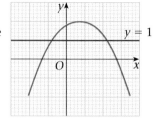

$$2 + x - x^2 = 1 \quad\text{——— Add 1 to each side}$$

The graph shows $\quad 2 + x - x^2 = y$

Comparing gives $\quad\quad\quad y = 1$

So draw the horizontal line $\quad y = 1$

This line meets the parabola at $x = -0.6$ and $x = 1.6$, so the solutions of $1 + x - x^2 = 0$ are $x = -0.6$ and $x = 1.6$ to 1 d.p.

(ii) To solve $2 - x - x^2 = 0$ using the graph:

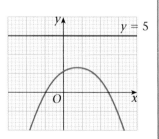

$$2 + x - x^2 = 2x \quad\text{——— Add 2x to each side}$$

The graph shows $\quad 2 + x - x^2 = y$

Comparing gives $\quad\quad\quad y = 2x$

So draw the line $\quad\quad\quad y = 2x$

This line meets the parabola at $x = -2$ and $x = 1$ so the solutions of $2 - x - x^2 = 0$ are $x = -2$ and $x = 1$.

(b) To solve $2x^2 - x = 6$ using the graph:

$$0 = 6 + x - 2x^2 \quad\text{——— Rearrange}$$

$$0 = 3 + \tfrac{1}{2}x - x^2 \quad\text{——— Divide each side by 2}$$

$$\tfrac{1}{2}x = 3 + x - x^2 \quad\text{——— Add } \tfrac{1}{2}x \text{ to each side}$$

$$\tfrac{1}{2}x - 1 = 2 + x - x^2 \quad\text{——— Subtract 1 from each side}$$

The graph shows $\quad y = 2 + x - x^2$

Comparing gives $\quad y = \tfrac{1}{2}x - 1$

So you would need to draw the line $y = \tfrac{1}{2}x - 1$ and find where it meets the parabola. The x-coordinates of the points of intersection give the solutions of the equation $2x^2 - x = 6$.

(c) To solve $x^2 - x + 3 = 0$ using the graph:

$$0 = -3 + x - x^2 \quad\text{——— Rearrange}$$

$$5 = 2 + x - x^2 \quad\text{——— Add 5 to each side}$$

The graph shows $\quad y = 2 + x - x^2$

Comparing gives $\quad y = 5$

The horizontal line $y = 5$ does not meet the parabola so you can deduce that $x^2 - x + 3 = 0$ has no real solutions.

(d) To solve $x^3 - x^2 - 2x + 6 = 0$ using the graph:

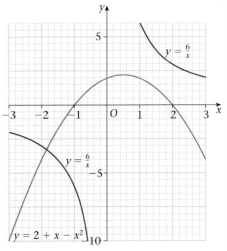

$$6 = 2x + x^2 - x^3 \quad\text{------- Rearrange}$$

$$6 = x(2 + x - x^2) \quad\text{------- Factorise}$$

$$\frac{6}{x} = 2 + x - x^2 \quad\text{-- Divide both sides by } x$$

The graph shows $y = 2 + x - x^2$

Comparing gives $y = \dfrac{6}{x}$

Draw the graph of the reciprocal function $y = \dfrac{6}{x}$.

Notice that it only meets the parabola at one point. This means that there is only one real solution of the equation $x^3 - x^2 - 2x + 6 = 0$.

The x-coordinate of this point of intersection gives the solution. So the one real solution of $x^3 - x^2 - 2x + 6 = 0$ is $x = -1.8$ to 1 d.p.

> **Alternative method:**
> You could rearrange $x^3 - x^2 - 2x + 6 = 0$ to get $2 + x - x^2 = 3x - x^3 - 4$.
>
> Then draw the curve $y = 3x - x^3 - 4$ and find its point of intersection with the parabola. Although this is valid, the aim is to rearrange so that the easiest graph to draw is obtained, preferably a straight line.

Solving equations from a given graph (or one to be drawn):

- Rearrange the equation to be solved to match the equation of the graph. Try to keep the rearrangement simple.
- Compare to find y. Note that y may be of the form $\dfrac{k}{x}$ if the equation to be solved has a higher power of x than the given graph function.
- Draw the line (or curve) from the previous step on the given axes.
- Read off the values of x at the points of intersection.

Exercise 19I

1 **(a)** On graph paper draw the graph of $y = x^2 - 4$ for $-3 \leqslant x \leqslant 3$.

(b) On the same axes, draw the graph of $y = 2 + \dfrac{1}{x}$ for $-3 \leqslant x \leqslant -\frac{1}{2}$ and $\frac{1}{2} \leqslant x \leqslant 3$.

(c) Use your graphs to find approximate solutions for

(i) $x^2 = 2$ **(ii)** $2 + \dfrac{1}{x} = 0.8$ **(iii)** $2 + \dfrac{1}{x} = x^2 - 4$

2 Part of the graph of the cubic function
$y = x^3 - 4x$ is shown.

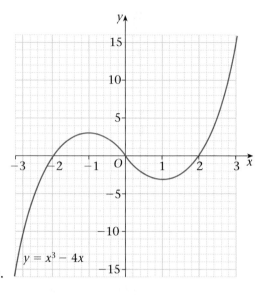

(a) Find the equation of the line which needs to be drawn on the graph to solve

    (i) $x^3 - 4x = 1$ (ii) $x^3 - 4x + 2 = 0$

    (iii) $x^3 - 6x = 1$ (iv) $x^3 - 2x = 4$

(b) Which one of the equations in (a) does not have three real solutions?
Explain your answer.

(c) The graph of $y = \dfrac{1}{x}$ (not drawn) intersects the given cubic curve $y = x^3 - 4x$ at points A and B. Find, in a form which includes an x^4 term, the equation which has the x-coordinates of A and B as two of its solutions.

3 (a) On graph paper, draw the graph of the parabola
$y = (6 - x)(2 + x)$ for $-2 \leqslant x \leqslant 6$.

(b) A container is in the shape of a cuboid of height x m.
The area of the base of the container is $(6 - x)(2 + x)$ m^2.
The volume of the container is 24 m^3.

By drawing the graph of an appropriate curve on your axes in (a) find, correct to the nearest tenth of a metre, the possible values for the height of the container.

4 (a) Copy and complete this table of values for $y = x^3 - 6x + 1$.

x	-3	-2	-1	0	1	2
y	-8					

(b) On graph paper, draw and label appropriate axes.
Plot the points represented by the values in your table. Join them with a smooth curve.

(c) (i) On the same axes, draw the line with equation
$y = 2x + 5$.

    (ii) Write down the x-coordinates of the two points of intersection of this line and the curve you have drawn in (b).

(d) Find the equation, expressed as simply as possible, which may be solved to give the x-coordinate of each point of intersection of the curve and the line.

(e) A line L is drawn parallel to the line $y = 2x + 5$ so that L intersects the curve $y = x^3 - 6x + 1$ at the point on the y-axis.

Write down the equation of line L.

(f) By forming and solving an algebraic equation, find the x-coordinates of the three points of intersection of L and the curve $y = x^3 - 6x + 1$.

Mixed exercise 19

1   The area of this rectangle is 9 cm^2.
Find the value of x.

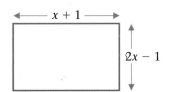

2   A rectangle measuring $(x + 3)$ cm by $(2x + 1)$ cm has a
rectangle measuring 2 cm by 4 cm cut from it.
The area left is 67 cm^2.
Work out the value of x.

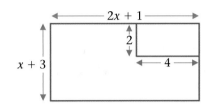

3   The length of the line joining $(2, 5)$ to $(x, -1)$ is 7 centimetres.
Work out the value of x.
Give your answer in the form $a + \sqrt{b}$.

4   Solve $x^2 + 6x = 4$.
Give your answer in the form $p + \sqrt{q}$, where p and q are integers.

5   (a)   Factorise $2x^2 - 35x + 98$.

(b)   Solve the equation $2x^2 - 35x + 98 = 0$.   [E]

6   Solve these simultaneous equations:

$$x^2 + y^2 = 29$$
$$y - x = 3$$ [E]

7   (a)   Solve $x^2 + x + 11 = 14$.
Give your solutions correct to 3 significant figures.

(b) $y = x^2 + x + 11$

The value of y is a prime number when $x = 0, 1, 2$ and 3.

The following statement is **not** true:

'$y = x^2 + x + 11$ is **always** a prime number when x is an
integer'.

Show that the statement is not true.   [E]

8 AT is a tangent at T to a circle, centre O.
$OT = x$ cm, $AT = (x + 5)$ cm, $OA = (x + 8)$ cm.

(a)   Show that $x^2 - 6x - 39 = 0$.

(b)   Solve the equation $x^2 - 6x - 39 = 0$ to find
the radius of the circle. Give your answer
correct to 3 significant figures.

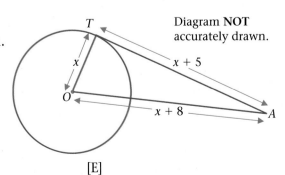

Diagram **NOT**
accurately drawn.

[E]

9 (a) On a grid, with $-12 \leqslant x \leqslant 12$ and $-12 \leqslant y \leqslant 12$, draw the graphs of $x^2 + y^2 = 100$ and $2y = 3x - 4$.

(b) Use the graphs to estimate the solutions of the simultaneous equations

$$x^2 + y^2 = 100$$

and $\qquad 2y = 3x - 4$

(c) For all the values of x, $x^2 + 6x = (x + 3)^2 - q$.

Find the value of q.

(d) One pair of integer values which satisfy the equation $x^2 + y^2 = 100$ is $x = 6$ and $y = 8$.

Find one pair of integer values which satisfy

$$x^2 + 6x + y^2 - 4y - 87 = 0.$$

[E]

Summary of key points

1 The quadratic equation $ax^2 + bx + c = 0$, with $a \neq 0$, has two solutions (or roots), which may be equal.

2 If $xy = 0$ then either $x = 0$ or $y = 0$.

3 If $y^2 = k$ then $y = \pm\sqrt{k}$

4 Completing the square: $x^2 + bx = \left(x + \dfrac{b}{2}\right)^2 - \left(\dfrac{b}{2}\right)^2$

5 The roots of the quadratic equation $ax^2 + bx + c = 0$, where $a \neq 0$, are given by the formula

$$x = \frac{\sqrt{-b \pm b^2 - 4ac}}{2a}$$

6 If $\dfrac{p}{r} = \dfrac{q}{r}$ then $p = q$.

7 When applied to problem solving, one of the roots of the quadratic equation is often not a solution to the problem and must be abandoned (with an explanation).

8 Solving a linear equation $(y = px + q)$ and a quadratic equation $(y = ax^2 + bx + c)$ simultaneously:

- Find y in terms of x from the linear equation (or x in terms of y).
- Substitute for y (or x) in the quadratic equation.
- Solve the resulting quadratic equation for x (or y).
- Substitute the values of x (or y) into the linear equation to find y (or x).

If the roots of the quadratic equation are equal, the line will be a tangent to the curve.

9 Solving equations from a given graph (or one to be drawn):

- Rearrange the equation to be solved to match the equation of the graph. Try to keep the rearrangement simple.
- Compare to find y. Note that y may be of the form $\dfrac{k}{x}$ if the equation to be solved has a higher power of x than the given graph function.
- Draw the line (or curve) from the previous step on the given axes.
- Read off the values of x at the points of intersection.

20 Presenting and analysing data 2

This chapter shows you how to use and interpret cumulative frequency diagrams, box plots and histograms.

20.1 Upper class boundaries

When you form a frequency distribution you usually group the information into 'classes' or 'intervals'.

In this frequency distribution, the values 10, 20, 30, 40 and 50 are the largest possible values in each of the intervals.

Marks in test	Frequency
$0 < b \leqslant 10$	5
$10 < b \leqslant 20$	8
$20 < b \leqslant 30$	12
$30 < b \leqslant 40$	7
$40 < b \leqslant 50$	4

> The largest possible value in an interval is called the **upper class boundary**.

The upper class boundary for an interval may depend upon whether the data is discrete or continuous.

Here are some examples for **discrete data**:

Number of bids in auction	Upper class boundary
5–9	9
10–14	14
15–19	19
20–25	25

Number of bids in auction	Upper class boundary
$5 < m \leqslant 9$	9
$9 < m \leqslant 14$	14
$14 < m \leqslant 19$	19
$19 < m \leqslant 25$	25

Number of bids in auction	Upper class boundary
$5 \leqslant m < 9$	8
$9 \leqslant m < 14$	13
$14 \leqslant m < 19$	18
$19 \leqslant m < 25$	24

> $5 \leqslant m < 9$ means 5, 6, 7 and 8. So 8 is the largest possible value in the interval.

Here are some examples for **continuous data**:

Weight of letter (g)	Upper class boundary
$20.5 < m \leqslant 30.5$	30.5
$30.5 < m \leqslant 40.5$	40.5
$40.5 < m \leqslant 50.5$	50.5
$50.5 < m \leqslant 60.5$	60.5

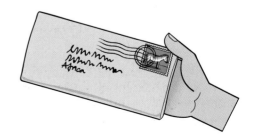

Weight of letter (g)	Upper class boundary
$20.5 \leqslant m < 30.5$	30.5
$30.5 \leqslant m < 40.5$	40.5
$40.5 \leqslant m < 50.5$	50.5
$50.5 \leqslant m < 60.5$	60.5

Weight of letter (g)	Upper class boundary
21–30	30.5
31–40	40.5
41–50	50.5
51–60	60.5

> If continuous data is grouped as 21–30, 31–40 and so on, then assume that the 21–30 interval goes from 20.5 to 30.5 with a midpoint of 25.5.

20.2 Cumulative frequency

The **cumulative frequency** is the total frequency up to a particular upper class boundary.

Here is an example for discrete data.

Example 1

The number of points scored in each of 40 rugby games was recorded. The results are summarised in the table.

Draw up a cumulative frequency table for this information.

Number of points	Frequency
0–10	2
11–20	5
21–30	9
31–40	12
41–50	8
51–60	4

The number of points scored in a game is discrete data, so the upper class boundaries are 10, 20, 30, ... The cumulative frequency table is:

Number of points	Cumulative frequency
0–10	2
0–20	$2 + 5 = 7$
0–30	$2 + 5 + 9 = 16$
0–40	$2 + 5 + 9 + 12 = 28$
0–50	$2 + 5 + 9 + 12 + 8 = 36$
0–60	$2 + 5 + 9 + 12 + 8 + 4 = 40$

Here is an example for continuous data.

Example 2

The table gives information about the lengths of 50 fossils.
Draw up a cumulative frequency table for this information.

Length (cm)	0–8	9–13	14–18	19–23	24–28	29–33
Frequency	5	10	16	9	7	3

The length of a fossil is continuous data, so the upper class
boundaries are 8.5, 13.5, 18.5, …
The cumulative frequency table is:

Length of fossil (cm)	Cumulative frequency
$0 \leqslant l < 8.5$	5
$0 \leqslant l < 13.5$	15
$0 \leqslant l < 18.5$	31
$0 \leqslant l < 23.5$	40
$0 \leqslant l < 28.5$	47
$0 \leqslant l < 33.5$	50

Fossil ammonites range from
under 2 cm to 300 cm in
diameter. These marine
creatures became extinct 65
million years ago.

Exercise 20A

1   Draw up a cumulative frequency table for each of the following sets of data.

(a)
Time watching TV (hours)	Frequency
0–3	3
4–7	5
8–11	8
12–15	3
16–18	1

(b)
Number of people on bus	Frequency
0–5	8
6–10	7
11–15	9
16–20	7
21–25	9

(c)
Age of mother at birth of baby (years)	Frequency
16–20	3
21–25	6
26–30	17
31–35	26
36–40	11
41–50	2

(d)
Daily temperature (°C)	Frequency
$-10 \leqslant t < 0$	12
$0 \leqslant t < 10$	86
$10 \leqslant t < 20$	185
$20 \leqslant t < 30$	79
$30 \leqslant t < 40$	3

2 Draw up a frequency table for this cumulative frequency table:

Weight of baby (kg)	Cumulative frequency
$1 \leqslant w < 2$	5
$1 \leqslant w < 3$	17
$1 \leqslant w < 4$	41
$1 \leqslant w < 5$	49
$1 \leqslant w < 6$	50

20.3 Cumulative frequency graphs

You can display data in a **cumulative frequency graph** by plotting the cumulative frequency against the upper class boundary for each class interval.

Example 3

The number of people queuing at a supermarket checkout was recorded at 10-minute intervals during one day. The table shows the frequency distribution.

Number in queue	0–9	10–19	20–29	30–39	40–49	50–59
Frequency	8	14	26	16	12	4

To draw the cumulative graph you must first draw up a cumulative frequency table. The numbers of people are discrete data, so the upper class boundaries are 9, 19, 29, 39, 49 and 59. The cumulative frequency table is:

Number in queue	Cumulative frequency
$0 \leqslant m \leqslant 9$	8
$0 \leqslant m \leqslant 19$	22
$0 \leqslant m \leqslant 29$	48
$0 \leqslant m \leqslant 39$	64
$0 \leqslant m \leqslant 49$	76
$0 \leqslant m \leqslant 59$	80

Draw the cumulative frequency graph by plotting (9, 8), (19, 22), (29, 48), ...

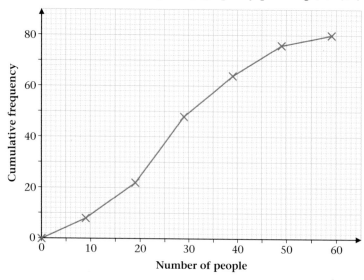

Cumulative frequency is always plotted on the vertical axis.

Here the points are joined by straight lines, giving a **cumulative frequency polygon.**

Plot the starting point (0, 0) even though it is not a value in the table.

A cumulative frequency polygon is used if the data is discrete. A cumulative frequency curve is used if the data is continuous.

Cumulative frequency diagrams may have different shapes.
Here are some examples:

1 In this table of data the values increase at a steady
rate, giving a straight line in the cumulative
frequency diagram.

Class interval	Frequency
0–10	10
11–20	10
21–30	10
31–40	10
41–50	10

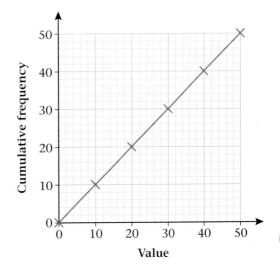

2 In this table the values are concentrated in the
middle. This is shown by the steepness of the curve
in the cumulative frequency diagram as it passes
through the class interval 21–30.

Class interval	Frequency
0–10	6
11–20	8
21–30	22
31–40	8
41–50	6

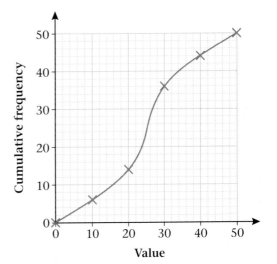

3 In this table there are few values in the class
interval 21–30. This is shown in the cumulative
frequency diagram by the less steep slope.

Class interval	Frequency
0–10	16
11–20	6
21–30	4
31–40	6
41–50	18

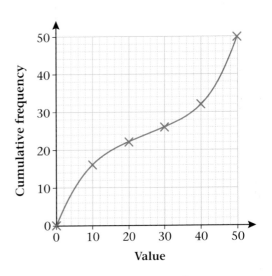

20.4 Using cumulative frequency graphs

A cumulative frequency graph can be used to *estimate* the median, the upper quartile and the lower quartile of a distribution.

Estimating the median

> The **median** is the value half way into the distribution.

To estimate the median from a cumulative frequency graph, find the $\frac{n}{2}$th value in the distribution.

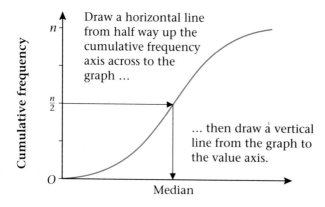

Draw a horizontal line from half way up the cumulative frequency axis across to the graph …

… then draw a vertical line from the graph to the value axis.

Estimating the quartiles

> The **lower quartile** is the value one quarter of the way into the distribution.

> The **upper quartile** is the value three quarters of the way into the distribution.

Here you find the $\frac{n}{4}$th value in the distribution.

Here you find the $\frac{3n}{4}$th value in the distribution.

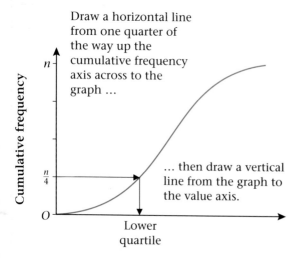

Draw a horizontal line from one quarter of the way up the cumulative frequency axis across to the graph …

… then draw a vertical line from the graph to the value axis.

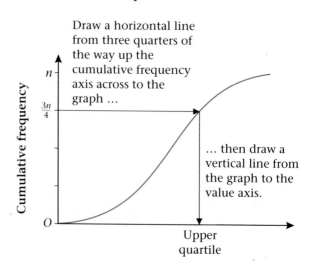

Draw a horizontal line from three quarters of the way up the cumulative frequency axis across to the graph …

… then draw a vertical line from the graph to the value axis.

The interquartile range

To estimate the interquartile range from a cumulative frequency graph, first estimate the upper and lower quartiles, then find:

> **interquartile range** = upper quartile − lower quartile

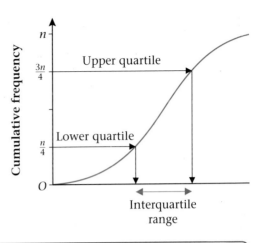

Example 4

The table gives information about the times (in seconds) between planes landing at an airport.

Find an estimate for

(a) the median

(b)   (i)   the lower quartile
      (ii) the upper quartile
      (iii) the interquartile range.

Time between planes (s)	Frequency
$60 \leqslant t < 100$	5
$100 \leqslant t < 140$	14
$140 \leqslant t < 180$	10
$180 \leqslant t < 220$	9
$220 \leqslant t < 260$	7
$260 \leqslant t < 300$	3

(a) The cumulative frequency table is on the right.

From this table the cumulative frequency graph can be drawn.

Time between planes (s)	Cumulative frequency
$60 \leqslant t < 100$	5
$60 \leqslant t < 140$	19
$60 \leqslant t < 180$	29
$60 \leqslant t < 220$	38
$60 \leqslant t < 260$	45
$60 \leqslant t < 300$	48

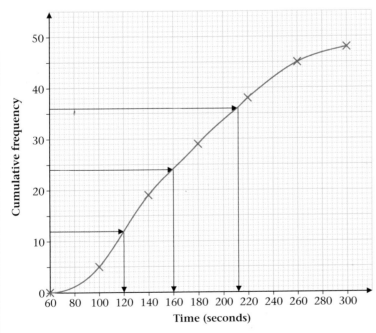

> Remember to plot the starting point (60, 0).

Read the required estimates from the graph:

Median $= \dfrac{48}{2}$th value = 24th value = 160 seconds

(b)   (i)   Lower quartile $= \dfrac{48}{4}$th value = 12th value = 120 seconds

(ii)   Upper quartile $= \dfrac{3(48)}{4}$th value = 36th value = 212 seconds

(iii)   Interquartile range = upper quartile − lower quartile
$$= 212 - 120 = 92 \text{ seconds}$$

Exercise 20B

1   The numbers of particles emitted each minute by a sample of uranium are summarised in the table.

Number of particles	Frequency
0–50	10
51–100	16
101–150	13
151–200	11
201–250	7
251–300	3

(a)   Draw a cumulative frequency graph for this data.

(b)   Use your graph to find an estimate for

(i)   the median   (ii)   the lower quartile

(iii)   the upper quartile   (iv)   the interquartile range.

2   A software company has a record of the age, at their last birthday, of each of its workers. This table shows the age distribution.

Age (years)	Frequency
$16 < a \leqslant 20$	6
$21 < a \leqslant 25$	9
$26 < a \leqslant 30$	14
$31 < a \leqslant 35$	4
$36 < a \leqslant 40$	2
$41 < a \leqslant 45$	1

(a)   Draw a cumulative frequency graph for this data.

(b)   Find an estimate for

(i)   the median   (ii)   the interquartile range.

3 The table gives information about the body temperatures of a random sample of people.

Use this data to find an estimate for

(a) the median body temperature

(b) the interquartile range

(c) the number of people with a body temperature less than 37 °C.

Body temperature (°C)	Cumulative frequency
<36.0	8
<36.3	23
<36.6	44
<36.9	78
<37.2	101
<37.5	115
<37.8	120

Activity – World temperatures

(a) From a newspaper, collect the highest temperatures (°C) recorded in cities around the world.

(b) Using suitable class intervals (e.g. 0–5, 6–10…), draw a cumulative frequency graph and find an estimate for the median temperature.

(c) How good is your answer as an estimate for the average temperature of the Earth? Explain your answer.

20.5 Using cumulative frequency graphs to solve problems

Cumulative frequency graphs can be used to solve a variety of problems.

Worked examination question

The speeds in miles per hour (mph) of 120 cars travelling on the A4 road were measured. The results are shown in the table.

(a) Draw a cumulative frequency graph to show these figures.

(b) Use your graph to find an estimate for the percentage of cars travelling at less than 42 miles per hour.

Speed (mph)	Cumulative frequency
$5 < s \leqslant 20$	2
$5 < s \leqslant 25$	10
$5 < s \leqslant 30$	28
$5 < s \leqslant 35$	50
$5 < s \leqslant 40$	84
$5 < s \leqslant 45$	106
$5 < s \leqslant 50$	116
$5 < s \leqslant 55$	120
$5 < s \leqslant 60$	120

(a) Here is the cumulative frequency graph.

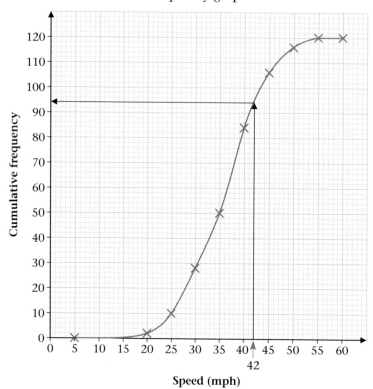

The starting point is (5, 0).

(b) The graph shows that 94 out of 120 cars are travelling at less than 42 mph.

$$\frac{94}{120} = 78.3\%$$

So 78.3% of the cars are travelling at less than 42 mph.

Exercise 20C

1 The marks in an examination are shown in the table.
The maximum mark obtainable was 100.

(a) Draw a cumulative frequency polygon.

(b) Use your graph to estimate the number of people who scored more than 67 marks.

(c) Work out an estimate for the mark that was exceeded by 60% of the people.

Mark range	Frequency
0–10	4
11–20	8
21–30	13
31–40	28
41–50	45
51–60	58
61–70	53
71–80	33
81–90	17
91–100	4

2 The table shows the frequency distribution of the masses of 50 students at a college.

Mass (kg)	$40 < m \leqslant 45$	$45 < m \leqslant 50$	$50 < m \leqslant 55$	$55 < m \leqslant 60$	$60 < m \leqslant 70$	$70 < m \leqslant 80$
Frequency	2	8	15	16	6	3

(a) Draw a cumulative frequency curve. Start your graph at (40, 0).

(b) Work out an estimate for the percentage of these students with a mass less than 53 kg.

(c) Find an estimate for the mass exceeded by 20% of the students.

3 The table gives information about the ages of people at a birthday party.

Age (years)	$0 \leqslant a < 5$	$5 \leqslant a < 10$	$10 \leqslant a < 15$	$15 \leqslant a < 25$	$25 \leqslant a < 35$
Frequency	3	9	7	3	5

Age (years)	$35 \leqslant a < 40$	$40 \leqslant a < 50$	$50 \leqslant a < 55$	$55 \leqslant a < 60$
Frequency	12	10	4	1

(a) Draw a cumulative frequency graph.

(b) Find an estimate for the age exceeded by the majority of the people.

(c) Find an estimate for the number of teenagers at the party.

20.6 Box plots

Here is the cumulative frequency graph from Example 4, for times between planes landing at an airport.

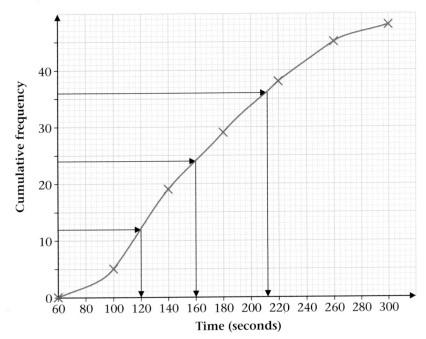

You can represent some of the important features of a cumulative frequency graph by drawing a **box plot** (or **box-and-whisker diagram**).

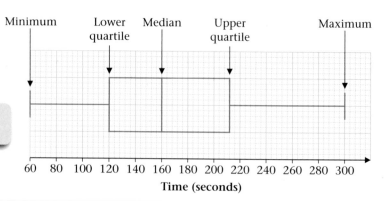

> A box plot needs a horizontal scale.

To draw a **box plot** you need five pieces of information:
- the lowest value
- the lower quartile
- the median
- the upper quartile
- the highest value.

From the box plot above, the actual lowest and highest values of the data are not known because the data is grouped. In this case you use the lowest and highest possible values for the data. The **lowest possible time** between planes is 60 seconds, and the **highest possible time** between planes is 300 seconds.

You can use a box plot to represent data given as a list or in a stem-and-leaf diagram.

Time between planes (s)	Frequency
$60 \leq t < 100$	5
$100 \leq t < 140$	14
$140 \leq t < 180$	10
$180 \leq t < 220$	9
$220 \leq t < 260$	7
$260 \leq t < 300$	3

___ **Example 5** ___

This stem and leaf diagram shows the age of each of the last 23 British prime ministers when taking office.

Draw a box plot to represent this information.

Age of prime minister (years)

4	3, 7, 8
5	3, 3, 3, 3, 5, 7, 7, 7, 7
6	0, 2, 2, 3, 4, 4, 5, 7, 8, 9
7	6

Key: 4 | 3 means 43.

The lowest value is 43

The lower quartile $\left(\text{the } \dfrac{23 + 1}{4} = \text{6th value} \right) = 53$

The median $\left(\text{the } \dfrac{23 + 1}{2} = \text{12th value} \right) = 57$

The upper quartile $\left(\text{the } \dfrac{3(23 + 1)}{4} = \text{18th value} \right) = 64$

The highest value is 76
So the box plot is:

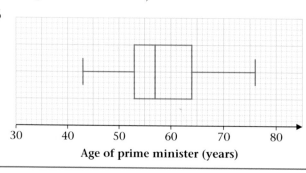

Age of prime minister (years)

Exercise 20D

1   This cumulative frequency diagram gives information about the
weights of 100 babies born at Bundledale Hospital.

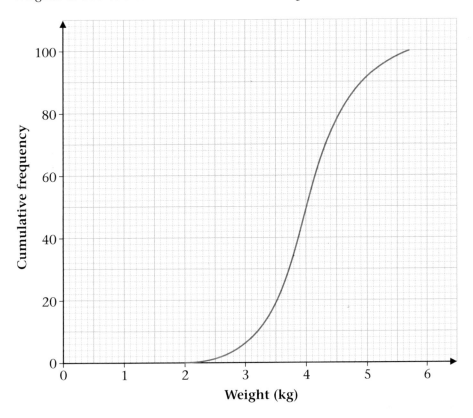

Draw a box plot to represent this information.

2   The number of consecutive press-ups by each of 27 members of a
sports club are summarised in this stem and leaf diagram:

Number of press-ups

0	5, 8
1	1, 1, 2, 3, 3, 3, 5, 7, 8, 8, 9, 9
2	0, 0, 2, 4, 4, 5, 5, 7, 7
3	0, 5, 7
4	8

Key: 4 | 8 means 48.

Draw a box plot to represent this information.

20.7 Comparing sets of data with box plots

Box plots can be very useful when comparing sets of data.

> When you compare the information in two or more box plots, always draw the diagrams lined up, one above the other.

Example 6

Deenita and Zac are writing party invitations. The table gives information about the times, in seconds, they take to write an invitation.

	Deenita	Zac
Lowest	20	11
Highest	58	72
Median	35	28
Lower quartile	26	20
Upper quartile	50	55

(a) Draw two box plots on the same scale to represent this information.

(b) Compare Deenita's and Zac's results.

(a)

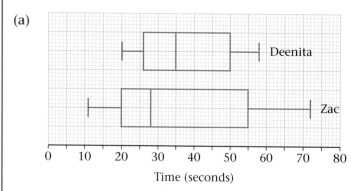

> Look at the maximum and minimum values to choose a suitable scale.

(b) Zac had a lower median value but his values are spread over a much wider range overall, and the interquartile range is greater than for Deenita. Deenita was much more consistent because both the range and the interquartile range are smaller.

Exercise 20E

1 These box plots give information about the results of two different classes for the same test.
Use the diagrams to compare the performance of the two classes.

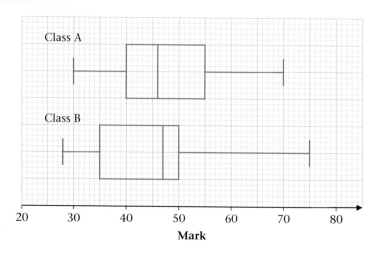

2   The cumulative frequency curves show the times taken by students to run 200 metres.

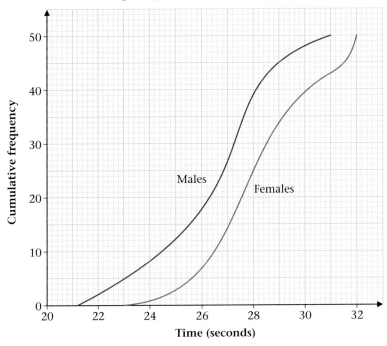

(a) Draw box plots to compare the times for males and females.

(b) Comment on the times for males and females.

Activity – The age of power

Find out the age of each of the last 23 US presidents when taking office.

Compare your results to those for the British prime ministers in Example 5.

20.8 Plotting data in a histogram

Equal-sized class intervals

The table shows how long an audience's laughter lasted, in seconds, for 54 jokes told by a stand-up comedian.

The data can be displayed in a histogram like this:

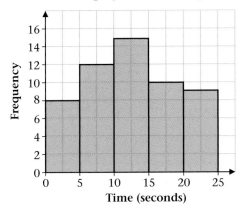

Duration of applause (s)	Frequency
$0 < t \leqslant 5$	8
$5 < t \leqslant 10$	12
$10 < t \leqslant 15$	15
$15 < t \leqslant 20$	10
$20 < t \leqslant 25$	9

In a **histogram** the areas of the rectangles are proportional to the frequencies they represent.

In this case all the class intervals are the same (5 seconds) so the rectangles all have the same width, and the height of each rectangle is proportional to its area and to the frequency.

Unequal class intervals

Duration of applause (s)	Frequency
$0 < t \leqslant 10$	20
$10 < t \leqslant 15$	15
$15 < t \leqslant 20$	10
$20 < t \leqslant 25$	9

When the class intervals are unequal it is incorrect to use the bar heights to represent frequencies.

If the stop-watch used to time the applause is not accurate for times of less than 5 seconds, the data can be recorded using different class intervals like this:

To draw the histogram you now construct a table with a column for **frequency density**, calculated by dividing the frequency by class width:

$$\textbf{frequency density} = \frac{\text{frequency}}{\text{class width}}$$

Duration of applause (s)	Frequency	Frequency density
$0 < t \leqslant 10$	20	$20 \div 10 = 2$
$10 < t \leqslant 15$	15	$15 \div 5 = 3$
$15 < t \leqslant 20$	10	$10 \div 5 = 2$
$20 < t \leqslant 25$	9	$9 \div 5 = 1.8$

This histogram displays the data:

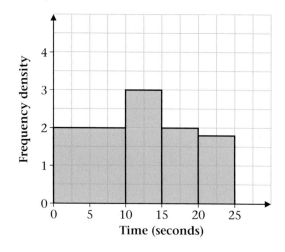

The vertical axis shows frequency density.

The areas of the rectangles are 20, 15, 10 and 9 respectively. They are in the correct proportion to the numbers of rounds of applause (frequencies).

Standard class intervals

The lengths of reign of the kings and queens of England are summarised in this table:

Length of reign (years)	Frequency
$0 < r \leqslant 10$	24
$10 < r \leqslant 20$	14
$20 < r \leqslant 30$	10
$30 < r \leqslant 40$	7
$40 < r \leqslant 50$	2
$50 < r \leqslant 60$	2
$60 < r \leqslant 70$	1

Windsor Castle is the oldest and largest inhabited castle in the world. It has been a royal home and fortress for over 900 years.

The last three class intervals ($40 < r \leqslant 50$, $50 < r \leqslant 60$ and $60 < r \leqslant 70$) have low frequencies compared to the others and will not show up so clearly on a histogram with class intervals of equal width. They will show up more clearly if the last four class intervals are combined:

Length of reign (years)	Frequency
$0 < r \leqslant 10$	24
$10 < r \leqslant 20$	14
$20 < r \leqslant 30$	10
$30 < r \leqslant 70$	12

You could now simply calculate frequency densities by using

$$\text{frequency density} = \frac{\text{frequency}}{\text{class width}}$$

but it is more convenient to use a **standard class interval**.

A standard class interval can be anything you wish, but in this case 10 years is sensible. So the class $30 < r \leqslant 70$ years is 4×10 years or 4 standard intervals.

The table is now:

Length of reign (years)	Class width in standard class intervals	Frequency	Frequency density
$0 < r \leqslant 10$	1	24	$24 \div 1 = 24$
$10 < r \leqslant 20$	1	14	$14 \div 1 = 14$
$20 < r \leqslant 30$	1	10	$10 \div 1 = 10$
$30 < r \leqslant 70$	4	12	$12 \div 4 = 3$

Here the frequency density column has been calculated using standard class intervals:

$$\text{frequency density} = \frac{\text{frequency}}{\text{class width in standard class intervals}}$$

This histogram shows the frequency density plotted against length of reign:

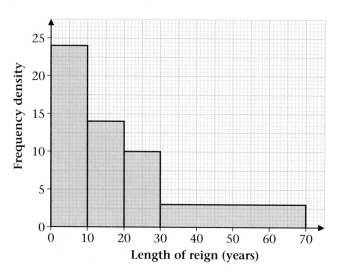

Handling continuous data

Sometimes data which is continuous is recorded as if it were discrete.

Example 7

Agnes drives a van making deliveries for mail order catalogues. She records the distance travelled for each delivery correct to the nearest mile:

3	12	18	13	16	18	19	15	17	16
15	7	20	22	15	18	14	19	17	6
14	17	21	25	18	14	12	17	8	19
13	25	15	14	5	16	16	12	11	7
19	12	26	18	19	10	11	26	7	16
33	13	8	22	16	18	11	38	31	21

(a) Produce a frequency table using intervals 0–9, 10–14, 15–19, 20–24 and 25–39.

(b) Use your completed frequency table to draw a histogram.

(a) As the distances are recorded to the nearest mile the class widths are 9.5, 5, 5, 5 and 15 respectively.

There is no advantage here in defining a standard class interval, so we will use:

$$\text{frequency density} = \frac{\text{frequency}}{\text{class width}}$$

This gives the following table:

> The data is continuous so the 0–9 class is actually 0–9.5, which has a width 9.5. The 10–14 class is actually 9.5–14.5, which has a width 5; and so on.

Distance travelled miles	Tally	Frequency	Frequency density
0–9	ⅢⅢ |||	8	$8 \div 9.5 = 0.84$
10–14	ⅢⅢ ⅢⅢ ⅢⅢ	15	$15 \div 5 = 3$
15–19	ⅢⅢ ⅢⅢ ⅢⅢ ⅢⅢ ⅢⅢ	25	$25 \div 5 = 5$
20–24	ⅢⅢ	5	$5 \div 5 = 1$
25–39	ⅢⅢ ||	7	$7 \div 15 = 0.47$

(b) Using the frequency density, you can draw the histogram.

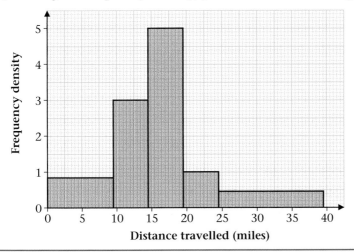

> Note that the histogram is drawn to the upper class boundaries: 9.5, 14.5, 19.5, 24.5 and 39.5.

Exercise 20F

1 This table shows the distribution of times, in seconds, required for students to memorise a list of 10 words.

Draw a histogram to display this data.

Time to memorise (s)	Frequency
$0 < t \leqslant 30$	3
$30 < t \leqslant 60$	9
$60 < t \leqslant 80$	10
$80 < t \leqslant 90$	8
$90 < t \leqslant 100$	6
$100 < t \leqslant 150$	12

2   This table shows the distribution of weights, in grams, of
    40 portions of rice served in a restaurant.

Weight (g)	Frequency
$0 \leqslant w < 30$	5
$30 \leqslant w < 45$	7
$45 \leqslant w < 60$	11
$60 \leqslant w < 75$	8
$75 \leqslant w < 105$	9

Draw a histogram to represent this information.

3   A group of people were asked to throw a tennis ball. The distances
    thrown, in metres, were recorded in a table:

Distance thrown (m)	Frequency
0–9	18
10–14	14
15–19	16
20–24	15
25–34	17
35–50	10

Draw a histogram to show this data.

4   In a survey of hand-lengths in centimetres these results were
    obtained:

20.1	13.8	17.9	14.6	20.3	16.3	21.5
21.6	17.9	18.3	17.7	21.8	18.4	22.1
22.1	20.3	18.7	18.7	23.1	19.1	19.9
21.7	21.0	21.8	18.4	19.7	18.7	24.7
21.5	22.3	20.9	17.1	18.8	19.6	21.6
20.2	23.1	15.8	18.2	22.0	20.8	22.0

> **Remember:**
> The interval 18–19 goes
> from 17.5 to 19.5, but
> does not include 19.5.

   (a) Produce a frequency table with class intervals of 13–17,
        18–19, 20–21 and 22–25.

   (b) Draw a histogram to show this data.

Activity – Distances travelled to school

(Go to www.heinemann.co.uk/hotlinks, insert the express code 4092P and click on this activity.)

Use the information in the Mayfield High School database to draw a histogram for the distances
between home and school of female KS4 students.

20.9 Interpreting histograms

Frequency = frequency density × class width

or

Frequency = frequency density × class width in standard class intervals

---

Example 8

The histogram below shows the distribution of the lifetimes of bees in a small colony.
Use the information to complete the frequency table.
One of the frequencies has already been entered in the table.

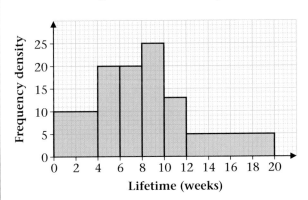

Lifetime (weeks)	Frequency density	Frequency
$0 \leqslant t < 4$	10	40
$4 \leqslant t < 6$	20	
$6 \leqslant t < 8$	20	
$8 \leqslant t < 10$		
$10 \leqslant t < 12$		
$12 \leqslant t < 20$		

The frequency densities can be read straight from the histogram.

The frequencies are obtained by multiplying the frequency density by the class width.

For the $0 \leqslant t < 4$ class interval:

$$\text{frequency} = \text{frequency density} \times 4$$
$$= 10 \times 4 = 40 \text{ bees}$$

For the class intervals in the middle:

$$\text{frequency} = \text{frequency density} \times 2$$

For the $12 \leqslant t < 20$ class interval:

$$\text{frequency} = \text{frequency density} \times 8$$
$$= 5 \times 8 = 40 \text{ bees}$$

Here is the completed table of frequencies:

Lifetime (weeks)	Frequency density	Frequency
$0 \leqslant t < 4$	10	$10 \times 4 = 40$
$4 \leqslant t < 6$	20	$20 \times 2 = 40$
$6 \leqslant t < 8$	20	$20 \times 2 = 40$
$8 \leqslant t < 10$	25	$25 \times 2 = 50$
$10 \leqslant t < 12$	13	$13 \times 2 = 26$
$12 \leqslant t < 20$	5	$5 \times 8 = 40$

Example 9

This histogram has been drawn using a standard class interval of 5 units.
Complete the frequency table.

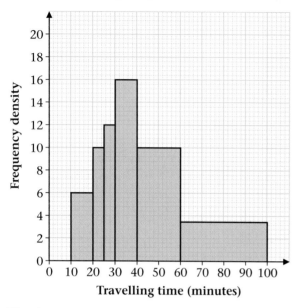

Travelling time (minutes)

Class interval	Frequency density	Frequency
$10 < t \leqslant 20$	6	
$20 < t \leqslant 25$	10	
$25 < t \leqslant 30$	12	
$30 < t \leqslant 40$	16	
$40 < t \leqslant 60$	10	
$60 < t \leqslant 100$	3.5	

The frequencies can be obtained by multiplying the frequency density by the class width in standard class intervals.

For the $10 < t \leqslant 20$ class interval:

width = 10 units = 2 standard class intervals

So

frequency = frequency density $\times$ 2
= $6 \times 2 = 12$ journeys

For the $20 < t \leqslant 25$ class interval:

width = 5 units = 1 standard class interval

So

frequency = frequency density $\times$ 1
= $10 \times 1 = 10$ journeys

and so on for the remaining class intervals.

Here is the completed table of frequencies:

Class interval	Frequency density	Frequency
$10 < t \leqslant 20$	6	$6 \times 2 = 12$
$20 < t \leqslant 25$	10	$10 \times 1 = 10$
$25 < t \leqslant 30$	12	$12 \times 1 = 12$
$30 < t \leqslant 40$	16	$16 \times 2 = 32$
$40 < t \leqslant 60$	10	$10 \times 4 = 40$
$60 < t \leqslant 100$	3.5	$3.5 \times 8 = 28$

A simpler method of showing frequencies is to draw a key to show what a standard area represents.

In this histogram the key shows the size of the area that represents 4 people.

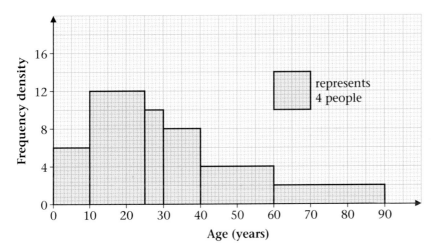

The histogram shows that there are
- 6 people aged between 0 and 10
- 18 people aged between 10 and 25
- 5 people aged between 25 and 30, and so on.

This method can be used with all histograms.

Exercise 20G

1 Students were asked to estimate the length of a line. Their responses are summarised in this histogram:

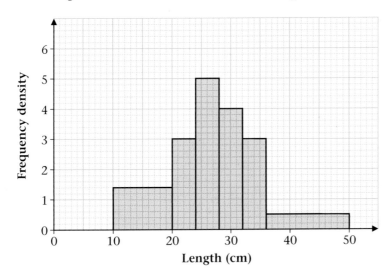

Estimate of length (cm)	Frequency
$10 < l \leqslant 20$	14
$20 < l \leqslant 24$	
$24 < l \leqslant 28$	
$28 < l \leqslant 32$	
$32 < l \leqslant 36$	
$36 < l \leqslant 50$	

Copy and complete the table.

2 The lifetimes of bulbs in an advertising hoarding were recorded. The information is summarised in this histogram:

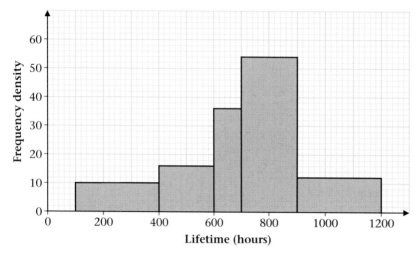

Lifetime (hours)	Frequency
$100 \leq t < 400$	30
$400 \leq t < 600$	30
$600 \leq t < 700$	33
$700 \leq t < 900$	15
$900 \leq t < 1200$	

The histogram was drawn using a standard class interval of 100 hours. Copy and complete the table.

3 The histogram shows the distribution of the weights of the letters in a mail bag. The histogram has been drawn using a standard class interval of 8 units.

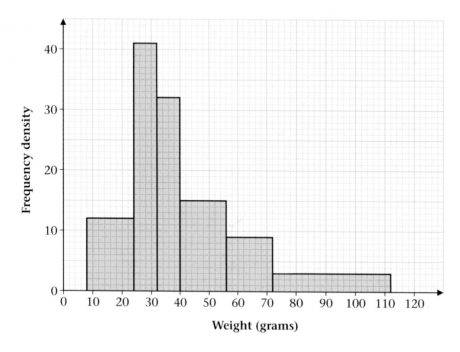

(a) Draw a frequency table.

(b) How many letters were in the mail bag?

(c) Find the percentage of letters that weighed 40 grams or more.

[E]

4   The histogram shows the distribution of the times taken to perform
    180 operations in a hospital.

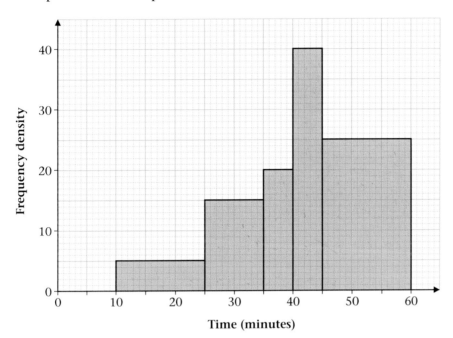

Time (minutes)

(a)   Copy and complete the table.

Time (minutes)	Frequency
$10 \leq t < 25$	~~5~~ 10
$25 \leq t < 35$	35
$35 \leq t < 40$	20
$40 \leq t < 45$	40
$45 \leq t < 60$	75

> Find the number of
> operations represented by
> 1 large square.

(b)   How many operations took less than 35 minutes?   [E]

Mixed exercise 20

1   This ordered stem and leaf diagram shows the times taken for
    each of 29 students to complete their maths homework.

Time to complete homework (minutes)

1	6, 7, 8, 8, 9
2	0, 1, 1, 1, 1, 3, 5, 5, 5, 6, 7, 8, 9
3	3, 5, 5, 5, 5, 7, 7, 8, 8
4	1, 2

Key: 4 | 1 means 41 minutes.

Draw a box plot to represent this information.

2 A supermarket manager wants to compare the performance of two checkout assistants. She gathers the information in the table for the time taken, in seconds, to serve each customer.

	Assistant A	Assistant B
Lowest	50	42
Highest	110	144
Median	80	76
Lower quartile	62	60
Upper quartile	90	84

(a) Choose an appropriate scale and draw two box plots for these results.

(b) The manager concluded that because Assistant B had a maximum of 144 seconds she was not performing as well as Assistant A.

Explain why the manager's conclusion might have been wrong.

3 The table shows information about the heights of 40 bushes.

Height (h cm)	Frequency
$170 \leqslant h < 175$	5
$175 \leqslant h < 180$	18
$180 \leqslant h < 185$	12
$185 \leqslant h < 190$	4
$190 \leqslant h < 195$	1

(a) Complete the cumulative frequency table.

Height (h cm)	Cumulative frequency
$170 \leqslant h < 175$	
$170 \leqslant h < 180$	
$170 \leqslant h < 185$	
$170 \leqslant h < 190$	
$170 \leqslant h < 195$	

(b) Draw a cumulative frequency graph for your table, using a scale of 2 cm for every 5 cm height.

(c) Use your graph to find an estimate for the median height of the bushes. [E]

4 40 boys each completed a puzzle. This cumulative frequency graph gives information about the times it took them to complete the puzzle.

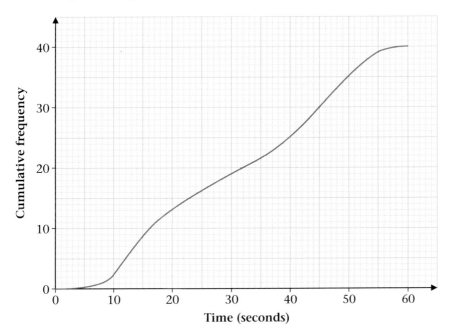

(a) Use the graph to find an estimate for the median time.

For the boys, the minimum time to complete the puzzle was 9 seconds, and the maximum time to complete the puzzle was 57 seconds.

(b) Use this information and the cumulative frequency graph to draw a box plot showing information about the boys' times. (Use graph paper and the same scale as in the diagram below.)

This box plot shows information about the times taken by 40 girls to complete the same puzzle.

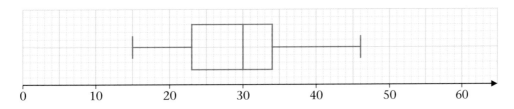

(c) Make two comparisons between the boys' times and the girls' times.   [E]

5   The incomplete table and histogram give some information
about the ages of the people who live in a village.

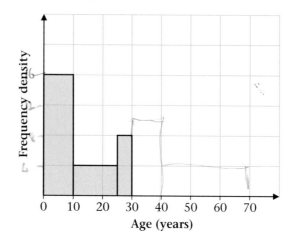

Age (x) in years	Frequency
$0 < x \leqslant 10$	160
$10 < x \leqslant 25$	60
$25 < x \leqslant 30$ 5	40
$30 < x \leqslant 40$ 10	100
$40 < x \leqslant 70$ 30	120

(a)   Copy and complete the frequency table, using the
information in the histogram.

(b)   Copy and complete the histogram.   [E]

6   The histogram gives information about the times, in minutes,
135 students spent on the internet last night.

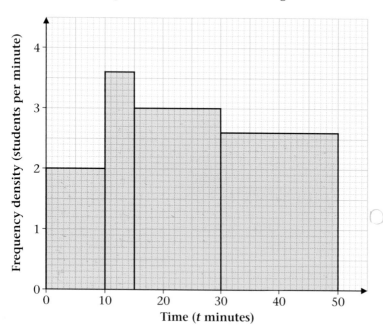

Time (t minutes)	Frequency
$0 < t \leqslant 10$	16 20
$10 < t \leqslant 15$	14.5 18
$15 < t \leqslant 30$	36
$30 < t \leqslant 50$	41
Total	135

$0 = wxF.$

5 2

Copy and complete the table, using the information in the
histogram.   [E]

7   One Monday, Tessa measured the times, in seconds, that individual birds spent on her bird table. She used this information to complete a frequency table.

Time (t seconds)	Frequency
$0 < t \leqslant 10$	8
$10 < t \leqslant 20$	16
$20 < t \leqslant 25$	15
$25 < t \leqslant 30$	12
$30 < t \leqslant 50$	6

(a)   Use the table to draw a complete histogram.

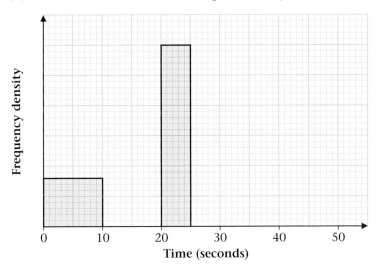

On Tuesday she conducted a similar survey and drew the following histogram from her results.

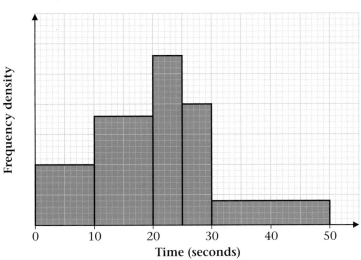

(b)   Use the histogram for Tuesday to complete a frequency table.

Time (t seconds)	Frequency
$0 < t \leqslant 10$	20 10
$10 < t \leqslant 20$	13
$20 < t \leqslant 25$	24 14
$25 < t \leqslant 30$	
$30 < t \leqslant 50$	

[E]

8 This back-to-back stem and leaf diagram gives information about the ages of oak trees in two forests:

Forest A Age in years		Forest B Age in years
	1	5
6, 4, 2	2	1, 9
7, 7, 2, 0	3	5, 6
6, 5, 5, 3, 1	4	1, 2, 2, 4, 8
8, 8, 5, 2, 2, 0	5	3, 6, 6, 7, 7, 9
9	6	1, 1, 2, 2, 3, 4, 5

Key: 9 | 6 | 1 means 69 years and 61 years.

(a) Draw box plots to compare this information.

(b) Comment on the ages of the oak trees in the two forests.

Summary of key points

1 The largest possible value in an interval is called the **upper class boundary**.

2 The **cumulative frequency** is the total frequency up to a particular upper class boundary.

3 You can display data in a **cumulative frequency graph** by plotting the cumulative frequency against the upper class boundary for each interval.

4 The **median** is the value half way into the distribution.

5 The **lower quartile** is the value one quarter of the way into the distribution.
The **upper quartile** is the value three quarters of the way into the distribution.

6 **Interquartile range** = upper quartile − lower quartile

7 To draw a **box plot** you need five pieces of information:
- the lowest value
- the lower quartile
- the median
- the upper quartile
- the highest value.

8 In a **histogram** the areas of the rectangles are proportional to the frequencies they represent.

9 **Frequency density** $= \dfrac{\text{frequency}}{\text{class width}}$

or

Frequency density $= \dfrac{\text{frequency}}{\text{class width in standard class intervals}}$

10 **Frequency** = frequency density × class width

or

Frequency = frequency density × class width in standard class intervals

21 Advanced trigonometry

This chapter extends the work on Pythagoras' theorem (Chapter 15) and basic trigonometry (Chapter 16) to finding areas, lengths and angles in triangles which do not contain a right angle. You will also learn to apply Pythagoras' theorem and trigonometry in three dimensions.

21.1 The area of a triangle

The diagram opposite shows a general triangle ABC. The lengths of its sides are a, b and c units.

The perpendicular from B to AC is drawn. It meets AC at X. Its length is h.

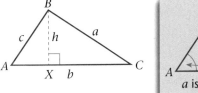

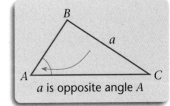

a is opposite angle A

The area of a triangle is $\frac{1}{2} \times$ base $\times$ height.
So the area of triangle ABC is $\frac{1}{2}bh$.

Using the **sine ratio**:

$$\sin C = \frac{h}{a} \quad \text{or} \quad h = a \sin C$$

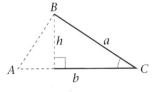

Substituting $h = a \sin C$ into area $ABC = \frac{1}{2}bh$ gives

area $ABC = \frac{1}{2} \times b \times a \sin C$, which you can write as

area $ABC = \frac{1}{2}ab \sin C$

Similarly, it can be shown that

area $ABC = \frac{1}{2}ac \sin B$

and

area $ABC = \frac{1}{2}bc \sin A$

Notice that, in all cases, to find the area of a triangle you need the **lengths of two sides** of the triangle and the **size of the angle between these sides**. Then the area is

$$\begin{array}{c} \text{area of} \\ \text{triangle} \end{array} = \begin{array}{c} \text{length of} \\ \text{first side} \end{array} \times \begin{array}{c} \text{length of} \\ \text{second side} \end{array} \times \begin{array}{c} \text{sine of angle} \\ \text{between} \\ \text{these sides} \end{array}$$

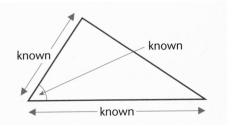

Example 1

Find the area of the triangle PQR.

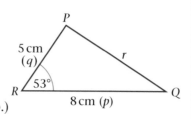

Using area $PQR = \frac{1}{2}pq \sin R$:

$$\text{area } PQR = \frac{1}{2} \times 5 \times 8 \times \sin 53°$$
$$= 20 \times 0.7986...$$
$$= 15.97 \text{ cm}^2 \text{ (to 2 d.p.)}$$

Example 2

A builder fences off a triangular plot of land XYZ.

$$XY = 43 \text{ m} \qquad XZ = 58 \text{ m} \qquad Y\hat{X}Z = 125°$$

Calculate the area of the plot.

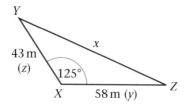

Using area $XYZ = \frac{1}{2}zy \sin X$:

$$\text{area } XYZ = \frac{1}{2} \times 43 \times 58 \times \sin 125°$$
$$= \frac{1}{2} \times 43 \times 58 \times 0.8192...$$
$$= 1021.5 \text{ m}^2 \text{ (to 1 d.p.)}$$

Example 3

The area of a triangle ABC is 52 cm^2.
All three angles of this triangle are acute.

$$AB = 14 \text{ cm} \qquad AC = 12 \text{ cm}$$

Calculate the angle A.

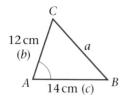

$$\text{Area} = \frac{1}{2} \times 14 \times 12 \times \sin A$$
$$52 = 84 \times \sin A$$

So, dividing both sides by 84:

$$\sin A = \frac{52}{84}$$
$$\sin A = 0.6190...$$

So $\qquad A = 42.5°$ (to 3 s.f.)

$A = \sin^{-1} 0.6190...$

Exercise 21A

1 Calculate the area of each of the triangles.
All lengths are in centimetres.

(a)

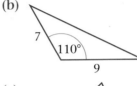

(b)

(c)

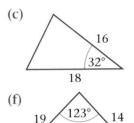

(d)

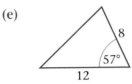

(e)

(f)

2 A farmer fences off a triangular field *PQR*.
PQ = 30 metres, *PR* = 34 metres and the angle at *P* is 75°.
Calculate the area of the field.

3 Calculate the area of this parallelogram:

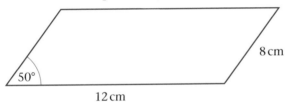

> The parallelogram can be treated as two triangles:

4 A triangle *ABC* has an area of 60 cm², *AB* = 15 cm and *AC* = 10 cm.

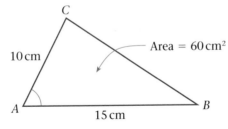

C

10 cm

Area = 60 cm²

A

15 cm

B

Calculate the size of the acute angle at *A*.

5 The area of triangle *PQR* is 40 cm².

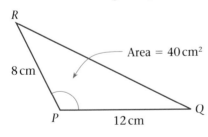

R

Area = 40 cm²

8 cm

P

12 cm

Q

> The sin⁻¹ function on a calculator gives the value of *P* between 0 and 90°.
> Remember that
> sin(180° − *x*) = sin *x*

PQ = 12 cm   *PR* = 8 cm

The angle at *P* is obtuse.

Calculate the angle at *P*.

21.2 The sine rule

The diagram shows a general triangle ABC.
The lengths of its sides are a, b and c.
The perpendicular from C to AB is drawn
and its length is labelled h. It meets AB at X.

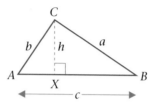

Now, using the sine ratio in triangle AXC:

$$\sin A = \frac{h}{b} \quad \text{or} \quad h = b \sin A$$

Also

$$\sin B = \frac{h}{a} \quad \text{or} \quad h = a \sin B$$

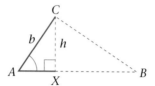

so $a \sin B = b \sin A$

So $\quad \dfrac{a}{\sin A} = \dfrac{b}{\sin B} \quad \text{or} \quad \dfrac{\sin A}{a} = \dfrac{\sin B}{b}$

This result is known as the **sine rule**.

Exercise 21B

Draw a triangle ABC.

Draw a perpendicular from B to AC.

Hence show that $\dfrac{a}{\sin A} = \dfrac{c}{\sin C}$.

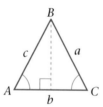

Using the sine rule

The full version of the **sine rule** is

$$\frac{a}{\sin A} = \frac{b}{\sin B} = \frac{c}{\sin C}$$

or

$$\frac{\sin A}{a} = \frac{\sin B}{b} = \frac{\sin C}{c}$$

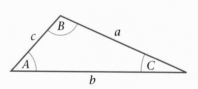

Example 4

Calculate the length of the side marked p.

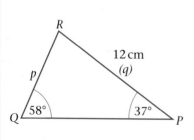

By the sine rule:

$$\frac{p}{\sin P} = \frac{q}{\sin Q}$$

$$\frac{p}{\sin 37°} = \frac{12}{\sin 58°}$$

$$p = \frac{12 \times \sin 37°}{\sin 58°}$$

$$= \frac{12 \times 0.6018\ldots}{0.8480\ldots}$$

$$= 8.52 \text{ cm (correct to 3 s.f.)}$$

Example 5

The angles in the triangle ABC are all acute.

Calculate the size of the angle at A.

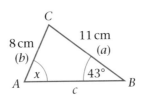

Using the sine rule:

$$\frac{\sin A}{a} = \frac{\sin B}{b}$$

$$\frac{\sin x}{11} = \frac{\sin 43°}{8}$$

So $\sin x = \dfrac{11 \times \sin 43°}{8}$

$$= \frac{11 \times 0.6820...}{8}$$

$$= 0.9377...$$

$$x = 69.7° \text{ (correct to 3 s.f.)}$$

Example 6

A lighthouse, L, is 40 km due North of a harbour, H.

A speedboat leaves H and travels on a bearing of 053° from H until it reaches a point P.

The point, P, lies on a bearing of 075° from L.

Calculate the distance travelled by the speedboat.

First make a sketch:

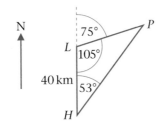

So the angle at P is $180° - (105° + 53°) = 22°$.

Now you can use the sine rule:

$$\frac{HP}{\sin 105°} = \frac{40}{\sin 22°}$$

So $HP = \dfrac{40 \times \sin 105°}{\sin 22°}$

$$= \frac{40 \times 0.9659...}{0.3746...}$$

$$= 103.14 \text{ km (to 2 d.p.)}$$

Exercise 21C

1 Calculate the lengths of the sides marked with letters.
  All lengths are in centimetres.

(a)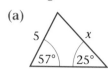

(b)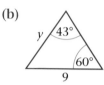

(c)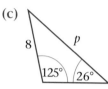

(d)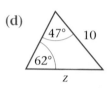

(e)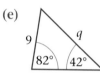

(f)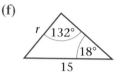

2 Calculate the sizes of the angles marked with letters.
  All of these angles are acute.

(a)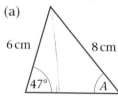

(b)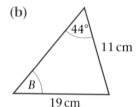

(c)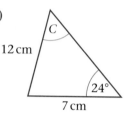

3 The diagram shows the relative positions of a port, P,
  a marker buoy, B, and a lighthouse, L.
  The bearing of L from P is 110°.
  The distance PL is 56 km.
  The bearing of B from P is 147°.
  The bearing of L from B is 030°.
  Calculate

  (a) the distance PB

  (b) the distance BL.

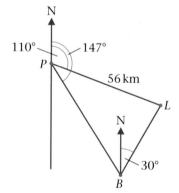

The ambiguous case (for an unknown angle)

In a triangle ABC, $AB = 9$ cm, $BC = 5$ cm and the angle at $A = 24°$.

Draw the base line AC, and $AB = 9$ cm, with the angle at A as 24°:

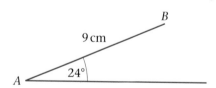

Imagine putting a compass point at B, then drawing a circle
centre B of radius 5 cm. This circle will intersect AC at two
possible points, marked as C and C'.
This shows that there are two possible angles at B, $A\hat{B}C$ or $A\hat{B}C'$.

You can compare the triangle to the general triangle *ABC*:

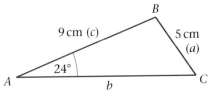

So, by the sine rule: $\dfrac{\sin C}{c} = \dfrac{\sin A}{a}$

You have $A = 24°, a = 5, c = 9$

So $\dfrac{\sin C}{9} = \dfrac{\sin 24°}{5}$

$\sin C = \dfrac{9 \times \sin 24°}{5}$

$= 0.7321\ldots$

$C = 47.06°$

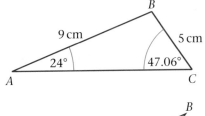

But $\sin(180° - x) = \sin x$, so we could also have

$C = 180° - 47.06°$

$C = 132.94°$

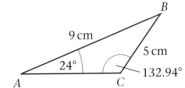

Exercise 21D

In a triangle *PQR*, *PQ* = 10 cm, *QR* = 8 cm and the angle *QPR* = 48°.

Calculate the two possible values for the angle *PQR*.

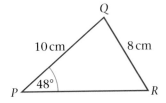

21.3 The cosine rule

ABC is a general triangle with sides *a*, *b* and *c*.
Side *a* is opposite angle *A*, and so on.
The perpendicular from *B* to *AC* meets *AC* at *P*.
This perpendicular has length *h* units.
Let *CP* = *x* units, so *PA* = (*b* − *x*) units.

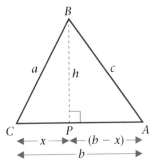

In triangle *BPA*, Pythagoras' theorem gives

$c^2 = h^2 + (b - x)^2$

$= h^2 + b^2 - 2xb + x^2$

$c^2 = h^2 + x^2 + b^2 - 2xb \qquad (1)$

But using Pythagoras' theorem again in triangle *BPC* gives

$a^2 = h^2 + x^2 \qquad (2)$

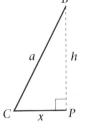

Substituting a^2 for $h^2 + x^2$ in (1):

$$c^2 = a^2 + b^2 - 2xb \qquad (3)$$

But using the cosine ratio on triangle BPC gives

$$\cos C = \frac{x}{a} \quad \text{or} \quad x = a \cos C$$

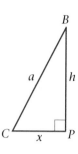

Substituting $a \cos C$ for x in (3) gives

$$c^2 = a^2 + b^2 - 2ab \cos C$$

This is the **cosine rule**.

You can rearrange this to give

$$2ab \cos C = a^2 + b^2 - c^2 \qquad \text{or}$$

$$\cos C = \frac{a^2 + b^2 - c^2}{2ab}$$

> In the GCSE exam the cosine rule will be given on a formula sheet. However, it might help you to remember it as:
>
> $$c^2 = \boxed{a^2 + b^2} - \boxed{2ab \cos C}$$
>
> $\boxed{\text{Pythagoras}} \;-\; \boxed{\text{a bit}}$

Example 7

Calculate the length of the side marked y.

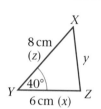

Using the cosine rule:

$$y^2 = x^2 + z^2 - 2xz \cos Y$$
$$= 6^2 + 8^2 - 2 \times 6 \times 8 \times \cos 40°$$
$$= 36 + 64 - 96 \times \cos 40°$$
$$= 100 - 96 \times 0.766...$$
$$y^2 = 100 - 73.54... \quad \text{or} \quad y^2 = 26.46...$$
$$y = \sqrt{26.46...} = 5.14 \text{ cm (to 3 s.f.)}$$

Example 8

Calculate the angle at A.

Using the cosine rule:

$$\cos A = \frac{c^2 + b^2 - a^2}{2cb}$$

$$= \frac{3^2 + 7^2 - 5^2}{2 \times 3 \times 7}$$

$$= \frac{9 + 49 - 25}{42} = \frac{33}{42} = 0.7857...$$

So $\qquad A = 38.2°$ (to 1 d.p.)

Exercise 21E

1   Calculate the lengths marked with letters. All lengths are in centimetres.

(a)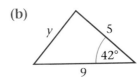

(b)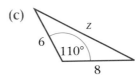

(c)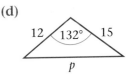

(d)

2   Calculate the angles marked with letters. All lengths are in centimetres.

(a)

(b)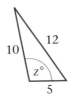

(c)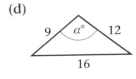

(d)

3   A builder ropes off a triangular plot of ground, *PQR*.
    The length of *PQ* = 42 m and the length of *PR* = 50 m.
    The angle *QPR* = 72°.

    Calculate the length of rope needed by the builder.

4   A lifeboat leaves its base, *S*, and travels 32 km due North to a
    lighthouse *L*. At *L* the lifeboat turns onto a bearing of 056°
    from *L* and travels for a further 45 km until it reaches a marker
    buoy, *B*. At *B* the lifeboat turns again and travels back in a
    straight line to *S*.

    Calculate the total distance travelled by the lifeboat.

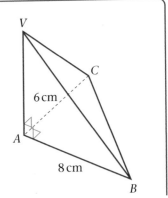

21.4 Trigonometry and Pythagoras' theorem in three dimensions

You need to be able to use Pythagoras' theorem and trigonometry in
three dimensions.

> For finding the longest
> diagonal in a cuboid see
> Section 15.8.

Example 9

The diagram shows a triangle-based pyramid *VABC* with *V* vertically
above *A*. The base *ABC* has a right angle at *A*.

   AB = 8 cm   *AC* = 6 cm   *VA* = 15 cm

Calculate

(a)   the length *BC*   (b)   the length *VB*

(c)   the length *VC*   (d)   angle *VBA*

(e)   angle *AVC*   (f)   angle *BVC*

(a) Using $BC^2 = AB^2 + AC^2$ by Pythagoras' theorem:

$$BC^2 = 8^2 + 6^2$$
$$= 64 + 36$$
$$= 100$$
$$BC = \sqrt{100}$$
$$= 10 \text{ cm}$$

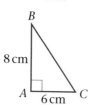

(b) Using $VB^2 = VA^2 + AB^2$:

$$VB^2 = 15^2 + 8^2$$
$$= 225 + 64$$
$$= 289$$
$$= \sqrt{289}$$
$$= 17 \text{ cm}$$

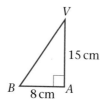

(c) Using $VC^2 = VA^2 + AC^2$:

$$VC^2 = 15^2 + 6^2$$
$$= 225 + 36$$
$$= 261$$
$$VC = \sqrt{261}$$
$$= 16.16... \text{ cm}$$

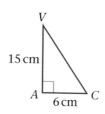

(d) Using the tangent ratio:

$$\tan V\hat{B}A = \tfrac{15}{8}$$
$$= 1.875$$
$$V\hat{B}A = 61.92°$$

$$\tan V\hat{B}A = \frac{\text{opp}}{\text{adj}} = \frac{AV}{AB}$$

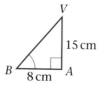

(e) Using the tangent ratio:

$$\tan A\hat{V}C = \tfrac{6}{15}$$
$$= 0.4$$
$$A\hat{V}C = 21.80°$$

$$\tan A\hat{V}C = \frac{\text{opp}}{\text{adj}} = \frac{AC}{AV}$$

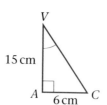

(f) Using the cosine rule:

$$\cos V = \frac{17^2 + 261 - 10^2}{2 \times 17 \times \sqrt{261}}$$

$VC^2 = 261$ (from part (c))

$$\cos V = \frac{c^2 + b^2 - v^2}{2cb}$$

$$= \frac{450}{549.3...}$$
$$= 0.8192...$$
$$V = 35°$$

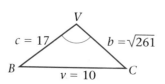

Exercise 21F

1 Here is a wedge *ABCDEF*. The base, *BCDE*, is a rectangle. The back face, *ABEF*, is also a rectangle. The angle between these two rectangles is 90°. *M* is the mid-point of *CD*.

Calculate

(a) the length *AC*   (b) the length *FC*

(c) the length *EC*   (d) angle *ACB*

(e) angle *ECF*   (f) angle *EMF*.

2 *VABCD* is a square-based pyramid. The vertex *V* is vertically above the mid-point, *M*, of the square base.

$$AB = 10 \text{ cm} \qquad VA = 16 \text{ cm}$$

Calculate

(a) the length of *BD*

(b) the length of *VM*

(c) the angle *VAM*

(d) the angle *AVM*.

3 *ABCV* is a pyramid with base *ABC*, right-angled at *A*.

$$AB = 24 \text{ cm} \qquad AC = 7 \text{ cm}$$

The vertex *V* is vertically above *A* with *AV* = 30 cm.
The point *M* lies on *BC* with angle *AMB* = 90°.

(a) Show that *BC* = 25 cm.

(b) Calculate the length of
     (i) *VB*   **(ii)** *VC*.

(c) Calculate the length *AM*.

(d) Calculate the angle *VMA*.

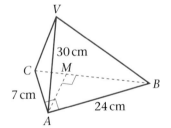

Mixed exercise 21

1 Calculate

(a) the area of triangle *ABC*

(b) the length of *BC*

(c) the angle *BCA*.

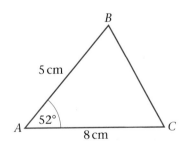

2   In triangle *ABC*, *AC* = 8 cm, *CB* = 15 cm and angle *ACB* = 70°.

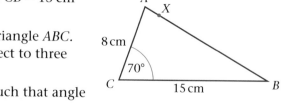

   (a) Calculate the area of triangle *ABC*.
        Give your answer correct to three
        significant figures.

   (b) *X* is the point on *AB* such that angle
        CXB = 90°.

        Calculate the length of *CX*.
        Give your answer correct to three significant figures.

3   *AB* = 3.2 cm,   *BC* = 8.4 cm.

   The area of triangle *ABC* is 10 cm².

   Calculate the perimeter of triangle *ABC*.
   Give your answer correct to three
   significant figures.

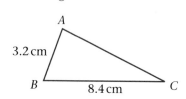

4   A yacht, *Y*, leaves a harbour, *H*, and travels 45 km due
   North until it reaches a marker buoy, *B*. At *B* the yacht
   turns onto a bearing of 290° from *B* and travels for a
   further 56 km until it reaches a lighthouse, *L*. At *L* it
   turns again and travels in a straight line back to *H*.

   Calculate

   (a) the total distance travelled by the yacht

   (b) the bearing of *L* from *H*

   (c) the shortest distance between *Y* and *B* on the
        yacht's return journey from *L* to *H*.

5   A man walks from his home *H* on a bearing of 060° for 3.2 miles
   until he reaches his friend's house, *F*. At *F* he turns onto a
   bearing of 280° from *F* and travels a further 4.1 miles to his
   sister's house, *S*. At *S* he turns again and walks in a straight line
   back home.

   Calculate

   (a) the distance he walks

   (b) the bearing of *H* from *S*

   (c) the shortest distance between the man and *F* on the
        man's return journey from *S* to *H*.

6   *VPQR* is a triangle-based pyramid. The vertex *V* is
   vertically above *P* with *VP* = 16 cm.

   In the triangular base, $Q\hat{P}R$ = 120°.

   Calculate

   (a) the length *QR*

   (b) the length *VQ*

   (c) the angle *QVR*.

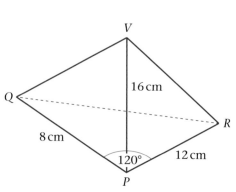

Summary of key points

1 For any triangle ABC, the **area of the triangle** is

$\frac{1}{2}ab \sin C = \frac{1}{2}ac \sin B = \frac{1}{2}bc \sin A$

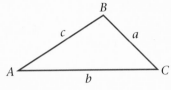

2 $\dfrac{\text{area of}}{\text{triangle}} = \dfrac{\text{length of}}{\text{first side}} \times \dfrac{\text{length of}}{\text{second side}} \times \dfrac{\text{sine of angle}}{\substack{\text{between} \\ \text{these sides}}}$

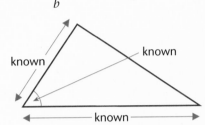

3 The **sine rule** is

$\dfrac{a}{\sin A} = \dfrac{b}{\sin B} = \dfrac{c}{\sin C}$ or $\dfrac{\sin A}{a} = \dfrac{\sin B}{b} = \dfrac{\sin C}{c}$

4 The **cosine rule** is

$c^2 = a^2 + b^2 - 2ab \cos C$ or $\cos C = \dfrac{a^2 + b^2 - c^2}{2ab}$

These formulae appear on the formula sheet which you will be given in the GCSE exam.

Remember also that

$\sin x = \sin(180° - x)$

This fact is not given on the sheet. You may need it for ambiguous cases related to the sine rule or for the area of a triangle.

22 Advanced mensuration

One of the problems involved in the development of trigonometry was finding the relationship between the angle a cannon could pivot through, the distance a cannon ball could travel, and the width of a passage into a harbour.

A natural extension of this work was to look at how to find:

- the length of an arc of a circle for an angle θ
- the area of a sector
- the area of a segment

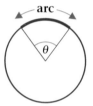

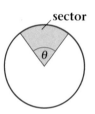

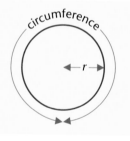

This is called a **minor** segment.

chord

major segment

To do this you need these formulae:

Circumference of a circle $= 2\pi r$

Area of a circle $= \pi r^2$

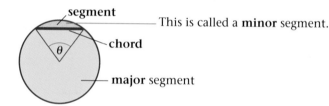

For more on circumference and area of a circle see Section 13.4.

22.1 Finding the length of an arc of a circle

In this diagram the arc length is $\frac{1}{4}$ of the circumference because the angle at the centre is 90°, which is $\frac{90}{360} = \frac{1}{4}$ of a whole turn.

In this diagram the angle at the centre is 60°. This is $\frac{60}{360} = \frac{1}{6}$ of a whole turn. So the arc length is $\frac{1}{6}$ of the circumference of the circle.

In this diagram the angle at the centre is 240°. This is $\frac{240}{360} = \frac{2}{3}$ of a whole turn. So the arc length is $\frac{2}{3}$ of the circumference of the circle.

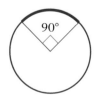

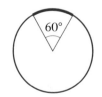

If the angle at the centre of a circle is θ then

$$\text{arc length} = \frac{\theta}{360} \text{ of the circumference}$$

$$= \frac{\theta}{360} \times 2\pi r$$

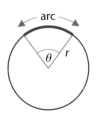

This gives the formula:

Arc length $= \dfrac{2\pi r\theta}{360} = \dfrac{\pi r\theta}{180}$

Example 1

Calculate the length of the arc AB.

$$\text{Arc length} = \frac{\pi \times 8 \times 72}{180}$$

$$= 10.05 \text{ cm (correct to 2 d.p.)}$$

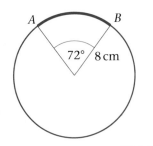

Example 2

The length of the arc PQ is 12 cm. Calculate the angle θ.

$$\text{Arc length} = \frac{\pi r\theta}{180}$$

so $\quad \dfrac{\pi \times \overset{1}{\cancel{10}} \times \theta}{\underset{18}{\cancel{180^\circ}}} = 12$

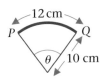

Multiply each side by 18 and divide each side by π:

$$\theta = \frac{18 \times 12}{\pi}$$

so $\quad \theta = 68.75° \text{ (correct to 2 d.p.)}$

Exercise 22A

1   Calculate each of these arc lengths:

(a)

(b)

(c)

(d)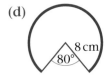

(e)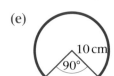

(f)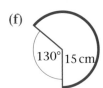

2   Calculate each of the angles marked θ.

(a)

(b)

(c)
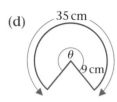

(d) 35 cm 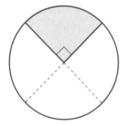 9 cm

(e) 45 cm

(f) 28 cm 5 cm

22.2 Finding the area of a sector of a circle

In this diagram the sector is $\frac{1}{4}$ of the whole circle, because the angle at the centre is 90°. This is $\frac{1}{4}$ of a whole turn or 360°. So the area of this sector is $\frac{1}{4}$ of the area of the circle.

In this diagram the angle at the centre is 60°. So the sector is $\frac{60}{360}$ or $\frac{1}{6}$ of the area of the whole circle. The area of this sector is $\frac{1}{6}$ of the area of the circle.

If the angle at the centre of a circle is θ then

$$\text{area of a sector} = \frac{\theta}{360} \text{ of area of circle}$$

$$= \frac{\theta}{360} \times \pi r^2$$

$$= \frac{\theta \times \pi \times r^2}{360}$$

This gives the formula:

Area of a sector $= \dfrac{\pi r^2 \theta}{360}$

Example 3

Calculate the area of this sector:

$$\text{Area} = \frac{\pi r^2 \theta}{360}$$

$r = 8$, so $r^2 = 64$

So $\text{area} = \dfrac{\pi \times 64 \times 75}{360}$

$= 41.89 \text{ cm}^2$ (correct to 2 d.p.)

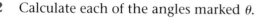
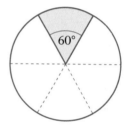

Example 4

AOB is a sector of a circle.
Angle $AOB = 130°$
Area of the sector $AOB = 200$ cm².

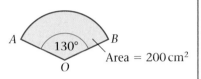

A 130° B
Area = 200 cm²
O

Calculate the radius OA of the circle
of which AOB is a sector.

$$\text{Area of sector} = \frac{\pi r^2 \theta}{360} = 200$$

Multiply each side by 360 and divide each side by $\pi\theta$:

$$r^2 = \frac{200 \times 360}{\pi \times \theta} = \frac{200 \times 360}{\pi \times 130}$$

$$= 176.2947$$

So $r = \sqrt{176.2947}$

$$= 13.28 \text{ cm (correct to 2 d.p.)}$$

Exercise 22B

1   Calculate the area of each of these sectors of circles:

   (a)   (b)

   40° 8 cm   70° 10 cm

   (c)   (d)

   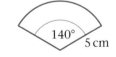
   140° 5 cm   65° 9 cm

   (e)   (f)

      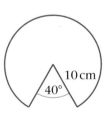
   8 cm 250°   10 cm 40°

2   OPQ is a sector of a circle centre O of radius 9 cm.
    The area of the sector OPQ is 51 cm².
    Calculate the size of the angle POQ.

    Area = 51 cm²   P   Q   9 cm   O

3   OXY is a sector of a circle centre O.
    The area of the sector OXY is 60 cm².
    The angle $XOY = 68°$.
    Calculate the length of the radius OX of the circle.

    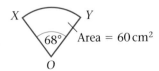

    X   Y   68° Area = 60 cm²   O

22.3 Finding the area of a segment of a circle

The area of the sector $OAB = \dfrac{\pi r^2 \theta}{360}$

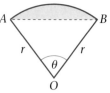

In triangle OAB: area of triangle $OAB = \frac{1}{2}r \times r \sin\theta$
$$= \tfrac{1}{2}r^2 \sin\theta$$

So the area of the shaded segment is the difference between these two areas:

> **Area of a segment** $= \dfrac{\pi r^2 \theta}{360} - \tfrac{1}{2}r^2 \sin\theta$

For any triangle:
Area $= \frac{1}{2}ab\sin C$

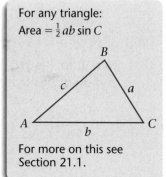

For more on this see Section 21.1.

___ Example 5 _____

Calculate the area of the shaded segment of the circle.

Area of segment $= \dfrac{\pi r^2 \theta}{360} - \tfrac{1}{2}r^2 \sin\theta$

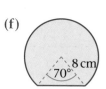

$$= \dfrac{\pi \times 6^2 \times 75}{360} - \tfrac{1}{2} \times 6^2 \times \sin 75°$$

$$= 23.562 - 17.387$$

$$= 6.18\,\text{cm}^2 \text{ (correct to 2 d.p.)}$$

Exercise 22C

1 Calculate the area of each shaded segment.

(a)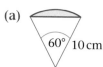

(b)

(c)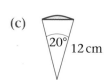

(d)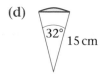

(e)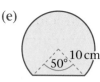

(f)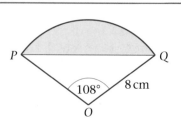

___ Worked examination question 1 _____

O is the centre of a circle, radius 8 cm.
P and Q are points on the circumference.
Angle $POQ = 108°$.

(a) Calculate the area of the sector POQ.

(b) Calculate the difference between the arc length PQ and the length of the chord PQ.

(c) Calculate the area of the shaded segment PQ.

Give your answers correct to three significant figures.

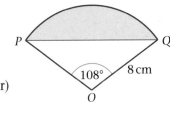

(a) Area of sector $= \dfrac{\pi r^2 \theta}{360}$

$r = 8$, so $r^2 = 64$, and $\theta = 108°$

So   area of sector $= \dfrac{\pi \times 64 \times 108}{180} = 60.318\,578\,95$ (by calculator)

$\qquad\qquad\qquad\qquad = 60.3\ \text{cm}^2$ (correct to 3 s.f.)

(b) Length of arc $PQ = \dfrac{\pi r \theta}{180} = \dfrac{\pi \times 8 \times 108}{180} = 15.0796\ldots$ cm

To find the length of the chord PQ mark the mid-point of PQ as M:

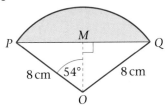

> You could also use the sine rule.
>
> $O\hat{P}Q = O\hat{Q}P = 36°$
> $\qquad$ (isosceles triangle)
>
> So $\quad \dfrac{8}{\sin 36°} = \dfrac{PQ}{\sin 108°}$
>
> $\qquad PQ = \dfrac{8 \sin 108°}{\sin 36°}$
>
> $\qquad\quad = 12.9442\ldots$

$\dfrac{PM}{8} = \sin 54°$, so $PM = 8 \times \sin 54° = 6.4721\ldots$ cm

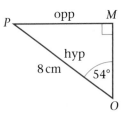

$PQ = 2 \times PM$, so $PQ = 2 \times 6.4721\ldots = 12.9442\ldots$ cm

The difference between the arc length PQ and the chord PQ is

$\qquad 15.0796\ldots - 12.9442\ldots = 2.1354\ldots$

$\qquad\qquad\qquad\qquad = 2.14$ cm (correct to 3 s.f.)

(c) The area of the shaded segment $PQ = \dfrac{\pi r^2 \theta}{360} - \tfrac{1}{2} r^2 \sin \theta$

$\qquad\qquad\qquad = 60.3185\ldots - \tfrac{1}{2} \times 64 \times \sin 108°$

$\qquad\qquad\qquad = 60.3185\ldots - 30.4338\ldots$

$\qquad\qquad\qquad = 29.8848\ldots$

$\qquad\qquad\qquad = 29.9\ \text{cm}^2$ (correct to 3 s.f.)

Exercise 22D

1 An arc of a circle of radius 15 cm has a length of 18 cm. Calculate (a) the angle θ (b) the area of the sector.

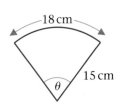

2 The diagram shows the landing area for a javelin competition. $OPQR$ is a sector of a circle, centre O, of radius 120 metres. Calculate

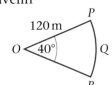

(a) the length of the chord PR

(b) the length of the arc PQR

(c) the area of the sector $OPQR$

(d) the area of the segment PQR.

3   A door is in the shape of a rectangle *ABCD* with a sector *OAD* of a circle.

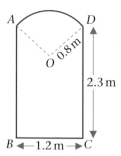

$DC = AB = 2.3$ m, $BC = AD = 1.2$ m and the radius of the circle is *OA* where $OA = OD = 0.8$ m.

Calculate

(a)   the perimeter of the door

(b)   the area of the door.

22.4 Finding volumes and surface areas

You need to know how to find the volumes and surface areas of a variety of 3-D objects.

The surface area of a cylinder

This cylinder has a circular base of radius *r* cm and a height of *h* cm.

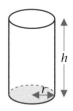

Its surface area is made up of the area of the curved surface plus the areas of the circular top and base.

The area of the circular base and top are both equal to πr^2. So the combined area of the top and base is $2\pi r^2$.

For the area of the curved surface, imagine unwrapping the cylinder. This creates a rectangle with length equal to the circumference of the circular base $2\pi r$ and width equal to the height *h*:

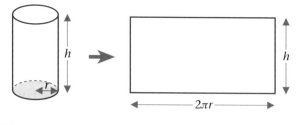

If you could unroll the surface it would look like this.

The area of the curved surface is $2\pi rh$.

So the total surface area of the cylinder in cm² is $2\pi rh + 2\pi r^2$.

> For a **cylinder** of height *h* with a circular base of radius *r*,
>
> **surface area** $= 2\pi rh + 2\pi r^2$

The volume of a cylinder

The **volume** of a cylinder is $\pi r^2 h$.

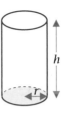

Example 6

Calculate the surface area and volume of a cylinder with circular base of radius 12 cm and height 30 cm.

$$\begin{aligned} \text{Surface area} &= 2\pi rh + 2\pi r^2 \\ &= 2 \times \pi \times 12 \times 30 + 2 \times \pi \times 12^2 \\ &= 2261.947 + 904.778 \\ &= 3166.725 = 3167 \text{ cm}^2 \text{ (to 4 s.f.)} \end{aligned}$$

$$\begin{aligned} \text{volume} &= \pi r^2 h \\ &= \pi \times 12^2 \times 30 \\ &= 13\,572 \text{ cm}^3 \text{ (to 5 s.f.)} \end{aligned}$$

Volume and surface area of prisms

A prism is a 3-D shape which has the same cross-section throughout its height.

These are prisms:

cuboid   triangular prism   pentagonal prism

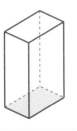
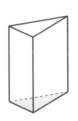
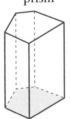

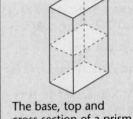

The **volume of any prism** is

   area of base × vertical height

     or

   area of cross-section × vertical height

The base, top and cross-section of a prism are all identical.

The **surface area of any prism** is

   2 × area of base + total area of vertical faces

Example 7

Here is a pentagonal prism.
The base *ABCDE* is a regular pentagon of side length 8 cm.
The height of the prism is 15 cm.

Calculate the volume of the prism.

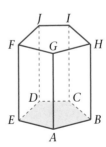

First calculate the area of the base *ABCDE*.

To do this split it into five congruent triangles and look at the area of one of these triangles, *AOB*.

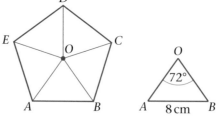

Angle $AOB = \dfrac{360°}{5} = 72°$

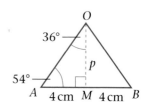

The area of *AOB* is $\frac{1}{2}AB \times p$ where p is the perpendicular distance *OM* from *O* to *AB*.

Angle *AOB* is 72°

so   angle *AOM* is $\frac{1}{2} \times 72° = 36°$

and   angle *OAM* is $90° - 36° = 54°$

Using the tangent ratio: $\dfrac{p}{AM} = \tan 54°$, so $p = AM \times \tan 54°$

So $p = 4 \times \tan 54°$

                        $= 5.5055...\,\text{cm}$

> For more about the tangent ratio see Section 16.1.

So the area of triangle *AOB* is

     $\frac{1}{2}AB \times p = AM \times p = 4 \times 5.5055... = 22.022...\,\text{cm}^2$

As *ABCDE* is made up of five identical triangles:

     area $ABCDE = 5 \times 22.022... = 110.1105...$

                  $= 110.11\,\text{cm}^2$ (to 2 d.p.)

The volume of the prism is area of base × vertical height

so   volume of prism $= 110.11... \times 15$

                     $= 1651.65...\,\text{cm}^3$

                     $= 1652\,\text{cm}^3$ (to 4 s.f.)

Exercise 22E

1   Find the volume and surface area of a cylinder of height 4 cm and circular base of radius 5 cm (to 3 s.f.).

2   A prism of height 10 cm has the cross-sectional shape of an equilateral triangle of side 6 cm. Find the volume and surface area of the prism (to 3 s.f.).

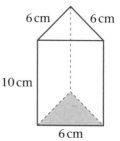

3   A waste bin is the shape of a cylinder. Its base has radius 0.5 m and its height is 1.2 m. Find its volume (to 3 s.f.).

4   A hexagonal prism has a vertical height of 32 cm. The base of the prism is a regular hexagon of side length 12 cm. Calculate the volume of the prism.

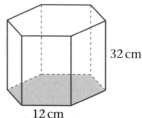

5 A prism has a vertical height of 30 cm.
The base of the prism is a regular
polygon with 10 sides.
Each side is of length 6 cm.

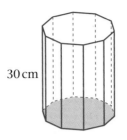

(a) Calculate the volume of the prism.

(b) Calculate the surface area.

30 cm

6 cm

22.5 The volume of a pyramid

These shapes are all **pyramids**:

triangle-based
pyramid

square-based
pyramid

hexagonal
pyramid

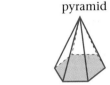

The base of any pyramid is a **polygon**.
The other edges are all straight lines which meet at a point,
usually called the vertex.

The area of a triangle is $\frac{1}{2}$ the
area of its smallest surrounding
rectangle.

In a similar way the volume of
a pyramid is $\frac{1}{3}$ the volume of
its smallest surrounding prism.

As the volume of a prism is
area of base × vertical height:

Volume of a pyramid $= \frac{1}{3} \times$ area of base × vertical height

A cylinder is a 'special prism'.

A **cone** is a special pyramid with a circular base.
Its smallest surrounding prism is a cylinder, so:

Volume of a cone $= \frac{1}{3} \times$ area of base × height
$= \frac{1}{3}\pi r^2 h$

h

r

___ **Example 8** _____

A pyramid $VABCD$ has a rectangular base $ABCD$.
The vertex V is 15 cm vertically above the mid-point M of
the base. $AB = 4$ cm and $BC = 9$ cm.

Calculate the volume of the pyramid.

The area of the base is $4 \times 9 = 36$ cm^2

so   volume of pyramid $= \frac{1}{3} \times$ area of base × vertical height
$= \frac{1}{3} \times 36 \times 15$
$= 180$ cm^3

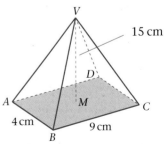

V

15 cm

D

A

M

C

4 cm

9 cm

B

Example 9

Here is a cone. The circular base has a diameter AB of length 10 cm. The **slant height** AV is of length 13 cm.

Calculate the volume of the cone.

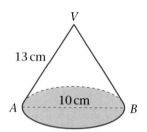

Before you can work out the volume, you need to work out the vertical height from the mid-point of AB to V.

Do this by applying Pythagoras' theorem to the triangular section ABV:

$$AV^2 = h^2 + MA^2$$
$$h^2 = AV^2 - MA^2$$
$$= 13^2 - 5^2$$
$$= 169 - 25 = 144$$
$$h = 12 \text{ cm}$$

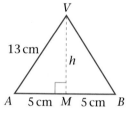

So   volume of cone $= \frac{1}{3}\pi r^2 h$
$$= \frac{1}{3} \times \pi \times 5^2 \times 12$$
$$= \frac{1}{3} \times \pi \times 25 \times 12$$
$$= 314.2 \text{ cm}^3 \text{ (correct to 1 d.p.)}$$

Example 10

A pyramid has a square base of side x cm and a vertical height 24 cm. The volume of the pyramid is 392 cm³.

Calculate the value of x.

$$\text{Volume} = \frac{1}{3} \times \text{area of base} \times \text{height}$$
$$= \frac{1}{3} \times x^2 \times h$$

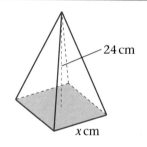

so $392 = \frac{1}{3} \times x^2 \times 24$
$$x^2 = \frac{3 \times 392}{24} = 49$$
$$x = 7$$

Exercise 22F

1   $VABCD$ is a square-based pyramid. The vertex V is 20 cm vertically above the mid-point of the horizontal square base $ABCD$, and $AB = 12$ cm.

Calculate the volume of the pyramid.

2   $VABC$ is a triangle-based pyramid.
The vertex V is vertically above the point B.
The base, ABC, is a triangle with a right angle at B.
$AB = 5$ cm, $BC = 7$ cm and $VC = 25$ cm.

Calculate the volume of the pyramid $VABC$.

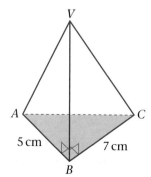

3 A cone has a circular base of radius r cm. The vertical height of the cone is 15 cm. The volume of the cone is 600 cm³. Calculate the value of r.

4 A solid metal cube of side length x cm is melted down and re-cast to make a cone. During this process none of the metal is lost and all of the metal is used to make the cone. The cone has a circular base of diameter 12 cm and a slant height of length 10 cm. Calculate

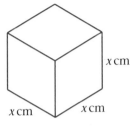

(a) the volume of the cone (b) the value of x.

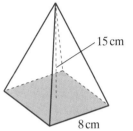

5 A solid pyramid has a square base of side length 8 cm. The vertical height of the pyramid is 15 cm. The pyramid is made by melting down and re-casting a metal cube of side length y cm. During the melting and re-casting process 10% of the metal is lost but all of the remainder is used to make the cone. Calculate the value of y.

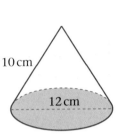

22.6 The surface area and volume of a sphere

For a sphere of radius r: **volume of sphere** $= \dfrac{4\pi r^3}{3}$

surface area $= 4\pi r^2$

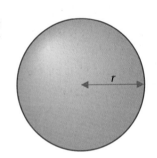

Example 11

Calculate the volume of a sphere of radius 5 cm.

$$\text{Volume of sphere} = \frac{4 \times \pi \times 5^3}{3}$$
$$= \frac{4 \times \pi \times 125}{3}$$
$$= 523.6 \text{ cm}^3 \text{ (correct to 1 d.p.)}$$

Example 12

The surface area of a sphere is 2000 cm².
Calculate the radius of the sphere.

Using surface area $= 4\pi r^2$:

$$4\pi r^2 = 2000$$

so $$r^2 = \frac{2000}{4\pi}$$

$$r^2 = \frac{500}{\pi}$$

$$r^2 = 159.15\ldots$$

So $$r = 12.62 \text{ cm (correct to 2 d.p.)}$$

Exercise 22G

1   Calculate the volume and surface area of a sphere
   (a)   of radius 8 cm
   (b)   of radius 7.2 cm
   (c)   of diameter 19 cm
   (d)   of diameter x cm.

2   A sphere has a volume of 5000 cm³.
   Calculate the radius of the sphere.

3   A cube of side x cm and a sphere of radius 6 cm have equal volumes.
   Calculate the value of x.

22.7 Areas and volumes of similar shapes

This rectangle measures 1 unit by 2 units: Enlarge it by scale factor 2:

It takes four of the smaller rectangles to fill the enlarged one:

area of enlarged rectangle = 4 × area of first rectangle

Here is the 1 by 2 rectangle again: Enlarge it by scale factor 3: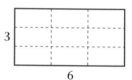

This time it takes nine (or 3^2) of the smaller rectangle to fill the larger one

area of enlarged rectangle = 9 × area of first rectangle

This relationship can be generalised for any enlargement:

> When a shape is enlarged by scale factor k, the area of the enlarged shape is k^2 × area of the original shape.

For example, if the scale factor is 5, the new area will be 5^2 (or 25) times the original area.

A similar result holds for 3-D shapes:

Here is a 1 by 1 by 2 cuboid: Enlarge it by scale factor 2: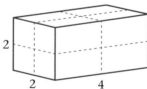

Eight (or 2^3) of the smaller cuboids are needed to fill the enlarged shape:

volume of enlarged cuboid = 8 × volume of first cuboid

This result generalises to:

> When a 3-D shape is enlarged by scale factor k, the volume of the enlarged shape is k^3 × volume of the original shape.

For example, if the scale factor is 5, the new volume will be 5^3 (or 125) times the original volume.

Example 13

A soft drinks company manufactures two **similar** bottles called *Standard* and *Super*.

The *Standard* bottle has volume 1000 cm³. The *Super* bottle has volume 1500 cm³. The height of the *Standard* bottle is 30 cm. Calculate the height of the *Super* bottle.

Because the two bottles are similar, *Super* is an enlargement of *Standard*.

To find the height of *Super* you need to calculate the scale factor of the enlargement. Let this scale factor be *k*.

So: volume of *Super* = $k^3 \times$ volume of *Standard*

$$k^3 = \frac{\text{volume of } Super}{\text{volume of } Standard}$$

$$= \frac{1500}{1000}$$

$$= 1.5$$

$$k = \sqrt[3]{1.5}$$

$$= 1.1447\ldots$$

Then the height of *Super* is scale factor ($k = 1.1447\ldots$) $\times$ height of *Standard*.

The height of *Super* is $1.1447\ldots \times 30 = 34.34$ cm (correct to 2 d.p.)

Example 14

The ratio of the radii of two spheres is $1 : 3$.
Calculate the ratio of:

(a) the surface areas of the spheres

(b) the volumes of the spheres.

Radius of larger sphere = 3 × radius of smaller sphere

(a) The large sphere is an enlargement of the small sphere, scale factor 3.

So surface area of larger sphere

$= 3^2 \times$ surface area of smaller sphere

The ratio of the surface areas is $1 : 9$.

(b) Volume of larger sphere

$= 3^3 \times$ volume of smaller sphere.

The ratio of the volumes is $1 : 27$.

Worked examination question 2

The diagram represents a box of chocolates.
This is a *Standard* size box.

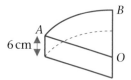

The lid and base of the box are sectors of a circle centre O.
The radius OA of the circle is 18 cm.
The angle $AOB = 75°$.
The box is 6 cm deep.

Calculate

(a)   the area of the face of the top of the box

(b)   the volume of the box.

The company manufacturing these boxes decides to manufacture
a *Large* size box which will be similar in shape to the *Standard* size box.
The volume of the *Large* size box is to be twice the volume of
the *Standard* size box.

(c)   Calculate the depth of the *Large* size box.

(a)   Area of the face of the top $= \dfrac{\pi \times 18^2 \times 75}{360}$

$$= \dfrac{\pi \times 324 \times 75}{360}$$

$$= 212.0575\ldots$$

$$= 212 \text{ cm}^2 \text{ (correct to 3 s.f.)}$$

(b)   Volume of box = area of top (or area of base) × height (or depth)

so   volume of the box $= 212.0575\ldots \times 6$

$$= 1272 \text{ cm}^3 \text{ (to the nearest whole number)}$$

(c)   Let the scale factor for the lengths be k.

The volume of *Large* box $= k^3 \times$ volume of *Standard* box

So $k^3 = \dfrac{\text{volume of } Large \text{ box}}{\text{volume of } Standard \text{ box}} = 2$

$k = \sqrt[3]{2} = 1.2599\ldots$

As the scale factor k is 1.2599...

depth of the *Large* box $= 1.2599\ldots \times$ depth of *Standard* box

$$= 1.2599\ldots \times 6 = 7.56 \text{ cm (correct to 2 d.p.)}$$

Exercise 22H

1   Washing liquid is sold in two sizes of similar bottles called *Basic* and *Extra*. The volume of a *Basic* bottle is $\frac{1}{2}$ litre and its height is 20 cm. The volume of an *Extra* bottle is 0.8 litres. Calculate the height of an *Extra* bottle.

2   A ball is a sphere of radius 12 cm.
The ball is inflated to increase its radius to 12.5 cm.
Show that the volume of air in the ball increases by approximately 13%.

3   The heights of two similar cylinders are in the ratio 2 : 3.
Calculate the ratio of

(a)   the surface area of the cylinders

(b)   the volumes of the cylinders.

4   A model car is a scale replica of the real car.
The length of the model car is $\frac{1}{50}$ of the length of the real car.

Calculate the fraction $\dfrac{\text{volume of the model car}}{\text{volume of the real car}}$

5   Beans are sold in two similar cylindrical cans called *Standard* and *Large*. The *Standard* can holds 500 grams of beans. The *Large* can holds 750 grams of beans. The height of a *Standard* can is 11 cm. Calculate

500 g   750 g

Standard   Large

(a)   the height of a *Large* can

(b)   the ratio of the areas of the circular bases of the cans.

22.8 Compound solids

You need to be able to find the volume of solids made from combining shapes you already know.

Example 15

This spinning top is made from a hemisphere and a cone.
Find its volume. Give your answer to 2 d.p.

Imagine splitting the shape into two sections.

A sphere of radius 3 cm has volume

$$\frac{4\pi(3)^3}{3} = 36\pi \text{ cm}^3$$

The hemisphere is half a sphere so it has volume $18\pi \text{ cm}^3$.

The volume of the cone is $\frac{1}{3}\pi r^2 h$

$$\frac{1}{3}\pi \times 3^2 \times 5 = 15\pi \text{ cm}^3$$

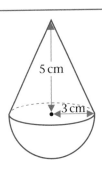

> You can leave your answers in terms of π until you have finished your working out.

So the volume of the entire spinning top is

$$18\pi + 15\pi = 33\pi = 103.67 \text{ cm}^3 \text{ (correct to 2 d.p.)}$$

Example 16

This hat is made from a cone with the top chopped off.
Find its volume correct to 2 d.p.

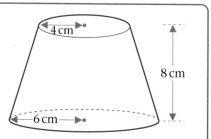

Imagine the hat was continued up to make a complete cone of vertical height h. If you know the value of h you can find the volume of the hat by subtracting the volume of the smaller cone from the volume of the larger one.

> This is known as a **truncated cone** or a **frustum**.

You can find h using similar triangles.

The shape $ABCD$ is a section from the hat.

Triangle ABE and triangle APD are similar, so

> See Section 3.6 for more about similar triangles.

$$\frac{h}{6} = \frac{8}{2}$$

$$h = 24 \text{ cm}$$

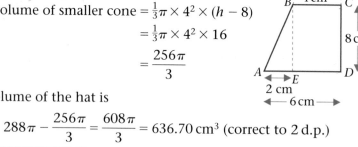

So   volume of larger cone $= \frac{1}{3}\pi r^2 h$

$$= \frac{1}{3}\pi \times 6^2 \times 24$$

$$= 288\pi$$

volume of smaller cone $= \frac{1}{3}\pi \times 4^2 \times (h-8)$

$$= \frac{1}{3}\pi \times 4^2 \times 16$$

$$= \frac{256\pi}{3}$$

So the volume of the hat is

$$288\pi - \frac{256\pi}{3} = \frac{608\pi}{3} = 636.70 \text{ cm}^3 \text{ (correct to 2 d.p.)}$$

Exercise 22I

1 This spinning top is made from a cube, a cylinder and a cone.

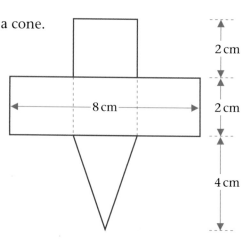

Find its volume, giving your answer to 2 d.p.

2 This box is made from a cube of
side 8 cm and a pyramid of
height 5 cm.
Find its volume.

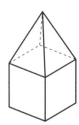

3 Find the volume of this truncated cone.
Give your answer to 2 d.p.

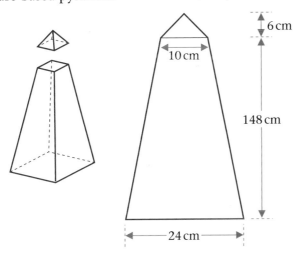

6 cm

3 cm

7 cm

4 The diagram represents a model of Cleopatra's Needle.
It is made from a truncated square-based pyramid and a smaller
square-based pyramid.

6 cm

10 cm

148 cm

24 cm

Cleopatra's Needle on the
Thames Embankment in
London is an original
Egyptian obelisk.

Find its volume, giving your answer to 2 d.p.

Mixed exercise 22

1 Here is a sector of a circle, *OAPB*, with centre *O*.

Calculate

(a) the length of the chord *AB*

(b) the length of the arc *APB*

(c) the area of the sector *OAPB*

(d) the area of the segment *APB*.

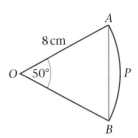

A

8 cm

O 50° P

B

2 Calculate the volume and surface area of a cylinder with circular
base of radius 12 cm and with vertical height 20 cm.

3 Calculate the volume of a sphere of diameter 8 cm.

4 The volume of a sphere of radius *r* cm is numerically equal to the
radius of that sphere.

Show that $r = \sqrt{\dfrac{3}{4\pi}}$.

5 *VABCD* is a pyramid with a rectangular base *ABCD*.

AB = 8 cm   *BC* = 12 cm
M is at the centre of *ABCD* and *VM* = 35 cm.
V is vertically above *M*.

Calculate the volume of *VABCD*.

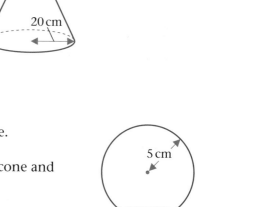

6 A cone has a circular base of radius 20 cm.
The slant height of the cone is 30 cm.

Calculate the volume of the cone.

7 Wine is sold in two similar bottles.
The volume of a small bottle of wine is 70 cl.
The volume of a larger bottle of wine is 100 cl.
The height of the larger bottle of wine is 30 cm.

Calculate the height of the smaller bottle of wine.

8 This trophy is made from a sphere, a truncated cone and
a cylinder.

Find its volume correct to 2 d.p.

9 X and Y are two geometrically similar solid shapes.
The total surface area of shape X is 450 cm².
The total surface area of shape Y is 800 cm².
The volume of shape X is 1350 cm³.
Calculate the volume of shape Y.   [E]

10 Cylinder A and cylinder B are mathematically similar.

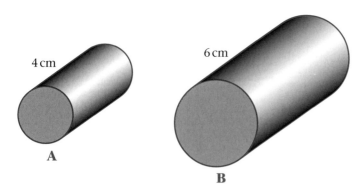

The length of cylinder A is 4 cm and the length of cylinder B is
6 cm. The volume of cylinder A is 80 cm³.
Calculate the volume of cylinder B.   [E]

Summary of key points

1   **Circumference** of a circle $= 2\pi r$
    Area of a circle $= \pi r^2$

2   **Arc length** of a circle $= \dfrac{\pi r \theta}{180}$

3   **Area of a sector** of a circle $= \dfrac{\pi r^2 \theta}{360}$

4   **Area of a segment** $= \dfrac{\pi r^2 \theta}{360} - \dfrac{1}{2}r^2 \sin\theta$

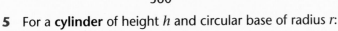

— arc

— sector

— segment

5   For a **cylinder** of height h and circular base of radius r:

   $$\textbf{surface area } = 2\pi rh + 2\pi r^2$$
   $$\textbf{volume } = \pi r^2 h$$

6   The **volume of any prism** is

   area of base $\times$ vertical height   or   area of cross-section $\times$ vertical height

7   **Volume of a pyramid** $= \frac{1}{3} \times$ area of base $\times$ vertical height

8   **Volume of a cone** $= \frac{1}{3} \times$ area of base $\times$ height $= \frac{1}{3}\pi r^2 h$

9   Cylinders and cones are special types of prisms and pyramids, respectively.

10 **Volume of a sphere** $= \dfrac{4\pi r^3}{3}$

   Surface area of a sphere $= 4\pi r^2$

11 When a shape is enlarged by a scale factor k to produce a similar shape:
   * area of enlarged shape $= k^2 \times$ area of original shape
   * volume of enlarged shape $= k^3 \times$ volume of original shape.

23 Exploring numbers 2

In this chapter you will investigate decimals and their fraction equivalents, and extend your knowledge of surds and upper and lower bounds.

23.1 Terminating and recurring decimals

When a fraction of the type $\frac{a}{b}$ is converted into a decimal, there are two different types of answer. The result will either be a **terminating decimal** such as 0.75 or 0.174 or a **recurring decimal** such as 0.3333... or 0.34343434....

> For more on recurring decimals see Section 4.6.

Terminating decimals

When the denominator, b, of the fraction $\frac{a}{b}$ is a factor of 10^n for some integer n (e.g. b is a factor of 10 or 100 or 1000 and so on), the division to evaluate the decimal will always finish at some stage. There will be an exact answer.

> A fraction is equivalent to a **terminating decimal** if the denominator is a factor of 10^n for some integer n.

> The only prime factors of 10 are 2 and 5. So the only prime factors of 10^n are 2 and 5, for any value of n.
> So $\frac{a}{b}$ is equivalent to a terminating decimal if b has prime factors of 2 and 5 only.

Example 1

Decide whether these fractions convert into terminating or recurring decimals.

(a) $\frac{1}{4}$ (b) $\frac{13}{64}$

Work out their decimal equivalents.

(a)   4 is a factor of $10^2 = 100$, so $\frac{1}{4}$ is a terminating decimal.
$\frac{1}{4} = 0.25$

(b)   Try dividing 64 into powers of 10.
64 is a factor of $10^6 = 1\,000\,000$, so $\frac{13}{64}$ is a terminating decimal.
$\frac{13}{64} = 0.203\,125$ •

```
       0.203 125
64)13.000 000
   12 8
      200
      192
       80
       64
      160
      128
       320
       320
```

To convert 0.25 to a fraction you multiply 0.25 by 100 and then divide this result by 100:

$$0.25 = \frac{0.25 \times 100}{100} = \frac{25}{100} = \frac{1}{4} \quad \text{—— Cancel by 25}$$

To convert a decimal which terminates after 2 places into a fraction you multiply the decimal by 10^2 and then divide this result by 10^2.

For a decimal which terminates after 3 decimal places, do the same but multiply and divide by 10^3. For example:

$$0.125 = \frac{0.125 \times 1000}{1000} = \frac{125}{1000} = \frac{1}{8} \quad\text{——— Cancel by 125}$$

Any terminating decimal with n decimal places can be converted into a fraction by multiplying the decimal by 10^n and dividing the result by 10^n.

> Cancel and write the fraction in its simplest form.

Recurring decimals

As a decimal, $\frac{1}{7}$ is

$$\frac{1}{7} = 0.142857142857\ldots \quad \text{or} \quad 0.\dot{1}4285\dot{7}$$

A **recurring decimal** has a repeating pattern in its digits.

Example 2

Write these fractions as decimals.

(a) $\frac{1}{3}$ (b) $\frac{1}{9}$ (c) $\frac{1}{6}$ (d) $\frac{1}{11}$ (e) $\frac{10}{21}$

(a) $\frac{1}{3} = 0.33333\ldots$ or $0.\dot{3}$

(b) $\frac{1}{9} = 0.11111\ldots$ or $0.\dot{1}$

(c) $\frac{1}{6} = 0.16666\ldots$ or $0.1\dot{6}$

(d) $\frac{1}{11} = 0.090909\ldots$ or $0.\dot{0}\dot{9}$ (a pattern every 2 decimal places)

(e) $\frac{10}{21} = 0.476190476190\ldots$ or $0.\dot{4}7619\dot{0}$ (a pattern every 6 decimal places)

Exercise 23A

1 Convert the following fractions into decimals. Indicate which of the decimals are terminating and which are recurring.

(a) $\frac{3}{4}$ (b) $\frac{3}{5}$ (c) $\frac{3}{7}$ (d) $\frac{3}{8}$ (e) $\frac{3}{10}$ (f) $\frac{3}{11}$

2 State, with a reason, whether the equivalent decimals for these fractions are terminating or recurring.

(a) $\frac{5}{6}$ (b) $\frac{5}{7}$ (c) $\frac{5}{8}$ (d) $\frac{5}{9}$ (e) $\frac{5}{12}$ (f) $\frac{5}{13}$ (g) $\frac{5}{16}$

3 Write out the recurring decimals equivalent to these fractions:

$$\frac{1}{7}, \ \frac{2}{7}, \ \frac{3}{7}, \ \frac{4}{7}, \ \frac{5}{7}, \ \frac{6}{7}$$

Explain the relationship between the pattern of the recurring decimals.

4 How many digits are there in the recurring pattern of digits in the decimal equivalent of the fraction $\frac{3}{17}$?

5   Find the recurring decimals which are equivalent to the fractions

$$\tfrac{1}{13}, \quad \tfrac{2}{13}, \quad \tfrac{3}{13}, \quad \dots, \quad \tfrac{10}{13}, \quad \tfrac{11}{13}, \quad \tfrac{12}{13}$$

Separate the recurring decimals into two distinct sets of numbers. Explain the connection between them.

6   By considering the process of long division to find a recurring decimal equivalent to a fraction $\frac{1}{m}$, such as $\frac{1}{17}$, explain

     (a)   why the decimal will recur

     (b)   why the number of digits in the recurring pattern is less than m.

23.2 Finding a fraction equivalent to a recurring decimal

The method is best illustrated with three examples.

Example 3

Find a fraction equivalent to 0.55555...

Let $x = 0.55555...$ (1)

Multiply equation (1) by 10:

     $10x = 5.55555...$ (2)

Subtract equation (1) from equation (2):

     $10x - x = 5$

So $9x = 5$

       $x = \tfrac{5}{9}$

The fraction equivalent to 0.55555... is $\tfrac{5}{9}$

> Pattern recurs every 1 decimal place, so multiply by 10^1.

Example 4

Find a fraction equivalent to 0.636363...

Let $x = 0.636363...$

Multiply by 100:

     $100x = 63.6363...$

Subtract:

     $100x - x = 63$

So $99x = 63$

       $x = \tfrac{63}{99} = \tfrac{7}{11}$

The fraction equivalent to 0.636363... is $\tfrac{7}{11}$

> Pattern recurs every 2 decimal places, so multiply by 10^2.

Example 5

Find a fraction equivalent to 0.103103103...

Let $\qquad x = 0.103103103...$

Multiply by 1000: $1000x = 103.103103103...$

Subtract: $\qquad 999x = 103$

So $\qquad x = \frac{103}{999}$

The fraction equivalent to 0.103103103... is $\frac{103}{999}$

> Pattern recurs every 3 decimal places, so multiply by 10^3.

Example 6

Write $1.1\dot{2}\dot{3}$ as a fraction in its simplest form.

Let $\qquad x = 1.1\dot{2}\dot{3}$

Multiply by 100: $100x = 112.3\dot{2}\dot{3}$

Subtract: $\qquad 99x = 111.2$

Multiply by 10: $990x = 1112$

So $x = \frac{1112}{990} = 1\frac{122}{990} = 1\frac{61}{495}$

Exercise 23B

1   Find the fractions which are equivalent to the recurring decimals:
   (a) 0.666666...
   (b) 0.777777...
   (c) 0.34343434...
   (d) 0.91919191...
   (e) 0.181818...
   (f) 0.125125125...
   (g) 0.513513513...
   (h) 0.100110011001...
   (i) 0.127912791279...
   (j) 0.089108910891...
   (k) 1.4133333...
   (l) 4.72121212...
   (m) $2.34\dot{5}$
   (n) $5.6\dot{2}\dot{7}$

2   Find the fraction which is equivalent to the recurring decimal 0.9999999... Explain the significance of your result.

23.3 Surds

The area of this square lawn is 20 m². The length of each side is x m.

$$x^2 = 20$$
$$x = \sqrt{20}$$

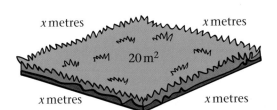

x metres · x metres · 20 m² · x metres · x metres

Using a calculator, you could write x as 4.47. This is correct to 2 decimal places but it is not an exact answer. If you wanted to say exactly what the number was you could just write $\sqrt{20}$.

A number written exactly using square roots is called a **surd**.
For example, $\sqrt{3}$ and $2 - \sqrt{5}$ are in surd form.

Unless you are told otherwise in a question, you can leave your answer as a surd.

Example 7

Solve the equation $x^2 - 8x + 11 = 0$, leaving your answer in surd form.

Complete the square:

$$(x - 4)^2 - 16 + 11 = 0$$
$$(x - 4)^2 - 5 = 0$$
$$(x - 4)^2 = 5$$
$$x - 4 = \pm\sqrt{5}$$

So $\qquad x = 4 + \sqrt{5}$ or $x = 4 - \sqrt{5}$

> For more about completing a square see Sections 19.2 and 19.3.

> $\sqrt{5}$ is the positive square root of 5. You have to use a $\pm$ sign to show both possible values.

Manipulating surds

Surds can be added, subtracted, multiplied and divided.

$$\sqrt{a \times b} = \sqrt{a} \times \sqrt{b}$$

$$\sqrt{\frac{a}{b}} = \frac{\sqrt{a}}{\sqrt{b}}$$

> You can work these rules out using the rules of indices. For example:
> $(ab)^n = a^n b^n$
> When $n = \frac{1}{2}$, $(ab)^{\frac{1}{2}} = a^{\frac{1}{2}} b^{\frac{1}{2}}$
> $\sqrt{ab} = \sqrt{a}\sqrt{b}$

Example 8

Simplify $\sqrt{12}$

$$\sqrt{12} = \sqrt{4 \times 3}$$
$$= \sqrt{4} \times \sqrt{3}$$
$$= 2\sqrt{3}$$

Sometimes you will be asked to simplify a fraction with a surd as the denominator.

Example 9

Simplify $\dfrac{1}{\sqrt{2}}$

Multiply the top and bottom by $\sqrt{2}$:

$$\frac{1}{\sqrt{2}} = \frac{1}{\sqrt{2}} \times \frac{\sqrt{2}}{\sqrt{2}} = \frac{\sqrt{2}}{2}$$

Simplified surds should never have a square root in the denominator.

Example 10

Rationalise $\dfrac{1}{3\sqrt{5}}$

> Simplifying an expression to remove the square root in the denominator is called 'rationalising'.

Multiply the top and bottom by $\sqrt{5}$:

$$\frac{1}{3\sqrt{5}} = \frac{1}{3\sqrt{5}} \times \frac{\sqrt{5}}{\sqrt{5}}$$

$$= \frac{\sqrt{5}}{3 \times 5} = \frac{\sqrt{5}}{15}$$

Example 11

Rationalise the denominator of $\dfrac{\sqrt{2}}{5 + \sqrt{2}}$

Multiply the top and bottom by $(5 - \sqrt{2})$:

$$\frac{\sqrt{2}}{5 + \sqrt{2}} = \frac{\sqrt{2}(5 - \sqrt{2})}{(5 + \sqrt{2})(5 - \sqrt{2})}$$

$$= \frac{5\sqrt{2} - 2}{23}$$

> $(5 + \sqrt{2})(5 - \sqrt{2})$
> $= 5 \times 5 - 5\sqrt{2} + 5\sqrt{2} - \sqrt{2}\sqrt{2}$
> $= 25 - 2$
> $= 23$

Exercise 23C

1 Solve the equation $x^2 = 30$, leaving your answer in surd form.

2 The area of a square is $40\,\text{cm}^2$. Find the length of one side of the square. Give your answer as a surd in its simplest form.

3 The lengths of the sides of a rectangle are
$$3 + \sqrt{5} \quad \text{and} \quad 3 - \sqrt{5} \text{ units}$$
Work out, in their most simplified forms,
(a) the perimeter of the rectangle
(b) the area of the rectangle.

4 Rationalise each of the following:

(a) $\dfrac{1}{\sqrt{7}}$ (b) $\dfrac{3}{\sqrt{5}}$ (c) $\dfrac{1}{\sqrt{17}}$

(d) $\dfrac{1}{\sqrt{11}}$ (e) $\dfrac{\sqrt{32}}{\sqrt{2}}$ (f) $\dfrac{\sqrt{5}}{3 + \sqrt{5}}$

(g) $\dfrac{\sqrt{7}}{2 - \sqrt{7}}$ (h) $\dfrac{\sqrt{3}}{2 - \sqrt{3}}$ (i) $\dfrac{\sqrt{11}}{\sqrt{11} - 7}$

5 Solve these equations, leaving your answers in surd form.
(a) $x^2 - 6x + 2 = 0$ (b) $x^2 + 10x + 14 = 0$

6 Show that $\dfrac{1}{2\sqrt{17}} = \dfrac{\sqrt{17}}{34}$

7 *ABCD* is a rectangle.

$AB = 2\sqrt{2}$ units $\qquad BC = \sqrt{3}$ units

Work out the length of the diagonal *AC*.
Give your answer in surd form.

8 The diagram represents a right-angled triangle *ABC*.

$AB = \sqrt{7} + 2$ units $\qquad AC = \sqrt{7} - 2$ units

Work out, leaving any appropriate answers in surd form,

(a) the area of triangle *ABC* **(b)** the length of *BC*.

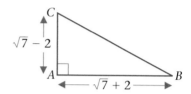

9 Simplify $\dfrac{\sqrt{3}}{\sqrt{2}}$

23.4 Rounding

When you round the measurements 2.14 cm, 2.15 cm and 2.16 cm to 1 decimal place

- 2.14 cm **rounds down** to 2.1 cm
- 2.16 cm **rounds up** to 2.2 cm.
- By convention, 2.15 cm **rounds up** to 2.2 cm.

However, you know that 2.15 cm is really exactly the same distance from both 2.1 cm and 2.2 cm.

The largest measurement that rounds down to 2.1 cm is 2.1499999… (recurring).

What is the value of $2.14\dot{9}$?

Let $x = 2.149999…$ (1)

Then using the method of Section 23.2:

$10x = 21.49999…$ (2)

Subtract equation (1) from equation (2):

$$\begin{array}{r} 10x = 21.49999… \\ x = 2.14999… \\ \hline 9x = 19.35 \end{array}$$

So $\qquad x = \dfrac{19.35}{9} = 2.15$

> So 2.1499… (recurring) is the same as 2.15

23.5 Upper bounds and lower bounds

All these numbers round down to 3.1 cm correct to 1 decimal place:

   3.11 cm, 3.136 cm, 3.1475 cm, 3.14 cm, 3.149 cm

A number is an upper bound of 3.1 cm if it is greater than or equal to all the numbers which round down to 3.1 cm. So upper bounds of 3.1 cm include the measurements

   3.2 cm, 3.16 cm, 3.155 cm, 3.152 cm, 3.151 cm and 3.15 cm

They are all greater than any number that rounds down to 3.1 cm.

Because 3.15 cm is the **smallest** measurement in this list, it is called the **least upper bound** of 3.1 cm.

Using similar arguments, the measurement 3.05 cm is the **greatest lower bound** of 3.1 cm.

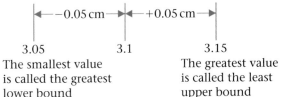

The smallest value is called the greatest lower bound

The greatest value is called the least upper bound

> The **greatest lower bound** and the **least upper bound** are the minimum and maximum possible values of a measurement or calculation.

> Sometimes these are called simply the **upper bound** and the **lower bound**.

Exercise 23D

In questions **1–7** write down the lower bound and the upper bound for each measurement, to the given degree of accuracy.

1   To the nearest unit: 4, 14, 104, 10, 100

2   To the nearest 10: 30, 50, 180, 3020, 10, 100

3   To 1 decimal place: 4.7, 2.9, 13.6, 0.3, 157.5, 10.0, 100.0

4   To 2 significant figures: 37, 50, 180, 3.2, 9.5, 9400, 10, 100

5   To the nearest 0.5 unit: 4.5, 7.5, 16.5, 3.0, 15.5, 10, 100

6   To the nearest 0.2 unit: 3.4, 3.2, 4.0, 9.4, 12.2, 24.6, 10.0, 100.0

7   To the nearest quarter unit: 3.75, 3.5, 4.25, 6.0, 15.5, 10.0, 100.0

8   Three students, James, Sunita and Sara, all try to find the size of their classroom.

   (a)   James estimates the lengths of the walls to be 7 m and 6 m to the nearest metre. Write the least upper bound and the greatest lower bound for the lengths of the classroom walls.

   (b)   Sunita measures the lengths of the classroom walls with her ruler. She writes '730 cm and 590 cm correct to 2 significant figures'. Write the least upper bound and the greatest lower bound for the lengths of the classroom walls.

(c) Sara measures the walls with a tape measure and obtains the lengths 7.32 m and 5.94 m correct to 2 decimal places. Write the least upper bound and the greatest lower bound for the lengths of the classroom walls.

(d) Explain whether the students' measurements are consistent.

9 The lap record for a racing circuit is 74.36 seconds correct to the nearest $\frac{1}{100}$ of a second.

(a) Write down the upper bound and the lower bound for the lap record.

(b) A racing driver has a recorded lap time of 74.357 seconds. Explain whether this is a new record.

23.6 Calculations involving upper and lower bounds – addition and multiplication

When two (or more) measurements are used in a calculation, the degree of accuracy of the measurements affects the degree of accuracy of the calculated result.

Example 12

Mr Robinson's garden is a rectangle with sides measured as 3.2 m and 5.9 m, both correct to 1 d.p. Calculate the greatest lower bound and the least upper bound for the area of the rectangle.

The largest area is found from the longest possible sides:

$3.25 \times 5.95 = 19.3375 \text{ m}^2$

The smallest area is found from the shortest possible sides:

$3.15 \times 5.85 = 18.4275 \text{ m}^2$

In addition and multiplication calculations use

• the two lower bounds to obtain the lower bound of the result
• the two upper bounds to obtain the upper bound of the result.

Exercise 23E

Using the upper and lower bounds for these measurements, find the upper and lower bounds for these quantities. The degree of accuracy of each measurement is given.

1 The area of a rectangle with sides 6 cm and 8 cm, both measured to the nearest centimetre.

2 The perimeter of a rectangle with sides 6 cm and 8 cm, both measured to the nearest centimetre.

3 The area of a square with side 7.5 cm measured to 2 s.f.

4   The perimeter of a square with side 7.5 cm measured to 2 s.f.

5   The area of a circle with radius 3 cm measured to 1 s.f.

6   The circumference of a circle with radius 0.34 m measured to 2 d.p.

7   The area of a circle with radius 41 mm measured to 2 s.f.

8   The area of a triangle with base length 8 cm and perpendicular height 6 cm, both measured to the nearest centimetre.

9   The area of a triangle with base length 8.0 cm and perpendicular height 6.0 cm, both measured to 2 s.f.

10   The area of a triangle with base length 18 cm and perpendicular height 16 cm, both measured to the nearest 2 cm.

11   The area of a rectangle with sides 3.75 cm and 1.5 cm, both measured to the nearest 0.25 cm.

12   The circumference of a circle with radius 3.5 cm measured to the nearest 0.5 cm.

13   The volume of a sphere with radius
    (a)   15 cm measured to the nearest 5 cm
    (b)   15 cm measured to 2 s.f.
    (c)   15.0 cm measured to 1 d.p.

23.7 Calculations involving upper and lower bounds – subtraction and division

The upper bound of the calculation results from a calculation using one upper bound and one lower bound.

Example 13

A piece of metal 32 cm long is cut from a 100 cm length. Both the measurements are correct to the nearest centimetre.
Find the least upper and greatest lower bounds of the length of the remaining piece.

   100 cm lies between 99.5 cm and 100.5 cm
   32 cm lies between 31.5 cm and 32.5 cm

The least upper bound for the remaining piece
   $= 100.5 - 31.5 = 69$ cm

The greatest lower bound for the remaining piece
   $= 99.5 - 32.5 = 67$ cm

> If in doubt, do these four calculations:
> $100.5 - 31.5 = $ (69)
> $100.5 - 32.5 = 68$
> $99.5 - 31.5 = 68$
> $99.5 - 32.5 = $ (67)

Example 14

The maximum and minimum temperatures, to the nearest °C, at Atlanta one day in 1993 were 27 °C and 8 °C, respectively. Calculate the range of temperatures.

27 °C represents a temperature between 26.5 °C and 27.5 °C
8 °C represents a temperature between 7.5 °C and 8.5 °C

Least upper bound of the range   = 27.5 − 7.5 = 20 °C

Greatest lower bound of the range = 26.5 − 8.5 = 18 °C

So the range of temperatures could be any value between 18 °C and 20 °C.

You need to think carefully how to combine the correct pair of initial bounds to find the upper and lower bounds of the result.

If you are in any doubt, do the calculation with the four different pairs of upper and lower bounds. Then choose the largest and the smallest values for the bounds.

Exercise 23F

1   A bar measures 44.8 cm long to 3 s.f. It is heated and expands to 45.7 cm. Calculate the upper bound and the lower bound of the expansion.

2   Two recordings of the same piece of music are measured, correct to the nearest minute, to be 34 minutes and 37 minutes. Calculate the least upper bound and the greatest lower bound for the difference in the times of the two recordings.

3   Calculate the least upper bound and the greatest lower bound for the difference in length of two pieces of paper. One measures 110 cm and the other 95 cm, both correct to 2 s.f.

4   Jeremy's time for a 100 m race is 14.7 seconds to the nearest 0.1 second. Calculate the upper bound and the lower bound of Jeremy's average speed in metres per second for the race

   (a)   assuming that the distance was accurately measured

   (b)   assuming that the distance was measured correct to the nearest metre.

5   Jennifer starts the day with a full tank of petrol. At the end of the day she has travelled 165 miles (correct to the nearest 5 miles). She fills the tank with 18 litres of petrol (correct to the nearest litre). Calculate the upper bound and the lower bound of the petrol consumption of Jennifer's car in

   (a)   miles per litre

   (b)   miles per gallon.

4.54 litres = 1 gallon

6　In a triangle ABC, angles A and B have been measured to be 35° and 72°, respectively. Length BC is 15 cm. All the measurements are correct to 2 s.f. Use the sine rule to calculate the upper bound and the lower bound of the length AC.

> For more about the sine rule see Section 21.2.

7　A ball has mass 130 g and volume 40 cm³, both correct to the nearest 10 units. Calculate the least upper bound and the greatest lower bound of the density of the ball.

> Density = $\dfrac{\text{mass}}{\text{volume}}$

8　Triangle DEF has sides of length 6 cm, 7 cm and 8 cm, all correct to the nearest centimetre. Use the cosine rule to calculate the upper bound and the lower bound of the smallest angle in the triangle.

> For more about the cosine rule see Section 21.3.

Mixed exercise 23

1　Convert the following fractions to decimals and indicate which of the decimals are terminating and which are recurring.

(a) $\frac{7}{8}$　　(b) $\frac{7}{9}$　　(c) $\frac{7}{10}$　　(d) $\frac{7}{11}$　　(e) $\frac{7}{12}$　　(f) $\frac{7}{13}$

2　Find a fraction equivalent to these decimal numbers:

(a)　13.747474...　　　(b)　6.753213213 21...

3　Simplify each of the following:

(a)　$\dfrac{1}{\sqrt{5}}$　　　　　　(b)　$\dfrac{5}{\sqrt{7}}$　　　　　　(c)　$\dfrac{1}{\sqrt{29}}$

(d)　$\dfrac{\sqrt{81}}{\sqrt{9}}$　　　　　　(e)　$\dfrac{\sqrt{48}}{\sqrt{2}}$　　　　　　(f)　$\dfrac{\sqrt{64}}{\sqrt{8}}$

4　Write the least upper bound and the greatest lower bound for

(a)　250 000 miles (correct to 2 s.f.)

(b)　6×10^6 cm (correct to 1 s.f.).

5　Explain the difference between these three lengths:

4 cm,　4.0 cm　and　4.00 cm

6　Find the least upper bound and the greatest lower bound of the shaded area. The two circles have radii of 5 cm and 3 cm, both correct to 1 s.f.

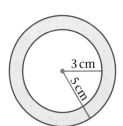

7　A cube has volume 55 mm³ correct to the nearest mm³. Calculate the least upper bound and the greatest lower bound of the length of a side of the cube.

8 **(a)** Find the value of $16^{\frac{1}{2}}$

(b) Given that $40 = k\sqrt{10}$, find the value of k.

(c) A large rectangular piece of card is $(\sqrt{5} + \sqrt{20})$ cm long and $\sqrt{8}$ cm wide.

A small rectangle $\sqrt{2}$ cm long and $\sqrt{5}$ cm wide is cut out of the piece of card.

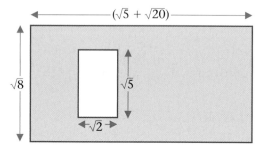

Express the area of the card that is left as a percentage of the area of the large rectangle. [E]

9 Elliot did an experiment to find the value of $g\,\text{m/s}^2$, the acceleration due to gravity.

He measured the time, T seconds, that a block took to slide L m down a smooth slope of angle $x°$.

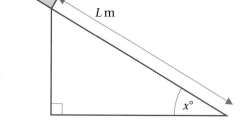

He then used the formula $g = \dfrac{2L}{T^2 \sin x°}$ to calculate an estimate for g.

$T = 1.3$ correct to 1 decimal place.
$L = 4.50$ correct to 2 decimal places.
$x = 30$ correct to the nearest integer.

(a) Calculate the lower bound and the upper bound for the value of g. Give your answers correct to 3 decimal places.

(b) Use your answers to part **(a)** to write down the value of g to a suitable degree of accuracy. Explain your reasoning.

Summary of key points

1 A fraction is equivalent to a **terminating decimal** if the denominator is a factor of 10^n for some integer n.

2 Any terminating decimal with n decimal places can be converted into a fraction by multiplying the decimal by 10^n and dividing the result by 10^n.

3 A **recurring decimal** has a repeating pattern in its digits.

4 A number written exactly using square roots is called a **surd**.
For example, $\sqrt{3}$ and $2 - \sqrt{5}$ are in surd form.

5 $\sqrt{a \times b} = \sqrt{a} \times \sqrt{b}$

6 $\sqrt{\dfrac{a}{b}} = \dfrac{\sqrt{a}}{\sqrt{b}}$

7   Simplified surds should never have a square root in the denominator.

8   The **greatest lower bound** and the **least upper bound** are the minimum and maximum possible values of a measurement or calculation.

9   In addition and multiplication calculations use
- the two lower bounds to obtain the lower bound of the result
- the two upper bounds to obtain the upper bound of the result.

24 Probability

Probability is about calculating or estimating what might happen in the future.

In this chapter you will find probabilities by considering all possible outcomes of an event, and estimate probabilities from experimental data.

24.1 Finding probabilities

> The **probability** of an event is expressed as a number from 0 to 1 inclusive.

That number can be expressed as a fraction, a decimal or a percentage. The larger the number, the greater the *likelihood* of the event happening.

> If an event is **impossible** its probability is 0.

> If an event is **certain** its probability is 1.

I will be the first person to swim to the North Pole.

The probability of you succeeding is 0!

You can calculate the probability of events for simple activities such as tossing coins and rolling dice.

> When there are n equally likely possible outcomes, the probability of each outcome $\frac{1}{n}$

Example 1

The numbers 1 to 10 are written on pieces of card which are then placed face downwards on a table. Ian picks up one of the cards at random. Calculate the probability that Ian's card will be

(a) a multiple of 3   (b) an even number   (c) a prime number.

> The key word **random** indicates that each of the possible outcomes is equally likely.

Number	1	2	3	4	5	6	7	8	9	10
Multiple of 3	✗	✗	✓	✗	✗	✓	✗	✗	✓	✗
Even	✗	✓	✗	✓	✗	✓	✗	✓	✗	✓
Prime	✗	✓	✓	✗	✓	✗	✓	✗	✗	✗

(a)   Three of the 10 numbers are multiples of 3. The probability of Ian selecting one of these is $\frac{3}{10}$ (or 0.3 or 30%).

> It is often worth drawing a rough diagram of all the possible outcomes.

(b)   Five of the 10 numbers are even, so the probability of Ian choosing one of these is $\frac{5}{10}$ (or $\frac{1}{2}$ or 0.5 or 50%).

(c)   Four of the 10 numbers are prime, so the probability of Ian choosing one of these is $\frac{4}{10}$ (or $\frac{2}{5}$ or 0.4 or 40%).

If all the outcomes are equally likely, the probability of an event can be calculated using the formula:

$$P(\text{event}) = \frac{\text{the number of ways the event can occur}}{\text{the total number of possible outcomes}}$$

> It may help your understanding if you read a probability like $\frac{3}{4}$ (three quarters) as 'three out of four possibilities'.

Example 2

Trevor and Kevin both answer this question:

There are eight snooker balls in a bag. Six of them are red, one is green and the other is yellow. Calculate the probability that a ball taken at random from the bag will be red.

Trevor's answer

> There are 3 possible outcomes:
> red, green and yellow
> $P(\text{red}) = \frac{6}{3} = 2$

Kevin's answer

> There are $6 + 1 + 1 = 8$ outcomes
> $P(\text{red}) = \frac{6}{8} = \frac{3}{4}$

> P(red) means 'the probability of selecting a red ball'. You should use this shorthand for expressing probabilities.

Who is correct? Explain your reasoning.

Kevin is correct. He realised that each of the snooker balls is a possible outcome and that altogether there are 8 balls and hence 8 possible outcomes. He could have left his answer as $\frac{6}{8}$ because the question did not ask him to cancel his answer. Equally correct answers are $\frac{3}{4}$, 0.75 and 75%.

Trevor made the common mistake of just counting the three possible colours. He was correct when he used the fact that there are 6 red balls as *the number of ways the event can occur*. He should have realised from his answer greater than 1 that he had made a mistake, because a probability is always between 0 and 1.

Exercise 24A

1 Nine playing cards are numbered 2 to 10. A card is selected from them at random. Calculate the probability that the card will be

(a) an odd number

(b) a multiple of 4.

2 Five strawberry, two orange and three blackcurrant flavoured sweets are placed in a box. One sweet is taken at random from the box. Calculate the probability that the sweet will be

(a) blackcurrant flavoured

(b) not orange flavoured.

3 A normal dice is rolled. Calculate the probability that the number on the uppermost face when it stops rolling will be

(a) 5   (b) odd   (c) prime   (d) not 6

> A normal dice is a cube. It is unbiased and has the numbers 1 to 6 on it. The numbers on opposite faces add up to 7.

4 A spinner is made from a regular octagon. It is labelled with three As, two Bs and three Cs. Each of the sides is equally likely to be resting on the table when it stops spinning. Calculate

(a) P(A resting on the table)

(b) P(B resting on the table)

(c) P(not C resting on the table).

5 A company is trying to prove that most cats prefer chicken flavoured cat food. Unfortunately, Alfred the cat doesn't know this and selects his food at random. The company puts out 10 saucers of rabbit, 6 saucers of sardine and only 4 saucers of chicken flavoured food. Calculate

(a) P(Alfred selects sardine flavoured food)

(b) P(Alfred selects rabbit flavoured flood)

(c) P(Alfred selects sardine or rabbit flavoured food)

(d) P(Alfred does not select chicken flavoured food).

(e) The answer to part (c) can be calculated from the answers to parts (a) and (b). Explain how.

(f) Explain how the answer to part (d) can be calculated if you are given that:
P(Alfred selects chicken flavoured food) = p

24.2 Mutually exclusive events

If two events cannot occur at the same time they are **mutually exclusive**.

> Events A and B are mutually exclusive when
> - if A happens then B does not
> - if B happens then A does not.

___ **Example 3** _____

A card is selected from a pack of 52 cards. What is the probability that the chosen card is either a 'king' or a '10'?

Choosing a 'king' and choosing a '10' are mutually exclusive events.

$$P(K) = \tfrac{4}{52}$$
$$P(10) = \tfrac{4}{52}$$
$$P(K \text{ or } 10) = \tfrac{4}{52} + \tfrac{4}{52} = \tfrac{8}{52}$$

When events A and B are mutually exclusive

$$P(A \text{ or } B) = P(A) + P(B)$$

Example 4

There are five red, three green and two yellow counters in a bag.
A counter is taken at random from the bag.
Calculate the probability that the counter will be

(a) red (b) green (c) yellow (d) red or green (e) not yellow.

(a) $P(\text{red}) = \frac{5}{10} = 0.5$ (b) $P(\text{green}) = \frac{3}{10} = 0.3$ (c) $P(\text{yellow}) = \frac{2}{10} = 0.2$

(d) A counter cannot be both red and green. So the events
'selecting a red' and 'selecting a green' are **mutually exclusive**.

So $P(\text{red or green}) = P(\text{red}) + P(\text{green}) = 0.5 + 0.3 = 0.8$

(e) The only possible colours which can occur are red, green and yellow. The sum of their
probabilities is 1. You are *certain* to get a red, a green or a yellow counter.

$$P(\text{red}) + P(\text{green}) + P(\text{yellow}) = 1$$

so $$P(\text{red}) + P(\text{green}) = 1 - P(\text{yellow})$$

so $$P(\text{red or green}) = 1 - P(\text{yellow})$$

If a counter selected is not yellow, it must be red or green.

So $P(\text{not yellow}) = 1 - P(\text{yellow})$
$$= 1 - 0.2$$
$$= 0.8$$

An obvious case of mutually exclusive events is 'selecting A' and 'not selecting
A'. In this situation there is a general statement that is often a useful short cut:

Event A and event 'not A' are mutually exclusive and cover all
possibilities and so $P(\text{not A}) = 1 - P(A)$

24.3 Lists and tables

You can make lists or tables to work out all the possibilities in experiments
where two or more objects are used to generate outcomes.

Example 5

Three coins are tossed. Calculate the probability of getting

(a) two heads and one tail   (b) at least one head.

Consider the three coins separately. Think of them as the first,
second and third coins even if they are tossed simultaneously.
Draw up a list of possible outcomes:

First coin	H	H	H	H	T	T	T	T
Second coin	H	H	T	T	H	H	T	T
Third coin	H	T	H	T	H	T	H	T

Each of the 3 coins can
land in 2 different ways –
head or tail. So to check
that you have found all the
possible outcomes use the
fact that there should be
2^3 outcomes, which is 8.
Write down the 8 different
possibilities systematically.
This helps to avoid missing
any or repeating some of
them.

(a) Three of the eight possible outcomes have 2 heads and 1 tail.

So $P(\text{2H and 1T}) = \frac{3}{8}$

(b) $P(\text{at least one head}) = 1 - P(\text{no heads})$
$$= 1 - \frac{1}{8} = \frac{7}{8}$$

You can see from the list that 7 of the 8 possible
outcomes have at least one head.

When there are many outcomes, it may be more efficient to draw a table.

___ **Example 6** ___

Two normal dice are rolled and the numbers on the tops are added together. Calculate the probability that the sum will be

(a)   10   (b) a multiple of 5   (c) not 7

> There are 2 dice and 6 ways each can land. The total number of possibilities is $6^2 = 36$.

Draw a sample space diagram for the sums.
There are 36 different ways that the two dice can land.

(a)   10 can occur in 3 ways so $P(10) = \frac{3}{36}$

(b)   5 can occur in 4 ways so $P(5) = \frac{4}{36}$

  $P(\text{multiple of 5}) = P(10) + P(5)$

  $= \frac{4}{36} + \frac{3}{36} = \frac{7}{36}$

(c)   7 can occur in 6 ways so $P(7) = \frac{6}{36}$

  $P(\text{not 7}) = 1 - P(7)$

  $= 1 - \frac{6}{36} = \frac{30}{36}$

Number on the first dice

+	1	2	3	4	5	6
1	2	3	4	5	6	7
2	3	4	5	6	7	8
3	4	5	6	7	8	9
4	5	6	7	8	9	10
5	6	7	8	9	10	11
6	7	8	9	10	11	12

Number on the second dice

___ **Example 7** ___

Arif and Moira are working together. Arif has a spinner which has an equal chance of landing on 1, 2, 3, 4 or 5. Moira has a normal dice. Arif spins his spinner and Moira rolls her dice. Calculate the probability that the two numbers they generate will:

(a)   have an odd sum

(b)   have a product which is a multiple of 3

(c)   be consecutive numbers.

> The total number of possibilities is the product of the possibilities on each piece of equipment. So the total number of possibilities is $5 \times 6 = 30$.

Draw a sample space diagram and
• put 'o' in every square where the sum of the two scores is an odd number
• put 't' in every square where the product of the two scores is a multiple of 3
• put 'c' in every square where the two scores are consecutive numbers.

(a)   There are 15 squares on the sample space diagram labelled 'o'.

  $P(\text{odd sum}) = \frac{15}{30}$

(b)   There are 14 squares labelled 't'.

  $P(\text{product is a multiple of 3}) = \frac{14}{30}$

(c)   There are 9 squares labelled 'c'.

  $P(\text{consecutive numbers}) = \frac{9}{30}$

Dice score

	1	2	3	4	5	6
1		o c	t	o		o t
2	o c		otc		o	t
3	t	otc	t	otc	t	o t
4	o		otc		o c	t
5		o	t	o c		otc

Spinner score

Do not cancel fractions in probability questions unless you are asked to. Adding or subtracting probabilities is easier with uncancelled fractions which have the same denominator.

Exercise 24B

1 A game is played using these two boards. In one 'go' a player spins both arrows. The sum of the numbers where the two arrows point is the score for the 'go'. In the diagram the score is 9.

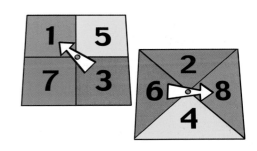

(a) Copy and complete the table showing all the possible scores for one 'go'.

+	1	3	5	7
2	3	5	7	9
4	5	7	9	11
6	7	9	11	13
8	9	11	13	15

(b) Write the probability that a player will score 11 in one 'go'. [E]

2 In a game, two normal dice, one red and the other blue, are thrown simultaneously. A score is found by adding together the numbers on the faces which finish uppermost.
Four players, Adam, Bethan, Christos and Dean, decide that Adam will win if the red dice shows 6, Bethan will win if the score is 5 or 6, Christos will win if the score is 2 or 3 or 4, and that Dean will win otherwise.
Find the probability, expressed as a fraction, that the game will be won by

(a) Adam   (b) Bethan   (c) Christos   (d) Dean. [E]

3 Four coins are tossed simultaneously. List all the possible ways in which the coins can land with heads or tails uppermost.
Calculate the probability that

(a) four heads will occur

(b) an equal number of heads and tails will occur

(c) at least one coin will land tail uppermost.

4 A bag contains one red and one white snooker ball. A second bag contains one brown, one green and one yellow ball. A third bag contains one blue, one pink and one black ball.
List all the possible sets of three balls which can arise when one ball is selected from each bag at random.
Calculate the probability that the three balls selected will include the red and the black ball.

5 Two normal dice are rolled. Calculate the probability that the two numbers on the upper faces will

(a) sum to a prime number

(b) have an odd difference

(c) have an even product

(d) have one which is a factor of the other.

6   A fair octahedral dice with the numbers 1 to 8 on its faces is rolled
    at the same time as a normal dice. The number on the upper face of
    the normal dice is used as the x-coordinate and the number from
    the upper face of the octahedral dice is used as the y-coordinate of a
    point on a grid. Calculate the probability that when the two dice
    are rolled the point generated will lie on the line:

(a) $x = 3$ (b) $y = 5$ (c) $y = 2x$

(d) $x + y = 8$ (e) $y = x + 2$ (f) $y + 2x = 7$

24.4 Relative frequency

So far you have calculated probabilities when all possible outcomes
are equally likely. Sometimes you can only estimate probabilities
based on experimental results.

Hannah and Ruth want to estimate the probability of a drawing pin
landing point up.

They carry out an experiment which consists of throwing the drawing
pin into the air 50 times.

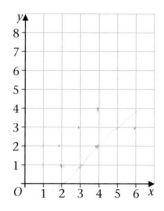

Result	Tally	Frequency
Point up	⅋⅋ ⅋⅋ ⅋⅋ |||	18
Point not up	⅋⅋ ⅋⅋ ⅋⅋ ⅋⅋ ⅋⅋ ⅋⅋ ||	32

They calculate an estimate for the probability using

$$\text{relative frequency} = \frac{\text{number of times event occurs}}{\text{total number of trials}}$$

The pin lands point up 18 times out of 50. The **relative frequency** of
the pin landing point up is $\frac{18}{50}$.

To impove the estimate they should carry out more trials or combine
their results with those of their classmates who have also done the
experiment.

To estimate a probability:
• carry out a large number of trials
• calculate the relative frequency.

> Sometimes you can
> estimate probability by
> experiment and then
> compare with the
> theoretical probability.

The larger the number of trials, the closer the relative frequency is to
the probability.

Activity – Playing cards

Select a card from a normal pack. Record whether it is odd, even, picture card or ace. Replace the card in
the pack.

Repeat 99 times (100 times in total).

From your experimental data estimate the probability of picking each type of card. Compare these
estimates with the theoretical probabilities.

Example 8

Some plastic trapezia like the one on the right are turned into spinners. Ten pairs of students each spin their spinner 20 times. They find that the relative frequencies of the 6 cm side stopping on the table are

$\frac{7}{20}$ $\frac{4}{20}$ $\frac{11}{20}$ $\frac{8}{20}$ $\frac{5}{20}$ $\frac{8}{20}$ $\frac{9}{20}$ $\frac{7}{20}$ $\frac{10}{20}$ $\frac{9}{20}$

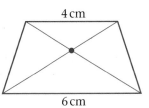
4 cm

6 cm

Calculate the relative frequency when the trials are combined.

Total number of trials = $10 \times 20 = 200$

Number of trials where event occurs = sum of numerators

$$= 78$$

Relative frequency $= \frac{78}{200} = \frac{39}{100} \approx \frac{8}{20}$

Two pairs actually got this result but some other pairs got results that were quite different from this. In real experiments, this does happen.

Exercise 24C

1   Roll an eraser and work out the relative frequency of it finishing in each of its possible positions after 100 trials.

2   Roll an object like a shoe or a small toy and work out the relative frequencies of it ending up in different positions. Estimate the probability of each of the possibilities and combine your results with other students' results to refine your estimates.

24.5 Estimating from experience

In some cases, even carrying out trials is impossible and so probabilities have to be estimated and justified using a reasonable argument.

Example 9

Hamwick rugby team have won each of their last six home games by at least 20 points.
St Debes have lost their last four away games quite heavily.
If St Debes play at Hamwick it is much more likely that Hamwick will win than that St Debes will. An estimate of Hamwick's probability of winning of 80% or more would be reasonable.

Challenge
Think of some more examples like this one.

24.6 Independent events

James tosses a coin and Sarah rolls a dice. James knows that the probability of his coin landing 'heads' is $\frac{1}{2}$.

The probability of Sarah's dice showing a 6 is $\frac{1}{6}$.

What is the probability of getting a head and a 6?

The sample space diagram shows that there are 12 possible outcomes.

$$P(\text{H and 6}) = \tfrac{1}{12}$$

Whether or not the coin lands 'heads' has no effect on whether the dice shows a 6 or any other score. The two events, getting a head and scoring a 6, are **independent**.

$$P(\text{H and 6}) = P(\text{H}) \times P(6)$$

Dice

		1	2	3	4	5	6
Coin	H						✓
	T						

Notice that
$P(\text{H}) = \tfrac{1}{2}$, $P(6) = \tfrac{1}{6}$ $P(\text{H and 6}) = \tfrac{1}{2} \times \tfrac{1}{6} = \tfrac{1}{12}$

Two events are **independent** when one does not affect the outcome of the other.

If events A and B are independent
$$P(\text{A and B}) = P(\text{A}) \times P(\text{B})$$

Worked examination question

Some college students have to take three examination papers, one in English, one in maths and one in science.
$P(\text{passing English}) = 0.7$, $P(\text{passing maths}) = 0.5$ and $P(\text{passing science}) = 0.8$
Assume these events are independent. Calculate the probability that a student will

(a) pass all three papers   (b) fail all three papers.

(a) $P(\text{passing all 3}) = 0.7 \times 0.5 \times 0.8 = 0.28$

(b) $P(\text{failing all 3}) = 0.3 \times 0.5 \times 0.2 = 0.03$

A, B and C are independent events, so
$P(\text{A and B and C})$
$= P(\text{A}) \times P(\text{B}) \times P(\text{C})$

Exercise 24D

1 When Nina and Zoe go to the shop the probability of Nina choosing a chocolate bar is $\tfrac{1}{3}$ and of choosing a toffee bar is $\tfrac{1}{5}$. The probability of Zoe choosing a chocolate bar is $\tfrac{1}{4}$ and of choosing a toffee bar is $\tfrac{1}{2}$. The girls choose independently of each other. Calculate the probability of
   (a) both choosing chocolate bars
   (b) both choosing toffee bars
   (c) one choosing a chocolate and the other choosing a toffee bar.

2 There are three tame mice, Roger, Susan and Timmy. They can choose to eat at five troughs (A, B, C, D and E) independently and at random.
   (a) What is the probability that Roger will eat from trough A?
   (b) Find the probability that Susan and Timmy will both eat
       (i) at trough A   (ii) at the same trough.   [E]

24.7 Probability trees

Tree diagrams can be used to illustrate possible outcomes for two or more independent events.

___ **Example 10** ___

Newton High School students can choose to wear or not wear a sweatshirt as part of the uniform. The probability that Ali will choose to wear a sweatshirt is $\frac{1}{3}$. The probability that Bethan will wear one is $\frac{2}{5}$ and the probability that Chris will wear one is $\frac{1}{2}$. Calculate the probability that:

> Each student has 2 choices: to wear or not to wear a sweatshirt. So there are $2^3 = 2 \times 2 \times 2$ or 8 paths.

(a) all three wear a sweatshirt

(b) exactly two of them wear a sweatshirt

(c) at least two of them wear a sweatshirt.

(a) The path to A includes the probabilities of each student wearing a sweatshirt.

$$P(\text{all wear sweatshirt}) = \frac{1}{3} \times \frac{2}{5} \times \frac{1}{2}$$
$$= \frac{2}{30}$$

> Their choices are independent so multiply the probabilities along the branches.

(b) Paths B, C and E all involve exactly two wearing sweatshirts.

$$P(\text{exactly 2 wear sweatshirts})$$
$$= (\tfrac{1}{3} \times \tfrac{2}{5} \times \tfrac{1}{2}) + (\tfrac{1}{3} \times \tfrac{3}{5} \times \tfrac{1}{2}) + (\tfrac{2}{3} \times \tfrac{2}{5} \times \tfrac{1}{2})$$
$$= \tfrac{2}{30} + \tfrac{3}{30} + \tfrac{4}{30}$$
$$= \tfrac{9}{30}$$

> The events A to H are mutually exclusive and add to 1.
> So P(B or C or E) = P(B) + P(C) + P(E)

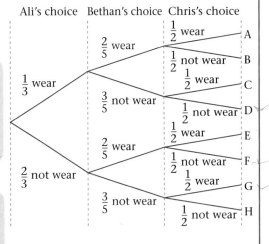

Ali's choice   Bethan's choice   Chris's choice

(c) P(at least 2) = P(all 3 or exactly 2)
$$= \frac{2}{30} + \frac{9}{30}$$
$$= \frac{11}{30}$$

> It does not matter which choice is put first in the tree because the probabilities are multiplied by each other and multiplication is a commutative operation.

Exercise 24E

1 Rick, Sher and Tessa always wear a coloured T-shirt when they go to the Youth Club. Amongst other colours they each have one, and only one, red T-shirt. They each pick a T-shirt at random from the pile in their drawer. Including the red one, there are three T-shirts in Rick's drawer, four in Sher's and five in Tessa's. Calculate the probability that:

(a) they will all go to the Youth Club in a red T-shirt

(b) none of them will go in a red T-shirt

(c) exactly two of them will go in a red T-shirt

(d) at least two of them will go in a red T-shirt.

2 A game at a school fair involves choosing a plastic cup from each of three boxes. The plastic cups each cover a coloured cube which the player cannot see. The first box contains 5 cups, covering 3 red and 2 blue cubes. The second box contains 4 cups, covering 3 red and 1 blue cube. The third box contains 3 cups, covering 2 red and 1 blue cube.

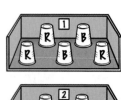

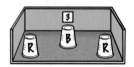

Vicky pays 20p for a go. If she selects 3 reds she loses her money. If she gets exactly 1 blue she gets 10p back. If she uncovers exactly 2 blue cubes she gets 30p back, and if all three of her choices cover blue cubes she receives 50p back.

Calculate the probability that Vicky will

(a) receive 50p

(b) receive 30p

(c) have less money at the end of the game than she had before paying her 20p.

Mixed exercise 24

1 A 50p, a 20p, a 10p and a 2p coin are tossed at the same time. List all the possible outcomes. Find the probability of obtaining

(a) exactly two heads (b) more than two heads.

2 One letter is selected at random from the word 'PARALLELOGRAM'. Write down the probability of selecting

(a) a P (b) an L (c) an A or an R.

3 The spinner shown has five equal sectors. How many 3s would you expect in 600 spins?

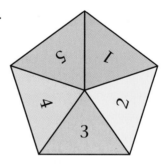

4 One ball is selected from a bag containing eight balls, of which x are black.

(a) Write down, in term of x, the probability of selecting a black ball.

(b) Write down, in terms of x, the probability of not selecting a black ball.

(c) When a further four black balls are added to the bag, the probability of selecting a black ball is doubled. Find x.

5 In an opinion poll some people were asked to state their favourite football team. The results are summarised in the pie chart.

If one person is chosen at random from the sample, find an estimate of the probability that their favourite football team is either Chelsea or Liverpool.

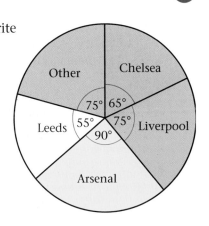

6 These spinners are spun together.

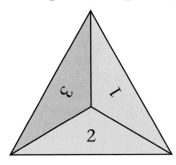 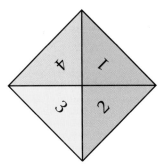

The numbers shown on the spinners are added. Display all the possible outcomes in a suitable way.
Find the probability of obtaining

(a) a total of 5

(b) the same number on both spinners

(c) an even number on one spinner and an odd number on the other spinner.

7 A tetrahedral dice (4 sides, numbered 1 to 4) and a normal dice (6 sides) are thrown at the same time. A 'win' occurs when the number on the tetrahedral dice is greater than or equal to the number on the normal dice. What is the probability of a 'win'?

8 Box A contains three blue and four red balls.
Box B contains two blue and seven red balls.
One ball is selected at random from box A and one ball is selected at random from box B. Find the probability that

(a) both balls are red

(b) both balls are the same colour.

9 A fruit machine has three independent reels and pays out a Jackpot of £50 when three raspberries are obtained. Each reel has 12 pictures. The first reel has four raspberries, the second has three raspberries and the third has two raspberries.
Find the probability of winning the Jackpot.

10 A biased dice is thrown 100 times. The results are shown in the table.

Number on dice	1	2	3	4	5	6
Frequency	14	16	15	12	11	32

Write down an estimate of the probability of throwing a 6 with this dice.

11 Jim is practising his golf. He drives a number of balls at a green. If he hits the green he records 'H', and if he misses the green he records 'M'. Here are his results for the first 17 drives:

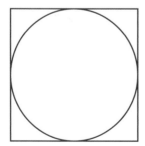

H M M H M H H M M M H H H H H H H

(a) Assuming that his drives are independent, write down an estimate of the probability that the next time he drives a ball he will hit the green.

(b) Is it reasonable to assume that his drives are independent? Give a reason for your answer.

12 The table shows how many boys and how many girls in a school are left handed and how many are right handed.

	Girls	Boys
Left handed	32	28
Right handed	143	137

(a) A student is selected at random from the school. Find the probability that the student is
  (i) a boy
  (ii) a left-handed girl

(b) A boy is selected at random from the school. Find the probability that he is right handed.

13 There are two parts to a driving test, the theory test and the practical test. You must pass the theory test before you take the practical test.
Sethina plans to take her driving test.
The probability that she will pass her theory test is 0.8
The probability that she will pass her practical test is 0.75

(a) Draw a tree diagram to represent this information.

(b) Calculate the probability that Sethina will pass her driving test.

14 The diagram shows a circle drawn inside a square.
A point is chosen at random from inside the square.

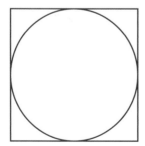

What is the probability that the point lies inside the circle?

Summary of key points

1 The **probability** of an event is expressed as a number from 0 to 1 inclusive.
   - If an event is **impossible** its probability is 0.
   - If an event is **certain** its probability is 1.

2 When there are n equally likely possible outcomes, the probability of each outcome is $\dfrac{1}{n}$

3 $\text{P(event)} = \dfrac{\text{the number of ways the event can occur}}{\text{the total number of possible outcomes}}$

4 If two events cannot occur at the same time they are **mutually exclusive**.

5 When events A and B are mutually exclusive
   $$\text{P(A or B)} = \text{P(A)} + \text{P(B)}$$

6 Event A and event 'not A' are mutually exclusive and cover all possibilities and so
   $$\text{P(not A)} = 1 - \text{P(A)}$$

7 $\text{Relative frequency} = \dfrac{\text{number of times event occurs}}{\text{total number of trials}}$

8 Two events are **independent** if one does not affect the outcome of the other.

9 If events A and B are independent
   $$\text{P(A and B)} = \text{P(A)} \times \text{P(B)}$$

25 Transformations of graphs

To draw a graph, you can draw up a table for values of x, plot the points and join them using a smooth curve. The range of values for x is chosen so as to include the main features of the graph.

You will now consider examples where this is not the case. For example, the graph on the right shows five points plotted for the equation $y = x^2 - 9x + 10$ for $-2 \leqslant x \leqslant 2$.

Without plotting any more points, consider what the graph looks like for values of x up to 10.

This chapter shows you how to use your knowledge of the graphs of some basic functions, and to apply suitable transformations (translations, reflections and stretches) to sketch the main features of more complicated graphs without the need to find and plot lots of points.

In previous chapters you have been using graphs with equations:

$$y = x^2, \quad y = x^3, \quad y = \frac{1}{x}$$

You will be using the graphs of these three equations along with the straight line $y = x$ in many examples during this chapter. For reference these four equations are called 'the basic functions'.

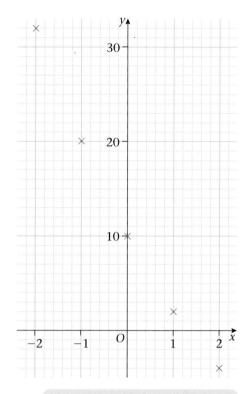

You will need tracing paper for many of the exercises in this chapter.

25.1 Function notation

A **function** is a rule which shows how one set of numbers relates to another set.

You can think of a function, f, as a machine which changes an input x into an output.

input x — f — output f(x)

When the function f is 'square' then, for an input x, the output is x^2 so $f(x) = x^2$.

f(x), which denotes the output, is read as 'f of x'.

In this case, for an input of 3 the output is 9. You write $f(3) = 3^2 = 9$.

input x — square — output x^2

Similarly $f(-4) = 16$, and $f(k + 1) = (k + 1)^2 = k^2 + 2k + 1$.

A function f can also be equivalent to a sequence of operations (rules).

For example, $f(x) = 2x^2 + 1$ is the output from this flow diagram:

input x — square — × 2 — + 1 — $2x^2 + 1$

Example 1

When $f(x) = 2x^2 + 1$, find

(a) $f(3)$ (b) $f(-4)$ (c) $f(0)$ (d) $f(-k)$

(a)   Replacing x by 3: $f(3) = 2(3)^2 + 1 = 19$

(b)   Replacing x by -4: $f(-4) = 2(-4)^2 + 1 = 33$

(c)   Replacing x by 0: $f(0) = 2(0)^2 + 1 = 1$

(d)   Replacing x by $-k$: $f(-k) = 2(-k)^2 + 1 = 2k^2 + 1$

Example 2

When $f(x) = x^2$, find

(a) $f(x + 1)$ (b) $f(2x)$ (c) $1 - f(-x)$ (d) $f(-\frac{1}{2}x) + 4$

(a)   Replacing x by $x + 1$: $f(x + 1) = (x + 1)^2 = x^2 + 2x + 1$

(b)   Replacing x by $2x$: $f(2x) = (2x)^2 = 4x^2$

(c)   Replacing x by $-x$: $1 - f(-x) = 1 - (-x)^2 = 1 - x^2$

(d)   Replacing x by $(-\frac{1}{2}x)$: $f(-\frac{1}{2}x) + 4 = (-\frac{1}{2}x)^2 + 4 = \dfrac{x^2}{4} + 4$

Example 3

Given that $f(x) = x^2 + 2$, find the values of x when $f(x) = 11$.

When $f(x) = 11$:

$\quad\quad 11 = x^2 + 2$

$\quad\quad\quad 9 = x^2$ ——————— Subtract 2 from both sides

$\quad\quad \pm 3 = x$ ——————— Take the square root of
$\quad\quad\quad\quad\quad\quad\quad\quad\quad\quad\quad\quad$ both sides

The plotted points for the graph of $y = x^2 - 9x + 10$ shown on page 460
came from the table:

x	-2	-1	0	1	2
y	32	20	10	2	-4

$f(-2) = (-2)^2 - 9(-2) + 10$
$\quad\quad = 4 + 18 + 10$
$\quad\quad = 32$

Using function notation, this is the same as the graph $y = f(x)$,
where $f(x) = x^2 - 9x + 10$, $f(-2) = 32$, $f(-1) = 20$ and so on.

Example 4

Here is part of the graph of $y = f(x)$, where $f(x) = x^2 + 1$

From the graph, find

(a)   the value of   (i) f(0)   (ii) f(2)

(b)   the values of x when $f(x) = 2$.

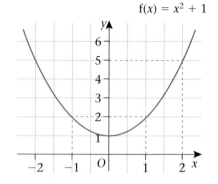

$f(x) = x^2 + 1$

(a)   (i)   f(0) is the output (y) when $x = 0$.
            From the graph, f(0) = 1.
       (ii)   From the graph, f(2) = 5.

(b) $f(x) = 2$; we need x when $y = 2$.
       From the graph, when $y = 2$, $x = -1$ or $x = 1$.
       So when $f(x) = 2$, $x = \pm 1$.

> If $f(x)$ is a function then $y = f(x)$ is the equation of the graph of the function.

Exercise 25A

1   When $f(x) = x^2 - 2$, find

   (a)   f(2)   (b)   f(4)   (c)   f(0)

   (d)   f(−1)   (e)   f(−3)   (f) $f(\tfrac{1}{4})$

2   When $f(x) = x^2$, find

   (a)   f(−x)   (b)   f(3x)   (c)   f(x + 2)

   (d)   f(x + 1) + 3   (e) $f\left(-\dfrac{x}{2} + 1\right) - 4$ (f)   5 − f(2x)

   (g)   f(kx + a) + b

3   Paul accidentally spilled some ink on his text book.
     His teacher says that he should have been able to
     answer the questions without using the equation
     for f(x). Explain why Paul did not need the equation
     for f(x). Then answer the question.

     Part of the graph of $y = f(x)$,

     where $f(x) = x^2 -$ ◆ is drawn.

     Find

     (a)   the value of
            (i) f(3)   (ii) f(1)   (iii) f(0)   (iv) f(−1)

     (b)   the values of x when $f(x) = 0$.

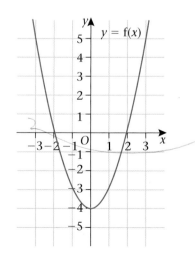

25.2 Applying vertical translations to graphs

Parabolas

The graphs of the parabolas $y = x^2$, $y = x^2 + 8$ and $y = x^2 - 15$ are shown below. Another parabola is also shown.

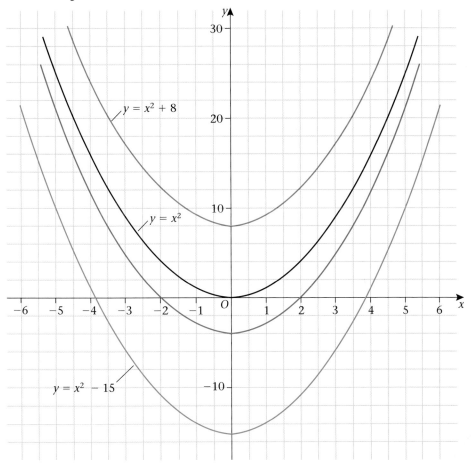

> These are all graphs of functions.

Step 1   Trace the graph of $y = x^2$.

Step 2   Slide your tracing paper vertically upwards along the y-axis so that your trace of $y = x^2$ coincides with the parabola $y = x^2 + 8$.
The tracing paper has moved *up* 8 units.
The graph of $y = x^2 + 8$ is the graph of $y = x^2$ **translated 8 units vertically** in the *positive y-direction*.

Step 3   Now put your tracing paper back on the parabola $y = x^2$.
This time slide your tracing paper vertically downwards so that it coincides with the parabola $y = x^2 - 15$.
The tracing paper has moved *down* 15 units.
The graph of $y = x^2 - 15$ is the graph of $y = x^2$ **translated 15 units vertically** in the *negative y-direction*.

The graph of $y = x^2 + a$ is the graph of $y = x^2$ translated a units vertically (in the *positive y-direction* if $a > 0$, in the *negative y-direction* if $a < 0$).

Exercise 25B

1 Using the graphs on the previous page, place your tracing paper on the parabola $y = x^2$ and slide it so that it coincides with the unlabelled parabola. What is the equation of this parabola?

2 Write down the equations of three different parabolas that always lie between the parabolas $y = x^2$ and $y = x^2 + 8$.

3 The parabola $y = x^2 + k$ always lies between the parabolas $y = x^2$ and $y = x^2 - 15$. Write down an inequality for k.

The Millennium Bridge in Gateshead is a parabola.

Other basic functions

The graphs below show how curves related to the basic functions $y = x$, $y = x^3$ and $y = \dfrac{1}{x}$ can be drawn by applying a vertical translation.

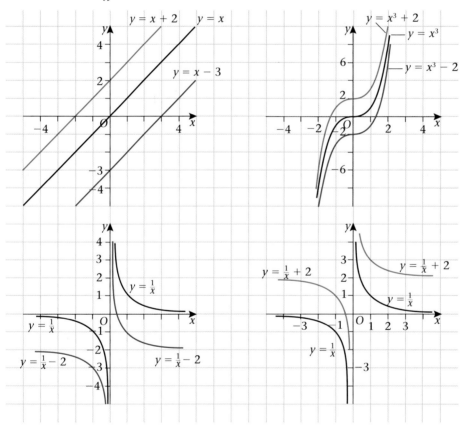

Example 5

Copy and complete these sentences:

(a) The graph of $y = x + 2$ is the graph of $y = x$ translated _____.

(b) The graph of $y = x^3 - 2$ is the graph of $y = x^3$ translated _____.

(c) The graph of $y = \dfrac{1}{x} + 2$ is the graph of $y = \dfrac{1}{x}$ _____.

(d) The graph of $y = x^3 + 2$ is the graph of $y = x^3$ _____.

(e) The graph of $y = \dfrac{1}{x} - 2$ is the graph of $y = \dfrac{1}{x}$ _____.

(a) The graph of $y = x + 2$ is the graph of $y = x$ translated 2 units vertically in the positive y-direction.

(b) The graph of $y = x^3 - 2$ is the graph of $y = x^3$ translated 2 units vertically in the negative y-direction.

(c) The graph of $y = \dfrac{1}{x} + 2$ is the graph of $y = \dfrac{1}{x}$ translated 2 units vertically in the positive y-direction.

(d) The graph of $y = x^3 + 2$ is the graph of $y = x^3$ translated 2 units vertically in the positive y-direction.

(e) The graph of $y = \dfrac{1}{x} - 2$ is the graph of $y = \dfrac{1}{x}$ translated 2 units vertically in the negative y-direction.

The graph of $y = x + a$ is the graph of $y = x$ translated a units vertically (in the *positive y*-direction if $a > 0$, in the *negative y*-direction if $a < 0$).

The graph of $y = x^3 + a$ is the graph of $y = x^3$ translated a units vertically (in the *positive y*-direction if $a > 0$, in the *negative y*-direction if $a < 0$).

The graph of $y = \dfrac{1}{x} + a$ is the graph of $y = \dfrac{1}{x}$ translated a units vertically (in the *positive y*-direction if $a > 0$, in the *negative y*-direction if $a < 0$).

Using function notation we can generalise all the results obtained so far:

For any function f, the graph of $y = f(x) + a$ is the graph of $y = f(x)$ translated a units vertically (in the *positive y*-direction if $a > 0$, in the *negative y*-direction if $a < 0$).

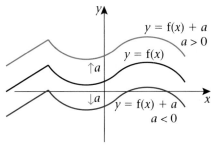

Example 6

Here is the graph of $y = f(x)$, where $f(x) = x^3 - x$, (in black).

(a) Sketch, on the same axes, the graph of $y = x^3 - x + 10$.

(b) Describe the transformation that gives $y = x^3 - x + 10$ from $y = x^3 - x$.

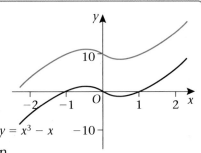

(a) $x^3 - x + 10 = f(x) + 10$
The graph of $y = x^3 - x + 10$ is shown in red.

(b) A vertical translation of 10 units in the positive y-direction.

Exercise 25C

1　In diagrams **(a)** to **(e)** the graphs of 'basic functions' are in black. In each part, write down the translations which have to be applied to these basic functions in order to obtain the graphs labelled A, B, C.

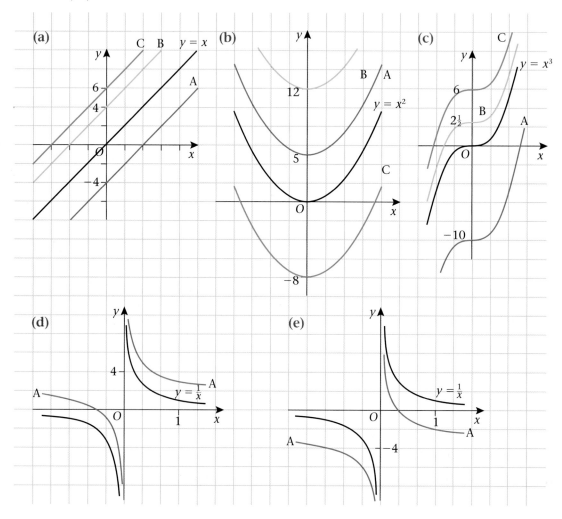

2　**(a)**　Trace the axes and sketch of the graph of $y = f(x)$ below.

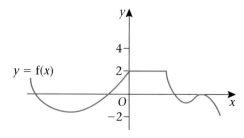

$y = f(x)$

(b)　On the same axes, sketch the graph of $y = f(x) + 2$.

(c)　Describe the transformation which gives $y = f(x) + 2$ from $y = f(x)$.

25.3 Applying horizontal translations to graphs

Parabolas

The graphs of the parabolas $y = x^2$, $y = (x + 2)^2$ and $y = (x - 4)^2$ are shown below. One other parabola is also shown.

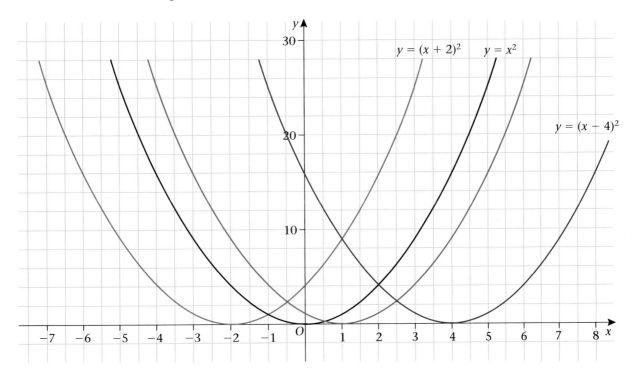

Use your tracing of the graph of $y = x^2$ (or trace a new one).

Put your tracing paper on the parabola $y = x^2$.

Slide your tracing paper horizontally along the x-axis so that your trace of $y = x^2$ coincides with the parabola $y = (x + 2)^2$.

You should have slid the tracing paper along 2 units in the *negative* x-direction.

The graph of $y = (x + 2)^2$ is the graph of $y = x^2$ **translated 2 units horizontally** in the *negative* x-direction.

Similarly, the graph of $y = (x - 4)^2$ is the graph of $y = x^2$ **translated 4 units horizontally** in the *positive* x-direction.

> The graph of $y = (x + a)^2$ is the graph of $y = x^2$ translated a units horizontally (in the *negative* x-direction if $a > 0$, in the *positive* x-direction if $a < 0$).

> The **vertex** of the parabola $y = (x + a)^2$ is at the point $(-a, 0)$.

Example 7

Sketch the graph of $y = x^2 + 6x + 9$. State the coordinates of the points where the graph crosses the axes.

Factorising $x^2 + 6x + 9$ gives $(x + 3)^2$
so $y = (x + 3)^2$.

The graph of $y = (x + 3)^2$ is the graph of $y = x^2$ translated 3 units horizontally in the *negative x*-direction.

The graph of $y = (x + 3)^2$ is a parabola with vertex $(-3, 0)$.
When $x = 0$, $y = (x + 3)^2$.
The parabola crosses the y-axis at the point $(0, 9)$ and touches the x-axis at the point $(-3, 0)$.

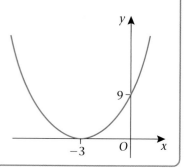

Exercise 25D

1   Using the graphs on the previous page, place your tracing paper on the parabola $y = x^2$ and slide it so that it coincides with the unlabelled parabola. What is the equation of this parabola?

2   Sketch the graph of (a) $y = (x + 1)^2$ (b) $y = (x + 5)^2$.

3   Sketch the graph of $y = x^2 - 4x + 4$.

Mark the coordinates where the graph crosses the axes.

Lines, cubics and reciprocals

Look back at the sketches of the lines on page 464.

The graph of the straight line $y = x - 3$ is the graph of $y = x$ translated *either* 3 units vertically down *or* 3 units horizontally in the positive x-direction.

The cubic and reciprocal graphs below have undergone transformations. You should examine these sketches carefully to understand the horizontal translations that have been applied.

Use tracing paper if necessary to help you.

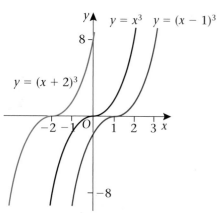

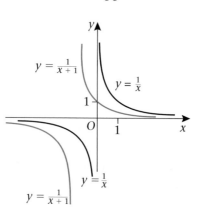

Example 8

Copy and complete these sentences:

(a) The graph of $y = \dfrac{1}{x+1}$ is the graph of $y = \dfrac{1}{x}$ _____.

(b) The graph of $y = (x+2)^3$ is the graph of $y = x^3$ _____.

(a) The graph of $y = \dfrac{1}{x+1}$ is the graph of $y = \dfrac{1}{x}$ translated 1 unit horizontally in the negative x-direction.

(b) The graph of $y = (x+2)^3$ is the graph of $y = x^3$ translated 2 units horizontally in the negative x-direction.

In general, for any function f, the graph of $y = f(x + a)$ is the graph of $y = f(x)$ translated a units horizontally (in the *negative x*-direction if $a > 0$, in the *positive x*-direction if $a < 0$).

Example 9

The graph of $y = f(x)$, where $f(x) = x(x + 2)$, is sketched below (in black).

(a) On the same axes sketch the graph of $y = x^2 + 4x + 3$.

(b) Describe the transformation which gives $y = x^2 + 4x + 3$ from $y = x(x + 2)$.

(a) Factorising $x^2 + 4x + 3$ gives $(x + 1)(x + 3)$ so we need to sketch the graph of $y = (x + 1)(x + 3)$.

Comparing $x(x + 2)$ with $(x + 1)(x + 3)$, we see that the x in the first expression has been replaced by $(x + 1)$ to get the second.

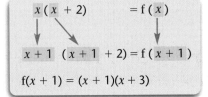

But $\quad f(x) = x(x + 2)$

so $\quad f(x + 1) = (x + 1)(x + 1 + 2) = (x + 1)(x + 3)$.

So sketching the graph of $y = (x + 1)(x + 3)$ is the same as sketching $y = f(x + 1)$.

The graph of $y = f(x + 1)$ is the graph of $y = f(x)$ translated 1 unit in the negative x-direction.

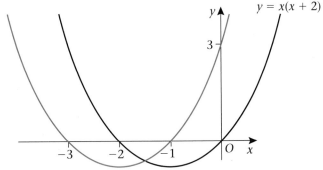

The graph of $y = x^2 + 4x + 3$ is sketched above (shown in red).

(b) A horizontal translation of 1 unit in the negative x-direction.

Exercise 25E

1   The graph of $y = x^3$ is transformed to $y = (x - 1)^3$.
     Write down the translation which has taken place.

2   The graph of $y = x(x + 2)$ is transformed to $y = (x - 1)(x + 1)$.
     Write down the transformation which has taken place.

3   The graph of $y = \dfrac{1}{x + 3} - 3$ is transformed to $y = \dfrac{1}{x + 3} + 4$.
     Write down the transformation which has taken place.

4   Graph A becomes $y = x^3 - 1$ following a vertical translation by
     3 units downwards. Write the equation of Graph A.

5   The graph of $y = x(x - 1)(x + 5)$ is translated by 4 units horizontally
     in the positive x-direction. Write down its new equation.

25.4 Applying double translations to graphs

This section shows you how to apply two translations and to sketch the
graphs of various functions.

Example 10

Sketch the graph of $y = (x - 2)^3 + 1$.

This graph can be built up from the basic curve $y = x^3$ like this:

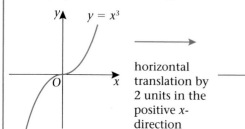
horizontal
translation by
2 units in the
positive x-
direction

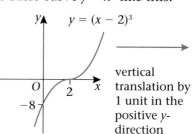
vertical
translation by
1 unit in the
positive y-
direction

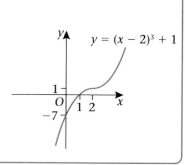

For **double translations**, the order does not matter. You should convince
yourself by sketching $y = x^3 + 1$ then applying the horizontal translation to
get the final result.

However, when combining **a translation with a different type of
transformation**, the order of the transformations becomes critical.

To sketch a quadratic function whose coefficient of x^2 is 1

Example 11

Sketch the graph of $y = x^2 + 4x - 1$.
State the coordinates of the vertex and the points at which the curve crosses the x-axis.

$y = x^2 + 4x - 1$

$= (x + 2)^2 - 4 - 1$ —————— Complete the square

$= (x + 2)^2 - 5$

Check:
$(x + 2)^2 - 5$
$= x^2 + 4x + 4 - 5$
$= x^2 + 4x - 1$

Start with the basic curve $y = x^2$:

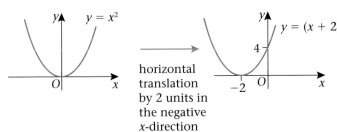

horizontal translation by 2 units in the negative x-direction

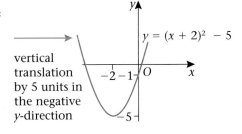

vertical translation by 5 units in the negative y-direction

The coordinates of the vertex are $(-2, -5)$.

To find where the graph crosses the x-axis solve the equation:

$$(x + 2)^2 - 5 = 0$$
$$(x + 2)^2 = 5$$
$$(x + 2) = \pm\sqrt{5}$$
$$x = -2 - \sqrt{5} \text{ or } -2 + \sqrt{5}$$

To sketch the graph of $y = x^2 + bx + c$, complete the square and apply a double translation to the parabola $y = x^2$.

In general, for any function f, the graph of $y = f(x + a) + b$ is the graph of $y = f(x)$ translated a units horizontally (in the *negative* x-direction if $a > 0$, in the *positive* x-direction if $a < 0$) followed by a translation of b units vertically (*upwards* if $b > 0$, *downwards* if $b < 0$).

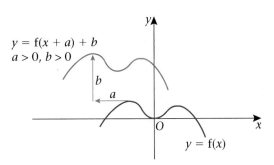

Exercise 25F

1   Sketch the graphs of the following quadratic functions. In each case give the coordinates of the vertex.

(a) $y = x^2 + 2x + 4$ (b) $y = x^2 - 6x + 4$

(c) $y = x^2 + 3x - 1$ (d) $y = x^2 - 4x - \frac{1}{2}$

2   Describe the transformations which transform the graph of $y = (x + 1)^2$ to the graph of:

(a) $y = (x + 7)^2$ (b) $y = (x - 3)^2 - 4$

3   Sketch the graphs of these quadratic functions:

(a) $y = x^2 - 6x + 2$ (b) $y = x^2 + 2x - 4$

Describe the transformations which give the graph in (b) from the graph in (a).

4   What is the equation of the graph obtained by applying these transformations to the graph of $y = x^2 - 16x$?

> a horizontal translation of 1 unit in the positive x-direction followed by a vertical translation of 3 units down

5   Sketch the graph of $y = 3 + \dfrac{1}{x + 2}$

Give the coordinates of the points where the graph and the asymptotes cross the axes.

25.5 Applying reflections to graphs

In the x-axis

The sketch graphs show $y = f(x)$ (in black) and $y = -f(x)$ (in red) for each of the four basic functions.

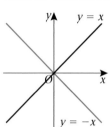 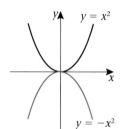 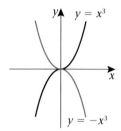 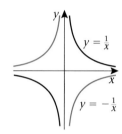

The graph of $y = -f(x)$ is obtained by changing the sign of y in $y = f(x)$.

Each red graph is obtained by reflecting each black graph in the x-axis.

> For any function f, the graph of $y = -f(x)$ is obtained by reflecting $y = f(x)$ in the x-axis.

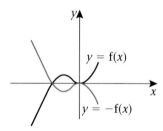

You can use tracing paper to help you reflect graphs in the x-axis:

- trace the function and the x-axis and mark the origin
- turn over the tracing paper and match up the origin and the x-axis. The trace now shows the reflection of the original function in the x-axis.

Example 12

A sketch of the parabola $y = f(x)$, where $f(x) = -x^2 + 4x$, is shown in black. Its vertex has coordinates (2, 4).

(a)   Find the equation of $y = -f(x)$.   (b)   Sketch the graph of $y = -f(x)$.

(c)   Describe the transformation which is applied to $y = f(x)$ to obtain $y = -f(x)$.

(d)   Write the coordinates of the vertex of $y = -f(x)$.

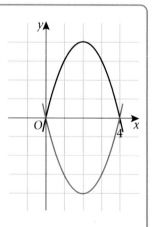

(a) $f(x) = -x^2 + 4x$
$y = -f(x) = -(-x^2 + 4x)$
So the equation of $y = -f(x)$ is $y = x^2 - 4x$.

(b)   The graph of $y = -f(x)$ is shown in red.

(c)   Reflection in the x-axis.   (d)   Vertex is (2, −4).

In the y-axis

The sketch graphs show $y = f(x)$ (in black) and $y = f(-x)$ (in red) for $f(x) = x + 1$ and $f(x) = x^3 + 8$.

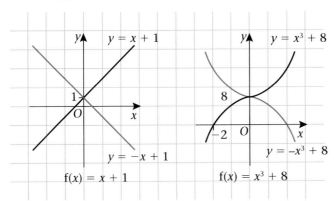

The graph of $y = f(-x)$ is obtained by changing the sign of x in $y = f(x)$.

Each red graph is obtained by reflecting each black graph in the y-axis.

For any function f, the graph of $y = f(-x)$ is obtained by reflecting $y = f(x)$ in the y-axis.

Example 13

This is a sketch of the parabola $y = f(x)$, where $f(x) = -x^2 + 6x$.

Its vertex has coordinates $(3, 9)$.

(a) Find the equation of $y = f(-x)$.

(b) Sketch the graph of $y = f(-x)$.

(c) Describe the transformation which is applied to $y = f(x)$ to obtain $y = f(-x)$.

(d) Write the coordinates of the vertex of $y = f(-x)$.

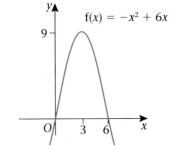

(a) $f(x) = -x^2 + 6x$
$y = f(-x) = -[(-x)^2] + 6(-x) = -[x^2] - 6x$
So the equation of $y = f(-x)$ is $y = -x^2 - 6x$.

(b)

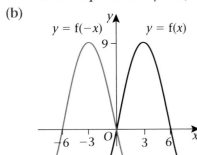

(c) Reflection in the y-axis.   (d) Vertex is $(-3, 9)$.

Exercise 25G

1 This is a sketch of $y = f(x)$, where $f(x) = x^2 - 2x$. The vertex of this parabola is $(1, -1)$.

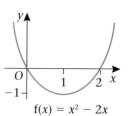
$$f(x) = x^2 - 2x$$

(a) Find the equation of $f(-x)$.

(b) Sketch the graph of $y = f(-x)$.

(c) Sketch the graph of $y = -x^2 + 2x$.

(d) Write the coordinates of the vertex of $y = -x^2 + 2x$.

2 A reflection is applied to the graph of $y = f(x)$, where $f(x) = x^3 + 2x^2$.

Find the equation of the new graph if

(a) the reflection is in the x-axis

(b) the reflection is in the y-axis.

3 (a) Copy this sketch of the graph of $y = f(x)$.

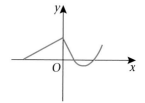

(b) On the same axes, sketch the graph of $y = -f(x)$.

(c) Describe fully the single transformation which when applied to the graph of $y = f(x)$ gives the graph of $y = -f(x)$.

4 (a) Copy this sketch of the graph of $y = f(x)$.

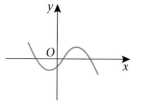

(b) On the same axes, sketch the graph of $y = f(-x)$.

(c) Describe fully the single transformation which when applied to the graph of $y = f(x)$ gives the graph of $y = f(-x)$.

25.6 Applying combined transformations

You can use a combination of transformations to sketch a function.

___ **Example 14** _____

Explain how to sketch the graph of $y = -x^2 + 6x + 3$ by applying transformations to the graph of $y = x^2$.
Illustrate your answer by sketches and state the coordinates of the vertex of $y = -x^2 + 6x + 3$.

> **Check:**
> $- (x - 3)^2 + 12$
> $= -(x^2 - 6x + 9) + 12$
> $= -x^2 + 6x - 9 + 12$
> $= -x^2 + 6x + 3$

$$-x^2 + 6x + 3 = -(x^2 - 6x) + 3$$

$$= -[(x - 3)^2 - 9] + 3 \quad\text{——— Complete the square}$$

$$= -(x - 3)^2 + 12$$

Let $f(x) = x^2$ so $\quad y = x^2 \qquad\qquad\qquad$ becomes $y = f(x)$

$\qquad\qquad 1 \quad y = (x - 3)^2 \qquad\qquad$ becomes $y = f(x - 3)$

$\qquad\qquad 2 \quad y = -(x - 3)^2 \qquad\qquad$ becomes $y = -f(x - 3)$

$\qquad\qquad 3 \quad y = -(x - 3)^2 + 12 \quad$ becomes $y = -f(x - 3) + 12$.

> For a combination of transformations, the order of the transformations is critical.

To transform the graph of $y = x^2$ to the graph of $y = -x^2 + 6x + 3$, apply the following transformations in this order:

1 horizontal translation of 3 units in the positive x-direction

2 reflection in the x-axis

3 vertical translation of 12 units in the positive y-direction.

> The transformations are in the order given on the previous page.

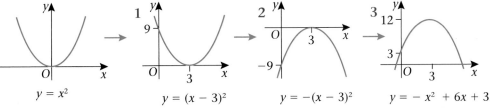

$y = x^2$ → $y = (x - 3)^2$ → $y = -(x - 3)^2$ → $y = -x^2 + 6x + 3$

The coordinates of the vertex of $y = -x^2 + 6x + 3$ are (3, 12).

Exercise 25H

1 List the transformations, in the correct order, which when applied to $y = x^2$ give the graphs of the following functions.

(a) $y = -x^2 + 6$ (b) $y = (x - 4)^2 + 2$ (c) $y = -x^2 + 4x - 3$

Sketch each graph.

25.7 Applying stretches to graphs

Stretch in the y-direction

The graphs show

$y = f(x)$ (in black) $y = 2f(x)$ (in red) $y = \frac{1}{2}f(x)$ (in green)

for the two functions $f(x) = x$ and $f(x) = x^2$.

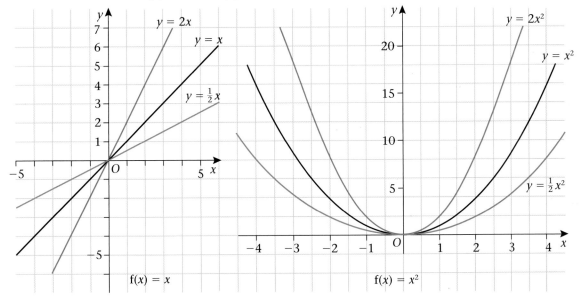

To obtain the graph of $y = 2f(x)$ from the graph of $y = f(x)$, multiply the y-coordinate of each point on the graph of $y = f(x)$ by 2 but leave the x-coordinate unaltered. For example, the point $(1, 1)$ becomes $(1, 2)$ and so on.

> A stretch is similar to an enlargement but in one direction only.

This has the effect of stretching the curve out in the direction of the y-axis by a scale factor 2. Points of the graph on the x-axis remain fixed.

Similarly, by comparing the curves $y = x^2$ and $y = \frac{1}{2}x^2$ we can see that the stretch required to get $y = \frac{1}{2}f(x)$ from $y = f(x)$ is again in the direction of the y-axis but with scale factor $\frac{1}{2}$.

> For any function f, the graph of $y = af(x)$, where a is a positive constant, is obtained from $y = f(x)$ by applying a stretch of scale factor a parallel to the y-axis.

Example 15

The graph of $y = f(x)$, where $f(x) = -x^2 + 6x - 5$, is shown below (in black) and labelled.

(a) Sketch the graphs of
    (i) $y = -2x^2 + 12x - 10$ (ii) $y = -\frac{1}{2}x^2 + 3x - \frac{5}{2}$ on the same axes.

(b) Describe the transformation which is applied to $y = -\frac{1}{2}x^2 + 3x - \frac{5}{2}$ to get $y = -2x^2 + 12x - 10$.

(a)   (i) $y = -2x^2 + 12x - 10 = 2(-x^2 + 6x - 5) = 2f(x)$

        Apply a stretch of scale factor 2 parallel to the y-axis.
        The graph of $y = -2x^2 + 12x - 10$ is shown in red.

> $(0, -5) \rightarrow (0, -10)$
> $(3, 4) \rightarrow (3, 8)$
> $(5, 0) \rightarrow (5, 0)$

  (ii) $y = -\frac{1}{2}x^2 + 3x - \frac{5}{2} = \frac{1}{2}(-x^2 + 6x - 5) = \frac{1}{2}f(x)$

        Apply a stretch of scale factor $\frac{1}{2}$ parallel to the y-axis.
        The graph of $y = -\frac{1}{2}x^2 + 3x - \frac{5}{2}$ is shown in green.

> $(0, -5) \rightarrow (0, -\frac{5}{2})$
> $(3, 4) \rightarrow (3, 2)$
> $(5, 0) \rightarrow (5, 0)$

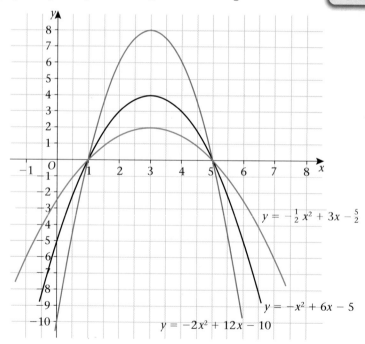

$y = -\frac{1}{2}x^2 + 3x - \frac{5}{2}$

$y = -x^2 + 6x - 5$

$y = -2x^2 + 12x - 10$

(b) The transformation applied to $y = \frac{1}{2}f(x)$ to get $y = 2f(x)$ is a stretch of scale factor 4 parallel to the y-axis.

Stretch in the x-direction

The graphs show

$\qquad$ $y = f(x)$ (in black) $\quad$ $y = f(2x)$ (in red) $\quad$ $y = f(\frac{1}{2}x)$ (in green)

for the two functions $f(x) = x + 1$ and $f(x) = x^2 - 9$.

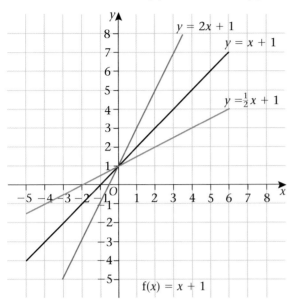

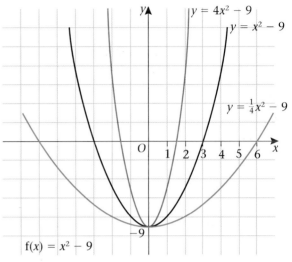

To get the graph of $y = f(2x)$ from the graph of $y = f(x)$, multiply the x-coordinate of each point on the graph of $y = f(x)$ by $\frac{1}{2}$ but leave the y-coordinate unaltered. For example, for the parabolas $(4, 7)$ becomes $(2, 7)$, $(2, -5)$ becomes $(1, -5)$ and so on.

This has the effect of stretching the curve out (in fact it is 'squashed') in the direction of the x-axis by scale factor $\frac{1}{2}$. Points on the graph on the y-axis remain fixed.

Similarly, by comparing the black and green curves you can see that the stretch required to get $y = f(\frac{1}{2}x)$ from $y = f(x)$ is again in the direction of the x-axis with scale factor 2.

> For any function f, the graph of $y = f(ax)$, where a is a positive constant, is obtained from $y = f(x)$ by applying a stretch of scale factor $\dfrac{1}{a}$ parallel to the x-axis.

Example 16

This is a sketch of $y = f(x)$, where $f(x) = 6x - x^2$.

The vertex of the parabola is $(3, 9)$.

(a) Write the equation of $y = f(2x)$.

(b) Sketch the graph of $y = f(2x)$. Give the coordinates of the vertex and the points where the graph crosses the axes.

(c) Sketch the graph of $y = 4f(x)$. Give the coordinates of the vertex and the points where the graph crosses the axes.

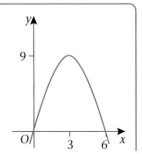

(a) $f(x) = 6x - x^2$
  $y = f(2x) = 6(2x) - (2x)^2 = 12x - 4x^2$ so $y = 12x - 4x^2$

(b)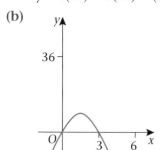

The vertex is $(\frac{3}{2}, 9)$.
The graph crosses
the axes at $(0, 0)$
and $(3, 0)$.

(c)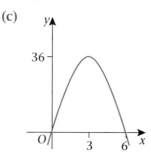

The vertex is $(3, 36)$.
The graph crosses
the axes at $(0, 0)$
and $(6, 0)$.

Example 17

Explain how to sketch the graph of $y = -4x^2 + 8x + 3$ by applying transformations
to the graph of $y = x^2$. Illustrate your answer by sketches and state the
coordinates of the vertex of $y = -4x^2 + 8x + 3$.

  $-4x^2 + 8x + 3 = -4(x^2 - 2x) + 3$

  $= -4[(x - 1)^2 - 1] + 3$ ———— Complete the square

  $= -4(x - 1)^2 + 7$

> Check:
> $-4(x - 1)^2 + 7$
> $= -4(x^2 - 2x + 1) + 7$
> $= -4x^2 + 8x - 4 + 7$
> $= -4x^2 + 8x + 3$

Let $f(x) = x^2$ so $y = x^2$ becomes $y = f(x)$

  1    $y = (x - 1)^2$ becomes $y = f(x - 1)$

  2    $y = 4(x - 1)^2$ becomes $y = 4f(x - 1)$

  3    $y = -4(x - 1)^2$ becomes $y = -4f(x - 1)$

  4    $y = -4(x - 1)^2 + 7$ becomes $y = -4f(x - 1) + 7$.

To transform the graph of $y = x^2$ to the graph of $y = -4x^2 + 8x + 3$, apply the following
transformations in this order:

1 horizontal translation of 1 unit in the positive x-direction

2 stretch of scale factor 4 parallel to the y-axis

3 reflection in the x-axis

4 vertical translation of 7 units in the positive y-direction.

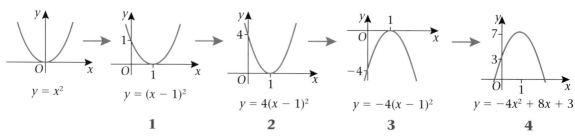

The vertex of $y = -4x^2 + 8x + 3$ is $(1, 7)$.

Exercise 25I

1 List the transformations, in the correct order, which when applied to $y = x^2$ give the graphs of the following functions.

(a) $y = 2x^2 + 4x - 1$

(b) $y = 3x^2 - 12x + 2$

(c) $y = x^2 - 4x + 2$

Sketch each graph.

2 List the transformations, in the correct order, which when applied to $y = x^3$ give the graphs of the following functions.

(a) $y = 4(x + 2)^3$

(b) $y = 8 - x^3$

(c) $y = 5 + (x - 1)^3$

(d) $y = (2x)^3 - 8$

Sketch each graph.

3 Explain how you would use your graph for question **1**(c) to sketch the graphs of these functions:

(a) $y = (x + 1)^2 - 4(x + 1) + 2$

(b) $y = x^2 + 4x + 2$

(c) $y = -x^2 + 4x$

25.8 Applying transformations to trigonometric functions

> For more on graphs of trigonometric functions see Chapter 16.

These are the graphs of $y = \sin x$ and $y = \cos x$ for $-360° \leqslant x \leqslant 360°$:

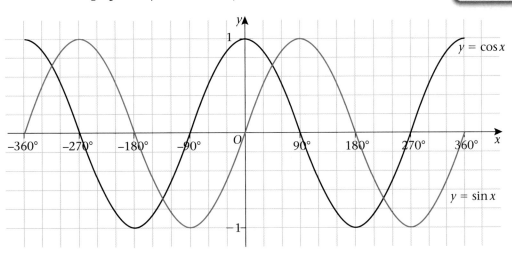

Example 18

Starting with $y = f(x)$, where $f(x) = \cos x$, apply a translation of 90° horizontally in the positive x-direction.

This leads to a trigonometrical identity. What is the identity?

$y = f(x)$ translated 90° in the positive x-direction becomes $y = f(x - 90°)$.

Applying this translation to $f(x) = \cos x$ you obtain $y = \cos(x - 90°)$.

But from the graphs, a translation of 90° horizontally in the positive x-direction on $y = \cos x$ gives the graph of $y = \sin x$.

Since $y = \cos(x - 90°)$ is the same graph as $y = \sin x$, we can deduce the trigonometrical identity:

$$\cos(x - 90°) = \sin x$$

Exercise 25J

1 Starting with $y = f(x)$, where $f(x) = \sin x$, apply a translation of 90° horizontally in the negative x-direction. This leads to a trigonometrical identity. What is the identity?

2 (a) What is the transformation which needs to be applied to $y = f(x)$, where $f(x) = \cos x$, to obtain $y = \cos 2x$?

   (b) Sketch the graph of $y = \cos 2x$ for $-360° \leqslant x \leqslant 360°$.

   (c) How many solutions of the equation $\cos 2x = \frac{1}{2}$ lie within the range $-360° \leqslant x \leqslant 360°$?

3 (a) What is the transformation which needs to be applied to $y = f(x)$, where $f(x) = \sin x$, to obtain $y = 2 \sin x$?

   (b) Sketch the graph of $y = 2 \sin x$ for $-360° \leqslant x \leqslant 360°$.

   (c) How many solutions of the equation $2 \sin x = 1$ lie within the range $-360° \leqslant x \leqslant 360°$?

   (d) Without any further sketches, state how many solutions of the equation $2 \sin 2x = 1$ lie within the range $-360° \leqslant x \leqslant 360°$. Explain your answer.

4 For $f(x) = \sin x$, sketch the graph of (a) $y = f(-x)$ (b) $y = -f(x)$.

5 The greatest value which $\cos x$ can have is $+1$.
   The lowest value that $\cos x$ can have is -1.
   Write down the greatest and lowest values for

   (a) $5 \cos x$ (b) $\cos 3x$ (c) $\cos(x + 60°)$

   (d) $-2 \cos x$ (e) $6 + 3 \cos 4x$

Mixed exercise 25

1 A sketch of the curve $y = \sin x$ for $0 \le x \le 360°$ is shown below.

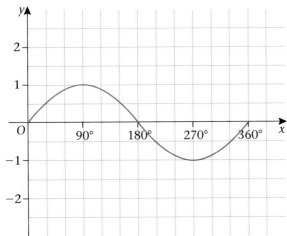

(a) Using the sketch above, or otherwise, find the equation of each of the following curves.

(i)

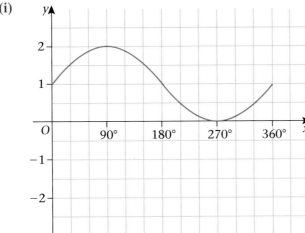

(ii)

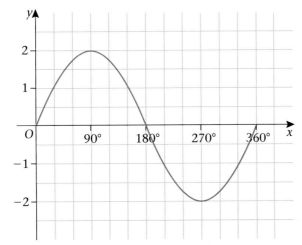

(b) Describe fully the sequence of two transformations that maps the graph of $y = \sin x$ on to the graph of $y = 3 \sin 2x$.

2

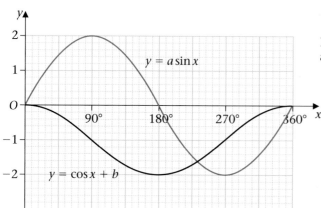

Diagram **NOT** accurately drawn

The diagram shows part of two graphs.

The equation of one graph is $y = a \sin x$

The equation of the other graph is $y = \cos x + b$

(a) Use the graphs to find the value of a and the value of b.

(b) Use the graphs to find the values of x in the range $0° \leqslant x \leqslant 720°$ for which $a \sin x = \cos x + b$

(c) Use the graphs to find the value of $a \sin x - (\cos x + b)$ when $x = 450°$.

3 This is a sketch of the curve with equation $y = f(x)$. It passes through the origin O.

The only vertex of the curve is at $A(2, -4)$.

(a) Write down the coordinates of the vertex of the curve with equation

    (i) $y = f(x - 3)$

    (ii) $y = f(x) - 5$

    (iii) $y = -f(x)$

    (iv) $y = f(2x)$.

(b) The curve with equation $y = x^2$ has been translated to give the curve $y = f(x)$.

Find $f(x)$ in terms of x.

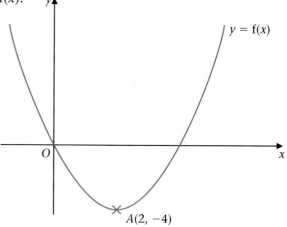

4 This is a sketch of the curve with equation $y = f(x)$.

The only maximum point of the curve $y = f(x)$ is $A(3, 6)$.

Write down the coordinates of the maximum point for curves with each of the following equations.

(a) $y = f(x + 2)$

(b) $y = f(x) + 4$

(c) $y = f(-x)$

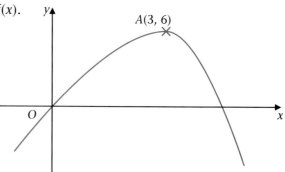

5   This is a sketch of the curve with equation $y = f(x)$.
The only vertex of the curve is at $P(2, -25)$.

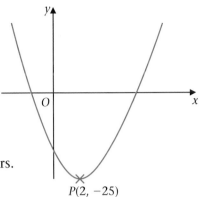

(a)   Write down the coordinates of the vertex for curves with each of the following equations.

   (i) $y = f(x - 2)$

   (ii) $y = f(x) + 22$

   (iii) $y = f(2x)$

(b) $f(x) = (2x + a)(2x - b)$, where a and b are positive integers.
The curve $y = f(x)$ passes through the point $(2, -25)$.
The curve $y = f(x + 4.5)$ passes through the point $(0, 0)$.

Find the value of a and the value of b.

Summary of key points

1   A **function** is a rule which shows how one set of numbers relates to another set.

2   The graph of $y = x^2 + a$ is the graph of $y = x^2$ translated a units vertically (in the *positive y*-direction if $a > 0$, in the *negative y*-direction if $a < 0$).

3   The graph of $y = x + a$ is the graph of $y = x$ translated a units vertically (in the *positive y*-direction if $a > 0$, in the *negative y*-direction if $a < 0$).

4   The graph of $y = x^3 + a$ is the graph of $y = x^3$ translated a units vertically (in the *positive y*-direction if $a > 0$, in the *negative y*-direction if $a < 0$).

5   The graph of $y = \dfrac{1}{x} + a$ is the graph of $y = \dfrac{1}{x}$ translated a units vertically (in the *positive y*-direction if $a > 0$, in the *negative y*-direction if $a < 0$).

6   For any function f, the graph of $y = f(x) + a$ is the graph of $y = f(x)$ translated a units vertically (in the *positive y*-direction if $a > 0$, in the *negative y*-direction if $a < 0$).

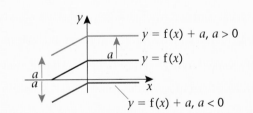

7   The graph of $y = (x + a)^2$ is the graph of $y = x^2$ translated a units horizontally (in the *negative x*-direction if $a > 0$, in the *positive x*-direction if $a < 0$).

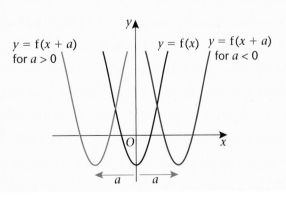

8 The **vertex** of the parabola $y = (x + a)^2$ is at the point $(-a, 0)$.

9 For any function f, the graph of $y = f(x + a)$ is the graph of $y = f(x)$ translated *a* units horizontally (in the *negative x-direction* if $a > 0$, in tthe *positive x-direction* if $a < 0$).

10 To sketch the graph of $y = x^2 + bx + c$, complete the square and apply a double translation to the parabola $y = x^2$.

11 For any function f, the graph of $y = f(x + a) + b$ is the graph of $y = f(x)$ translated *a* units horizontally (in the *negative x-direction* if $a > 0$, in the *positive x-direction* if $a < 0$) followed by a translation of *b* units vertically (*upwards* if $b > 0$, *downwards* if $b < 0$).

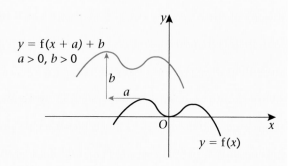

12 For any function f, the graph of $y = -f(x)$ is obtained by reflecting $y = f(x)$ in the *x*-axis.

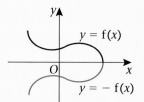

13 For any function f, the graph of $y = f(-x)$ is obtained by reflecting $y = f(x)$ in the *y*-axis.

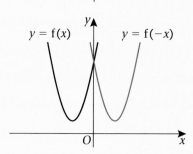

14 For any function f, the graph of $y = af(x)$, where *a* is a positive constant, is obtained from $y = f(x)$ by applying a stretch of scale factor *a* parallel to the *y*-axis.

15 For any function f, the graph of $y = f(ax)$, where *a* is a positive constant, is obtained from $y = f(x)$ by applying a stretch of scale factor $\dfrac{1}{a}$ parallel to the *x*-axis.

26 Circle theorems

In this chapter you will learn about calculating angles and using angle properties related to circles, and about proof.

We shall start by looking at a collection of important mathematical results known as **circle theorems**.

Proofs that may be required in the GCSE exam are given in Section 26.3.

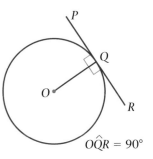

26.1 Circle theorems

Axiom

> An axiom is a given fact that can be used to prove a theorem.

The angle between a tangent and a radius is 90°.

$O\hat{Q}R = 90°$

Example 1

In the diagram on the right, given that PT is a tangent to the circle, find angle PTQ (x).

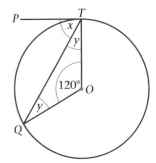

$$OT = OQ = \text{radius}$$

So triangle QOT is isosceles.

So   angle OQT = angle OTQ (call each y)

Then $y + y + 120° = 180°$ (angles in a triangle)

$$2y + 120° = 180°$$
$$2y = 60°$$
$$y = 30°$$

But $x + y = 90°$ (PT is a tangent, OT a radius)

So $x = 90° - y$
$$= 90° - 30°$$
$$= 60°$$

Theorem 1

The lengths of the two tangents from a point to a circle are equal.

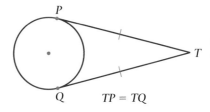

$TP = TQ$

> For a proof of Theorem 1 see Section 26.3.

Example 2

O is the centre of the circle.
TA and TB are the tangents from T to the circle.
Angle $ATB = 64°$
Calculate the size of

(a)   angle BAT (b)   angle OAB.

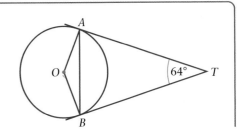

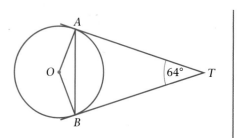

(a) The tangents TA and TB are equal, so ATB is isosceles.

$$B\hat{A}T + T\hat{B}A = 180° - 64° = 116°$$
(sum of angles of triangles)

$$B\hat{A}T = T\hat{B}A = 116° \div 2 = 58°$$
(base angles of isosceles triangle)

(b) $O\hat{A}T = 90°$ (radius perpendicular to tangent)

So $O\hat{A}B = 90° - 58° = 32°$

Theorem 2

The perpendicular from the centre of a
circle to a chord bisects the chord.

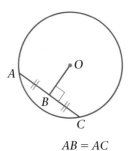

$AB = AC$

For a proof of Theorem 2
see Section 26.3.

Example 3

Find the values of a and b.

$$A\hat{O}B = 360° - 230° = 130°$$
$$a = \tfrac{1}{2} \text{ of } A\hat{O}B \text{ (by symmetry)}$$
so $\quad a = \tfrac{1}{2} \text{ of } 130°$
$$= 65°$$

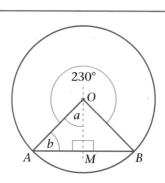

In the triangle AMO:
$$b + a + 90° = 180°$$
$$b = 180° - 90° - a$$
$$= 90° - 65°$$
$$= 25°$$

Example 4

The distance of the chord AB from the centre O
is 5 cm. $AB = 12$ cm.

Calculate the radius of the circle.

$$AM = \tfrac{1}{2}AB = \tfrac{1}{2} \times 12 = 6 \text{ cm}$$

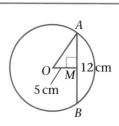

Using Pythagoras' theorem in triangle OAM:
$$OA^2 = 25 + 36$$
$$OA = \sqrt{61}$$
$$= 7.81 \text{ cm}$$

Exercise 26A

1 N is the middle point of the chord *PQ*.

$Q\hat{O}N$ is 52°

Calculate angle *NQO*.

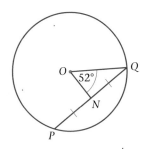

2 *BC* is the tangent at *B* to the circle.

$O\hat{B}A = 24°$

Calculate

(a) $A\hat{B}C$

(b) $O\hat{A}B$

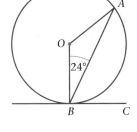

3 *QP* and *QR* are the tangents from *Q* to the circle.

$Q\hat{P}R = 73°$

Calculate

(a) $P\hat{Q}R$

(b) $O\hat{P}R$

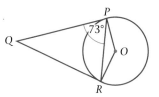

26.2 More circle theorems

Theorem 3

The angle subtended at the centre of a circle is twice the angle at the circumference.

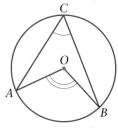

$A\hat{O}B = 2 \times A\hat{C}B$

> For a proof of Theorem 3 see Section 26.3.

> The subtended angle is the angle formed at *O* by the straight lines *AO* and *BO*.

___ **Example 5** _____

Calculate the size of angles *ACB* and *ADB*.

Angle *ACB*	$= \frac{1}{2}$ angle *AOB*
	$= 40°$
Reflex angle *AOB*	$= 360° - 80°$
	$= 280°$
Angle *ADB*	$= \frac{1}{2}$ reflex angle *AOB*
	$= 140°$

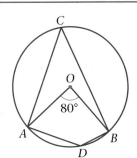

From Theorem 3 it follows that:

Theorem 4

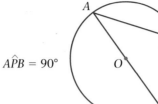

The angle in a semicircle is a right angle.

$A\hat{P}B = 90°$

For a proof of Theorem 4 see Section 26.3.

Example 6

In the diagram, calculate the size of angle AOX.

Angle $BDA = 90°$ (angle in semicircle)

Angle $BAD = 32°$ (third angle of triangle)

Angle $AOX = 32°$ (alternate angles)

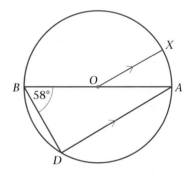

Theorem 5

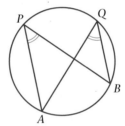

Angles in the same segment are equal.

$A\hat{P}B = A\hat{Q}B$

For a proof of Theorem 5 see Section 26.3.

Example 7

Calculate the angle a.

    $a = 54°$ (angles in the same segment are equal)

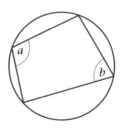

Theorem 6

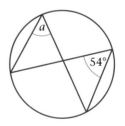

Opposite angles of a cyclic quadrilateral are supplementary.

$a + b = 180°$

For a proof of Theorem 6 see Section 26.3.

Points which lie on the circumference of the same circle are called **concyclic**. If a circle can be drawn through all four corners of a quadrilateral, then the quadrilateral is known as **cyclic**.

Two angles are **supplementary** if their sum is 180°.

Example 8

Find the sizes of angles *ABC* and *BAC*.

Angle *ABC* = 146° (opposite angles of a cyclic quadrilateral)

Angle *BAC* = $\frac{1}{2}(180° - 146°)$ (triangle *ABC* is isosceles)

$\qquad = \frac{1}{2} \times 34°$

$\qquad = 17°$

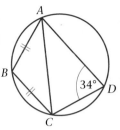

Theorem 7

The angle between a tangent and a chord is equal to the angle in the alternate segment.

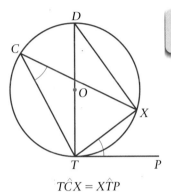

For a proof of Theorem 7 see Section 26.3.

This theorem is called the **alternate segment theorem** because the angle a chord makes with a tangent is equal to the angle subtended by the chord at the circumference in the alternate segment to the tangent.

$$T\hat{C}X = X\hat{T}P$$

Example 9

Calculate the sizes of angles *TQX* and *TXQ*.

Angle *TQX* = 36° (alternate segment)

Angle *TXQ* = $\frac{1}{2}(180° - 36°)$ (triangle *TXQ* is isosceles)

$\qquad = \frac{1}{2} \times 144°$

$\qquad = 72°$

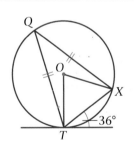

Exercise 26B

1 *PQ* is a diameter of the circle. *TS* is the tangent at *P*.
$P\hat{R}Q = 90°$ and $P\hat{Q}R = 34°$.
Calculate

(a) $R\hat{P}Q$ (b) $R\hat{P}T$

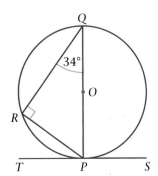

2 *DE* is the tangent at *B* to the circle.
AO is parallel to *BC*. $O\hat{B}A = 56°$.
Calculate

(a) $A\hat{B}D$ (b) $O\hat{A}B$ (c) $A\hat{O}B$

(d) $O\hat{B}C$ (e) $C\hat{B}E$

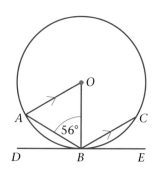

3 *PQ* is parallel to *SR*.
Calculate *a*, *b*, *c* and *d*.

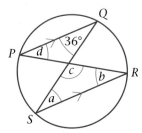

4 *TPS* is the tangent to the circle at *P*.

$T\hat{P}Q = 47°$.

Calculate the size of

(a) $P\hat{Q}R$

(b) $Q\hat{R}P$

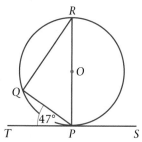

5 *AD* is parallel to *BC*. $A\hat{D}C = 102°$.
Calculate the size of

(a) $A\hat{D}E$ (b) $A\hat{B}C$

(c) $D\hat{C}B$ (d) $B\hat{A}D$

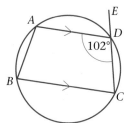

6 *PT* is a tangent to the circle.
Find the angles *OPT* and *OPA*.

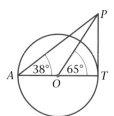

7 *M* is the mid-point of *AC*.
Calculate the lengths of *AC* and *OM*.

8 M is the mid-point of PQ.
Calculate angle POM and angle OPM.

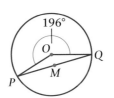

9 Calculate SO and the area of OSPT.

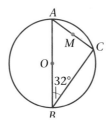

10 M is the mid-point of AC.
Calculate the size of angle MOC.

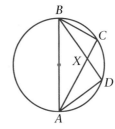

11 AB is a diameter.
Angle ABC = 54° and angle ABD = 32°.
Calculate the sizes of angles CXD and CAD.

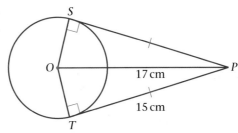

12 TQ is a diameter. PT is a tangent.
Find, in terms of x,
(a) QP̂T   (b) RT̂P

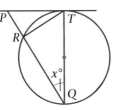

13 AB is a diameter.
Calculate the area of triangle ABD.

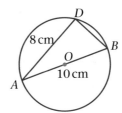

14 Calculate angles d, e and f.

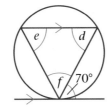

15 Calculate the sizes of angles *TBA* and *TCA*.

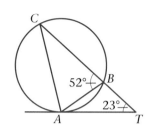

16 Calculate the length of *AB*.

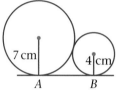

17 *PT* is a tangent to the circle at *T*.
Calculate the size of angle *TXS*.

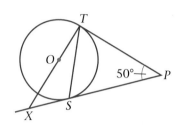

18 A bridge is built in the shape of
the arc of a circle, centre *O*.
The width, *AB*, is 12 m and the
height is 4 m. Calculate the
distance *AO*.

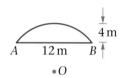

19 Draw accurately a triangle *ABC* which has all its angles acute.
Draw the perpendicular bisectors of the sides *AC* and *AB*.
These meet at the point *O*. Explain why *O* is the centre of the
circle which passes through the three corners of the triangle.

26.3 Proving circle and geometrical theorems

This section shows the proofs you may need for your GCSE exam.

When writing geometrical proofs you need to give reasons for the
steps that you take. This usually means stating reasons or theorems
that justify the steps in your proof.

Proof of Theorem 1

The lengths of the two tangents from a point to a circle are equal.

In triangles *OPS* and *OPT*:

    OS = *OT*   (radii)

    OP is common to both

    angle *OSP* = angle *OTP* = 90°   (tangent and radius)

So triangles *OPS* and *OPT* are congruent. (RHS)

So   *PS* = *PT*

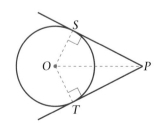

Proof of Theorem 2

The perpendicular from the centre of a circle to a chord bisects the chord.

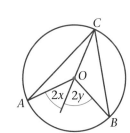

M is the point where the perpendicular from O meets the chord AB.
In triangles OAM and OBM:

$OA = OB$ (radii)

OM is common to both

angle OMA = angle $OMB = 90°$

So triangles OAM and OBM are congruent.   (RHS)

So $AM = MB$

Proof of Theorem 3

The angle subtended at the centre of a circle is twice the angle at the circumference.

You need to prove that $A\hat{O}B = 2 \times A\hat{C}B$.

Call the two parts of $A\hat{O}B$ $2x$ and $2y$, as shown.

So $A\hat{O}B = 2x + 2y$

Now $A\hat{O}C = 180° - 2x$ (straight line)

$O\hat{A}C + O\hat{C}A = 2x$ (exterior angle of triangle = sum of two
                                       interior opposite angles)

$O\hat{A}C = O\hat{C}A = x$ (triangle AOC is isosceles)

Similarly, $O\hat{B}C = O\hat{C}B = y$

So $A\hat{C}B = x + y$

That is, $A\hat{O}B = 2 \times A\hat{C}B$

Since nothing has been assumed about the values of x and y, this result must be generally true for all diagrams like the one above.

The theorem is also true if C is on the minor arc, as shown on the right.

To prove, call the two parts of the the reflex angle AOB $2x$ and $2y$, as before.

$A\hat{O}C = 180° - 2x$ (straight line)

So $O\hat{A}C = O\hat{C}A = x$ (triangle OAC is isosceles)

Similarly $O\hat{C}B = O\hat{B}C = y$

So $A\hat{C}B = x + y$

That is, reflex angle $AOB = 2 \times A\hat{C}B$

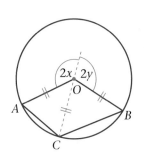

There is another case to consider. Here C is on the major arc but relatively close to B. The theorem is still true but the proof is not demonstrated here.

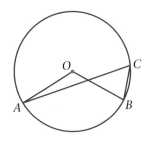

Proof of Theorem 4

> The angle in a semicircle is a right angle.

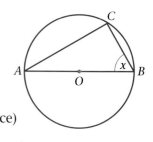

In triangle ABC:

angle $AOB = 180°$ (straight line)

So angle $ACB = \frac{1}{2} \times 180°$ (angle at centre is twice angle at circumference)

$= 90°$

Proof of Theorem 5

> Angles in the same segment are equal.

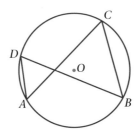

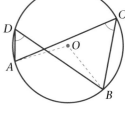

Join AO and OB.

angle $ACB = \frac{1}{2}$ angle AOB (angle at centre is twice angle

and   angle $ADB = \frac{1}{2}$ angle AOB at circumference)

So   angle $ACB =$ angle ADB

Proof of Theorem 6

> Opposite angles of a cyclic quadrilateral are supplementary.

$2x + 2y = 360°$

So $x + y = 180°$

angle $BAD = x$ (angle at centre is twice angle at circumference)

angle $BCD = y$ (angle at centre is twice angle at circumference)

angle $BAD +$ angle $BCD = x + y = 180°$

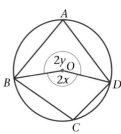

> You can show that $A\hat{B}C + C\hat{D}A = 180°$ by drawing AO and CO.

Proof of Theorem 7

> The angle between a tangent and a chord is equal to the angle in the alternate segment.

TD is the diameter of the circle, so $T\hat{X}D = 90°$.

(angle in a semicircle is a right angle)

$T\hat{D}X + D\hat{T}X = 90°$ (1)   (angle sum of triangle is 180°)

PT is a tangent so $P\hat{T}D$ is 90°.

Thus $P\hat{T}X + D\hat{T}X = 90°$ (2)

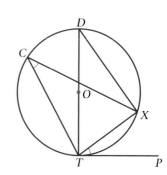

From (1) and (2):

$T\hat{D}X + D\hat{T}X = P\hat{T}X + D\hat{T}X = 90°$

So $T\hat{D}X = P\hat{T}X$

C is any other point on the circumference.

Now $T\hat{C}X = T\hat{D}X$ (angles in the same segment)

Thus $P\hat{T}X = T\hat{C}X$

Example 10

O is the centre of the smaller circle. *FEGH* is a straight line.
Prove that:

(a)   angle *DGH* = 2 × angle *DFE*   (b)   *FG* = *DG*

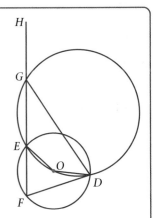

(a)   In the small circle, angle *DOE* = 2 × angle *DFE*
                          (angle at centre is twice angle at circumference)

   In the large circle, angle *DOE* = 180° − angle *DGE*
                          (opposite angles of a cyclic quadrilateral)

   Thus 180° − angle *DGE* = 2 × angle *DFE*

   But   180° − angle *DGE* = angle *DGH*   (*FEGH* is a straight line)

   So   angle *DGH* = 2 × angle *DFE*

(b)   Let angle *DFE* = x.
   Then from **(a)** angle *DGE* = 180° − 2x

   Thus   angle *FDG* = x   (angle sum of triangle is 180°)
   and   *FG* = *DG*   (*DFG* is an isosceles triangle)

Exercise 26C

1   O is the centre of the circle. *AOC* is a straight line.
   TB and *TC* are tangents.
   Prove that the triangles *AOB* and *BTC* are similar.

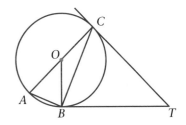

2   *PQ* and *RS* intersect at right angles. *PY* is a tangent.
   RXSY is a straight line so that *YX* = *XR*. *QSZ* is a straight line.
   Prove that:

   (a)   angle *YPX* = angle *XSQ*

   (b)   angle *PZQ* = 90°

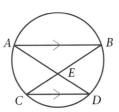

3   *AB* is parallel to *CD*. Prove that

   (a)   angle *AEC* = 2 × angle *ABC*

   (b)   *AE* = *BE*

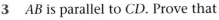

4   *ADX* and *BCX* are straight lines.
   The straight line *XFE* bisects angle *AXB*.
   Prove that angle *DFE* = angle *FEA*.

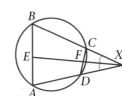

5   Prove that angle ADC = angle BCD.

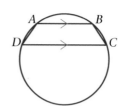

6 $ABCD$ is a cyclic quadrilateral with $AB = AD$.
Prove that the line AC bisects the angle DCB.

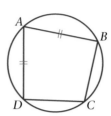

7 $ABCD$ is a parallelogram. Prove that $AE = AD$.

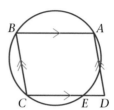

8 ADE is a straight line. FD bisects angle EDC.
Prove that FB bisects angle ABC.

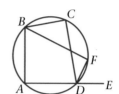

9 AB is parallel to the straight line CDE.
AF bisects angle DAG. Show that $AF = FE$.

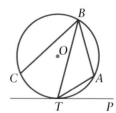

10 C is the point on the circumference such that BT bisects
angle CBA. Prove that angle COA = $4 \times$ angle ATP.

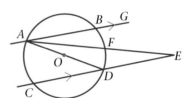

Mixed exercise 26

1   For each of the diagrams below, calculate the length of the
chord AB.

(a)

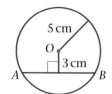

(b)

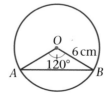

2 Angle *ABC* = 42°.
Name two other angles in the diagram with size 42°.

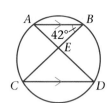

3 *QS* is a diameter. Triangle *PQS* is isosceles.
Calculate the size of angle *PQR*.

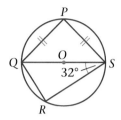

4 *PR* is a diameter of the semicircle.
PRY is a straight line.
Find the sizes of these angles:

(a) *P*Q̂*R*   (b) *Q*R̂*P*   (c) *Q*R̂*X*

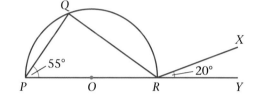

5 Calculate the sizes of these angles:

(a) *C*Â*D*   (b) *A*D̂*B*   (c) *A*Ĉ*D*

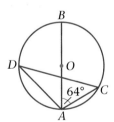

6 *AOB* is a diameter.
Calculate the size of angle *ADC*.

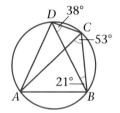

7 *AOB* is a diameter. Calculate the sizes of angles *OBC* and *OCA*.

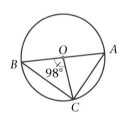

8 Calculate the radius of the circle and the length of the chord *PQ*.

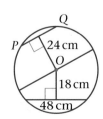

9 Triangle *ABC* is isosceles. Calculate the sizes of these angles:

(a) $B\hat{A}D$ (b) $A\hat{C}B$ (c) $A\hat{D}B$

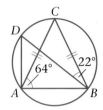

10 The straight lines *PSX* and *PTY* are tangents to the circle with centre *O*.

Given that the radius of the circle is *r*, calculate the length of *PB* in terms of *r*.

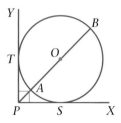

11 *PQRS* is a cyclic quadrilateral.
PSB and *QRB* are straight lines. *PQA* and *SRA* are straight lines.
Calculate the values of *x* and *y*.

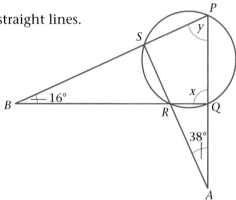

12 Triangle *AOB* is equilateral and *DA* is a tangent.
Calculate the sizes of these angles:

(a) $O\hat{A}B$ (b) $O\hat{A}D$ (c) $B\hat{D}A$

(d) $C\hat{A}B$ (e) $A\hat{C}B$ (f) Calculate the length of *AC*.

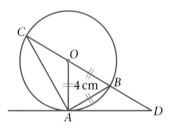

13 *TA* is a tangent to the circle, centre *O*.
DOB is a diameter of the circle.
Find the sizes of these angles:

(a) $D\hat{B}A$ (b) $B\hat{D}A$ (c) $C\hat{B}D$

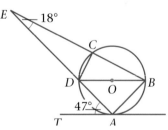

14 *AB* is a tangent to the circle. *DOE* is a diameter.
Find the sizes of these angles:

(a) $D\hat{O}B$ (b) $B\hat{E}D$

(c) Given that *AB* = 5.8 cm, calculate the radius.

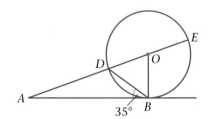

15 Prove that

(a) the tangent at *A* bisects the line *XY*

(b) angle *XAY* = 90°

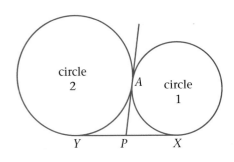

16 *BCE* and *ADE* are straight lines. *AB* and *DC* are parallel lines.
Calculate the sizes of these angles:

(a) *A*Ĉ*B*   (b) *D*Ĉ*A*   (c) *C*B̂*A*

(d) Show that triangle *EBA* is isosceles.

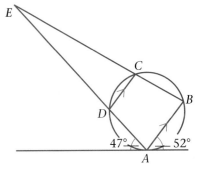

17 The circle, centre *I*, is the *inscribed* circle of triangle *ABC*.
Calculate the size of angle *EDF*.

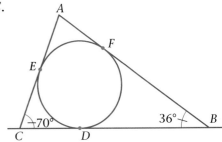

18 (a) Calculate the sizes of angles *ACB* and *CBA*.
Give reasons for your answers.

(b) Prove that triangle *ABC* is isosceles.

(c) Prove that triangles *PBR* and *ACR* are similar.

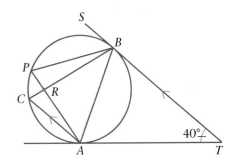

19 *C* is the centre of the large circle and *O* is the centre
of the small circle. *PT* is a tangent to the small circle.
PCAO is a straight line. The small circle has radius
1 cm. Calculate the radius of the large circle.

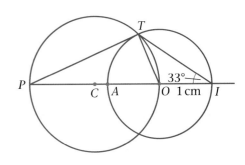

20 (a) Draw any acute-angled triangle. An **altitude** of a triangle is a straight line drawn from a corner to meet the opposite side at right angles. Draw the three altitudes of the triangle. Write what you notice.

(b) *AFB*, *AEC*, *BHE*, *AHD* and *FHC* are straight lines. *AOH* is a diameter. Find, in terms of *x* or *y* or both, the size of angle *BHD*. Prove that *FBCE* is a cyclic quadrilateral. Hence, prove that *AH* is perpendicular to *BC*.

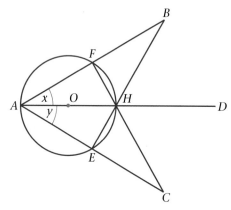

(c) What is the relationship between parts (a) and (b)?

Summary of key points

1 The angle between a tangent and a radius is 90°.

2 The lengths of the two tangents from a point to a circle are equal.

3 The perpendicular bisector from the centre of a circle to a chord bisects the chord.

4 The angle subtended at the centre of a circle is twice the angle at the circumference.

5 The angle in a semicircle is a right angle.

6 Angles in the same segment are equal.

7 Opposite angles of a cyclic quadrilateral are supplementary.

8 The angle between a tangent and a chord is equal to the angle in the alternate segment.

27 Vectors

In this chapter you will use vectors to represent translations and prove geometrical results.

27.1 Translations

A **translation** moves all the points on a shape the same distance and in the same direction. The shape is neither turned nor rotated.

For more on translations see Section 7.1.

Translations can be described using **column vectors**: $\begin{pmatrix} x \\ y \end{pmatrix}$

$\begin{pmatrix} x \\ y \end{pmatrix}$ displacement in the x-direction / displacement in the y-direction

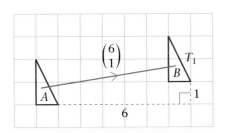
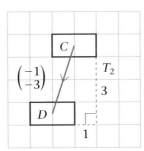

The translation T_1 is given by the column vector $\begin{pmatrix} 6 \\ 1 \end{pmatrix}$, meaning 6 units in the positive x-direction and 1 unit in the positive y-direction.

T_2 is given by $\begin{pmatrix} -1 \\ -3 \end{pmatrix}$, meaning 1 unit in the negative x-direction and 3 units in the negative y-direction.

The values of x and y in a column vector are called **components**.

The translation that takes A to B can be written as a translation vector $\overrightarrow{AB}$.

In the examples above,

$$\overrightarrow{AB} = \begin{pmatrix} 6 \\ 1 \end{pmatrix} \qquad \overrightarrow{CD} = \begin{pmatrix} -1 \\ -3 \end{pmatrix}$$

The arrow shows the direction of the translation:
A to B = $\overrightarrow{AB}$
B to A = $\overrightarrow{BA}$

Example 1

A is the point $(2, -3)$ and B is the point $(5, 2)$. $\overrightarrow{AC} = \begin{pmatrix} -1 \\ 3 \end{pmatrix}$

(a) Write the column vector $\overrightarrow{AB}$.

(b) Write the column vector $\overrightarrow{BA}$.

(c) Write the coordinates of C.

(a) $\overrightarrow{AB} = \begin{pmatrix} 3 \\ 5 \end{pmatrix}$

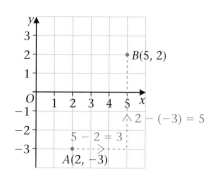

To get from A to B you move
3 in the positive x-direction
5 in the positive y-direction

(b) $\overrightarrow{BA} = \begin{pmatrix} -3 \\ -5 \end{pmatrix}$

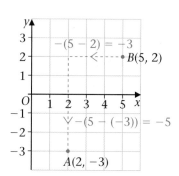

To get from B to A you move
3 in the negative x-direction
5 in the negative y-direction

(c) $\overrightarrow{AC} = \begin{pmatrix} -1 \\ 3 \end{pmatrix}$

So from A you move -1 unit
in the x-direction and 3 units
in the y-direction.
So C is at $(1, 0)$.

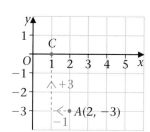

Example 2

$ABCD$ is a rectangle.

(a) Write down the column vectors for all sides.

(b) What do you notice about
(i) $\overrightarrow{BC}$ and $\overrightarrow{AD}$ (ii) $\overrightarrow{AB}$ and $\overrightarrow{DC}$?

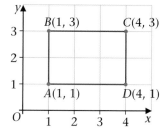

(a) $\overrightarrow{AB} = \begin{pmatrix} 0 \\ 2 \end{pmatrix}$ $\overrightarrow{BC} = \begin{pmatrix} 3 \\ 0 \end{pmatrix}$ $\overrightarrow{CD} = \begin{pmatrix} 0 \\ -2 \end{pmatrix}$ $\overrightarrow{AD} = \begin{pmatrix} 3 \\ 0 \end{pmatrix}$

(b) (i) $\overrightarrow{BC} = \begin{pmatrix} 3 \\ 0 \end{pmatrix} = \overrightarrow{AD}$

   (ii) $\overrightarrow{AB} = \begin{pmatrix} 0 \\ 2 \end{pmatrix}$ $\overrightarrow{DC} = -\overrightarrow{CD} = \begin{pmatrix} 0 \\ 2 \end{pmatrix}$

     $\overrightarrow{AB} = \overrightarrow{DC}$

If $\overrightarrow{AD} = \overrightarrow{BC}$, then AD and BC are parallel and have equal length.

Exercise 27A

1 *A* is the point (4, 6), *B* is the point (2, 5) and *C* is the point $(-1, 0)$.
Write the column vectors

    (a) $\overrightarrow{AB}$ (b) $\overrightarrow{BC}$ (c) $\overrightarrow{CA}$ (d) $\overrightarrow{AC}$

2 *A* is the point (2, 4), *B* is the point (3, 5) and *C* is the point (4, 2).
The vector $\overrightarrow{CD}$ is parallel to the vector $\overrightarrow{BA}$. The point *D* lies on
the *y*-axis. Find the coordinates of *D*.

3 (a) The point *A*(0, 2) is reflected in the line $x = 3$ to give the
point *B*. The point *B* is reflected in the line $x = 6$ to give *C*.
Describe the single transformation that maps *A* to *C*.

    (b) Generalise your result in the case where there are successive
reflections in the lines $x = a$ and $x = b$, with $a < b$.

4 *A*(2, 3), *B*(5, 7) and *C*(9, 5) are three points.

    (a) Find the column vector $\overrightarrow{BC}$.

    (b) *ABCD* is a parallelogram with *BC* parallel to *AD*.
Use your answer to (a) to find the coordinates of *D*.

    (c) Show that $\overrightarrow{AB}$ and $\overrightarrow{DC}$ have the same column vector.

27.2 Vectors

Another way of writing a translation is using bold type single letters
such as **a** and **b**. By hand write them underlined.

Translations described in this way are called **vectors**. The vectors **a**
and **b** are shown here. The lines with arrows are called **directed line
segments** and show a unique **length** and **direction** for each of
vectors **a** and **b**.

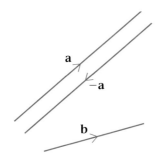

A **vector** defined as **a** has a unique length and direction.
The vector with the same length but opposite direction is −**a**.

Addition of vectors

Look at the diagram below.

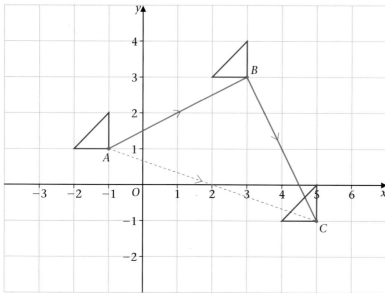

The vectors $\overrightarrow{AB}$ and $\overrightarrow{BC}$ can be written $\begin{pmatrix} 4 \\ 2 \end{pmatrix}$ and $\begin{pmatrix} 2 \\ -4 \end{pmatrix}$.

$\overrightarrow{AB} + \overrightarrow{BC}$ can be interpreted as the result of two successive translations, equivalent to $\overrightarrow{AC}$.

$$\begin{pmatrix} 4 \\ 2 \end{pmatrix} + \begin{pmatrix} 2 \\ -4 \end{pmatrix} = \begin{pmatrix} 4+2 \\ 2-4 \end{pmatrix} = \begin{pmatrix} 6 \\ -2 \end{pmatrix}$$

$$\overrightarrow{AB} + \overrightarrow{BC} = \overrightarrow{AC}$$

Here vectors $\overrightarrow{AB}$ and $\overrightarrow{BC}$ have been added to give the vector $\overrightarrow{AC}$.

For any two vectors **a** and **b** it is possible to add them by placing them 'nose to tail':

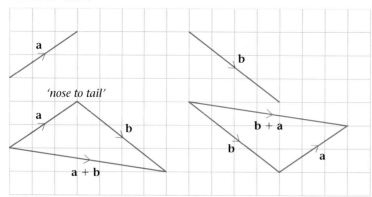

> Note that
> $\mathbf{a} + \mathbf{b} = \mathbf{b} + \mathbf{a}$
> $\mathbf{b} + \mathbf{a}$ gives a vector of the same length and direction as $\mathbf{a} + \mathbf{b}$.

This is called the **triangle law of addition**:

$\mathbf{a} + \mathbf{b} = \mathbf{b} + \mathbf{a}$

For example: $\begin{pmatrix} 3 \\ 2 \end{pmatrix} + \begin{pmatrix} 4 \\ -3 \end{pmatrix} = \begin{pmatrix} 4 \\ -3 \end{pmatrix} + \begin{pmatrix} 3 \\ 2 \end{pmatrix} = \begin{pmatrix} 7 \\ -1 \end{pmatrix}$

Subtraction of vectors

$\mathbf{p} - \mathbf{q}$ can be interpreted as $\mathbf{p} + (-\mathbf{q})$.

Then, using the triangle law of addition:

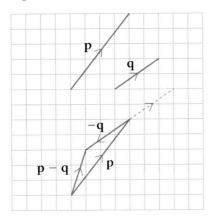

Using column vectors, this means:

$$\binom{4}{5} - \binom{3}{2} = \binom{4}{5} + \binom{-3}{-2} = \binom{4-3}{5-2} = \binom{1}{3}$$

Multiplication of vectors by scalars

Geometrically, the expression $2\overrightarrow{AB}$ means the directed line segment parallel to $\overrightarrow{AB}$, but twice the length. In other words, when vector $\overrightarrow{AB}$ is multiplied by 2 the result is $2\overrightarrow{AB}$.

A scalar is a number, e.g. $3, 2, \frac{1}{2}$...

If a vector **a** is multiplied by a scalar k then the vector $k\mathbf{a}$ is parallel to **a** and is equal to k times **a**.

For the column vector $\binom{p}{q}$, $k \times \binom{p}{q} = \binom{kp}{kq}$

For example, $3\binom{2}{1} = \binom{3 \times 2}{3 \times 1} = \binom{6}{3}$

Exercise 27B

1 $\overrightarrow{AB} = \binom{3}{1}$

Write down the column vector for

(a) $3\overrightarrow{AB}$ (b) $5\overrightarrow{AB}$ (c) $-2\overrightarrow{AB}$ (d) $\overrightarrow{BA}$ (e) $k\overrightarrow{AB}$

2 In the diagram $\overrightarrow{OA} = \mathbf{a}$, $\overrightarrow{AM} = \mathbf{b}$.

M is the mid-point of AC.

Write down, in terms of **a** and **b**

(a) $\overrightarrow{AC}$ (b) $\overrightarrow{OM}$ (c) $\overrightarrow{OC}$

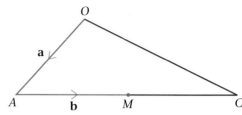

3 $\overrightarrow{AB}$ is the column vector $\begin{pmatrix} 4 \\ -3 \end{pmatrix}$. $\overrightarrow{BC}$ is the column vector $\begin{pmatrix} 2 \\ 4 \end{pmatrix}$.

Find the column vector $\overrightarrow{AC}$.

Draw a diagram to show your answer.

4 $\overrightarrow{AB} = \begin{pmatrix} 1 \\ 4 \end{pmatrix}$ $\overrightarrow{BC} = \begin{pmatrix} 2 \\ -3 \end{pmatrix}$ $\overrightarrow{CD} = \begin{pmatrix} -5 \\ 2 \end{pmatrix}$

(a) Find the column vector for $\overrightarrow{AD}$.

Draw a diagram to show this.

(b) Show that $\overrightarrow{AC} = \overrightarrow{DB}$.

5 M is the mid-point of OA.

$ON = \frac{1}{3}OC$

$\overrightarrow{OA} = \mathbf{a}$

$\overrightarrow{MN} = \mathbf{b}$

Find $\overrightarrow{OC}$ in terms of **a** and **b**.

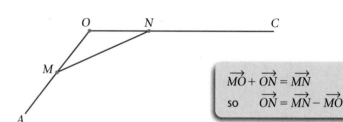

$$\overrightarrow{MO} + \overrightarrow{ON} = \overrightarrow{MN}$$

so $\quad \overrightarrow{ON} = \overrightarrow{MN} - \overrightarrow{MO}$

27.3 Vector algebra

The rules of algebra that you already know can be applied to vectors, providing you do not multiply or divide one vector by another.

> For any vectors **a**, **b** and c and any scalars p, q and k:
> - $\mathbf{a} + \mathbf{b} = \mathbf{b} + \mathbf{a}$
> - $\mathbf{a} + \mathbf{b} = \mathbf{c}$ can be written as $\mathbf{a} = \mathbf{c} - \mathbf{b}$
> - $\mathbf{a} + (\mathbf{b} + \mathbf{c}) = (\mathbf{a} + \mathbf{b}) + \mathbf{c}$
> - $k(\mathbf{a} + \mathbf{b}) = k\mathbf{a} + k\mathbf{b}$
> - $(p + q)\mathbf{a} = p\mathbf{a} + q\mathbf{a}$

You can use these rules to solve vector equations. For example

$$2\mathbf{a} = \mathbf{b} \quad \rightarrow \quad \mathbf{a} = \frac{\mathbf{b}}{2}$$

Two column vectors are equal if they represent the same translation.

If $\quad \begin{pmatrix} a \\ b \end{pmatrix} = \begin{pmatrix} c \\ d \end{pmatrix}$

then $\quad a = c$ and $b = d$

> If two vectors **a** and **b** are parallel, then $\mathbf{a} = k\mathbf{b}$ for some scalar k.

Example 3

$$a = \begin{pmatrix} 3 \\ -2 \end{pmatrix} \quad \text{and} \quad b = \begin{pmatrix} -1 \\ -3 \end{pmatrix}$$

(a)   Write as column vectors   (i) $a - b$ (ii) $3a$ (iii) $3a - 2b$

(b)   Find the vector x such that $a + 2x = b$

(c)   Find a vector c such that $c + 3a$ is parallel to $\begin{pmatrix} 3 \\ -2 \end{pmatrix}$.

(a)   (i) $a - b = a + (-b) = \begin{pmatrix} 3 \\ -2 \end{pmatrix} + -\begin{pmatrix} -1 \\ -3 \end{pmatrix} = \begin{pmatrix} 3 - -1 \\ -2 - -3 \end{pmatrix} = \begin{pmatrix} 4 \\ 1 \end{pmatrix}$

> Remember:
> $a - b = a + (-b)$

  (ii) $3a = 3\begin{pmatrix} 3 \\ -2 \end{pmatrix} = \begin{pmatrix} 9 \\ -6 \end{pmatrix}$

  (iii) $3a - 2b = \begin{pmatrix} 9 \\ -6 \end{pmatrix} - 2\begin{pmatrix} -1 \\ -3 \end{pmatrix} = \begin{pmatrix} 9 \\ -6 \end{pmatrix} - \begin{pmatrix} -2 \\ -6 \end{pmatrix} = \begin{pmatrix} 9 \\ -6 \end{pmatrix} + \begin{pmatrix} 2 \\ 6 \end{pmatrix} = \begin{pmatrix} 11 \\ 0 \end{pmatrix}$

(b) $a + 2x = b$

So $2x = b - a$

$$x = \frac{b - a}{2} = \frac{1}{2}\begin{pmatrix} -1 - 3 \\ -3 - -2 \end{pmatrix} = \frac{1}{2}\begin{pmatrix} -4 \\ -1 \end{pmatrix} = \begin{pmatrix} -2 \\ -\frac{1}{2} \end{pmatrix}$$

(c) $c + \begin{pmatrix} 9 \\ -6 \end{pmatrix} = \begin{pmatrix} 3 \\ -2 \end{pmatrix}$

Writing c as $\begin{pmatrix} p \\ q \end{pmatrix}$:

$$\begin{pmatrix} p \\ q \end{pmatrix} + \begin{pmatrix} 9 \\ -6 \end{pmatrix} = \begin{pmatrix} 3 \\ -2 \end{pmatrix}$$

So $p + 9 = 3$ $\rightarrow$ $p = -6$

and $q - 6 = -2$ $\rightarrow$ $q = 4$

So c is $\begin{pmatrix} -6 \\ 4 \end{pmatrix}$.

Exercise 27C

1   $a = \begin{pmatrix} 2 \\ 1 \end{pmatrix}$ and $b = \begin{pmatrix} -4 \\ 3 \end{pmatrix}$

Calculate   (a) $a + b$ (b) $2a$ (c) $2a - 3b$ (d) $2(a - b)$

(e)   Find a vector c such that $a + c$ is parallel to $\begin{pmatrix} 4 \\ 4 \end{pmatrix}$.

2   $a = \begin{pmatrix} 3 \\ 1 \end{pmatrix}$ and $b = \begin{pmatrix} 2 \\ 2 \end{pmatrix}$

Draw diagrams to show that

(a) $2(a + b) = 2a + 2b$ (b) $(2 + 3)a = 2a + 3a$

3   P is the point $(1, 3)$, Q is the point $(2, 4)$ and R is the point $(5, 4)$.
S is the point such that $\overrightarrow{PQ} = \overrightarrow{SR}$. Find the coordinates of S.

4   $a = \begin{pmatrix} -1 \\ 2 \end{pmatrix}$ and $b = \begin{pmatrix} -3 \\ 4 \end{pmatrix}$

Calculate **x**, given that **a** + **x** = **b**

5   $c = \begin{pmatrix} 2 \\ -1 \end{pmatrix}$ and $d = \begin{pmatrix} 4 \\ -3 \end{pmatrix}$

Calculate **x** given that 2**x** + **c** = **d**

6   $e = \begin{pmatrix} 4 \\ 1 \end{pmatrix}$ and $f = \begin{pmatrix} -2 \\ 3 \end{pmatrix}$

Calculate **x** given that 2**e** − **x** = **f**

27.4 Finding the magnitude of a vector

The **magnitude** of a vector is the length of the directed line segment representing it.

If the vector is expressed in column form, you can use Pythagoras' theorem to find the magnitude.

Example 4

Find the magnitude of the vector $\overrightarrow{AB} = \begin{pmatrix} 6 \\ -8 \end{pmatrix}$.

Draw $\overrightarrow{AB}$ as the hypotenuse of a right-angled triangle:

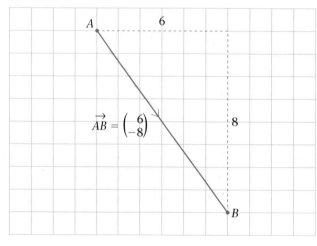

By Pythagoras' theorem, the length AB is given by

$$AB^2 = 6^2 + 8^2$$
$$= 36 + 64$$
$$= 100$$
$$AB = 10$$

The magnitude of $\overrightarrow{AB}$ is 10 units.

In general, the magnitude of the vector $\begin{pmatrix} x \\ y \end{pmatrix}$ is $\sqrt{x^2 + y^2}$

Exercise 27D

1   Find the magnitude of each of these vectors:

(a) $\begin{pmatrix} -4 \\ 3 \end{pmatrix}$ (b) $\begin{pmatrix} 7 \\ 24 \end{pmatrix}$ (c) $\begin{pmatrix} -6 \\ -8 \end{pmatrix}$ (d) $\begin{pmatrix} 7 \\ 11 \end{pmatrix}$

2   $a = \begin{pmatrix} 5 \\ 9 \end{pmatrix}$ and $b = \begin{pmatrix} 3 \\ 3 \end{pmatrix}$

Work out the magnitude of

(a) a   (b) b   (c) a + b   (d) a − b

3   $c = \begin{pmatrix} 2 \\ 3 \end{pmatrix}$ and $d = \begin{pmatrix} -3 \\ 4 \end{pmatrix}$

Work out the magnitude of

(a) c   (b) 2c   (c) 2d   (d) c − d

27.5 Linear combinations of vectors

One useful and important application of vectors is to produce special combinations of two given vectors.

> Combinations of the vectors **a** and **b** of the form $p\mathbf{a} + q\mathbf{b}$, where p and q are scalars, are called **linear combinations** of the vectors **a** and **b**.

Generally, any two non-parallel vectors can be combined to give a single vector in a different direction.

Example 5

$$a = \begin{pmatrix} 3 \\ 2 \end{pmatrix} \quad \text{and} \quad b = \begin{pmatrix} 1 \\ 3 \end{pmatrix}$$

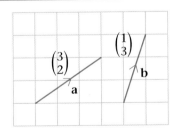

(a)   Find scalars p and q such that $p\mathbf{a} + q\mathbf{b}$ is parallel to the x-axis.

Vectors parallel to the x-axis are of the form $\begin{pmatrix} x \\ 0 \end{pmatrix}$.

(b)   Find scalars r and s such that $r\mathbf{a} + s\mathbf{b}$ is parallel to the y-axis.

Vectors parallel to the y-axis are of the form $\begin{pmatrix} 0 \\ y \end{pmatrix}$.

(a) $pa + qb = p\begin{pmatrix} 3 \\ 2 \end{pmatrix} + q\begin{pmatrix} 1 \\ 3 \end{pmatrix}$

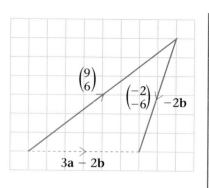

The y-component of vector $pa + qb$ must be zero because the vector lies in the direction of the x-axis.

So $2p + 3q = 0$

Two possible values of p and q which satisfy the equation are

$\qquad p = 3, \quad q = -2$

$3a - 2b$ is parallel to the x-axis.

(b) $ra + sb = r\begin{pmatrix} 3 \\ 2 \end{pmatrix} + s\begin{pmatrix} 1 \\ 3 \end{pmatrix}$

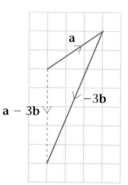

The x-component must be zero because the vector is in the direction of the y-axis.

So $3r + s = 0$

Possible values of r and s are

$\qquad r = 1, \quad s = -3$

$a - 3b$ is parallel to the y-axis.

For any two vectors you can find a linear combination parallel to any required direction.

Example 6

Find a linear combination of vectors $a = \begin{pmatrix} 3 \\ 2 \end{pmatrix}$ and $b = \begin{pmatrix} 1 \\ 3 \end{pmatrix}$ which is equal to vector $c = \begin{pmatrix} 9 \\ 13 \end{pmatrix}$

$$pa + qb = c$$

$$p\begin{pmatrix} 3 \\ 2 \end{pmatrix} + q\begin{pmatrix} 1 \\ 3 \end{pmatrix} = \begin{pmatrix} 9 \\ 13 \end{pmatrix}$$

$$\begin{pmatrix} 3p \\ 2p \end{pmatrix} + \begin{pmatrix} q \\ 3q \end{pmatrix} = \begin{pmatrix} 9 \\ 13 \end{pmatrix}$$

Using simultaneous equations: $3p + q = 9$ (1)

$\qquad\qquad\qquad\qquad\qquad\quad 2p + 3q = 13$ (2)

Multiply equation (1) by 3: $9p + 3q = 27$ (3)

$\qquad\qquad\qquad\qquad\qquad\quad 2p + 3q = 13$ (2)

Subtract (2) from (3): $7p = 14$

$\qquad\qquad\qquad\qquad\qquad\qquad\quad p = 2$

Substitute $p = 2$ in (1): $6 + q = 9$

$\qquad\qquad\qquad\qquad\qquad\qquad\quad q = 3$

The linear combination is $2a + 3b$

Exercise 27E

1   $\mathbf{a} = \begin{pmatrix} 2 \\ -1 \end{pmatrix}$ and $\mathbf{b} = \begin{pmatrix} 1 \\ 1 \end{pmatrix}$

Given that $2\mathbf{a} + p\mathbf{b}$ is parallel to the x-axis, find the value of p.

2   $\mathbf{a} = \begin{pmatrix} 2 \\ -1 \end{pmatrix}$ and $\mathbf{b} = \begin{pmatrix} 1 \\ 1 \end{pmatrix}$

Given that $\mathbf{a} + q\mathbf{b}$ is parallel to the y-axis, find the value of q.

3   Find the values of x and y, given that $2\begin{pmatrix} 1 \\ y \end{pmatrix} + \begin{pmatrix} x \\ y \end{pmatrix} = \begin{pmatrix} 2 \\ -6 \end{pmatrix}$

4   Find the values of the scalars p and q, given that

$$p\begin{pmatrix} 2 \\ 1 \end{pmatrix} + q\begin{pmatrix} -1 \\ 2 \end{pmatrix} = \begin{pmatrix} 7 \\ -2 \end{pmatrix}.$$

5   P is a variable point which moves so that the vector $\overrightarrow{OP}$ is given by

$$\overrightarrow{OP} = \begin{pmatrix} 2 \\ 6 \end{pmatrix} + t\begin{pmatrix} 1 \\ 1 \end{pmatrix}$$

Calculate the coordinates of P, for values of t from 0 to 5. Plot these coordinates.
What is the path of P as t varies? Give the equation of the path in the form $y = mx + c$.

6   A is the point $(2, 1)$, B is the point $(8, 4)$ and C is the point $(6, 6)$.

(a)   Calculate $\overrightarrow{AB}$.

(b)   Write an expression in terms of k for $\overrightarrow{AB} + k\overrightarrow{BC}$.

(c)   The line BC is extended to a point D where AD is parallel to the y-axis.

Find the value of k and the coordinates of D.

27.6 Position vectors

The column vector $\begin{pmatrix} x \\ y \end{pmatrix}$ denotes a translation.

There are an infinite number of points which are related by such a translation.

Look at the diagram, which shows several pairs of points linked by the same vector. The vector which translates

O to P, $\overrightarrow{OP}$, is a special vector, called the **position vector** of P.

It is called this because it fixes the position of point P relative to a fixed reference point, which is usually the origin.

In this case, $\overrightarrow{OP} = \begin{pmatrix} x \\ y \end{pmatrix}$.

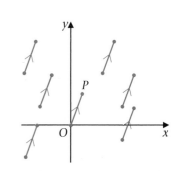

The **position vector** of a point P is $\overrightarrow{OP}$, where O is usually the origin.

Example 7

P is the point $(2, 3)$. $\overrightarrow{PQ} = \begin{pmatrix} -1 \\ 2 \end{pmatrix}$.

Find the position vector of Q.

The position vector of Q is

$$\overrightarrow{OQ} = \overrightarrow{OP} + \overrightarrow{PQ}$$

$$= \begin{pmatrix} 2 \\ 3 \end{pmatrix} + \begin{pmatrix} -1 \\ 2 \end{pmatrix}$$

$$= \begin{pmatrix} 1 \\ 5 \end{pmatrix}$$

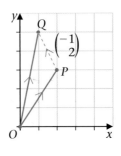

Example 8

A is the point $(3, 5)$ and B is the point $(6, -1)$. C is the point on AB such that $AC = \frac{1}{2} CB$. Find the position vector of the point C.

The position vector of C is $\overrightarrow{OC}$. $\overrightarrow{OC} = \overrightarrow{OA} + \overrightarrow{AC}$

$$\overrightarrow{OA} = \begin{pmatrix} 3 \\ 5 \end{pmatrix}$$

$$\overrightarrow{AB} = \begin{pmatrix} 3 \\ -6 \end{pmatrix}$$

$$\overrightarrow{AC} = \tfrac{1}{3} \overrightarrow{AB}$$

$$= \tfrac{1}{3} \begin{pmatrix} 3 \\ -6 \end{pmatrix}$$

$$= \begin{pmatrix} 1 \\ -2 \end{pmatrix}$$

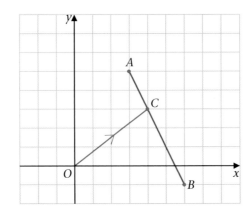

$$\overrightarrow{OC} = \overrightarrow{OA} + \overrightarrow{AC} = \begin{pmatrix} 3 \\ 5 \end{pmatrix} + \begin{pmatrix} 1 \\ -2 \end{pmatrix}$$

$$= \begin{pmatrix} 4 \\ 3 \end{pmatrix}$$

This problem is solved again in Example 9, using vector notation.

All the properties established for vectors earlier hold for position vectors.

Here are two very useful results.

If A and B have position vectors **a** and **b** respectively, then the vector

$\overrightarrow{AB} = \mathbf{b} - \mathbf{a}$ (Note the reversal of letters.)

From the diagram:

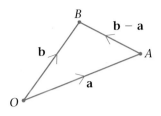

$$\overrightarrow{AB} = \overrightarrow{AO} + \overrightarrow{OB}$$
$$= -\overrightarrow{OA} + \overrightarrow{OB}$$
$$= -\mathbf{a} + \mathbf{b}$$
$$= \mathbf{b} - \mathbf{a}$$

___ **Example 9** ___

If the position vector of A is $\mathbf{a}$ and the position vector of B is $\mathbf{b}$ find $\mathbf{c}$, the position vector of C, in terms of $\mathbf{a}$ and $\mathbf{b}$ so that $AC = \frac{1}{2}CB$.

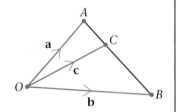

$$\mathbf{c} = \overrightarrow{OC} = \overrightarrow{OA} + \overrightarrow{AC}$$
$$\overrightarrow{AB} = \mathbf{b} - \mathbf{a}$$
$$\overrightarrow{AC} = \tfrac{1}{3}(\mathbf{b} - \mathbf{a})$$
$$\mathbf{c} = \overrightarrow{OA} + \overrightarrow{AC} = \mathbf{a} + \tfrac{1}{3}(\mathbf{b} - \mathbf{a})$$
$$= \tfrac{2}{3}\mathbf{a} + \tfrac{1}{3}\mathbf{b}$$

Check that this gives the same answer as Example 8.

If A and B have position vectors $\mathbf{a}$ and $\mathbf{b}$ respectively, then the position vector of the mid-point, M, of the line joining A to B is
$$\overrightarrow{OM} = \mathbf{m} = \tfrac{1}{2}(\mathbf{a} + \mathbf{b})$$

From the diagram:

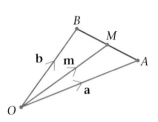

$$\overrightarrow{AM} = \tfrac{1}{2}\overrightarrow{AB}$$
$$\overrightarrow{OM} = \overrightarrow{OA} + \overrightarrow{AM}$$
$$= \mathbf{a} + \tfrac{1}{2}(\mathbf{b} - \mathbf{a})$$
$$= \mathbf{a} + \tfrac{1}{2}\mathbf{b} - \tfrac{1}{2}\mathbf{a}$$
$$= \tfrac{1}{2}(\mathbf{a} + \mathbf{b})$$

Exercise 27F

1   A is the point $(1, 2)$ and B is the point $(3, 6)$.
   Write the position vector of the mid-point of AB.

2   C is the point $(2, -2)$ and D is the point $(8, 4)$. E lies on the line
   CD, such that $CE = \frac{1}{2}ED$.
   Find the coordinates of E.

3   A is the point $(2, 3)$ and B is the point $(4, 7)$. C lies on the
   extension of the line AB, such that $BC = \frac{1}{2}AB$.
   Find the coordinates of C.

4   $PQRS$ is a parallelogram, with $P = (1, 1)$, $Q = (5, 3)$ and $R = (7, 7)$.
   Find the position vector of S and of the mid-point of PR.

27.7 Proving geometrical results

Example 10

A and B are the mid-points of the sides OX and OY of a triangle.
Prove that the line XY is parallel to the line AB and is twice the length of AB.

Use letters **a** and **b** for the position vectors of A and B, respectively.

$$\overrightarrow{OX} = 2\mathbf{a}$$
$$\overrightarrow{OY} = 2\mathbf{b}$$
$$\overrightarrow{XY} = \overrightarrow{OY} - \overrightarrow{OX}$$
$$= 2\mathbf{b} - 2\mathbf{a}$$
$$= 2(\mathbf{b} - \mathbf{a})$$

But $\overrightarrow{AB} = \mathbf{b} - \mathbf{a}$

so $\overrightarrow{XY} = 2\overrightarrow{AB}$

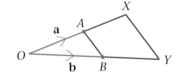

So XY is twice the length of AB and XY is parallel to AB.

Example 11

Show that the diagonals of a parallelogram bisect one another.

In the parallelogram $OACB$, $\overrightarrow{OA} = \mathbf{a}$ and $\overrightarrow{OB} = \mathbf{b}$.

Opposite sides of a parallelogram are parallel and equal so $\overrightarrow{AC} = \mathbf{b}$.

$$\overrightarrow{OC} = \overrightarrow{OA} + \overrightarrow{AC}$$
$$= \mathbf{a} + \mathbf{b}$$

If D is the mid-point of OC:

$$\overrightarrow{OD} = \tfrac{1}{2}\overrightarrow{OC}$$
$$= \tfrac{1}{2}(\mathbf{a} + \mathbf{b})$$

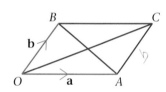

But $\tfrac{1}{2}(\mathbf{a} + \mathbf{b})$ is the position vector of the mid-point of AB.
So mid-point of AB = mid-point of OC, which means that OC bisects AB.

> Remember the result for two vectors **a** and **b**:
>
> the position vector of the mid-point of AB is $\tfrac{1}{2}(\mathbf{a} + \mathbf{b})$

Example 12

$OABC$ is a quadrilateral, with $\overrightarrow{OA} = \mathbf{a}$, $\overrightarrow{OB} = \mathbf{b}$ and $\overrightarrow{OC} = \mathbf{c}$.
P, Q, R and S are the mid-points of sides OA, AB, BC and CO.
Show that $PQRS$ is a parallelogram.

From the diagram,

$$\overrightarrow{OP} = \tfrac{1}{2}\mathbf{a} \text{ and } \overrightarrow{OS} = \tfrac{1}{2}\mathbf{c}$$
$$\overrightarrow{OQ} = \tfrac{1}{2}(\mathbf{a} + \mathbf{b}) \text{ and } \overrightarrow{OR} = \tfrac{1}{2}(\mathbf{b} + \mathbf{c})$$

Now $\overrightarrow{PQ} = \overrightarrow{OQ} - \overrightarrow{OP} = \tfrac{1}{2}\mathbf{b}$ and $\overrightarrow{SR} = \overrightarrow{OR} - \overrightarrow{OS} = \tfrac{1}{2}\mathbf{b}$

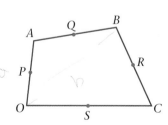

Thus PQ has equal length and is parallel to SR.
So QR is equal and parallel to PS, and $PQRS$ is a parallelogram.

Exercise 27G

1   $\overrightarrow{OX} = \mathbf{x}$ and $\overrightarrow{OY} = \mathbf{y}$. X and and Y are $\frac{2}{3}$ of the way along OA and
    OB, respectively. Write the vector $\overrightarrow{XY}$ in terms of $\mathbf{x}$ and $\mathbf{y}$.
    Write the geometrical relationship between the line XY and the
    line AB.

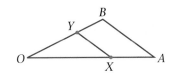

2   Express $\overrightarrow{AB}$ in terms of $\mathbf{a}$ and $\mathbf{b}$, where $\overrightarrow{OA} = \mathbf{a}$ and $\overrightarrow{OB} = \mathbf{b}$.
    Given that $AC = 2OA$ and $BD = 2OB$, and that OAC and OBD are
    straight lines, express $\overrightarrow{CD}$ in terms of $\mathbf{a}$ and $\mathbf{b}$.
    What is the geometrical relationship between AB and CD?

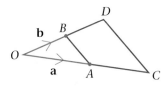

3   (a) X is the mid-point of AB and Y is the mid-point of OA.
          G is the point which is $\frac{2}{3}$ of the way along the line OX.
          Find the position vector of G.

    (b)   A point H is chosen to be $\frac{2}{3}$ of the way along the line BY.
          Find the position vector of H.
          What geometrical fact can you conclude from your
          answers?

    (c)   The line AG, when extended, cuts the line OB at D.
          Find the position vector of D.

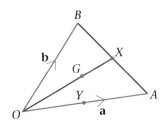

4   $OCDB$ is a trapezium, with OC parallel to BD. A is the
    mid-point of OC and E is the mid-point of BD. X is the
    point of intersection of AE and BC. $BD = 4\mathbf{a}$.

    (a)   Show that $\overrightarrow{OX}$ can be written in the form
          $(1 + k)\mathbf{a} + k\mathbf{b}$, where k is a scalar.

    (b)   Show that $\overrightarrow{OX}$ can be written in the form
          $(2 - 2m)\mathbf{a} + m\mathbf{b}$, where m is a scalar.

    (c)   Find the values of k and m, and hence the position vector of X.

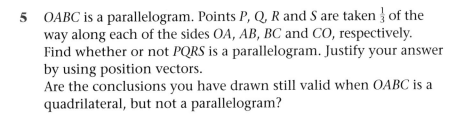

Diagram NOT
accurately
drawn

5   $OABC$ is a parallelogram. Points P, Q, R and S are taken $\frac{1}{3}$ of the
    way along each of the sides OA, AB, BC and CO, respectively.
    Find whether or not $PQRS$ is a parallelogram. Justify your answer
    by using position vectors.
    Are the conclusions you have drawn still valid when $OABC$ is a
    quadrilateral, but not a parallelogram?

6   Find, in terms of $\mathbf{a}$ and $\mathbf{b}$
    (a) $\overrightarrow{AB}$ (b) $\overrightarrow{AF}$ (c) $\overrightarrow{AE}$
    (d)   The mid-points of each of the sides of the regular hexagon
          $ABCDEF$ are joined to give a second hexagon. Find the
          position vectors of each of the corners of this second
          hexagon. This process is repeated to produce a third
          hexagon. What is the relationship between the first and third
          hexagons? Justify your answer.

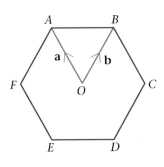

Mixed exercise 27

1   Give the column vector that
    describes the translation which maps

    (a)   *A* to *B*

    (b)   *B* to *A*

    (c)   *B* to *C*

    (d)   *C* to *B*

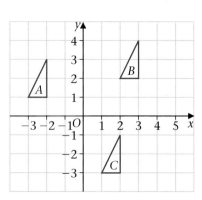

2   Write these as column vectors:

    (a) $\overrightarrow{OA}$ (b) $\overrightarrow{OB}$

    (c) $\overrightarrow{OA} + \overrightarrow{OB}$ (d) $\overrightarrow{OA} - \overrightarrow{OB}$

    (e) $2\overrightarrow{OA}$ (f) $\overrightarrow{OB} + \overrightarrow{AO}$

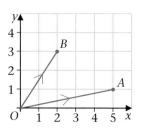

3   *B* is the point such that *OABC* is a parallelogram.
    Copy the diagram and plot the point *B*.

    Write the vector $\overrightarrow{OB}$.

    Write the vector $\overrightarrow{OM}$ where *M* is the mid-point of *OB*.

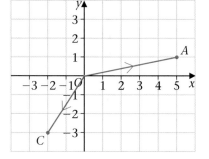

4   $\overrightarrow{OA} = \begin{pmatrix} 2 \\ 1 \end{pmatrix}$ and $\overrightarrow{OB} = \begin{pmatrix} 5 \\ 6 \end{pmatrix}$ Write down $\overrightarrow{AB}$.

5   $\overrightarrow{XY} = \begin{pmatrix} 2 \\ 3 \end{pmatrix}$ and $\overrightarrow{YO} = \begin{pmatrix} -1 \\ 4 \end{pmatrix}$ Write down $\overrightarrow{OX}$.

6   Simplify   (a) $\begin{pmatrix} 2 \\ 1 \end{pmatrix} + \begin{pmatrix} 1 \\ -1 \end{pmatrix}$ (b) $\begin{pmatrix} 2 \\ 1 \end{pmatrix} - \begin{pmatrix} 1 \\ -1 \end{pmatrix}$

7   Simplify $3\begin{pmatrix} 2 \\ 1 \end{pmatrix} - 2\begin{pmatrix} -2 \\ 1 \end{pmatrix}$

8   Given that $\begin{pmatrix} 2 \\ x \end{pmatrix} + \begin{pmatrix} x \\ y \end{pmatrix} = \begin{pmatrix} 5 \\ 6 \end{pmatrix}$, find the values of *x* and *y*.

9   Solve the vector equation $\begin{pmatrix} a \\ 2 \end{pmatrix} - \begin{pmatrix} -3 \\ b \end{pmatrix} = \begin{pmatrix} 4 \\ b \end{pmatrix}$

10 $\mathbf{a} = \begin{pmatrix} 3 \\ 0 \end{pmatrix}$, $\mathbf{b} = \begin{pmatrix} 0 \\ 2 \end{pmatrix}$, $\mathbf{c} = \begin{pmatrix} 3 \\ 1 \end{pmatrix}$

    Find the values of *p* and *q* such that $p\mathbf{a} + q\mathbf{b} = \mathbf{c}$

11 *ABCD* is a parallelogram.
    Express, in terms of **a** and **b**

    (a) $\overrightarrow{AC}$ (b) $\overrightarrow{DA}$ (c) $\overrightarrow{DB}$

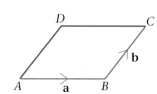

12 Given that $\overrightarrow{OC} = \frac{2}{5}\overrightarrow{OB}$, express, in terms of **a** and **b**:

(a) $\overrightarrow{CB}$ (b) $\overrightarrow{BA}$

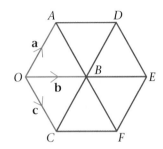

D is the point on *BA* so that $BD:DA = 3:2$

Express, in terms of **a** and **b**

(c) $\overrightarrow{BD}$ (d) $\overrightarrow{CD}$

What can you conclude about *CD* and *OA*?

13 *OADEFC* is a regular hexagon and *B* is the point of intersection of the diagonals.

$\overrightarrow{OA} = \mathbf{a}$, $\overrightarrow{OB} = \mathbf{b}$ and $\overrightarrow{OC} = \mathbf{c}$

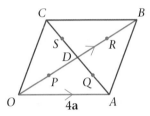

(a) Express, in terms of **a**, **b** or **c** as appropriate, the vector $\overrightarrow{OD}$.

(b) Find two different expressions for the vector $\overrightarrow{OE}$.

(c) Write down an equality involving **a**, **b** and **c**.

14 *OABC* is a parallelogram with $\overrightarrow{OA} = 4\mathbf{a}$ and $\overrightarrow{OB} = 4\mathbf{b}$.
The diagonals intersect at *D*.
P, *Q*, *R* and *S* are the mid-points of *OD*, *AD*, *DB* and *DC* respectively.
Show, using vector algebra, that *PQRS* is a parallelogram.

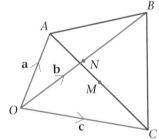

15 $\overrightarrow{OA} = \mathbf{a}$, $\overrightarrow{OB} = \mathbf{b}$ and $\overrightarrow{OC} = \mathbf{c}$. *N* is the mid-point of *OB* and *M* is the mid-point of *AC*. Express

(a) $\overrightarrow{AB}$ in terms of **a** and **b**

(b) $\overrightarrow{ON}$ in terms of **b**

(c) $\overrightarrow{AC}$ in terms of **a** and **c**

(d) $\overrightarrow{AM}$ in terms of **a** and **c**

(e) $\overrightarrow{OM}$ in terms of **a** and **c**

(f) $\overrightarrow{NM}$ in terms of **a**, **b** and **c**.

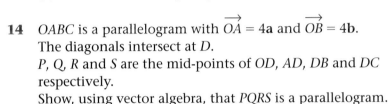

16 $\overrightarrow{OA} = \mathbf{a}$ and $\overrightarrow{OB} = \mathbf{b}$. *X* is the point such that $\overrightarrow{OX} = \frac{3}{2}\overrightarrow{OA}$ and *Y* is the point such that $\overrightarrow{OY} = 3\overrightarrow{OB}$.

Express, in terms of **a** and **b**

(a) $\overrightarrow{OX}$ (b) $\overrightarrow{OY}$ (c) $\overrightarrow{YX}$

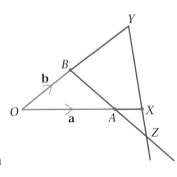

(d) The lines *YX* and *BA* are extended to meet at *Z*.
Explain why the position vector of any point on *YX* extended can be written as $3\mathbf{b} + p(\frac{3}{2}\mathbf{a} - 3\mathbf{b})$ and that the position vector of any point on *BA* extended can be written as $\mathbf{b} + q(\mathbf{a} - \mathbf{b})$.
Hence, or otherwise, find the position vector of *Z*.

17 $\overrightarrow{AB} = \begin{pmatrix} 1 \\ -4 \end{pmatrix}$, $A = (2, 3)$, $C = (-1, 15)$

Find the values of *x* and *y* if $x\overrightarrow{OA} - y\overrightarrow{OB} = \overrightarrow{OC}$

Summary of key points

1 Translations can be described using **column vectors**: $\begin{pmatrix} x \\ y \end{pmatrix}$

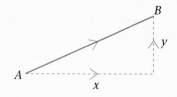

2 The values of x and y in a column vector are called **components**.

3 The translation that takes A to B can be written as a translation vector $\overrightarrow{AB}$.

4 If $\overrightarrow{AD} = \overrightarrow{BC}$, then AD and BC are parallel and have equal length.

5 A vector described as **a** has a unique length and direction.
The vector with same length but opposite direction is $-\mathbf{a}$.

6 For any two vectors **a** and **b**:
- $\mathbf{a} + \mathbf{b} = \mathbf{b} + \mathbf{a}$
- $\mathbf{a} - \mathbf{b} = \mathbf{a} + (-\mathbf{b})$

7 If a vector **a** is multiplied by a scalar k then the vector $k\mathbf{a}$ is parallel to **a** and is equal to k times **a**.

8 For any vectors **a**, **b** and **c** and any scalars p, q and k:
- $\mathbf{a} + \mathbf{b} = \mathbf{c}$ can be written as $\mathbf{a} = \mathbf{c} - \mathbf{b}$
- $k\mathbf{a} = k \times \mathbf{a}$
- $\mathbf{a} + (\mathbf{b} + \mathbf{c}) = (\mathbf{a} + \mathbf{b}) + \mathbf{c}$
- $k(\mathbf{a} + \mathbf{b}) = k\mathbf{a} + k\mathbf{b}$
- $(p + q)\mathbf{a} = p\mathbf{a} + q\mathbf{a}$

9 If two vectors **a** and **b** are parallel, then $\mathbf{a} = k\mathbf{b}$ for some scalar k.

10 The **magnitude** of a vector is the length of the directed line segment representing it.

11 In general, the magnitude of a vector $\begin{pmatrix} x \\ y \end{pmatrix}$ is $\sqrt{x^2 + y^2}$

12 Combinations of the vectors **a** and **b** of the form $p\mathbf{a} + q\mathbf{b}$, where p and q are scalars, are called **linear combinations** of the vectors **a** and **b**.

13 For any two vectors you can find a linear combination parallel to any required direction.

14 The **position vector** of a point P is $\overrightarrow{OP}$, where O is usually the origin.

15 If A and B have position vectors **a** and **b** respectively, then the vector $\overrightarrow{AB} = \mathbf{b} - \mathbf{a}$.

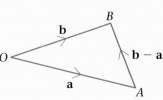

16 If A and B have position vectors **a** and **b** respectively, then the position vector of the mid-point, M, of the line joining A to B is $\overrightarrow{OM} = \mathbf{m} = \frac{1}{2}(\mathbf{a} + \mathbf{b})$.

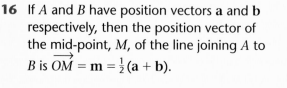

28 Introducing modelling

Simplifying a 'real-life' problem into a model which can be solved mathematically is called **mathematical modelling**. Estimating the height of a large waterfall using trigonometry is an example of mathematical modelling.

This chapter shows how exponential functions and trigonometric functions may be used in such models. It also shows how experimental data can be analysed to obtain relationships between two variables.

Mathematical model

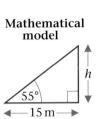

28.1 Modelling using exponential functions

2^x, $(0.5)^y$ and 3^{2t} are all examples of exponential functions.

> The function a^x, where a is a positive constant and x is a variable, is called an **exponential function**.

___
Example 1
___

Ben invests £2000 at Town Bank.
Town Bank pays 9% compound interest per annum, added yearly.
Ben does not intend to withdraw any interest.
(a) Find a formula which models this problem to give the value of the investment after t years.
(b) Use your formula to find when the investment is first worth more than £6000.

(a) Interest for the first year is $0.09 \times £2000$

Value of the investment after 1 year $= £2000 + 0.09 \times £2000$
$\qquad\qquad = (1 + 0.09) \times £2000$
$\qquad\qquad = 1.09 \times £2000$

> 9% = 0.09

Interest for the second year $= 0.09 \times (1.09 \times £2000)$
Value of the investment after 2 years $= (1.09 \times £2000) + 0.09 \times (1.09 \times £2000)$
$\qquad\qquad = (1 + 0.09)(1.09 \times £2000)$
$\qquad\qquad = 1.09 \times (1.09 \times £2000)$
$\qquad\qquad = (1.09)^2 \times £2000$

Similarly, the value of the investment after 3 years
$\qquad\qquad = (1.09)^3 \times £2000$

and the value of the investment after 4 years
$\qquad\qquad = (1.09)^4 \times £2000$

This is a sequence which gives the formua for the value, V, of the investment after t years as,
$\qquad V = (1.09)^t \times £2000$

(b) We need the smallest integer value of t such that $(1.09)^t \times £2000 > £6000$

so
$$(1.09)^t > 3$$

Using the method of trial and improvement and a calculator, try:

$t = 12$ $(1.09)^{12} = 2.8126...$ ——too small

$t = 13$ $(1.09)^{13} = 3.0658...$

So the value of the investment is over £6000 for the first time after 13 years.

The graph shows the value of the investment V after t years for Example 1.

This is an example of **exponential growth**.

The value of the investment is said to **grow exponentially** with a **multiplier** of 1.09

If $a > 1$ then a^x is an example of **exponential growth** with a multiplier of a.

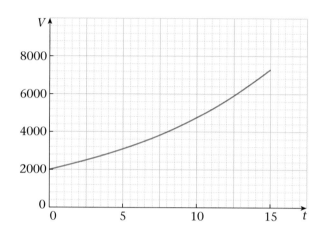

Example 2

A student published an article in which she reported that the number of birds of prey nesting in an area had been decreasing by 8% per year since 1990. There were 600 birds in 1990.

Assuming this same rate of decrease, find a formula for the number, N, of these birds of prey that will nest in the same area t years after 1990.

Decrease of birds during 1991 $= 0.08 \times 600$

Number of birds in 1991 $= 600 - 0.08 \times 600$

$= 0.92 \times 600$

Decrease of birds during 1992 $= 0.08 \times (0.92 \times 600)$

Number of birds in 1992 $= (0.92 \times 600) - 0.08 \times (0.92 \times 600)$

$= (1 - 0.08)(0.92 \times 600)$

$= (0.92)^2 \times 600$

Similarly, the number of birds 3 years after 1990 $= (0.92)^3 \times 600$

and   the number of birds 4 years after 1990 $= (0.92)^4 \times 600$

So   the number of birds t years after 1990 $= (0.92)^t \times 600$

$$N = (0.92)^t \times 600$$

The graph shows the number of nesting birds of prey, N, t years after 1990 for Example 2.

This is an example of **exponential decay**.

The number of birds is said to **decay exponentially** with a **multiplier** of 0.92.

> If $0 < a < 1$ then a^x is an example of **exponential decay** with a multiplier of a.

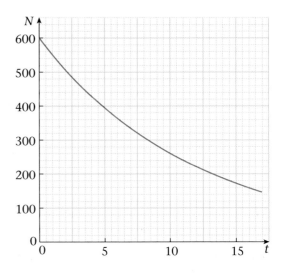

Exercise 28A

1   Rina invests £1500 at Shirebank. Shirebank pays 6% compound interest per annum, added yearly. Rina does not intend to withdraw any interest.

   (a)   Find a formula which models this problem to give the value of the investment after t years.

   (b)   Use your formula to find when the investment is first worth at least £3750.

2   A ball, dropped from a height x metres, rebounds to a height $0.6x$ metres. If the ball is dropped from a height of 8 metres, find

   (a)   a formula which models this problem to give the height of rebound after the nth bounce

   (b)   the total vertical distance the ball has travelled just before its third bounce.

3   The time taken for a mass of radioactive uranium to halve is 24 days. How long does it take for 40 mg to reduce to

   (a)   10 mg

   (b)   2.5 mg?

4   Winston's debt of £40 000 decreases by 20% per year.

   (a)   What is his debt after

      (i)   1 year

      (ii)   4 years

      (iii) n years?

   (b)   After how many years will the debt first fall below £100?

28.2 Modelling using trigonometric functions

For problems where the motion repeats itself after a certain time, the sine and cosine functions are the most appropriate.

> For the graphs of the sine and cosine functions, see Chapter 16.

Example 3

A travelling fair comes to a Cheshire town every May. The fair includes a big wheel which is constructed and tested. The diameter of the wheel is 20 m. Its centre is 12 m above the ground.

The wheel is first tested using one chair. The wheel rotates anticlockwise and is timed for each revolution once it reaches its working speed.

The timing starts as the chair moves upwards through the point level with the centre of the wheel.

The wheel rotates once every 36 seconds.

(a) Find the constants p and q so that $y = p + q \sin(10t)°$ is a suitable model for the height, y metres, of the chair above the ground t seconds after timing starts.

(b) Find the times during the first minute when the chair is 17 m above the ground.

(c) Sketch the graph of y against t for $0 \leq t \leq 54$.

(a) The diagram represents the big wheel, with A the position of the chair when the timing starts and B, the highest point reached by the chair shown.

$$y = p + q \sin(10t)°$$

When $t = 0$, $y = 12$, giving
$$12 = p + q \sin 0°$$
$$12 = p + q(0)$$
$$12 = p$$

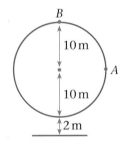

The chair reaches B when the wheel completes a $\frac{1}{4}$ of a revolution. The time taken, t, is $\frac{1}{4} \times 36 = 9$ seconds.

$$y = p + q \sin(10t)°$$

When $t = 9$, $y = 22$, giving
$$22 = p + q \sin 90°$$
$$22 = p + q(1)$$

But $p = 12$, so
$$22 = 12 + q$$
$$10 = q$$

(b) Use $y = 12 + 10 \sin(10t)°$ to find the value of t when $y = 17$:
$$17 = 12 + 10 \sin(10t)°$$
$$5 = 10 \sin(10t)°$$
$$\sin(10t)° = 0.5$$

So $(10t)° = 30°, 180° - 30°, 360° + 30°, 540° - 30°, 720° + 30°, \ldots$

$t = 3, 15, 39, 51, 75, \ldots$

The required times are 3 s, 15 s, 39 s and 51 seconds.

(c)

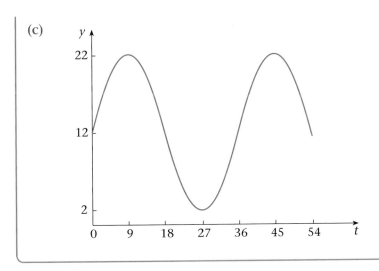

$y = 12 + 10 \sin (10t)$ is the graph of $y = \sin t$
- stretched by 10 parallel to the y-axis
- stretched by $\frac{1}{10}$ parallel to the x-axis and
- translated 12 units parallel to the y-axis.

Check the formula for y:
when $t = 18$, $y = 12$
when $t = 27$, $y = 2$
when $t = 36$, $y = 12$.

In coastal areas the depth of the water depends on the time of day.

The depth of the water is greatest at high tide and least at low tide. Sailors need to know the times when these two extremes occur.

This situation can be modelled using trigonometric functions.

Example 4

Martin believes that the depth of water, d metres, at the end of a jetty t hours after low tide can be modelled by a formula of the form

$d = a + b \cos(kt)°$

where a, b and k are constants.

He measures the depth of water at low tide as 2 metres.

(a) Assuming that low tides occur every 12 hours, show that $k = 30$.

Martin also measures the depth of water at high tide as 6 metres.

(b) Calculate the values of a and b.

Martin needs at least 3 metres of water at the end of the jetty to sail his boat.

(c) Given that low tide on a particular day was at 09:00, find the earliest time that Martin could sail his boat.

(a) Low tide is when $t = 0$ and $t = 12$

$$d = a + b \cos(kt)°$$

When $t = 0$, $d = 2$ $2 = a + b \cos 0°$ so $2 = a + b$

When $t = 12$, $d = 2$ $2 = a + b \cos(12k)°$

$k \neq 0$ because the depth of water is not constant.

This gives $\cos(12k)° = 1$

$$12k = \cos^{-1} 1 = 0 \text{ or } 360$$

$$k = 30$$

(b) High tide is when $t = 6$

$$d = a + b\cos(30t)°$$

When $t = 6$, $d = 6$ $6 = a + b\cos 180°$ so $6 = a - b$

But $\underline{2 = a + b}$

$\;\; 8 = 2a$

so $a = 4$ and $b = -2$

(c) Use $d = 4 - 2\cos(30t)°$ to find the value of t when $d = 3$:

$$d = 4 - 2\cos(30t)°$$

When $d = 3$ $3 = 4 - 2\cos(30t)°$

$$2\cos(30t)° = 1$$

$$\cos(30t)° = 0.5$$

$$30t = 60 \quad \text{so } t = 2$$

The **earliest** time that Martin can sail his boat is 11:00.

Exercise 28B

1   The depth, d metres, of water at the end of a jetty t hours after noon is modelled by the formula

$$d = 4 + 2.5\cos(30t)°$$

   (a)   Find the depth of water at
      (i) noon   **(ii)** 2 pm   **(iii)** 3 pm   **(iv)** 6 pm   **(v)** midnight.

   (b)   Find the first time, correct to the nearest minute, when the depth of water is 6 metres.

   (c)   Sketch the graph of d against t for $0 \leqslant t \leqslant 12$.

2   One end of a spring is fixed to a wall at point P. A mass M, which lies on a table, is attached to the other end. PM is horizontal. Damien pushes the mass towards P and releases it. He models the distance, y cm, of the mass from P at time t seconds after releasing it by the formula

$$y = 15 - 5\cos(45t)°$$

   (a)   Find the distance of the mass from the wall when
      (i) $t = 2$ **(ii)** $t = 4$ **(iii)** the mass is released.

   (b)   Sketch the graph of y against t for $0 \leqslant t \leqslant 8$.

3 t hours after midnight, the depth of water, d metres, at the entrance of a harbour is modelled by the formula:

$$d = 6 + 3\sin(30t)°$$

   (a)   What is the depth of water at **(i)** 1 am   **(ii)** noon?

   (b)   What is the depth of water at low tide?

   (c)   Find the times of high tide during a complete day.

   (d)   Sketch the graph of d against t for $0 \leqslant t \leqslant 24$.

4 The diameter of a big wheel is 16 m. Its centre is 9 m above the ground. The wheel rotates clockwise.

Mandy rides on the big wheel and starts to time it when her chair reaches its highest point. The wheel rotates once every 20 seconds.

(a) Find the constants p and q so that

$$y = p + q \cos(18t)°$$

is a suitable model for the height of the chair, y metres, above the ground t seconds after timing starts.

(b) Find the times during the first half minute when the chair is 13 m above the ground.

(c) Find the times during the first half minute when the chair is 5 m above the ground.

(d) Sketch the graph of y against t for $0 \leqslant t \leqslant 30$.

28.3 Using a line of best fit to obtain a relationship

Scientists frequently collect data from scientific experiments involving two quantities.

These experiments can be costly so the scientists try to use the data to obtain a relationship between the two quantities. They can then apply this relationship to other values of one of the variables and predict other results without having to carry out further experiments.

Experimental data is subject to errors in the measuring instruments so points are unlikely to lie exactly on a straight line when plotted.

> For more on lines of best fit see Section 10.12.

A **line of best fit** is used.

Example 5

This scatter diagram shows the results of a scientific experiment involving two variables x and y.

The scientist has drawn in the line of best fit.

(a) Find the equation of the line of best fit.

(b) Assuming that this line is valid for larger values of x, find the value of y when $x = 52$.

> You cannot always assume that the results will be true for values outside the range of the data.

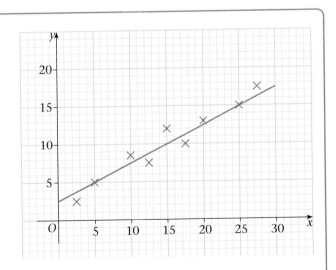

(a) The equation of a straight line is $y = mx + c$.

(25, 15) and (5, 5) are two points on the line of best fit.

> You could use *any* two points on the line.

Gradient $m = \dfrac{15 - 5}{25 - 5} = \dfrac{10}{20} = 0.5$

The equation of the line will take the form $y = 0.5x + c$

The point (5, 5) lies on the line

so $\quad 5 = 0.5(5) + c \Rightarrow c = 2.5$

The equation of the line of best fit is $y = 0.5x + 2.5$

> c could also have been read off from the graph as the y-intercept. In this example, reading off between squares may not be very accurate.

(b) $\qquad\qquad y = 0.5x + 2.5$

When $x = 52,\quad y = 0.5(52) + 2.5 = 26 + 2.5$

So $\qquad\qquad y = 28.5 \qquad$ when $x = 52$

To test the formula $y = px + q$, plot y against x.
If the points lie approximately on a straight line then p is the gradient of the line of best fit and q is the intercept on the vertical axis.

Exercise 28C

The scatter diagrams in questions **1–3** show the results of experiments involving two variables x and y. The line of best fit is drawn on each.

1 (a) Find the equation of the line of best fit.

(b) Assuming that this line is valid for larger values of x, find the value of y when $x = 48$.

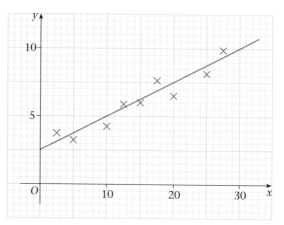

2 (a) Find the equation of the line of best fit.

(b) Assuming that this line is valid for larger values of x, find the value of y when $x = 18$.

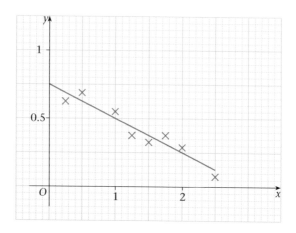

3 (a) Find the equation of the line of best fit.

(b) Assuming that this line is valid for larger values of x,
find the value of y when $x = 75$.

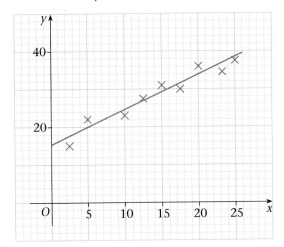

28.4 Reducing equations to linear form

Not all scientific data, when plotted as two variables, will lead to a
straight line.

> Y and X are the variables transformed to form a linear relationship.

This section shows how to rewrite some relationships in a linear form.
We shall use the general equation of a straight line as $Y = mX + c$,
where Y and X are the two variables, m is the gradient of the line and
c is the intercept on the y-axis.

We shall also use letters p and q to denote constants.

> Note: $Y = mX + c$ has three terms, two of which contain variables, that is Y and mX, and a third term, c, which does not contain a variable.

Equations of the form $y = px^2 + q$

Like $Y = mX + c$, $y = px^2 + q$ also has three terms, two with variables
and one without.

Comparing the two

$$Y = mX + c$$
$$\downarrow \quad \downarrow\downarrow \quad \downarrow$$
$$y = px^2 + q$$

> If $Y = mX + c$ passes through the origin then $c = 0$.

> To test the formula $y = px^2 + q$, plot y against x^2.
> If the points lie approximately on a straight line then p is the gradient
> of the line of best fit and q is the intercept on the vertical axis.

The gradient of the line gives the value of p and the intercept on the
y-axis gives the value of q.

Example 6

In an experiment these values of the variables V and R were obtained:

V	5	10	15	20	25
R	140	166	212	280	365

The variables V and R are thought to satisfy a relationship of the form $R = pV^2 + q$.

(a) Draw a graph to test this.

(b) Use your graph to estimate the values of the constants p and q.

(c) Use your relationship to find R when $V = 18$.

(a) Comparing $R = pV^2 + q$ to $Y = mX + c$

$$Y = mX + c$$
$$\downarrow \quad \downarrow\downarrow \quad \downarrow$$
$$R = pV^2 + q$$

Plotting R on the vertical axis and V^2 on the horizontal axis, should lead to an approximate straight line if $R = pV^2 + q$ is the correct relationship. The gradient of the line gives p and the intercept on the R-axis gives q.

V^2	25	100	225	400	625
R	140	166	212	280	365

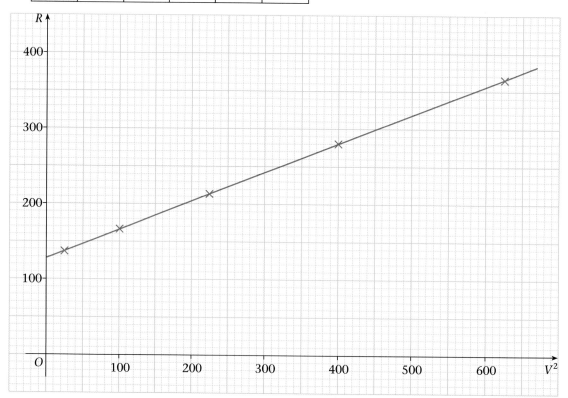

(b)   Taking the points on the line as (100, 166) and (400, 280):

> You could take any two points **on the line**.

$$\text{the gradient, } m = p = \frac{280 - 166}{400 - 100} = \frac{114}{300} = 0.38$$

and the y-intercept $q = 128$

So $R = 0.38V^2 + 128$

(c)   When $V = 18$, $R = 0.38 \times 18^2 + 128$

$$= 251.12$$

$$R = 251 \text{ (to 3 s.f.)}$$

Equations of the form $y = px^2 + qx$

In the equation $y = px^2 + qx$, all three terms contain variables so we cannot compare this with $Y = mX + c$ directly.

Dividing each term of $y = px^2 + qx$ by x gives $\dfrac{y}{x} = px + q$

> $y = px^2 + qx$
>
> $\dfrac{y}{x} = \dfrac{px^2}{x} + \dfrac{qx}{x}$
>
> $\dfrac{y}{x} = px + q$

Comparing $\dfrac{y}{x} = px + q$ with $Y = mX + c$

$$Y \;=\; mX + c$$

$$\downarrow \quad \downarrow\downarrow \quad \downarrow$$

$$\frac{y}{x} \;=\; px^2 + q$$

To test the formula $y = px^2 + qx$, plot $\dfrac{y}{x}$ against x.

If the points lie approximately on a straight line then p is the gradient of the line of best fit and q is the intercept on the vertical axis.

The gradient of the line gives p and the intercept on the $\dfrac{y}{x}$-axis gives q.

Example 7

An object is fired vertically upwards and its height, h metres, above the firing point is recorded t seconds later. The table shows the results.

t	1	2	3	4	5
h	27	44	48	44	27

(a)   Plot $\dfrac{h}{t}$ against t.

(b)   Explain why your graph verifies that $h = pt^2 + qt$, and use your graph to estimate the values of the constants p and q.

(c)   Use your relationship to find h when $t = 5.5$

(a)

t	1	2	3	4	5
$\dfrac{h}{t}$	27	22	16	11	5.4

(b) The graph of $\dfrac{h}{t}$ against t gives a straight line so $\dfrac{h}{t} = pt + q$, where p is the gradient and q is the intercept on the vertical axis. Multiplying both sides of the equation by t leads to $h = pt^2 + qt$.

$p =$ gradient of the line $= \dfrac{-11}{2} = -5.5$

$q =$ intercept on the vertical axis $= 33$

(c) $\qquad h = -5.5t^2 + 33t$

When $t = 5.5$,

$\qquad h = -5.5(5.5)^2 + 33(5.5)$

$\qquad = 15.125$ metres

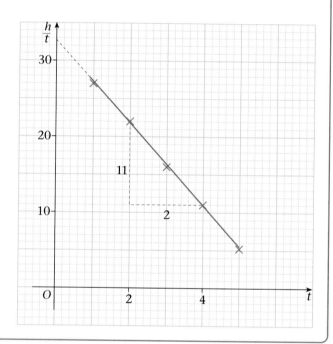

Exercise 28D

1

x	11	20	27	36	45
y	8	10.2	12.4	15	17

The table shows corresponding values of the variables x and y.
Peter believes that they satisfy a relationship of the form $y = ax + b$.

(a) Draw a graph to test whether Peter is correct.

(b) Use your graph to estimate the values of the constants a and b.

2

x	1	2	3	4	5
y	3.5	9.3	19	33	52

y is approximately equal to $ax^2 + b$.
Plot y against x^2 and use the graph to estimate a and b.

3 Water is squirted horizontally from a hosepipe.
   The height of the water is y metres at a distance x metres
   from the hosepipe. Measurements of x and y are:

x	0	1	2	3	5	6
y	7.2	7.0	6.4	5.5	2.2	0.1

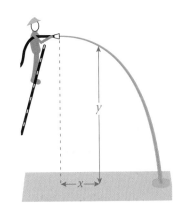

It is thought that the relationship between x and y is
of the form $y = ax^2 + b$. Plot y against x^2 and use the
graph to estimate a and b.

4

x	1	2	3	4	5
y	2.2	8.4	18.5	32.8	51.1

y is approximately equal to $ax^2 + bx$.

Plot $\dfrac{y}{x}$ against x and use the graph to
estimate a and b.

5

x	1	2	3	4	5
y	2.5	4.1	4.6	4.0	2.5

y is approximately equal to $ax^2 + bx$.

Plot $\dfrac{y}{x}$ against x and use the graph to
estimate a and b.

6

x	1	2	3	4	5
y	5.52	4.03	1.53	-1.96	-6.48

y is approximately equal to $ax^2 + b$.
Use a graph to estimate a and b.

28.5 Finding the constants in an exponential relationship

To draw the graph of $y = 3^x$, you can complete
a table of values of x by finding the corresponding
values of y.

x	0	1	2	3
y	1	3	9	27

You can then plot these points and join them
with a smooth curve.

The reverse process is also valid.

> If a point lies on a curve then the coordinates
> of the point satisfy the equation of the curve.

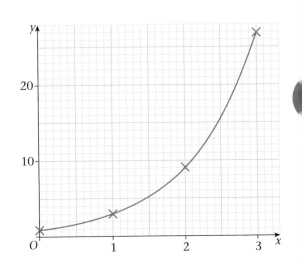

Example 8

This sketch shows part of the graph of $y = pq^x$.

It is known that the points $(0, 5)$, $(2, k)$ and $(3, 40)$ lie on this curve.

Use the sketch to find the values of p, q and k.

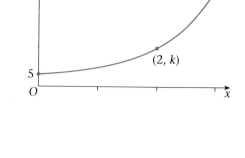

Point $(0, 5)$ lies on the curve $y = pq^x$,

so $\qquad 5 = pq^0$

$\qquad\qquad 5 = p \times 1 \qquad$ so $\qquad p = 5$

Point $(3, 40)$ lies on the curve $y = 5q^x$,

so $\qquad 40 = 5q^3 \qquad$ so $\qquad q^3 = 8$

$\qquad\qquad\qquad\qquad\qquad\qquad q = 2$

So $\qquad\qquad\qquad\qquad\qquad y = 5 \times 2^x$

Point $(2, k)$ lies on the curve $\quad y = 5 \times 2^x$, $\qquad$ so $\qquad k = 5 \times 2^2$

$\qquad\qquad\qquad\qquad\qquad\qquad\qquad\qquad\qquad\qquad\qquad k = 20$

Exercise 28E

1 This sketch shows part of the graph of $y = pq^x$.
Use the sketch to find the values of p, q and k.

2 The sketch below shows part of the graph of $y = pq^x$.
Use the sketch to find the values of p, q and k.

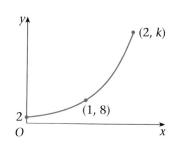

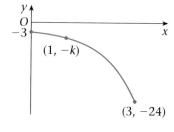

3 This sketch shows part of the graph of $y = a + b^x$.
Use the sketch to find the values of a, b and k.

4 The point $(1, 3)$ lies on the curve $y = a^{-x}$.
Calculate the value of a.

5 The point $(2, 2\frac{1}{4})$ lies on the curve $y = a^{-x}$.
Calculate the two possible values of a.

6 The three points $(0, 5)$, $(1, 4\frac{1}{3})$ and $(2, k)$ lie on the curve $y = a + b^{-x}$.
Calculate the values of a, b and k.

7 The point $(1, 8)$ lies on the curve $y = a^{-3x}$.
Calculate the value of a.

8 **Challenge**
   The points (2, 32) and (5, 2048) both lie on the curve $y = pq^x$.
   (a) Find the values of the constants p and q.
   (b) Given also that the point $(k, 128)$ lies on this curve, find the value of k.

Summary of key points

1 The function a^x, where a is a positive constant and x is a variable, is called an **exponential function**.

2 • If $a > 1$ then a^x is an example of **exponential growth** with a multiplier of a.
   • If $0 < a < 1$ then a^x is an example of **exponential decay** with a multiplier of a.

3 A point lies on a curve if the coordinates of the point satisfy the equation of the curve.

4 To test the formula $y = px + q$, plot y against x. If the points lie approximately on a straight line then p is the gradient of the line of best fit and q is the intercept on the vertical axis.

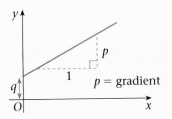

5 To test the formula $y = px^2 + q$, plot y against x^2. If the points lie approximately on a straight line then p is the gradient of the line of best fit and q is the intercept on the vertical axis.

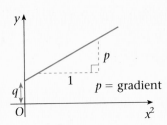

6 To test the formula $y = px^2 + qx$, plot $\dfrac{y}{x}$ against x.
   If the points lie approximately on a straight line then p is the gradient of the line of best fit and q is the intercept on the vertical axis.

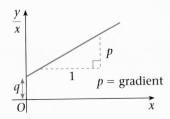

7 If a point lies on a curve then the coordinates of the point satisfy the equation of the curve.

29 Conditional probability

In this chapter you will find probabilities of events that are not independent.

29.1 Dependent and independent events

Imagine a bag containing eight snooker balls. Five are red and three are other colours (not red). If you pick two balls out of the bag at random, what is the probability that they will both be red?

The answer will depend on whether or not you put the first ball back in the bag before you pick the second ball out.

If you put the first ball back, there will be the same number of red balls to choose from second time round as there were first time. Then the probability of picking a second red ball *does not* depend on whether the first ball picked was red or not. It is **independent** of the colour of the first ball chosen.

If the first ball is red and you do not put it back, there will be one less red ball to pick out. In this case the probability of picking a second red ball depends on whether the first ball picked out was red or not. It is **dependent** on the colour of the first ball chosen.

These tree diagrams show the probability of picking two red balls in these two different situations:

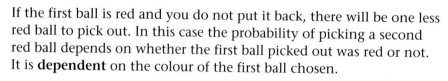

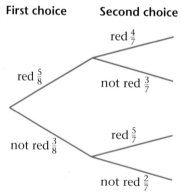

Independent events – first ball replaced before second ball picked

First choice **Second choice**

red $\frac{5}{8}$

red $\frac{5}{8}$

not red $\frac{3}{8}$

not red $\frac{3}{8}$

red $\frac{5}{8}$

not red $\frac{3}{8}$

Here the probability of picking two red balls is

$$\frac{5}{8} \times \frac{5}{8} = \frac{25}{64}$$

or

$$P(red) \times P(red) = P(red \text{ and } red)$$

Dependent events – first ball not replaced before second ball picked

First choice **Second choice**

red $\frac{4}{7}$

red $\frac{5}{8}$

not red $\frac{3}{7}$

not red $\frac{3}{8}$

red $\frac{5}{7}$

not red $\frac{2}{7}$

If a red ball is picked first and not replaced, there are only seven balls left to choose from. Only four of these are red, so for the second ball

$$P(red) = \tfrac{4}{7} \text{ and } P(not\ red) = \tfrac{3}{7}$$

If a 'not red' ball is picked first and not replaced, there are still seven balls left to choose from and five of these are red, so for the second ball

$$P(red) = \tfrac{5}{7} \text{ and } P(not\ red) = \tfrac{2}{7}$$

So when the first ball is not replaced, the probability of two reds being picked is

$$\frac{5}{8} \times \frac{4}{7} = \frac{20}{56} = \frac{5}{14}$$

Conditional probability is the name for the probability of an event that is dependent on a previous event. The probability depends on the conditions before the event.

___ **Worked examination question** _____

A box contains ten discs. Four of the discs are green and six are red. Two discs are removed at random from the box. By drawing a tree diagram, or otherwise, calculate the probability that both discs will be red.

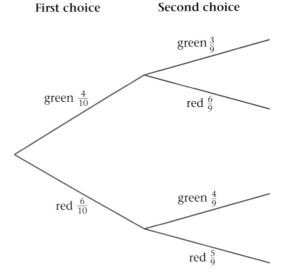

First choice   **Second choice**

green $\frac{4}{10}$

green $\frac{3}{9}$

red $\frac{6}{9}$

red $\frac{6}{10}$

green $\frac{4}{9}$

red $\frac{5}{9}$

So the probability of both discs being red is:

$$\frac{6}{10} \times \frac{5}{9} = \frac{30}{90} = \frac{1}{3}$$

Any other exactly correct equivalent fraction, decimal or percentage would also gain full marks. Answers such as 0.3, 0.33, 0.333, 30% or 33% are not sufficiently accurate. The answer must be exact: 0.3̇, $\frac{1}{3}$ or 33$\frac{1}{3}$%.

Exercise 29A

1 Leo's cat has a litter of kittens: five female and two male.
The vet examines them randomly one by one.
Draw a tree diagram and use it to calculate the probability that the first three kittens examined:

(a) will all be male

(b) will all be female

(c) will include at least one of each sex

(d) will include both males.

Read through the whole question before you start to draw the tree diagram because the choices to be made at each stage do not always become clear until the end.

2 Nine playing cards numbered 2 to 10 are shuffled thoroughly.
The top three cards are turned face up on a table.
Draw a tree diagram and use it to calculate the probability that the numbers on the three cards

(a) will all be even   (b) will all be odd

(c) will have an even product.

3   Ten identical plastic counters numbered 1 to 10 are placed in a box. Meena picks three of them at random. She wins if the product of the numbers she picks is a multiple of 3.
Draw a tree diagram to illustrate the winning outcomes and their probabilities and use it to calculate the probability that Meena will win when she picks three counters.

4   One hundred raffle tickets numbered 1 to 100 are used in a tombola to decide who wins a prize. People who pick a ticket with a units digit of 0 win. The first person to try his luck picks two tickets.
By drawing a tree diagram, or otherwise, calculate the probability that he will win

(a)   no prize

(b)   at least one prize.

5   After a holiday abroad Janet has 10 coins in her purse. Five are €1 coins, three are €2 coins and two are £1 pieces. All the coins are similar sizes. Janet takes three coins at random from her purse. Calculate the probability that the three coins will

(a)   all be €1 coins

(b)   not include a €1 coin

(c)   be one of each type.

29.2 Paths through tree diagrams

Large tree diagrams can take some time to draw.

> You only need to draw paths through a tree diagram which are needed to solve the problem set. Make sure you include all the paths you need.

> An incomplete tree showing only the relevant branches and leading to a correct answer is sufficient for full marks in the GCSE exam.

In the snooker ball problem at the start of Section 29.1, if you pick two balls at random, the probability that at least one of them is red includes every event except that of picking no reds at all. These two groups of events are **mutually exclusive** so

P(at least one red) = 1 − P(no reds at all)

> For more about mutually exclusive events see Section 24.2.

It may be easier to find P(at least one red) by calculating 1 − P(no reds at all).

> It may be quicker to calculate the probability of a mutually exclusive event than to draw a tree diagram.

Example 1

There are ten sweets in a box: three are orange, four are lime and three are lemon flavoured. Three sweets are taken randomly from the box and eaten.

Calculate the probability that those eaten are not all the same flavour.

$$P(\text{not all the same flavour}) = 1 - P(\text{all the same flavour})$$

The paths through the tree for all three sweets the same flavour are:

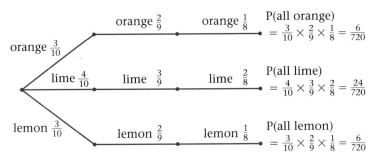

orange $\frac{2}{9}$ orange $\frac{1}{8}$ P(all orange)
$$= \frac{3}{10} \times \frac{2}{9} \times \frac{1}{8} = \frac{6}{720}$$

orange $\frac{3}{10}$

lime $\frac{4}{10}$ lime $\frac{3}{9}$ lime $\frac{2}{8}$ P(all lime)
$$= \frac{4}{10} \times \frac{3}{9} \times \frac{2}{8} = \frac{24}{720}$$

lemon $\frac{3}{10}$ lemon $\frac{2}{9}$ lemon $\frac{1}{8}$ P(all lemon)
$$= \frac{3}{10} \times \frac{2}{9} \times \frac{1}{8} = \frac{6}{720}$$

$$P(\text{all the same flavour}) = \frac{6}{720} + \frac{24}{720} + \frac{6}{720} = \frac{36}{720} = \frac{1}{20}$$

$$P(\text{not all the same flavour}) = 1 - \frac{1}{20} = \frac{19}{20}$$

> Notice that when the products of the fractions were found they were not simplified straight away because they had to be added later. This usually saves time in an exam.

Exercise 29B

In this exercise try drawing only those parts of each tree diagram that you need to answer the questions.

1   Shelley has 15 full one-litre tins of paint. Six contain Sunset Red, five contain Sunflower Yellow and the other four contain Midnight Blue. Shelley selects three tins at random, opens them and empties them into a ten-litre bucket.
Calculate the probability that when she mixes the paint in the bucket the colour produced will not be exactly one of the original three colours.

2   Twenty-six tiles, each with one of the letters of the alphabet on, are placed on a table so that none of the letters shows. Winston picks four at random. Calculate the probability that:
   (a)   none of the letters picked will be a vowel (a, e, i, o, u)
   (b)   at least one of the letters picked will be a vowel.

3   Courtney has seven plastic letters: three As, two Rs and two Ts. She picks three at random and puts them on the table in a row, the first at the left and so on. Calculate the probability that
   (a)   the letters will spell   (i) ART   (ii) RAT   (iii) TAR
   (b)   the three letters will all be different.

4 Helen has six pens in her pencil case. Three are blue, two are black and the other is red. She takes out three pens at random and puts them on her desk.
Calculate the probability that there will be one pen of each colour on the desk.

5 Under his sofa David has a pair of black shoes, a pair of brown shoes and an odd red shoe. He cannot see them and can only just reach them, so has to select them at random. He does not replace any shoe he pulls out.
Calculate the probability that:

(a) the first two shoes selected will form a pair

(b) David will obtain a pair of shoes by pulling out exactly three shoes, but not less than three

(c) David will obtain a pair after pulling out three shoes at most.

29.3 Probability and human behaviour

The probability that you will win a game can be affected by your previous success or failure.

For example, you might become demoralised by losing a game often and so be even more likely to lose. Or, if you fail a test or task the first time, you may be more likely to pass it the second time, as you may practise more. Tree diagrams can be used to represent situations like these as well.

___ **Example 2** _____

Steve is attempting to pass GCSE mathematics. The probability of him passing on his first attempt is 50%. If he fails on his first attempt then the probability of him passing on his second attempt will be 70%. Calculate the probability that he passes on his first or second attempt.

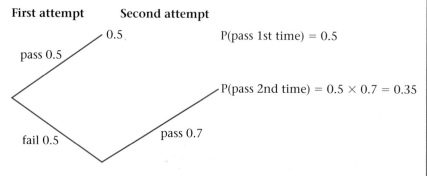

First attempt   **Second attempt**

pass 0.5   0.5   P(pass 1st time) = 0.5

P(pass 2nd time) = 0.5 × 0.7 = 0.35

fail 0.5   pass 0.7

So the probability that Steve passes on his first or second attempt is 0.5 + 0.35 = 0.85 or 85%.

Remember:
A probability can be written as a fraction, a percentage or a decimal.

For ease of calculation, convert the percentages to decimals before entering them on the probability tree. Draw only the relevant parts of the tree.

Exercise 29C

1 Rifat is playing darts. She is trying to score 180 points by throwing all three darts into the treble 20. The probability of her first dart landing in the treble 20 is 10%. If her first dart lands in the treble 20 then the probability of her second dart landing there as well is 5%. Having successfully thrown two darts into the treble 20, the probability of the third dart landing there as well is 3%.
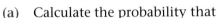

   (a) Calculate the probability that
       (i)   she will score 180 with her three darts
       (ii) she will not score 180 with her three darts.
   (b) Explain why it is impossible to calculate the probability of any two of her three darts landing in the treble 20 from the information given.

2 The probability of a letter posted first class on a Monday being delivered on Tuesday is 0.8. If it is not delivered on Tuesday the probability of it being delivered on Wednesday is 0.9. If it still has not been delivered then the probability of it being delivered on Thursday is 0.95.
   Draw the relevant parts of a tree diagram to illustrate this situation and use it to calculate the probability that the letter will be delivered before Friday. Assume that the days mentioned all fall within the same week.

3 The probability that Mr Patel will have his free newspaper *The Advertiser* delivered to his house in any week is 0.8. If he gets *The Advertiser* the probability that he will also buy the local newspaper *The Herald* is 0.15. If he does not get *The Advertiser* the probability that he will buy *The Herald* is 0.55.
   Draw a full tree diagram to illustrate the situation and use it to calculate the probability that he will only get one of the newspapers in any particular week.

4 When Ian and Bill go to buy ice-cream cornets, Ian always chooses first. The probability that Ian will choose a chocolate flavoured cornet is 70%. If Ian chooses chocolate the probability that Bill will choose chocolate as well is 80%. If Ian does not choose chocolate then the probability that Bill will not choose chocolate is 60%.
   Draw a tree diagram and use it to calculate the probability that
   (a) both boys choose to have chocolate flavoured cornets
   (b) neither boy chooses a chocolate flavoured cornet
   (c) the boys choose cornets of different flavours.

5 Majid goes to school by bus every day. The probability that the bus will be on time is 0.75. If the bus turns up on time the probability that Majid will get on it is 0.8, but if it is late the probability of him getting on falls to 0.65. The bus is never early.
   Draw a tree diagram to illustrate this situation and include on it all the appropriate probabilities. Use the diagram to calculate the probability that on any particular school day Majid will get on the bus.

6 A box contains twelve discs. Five of the discs are green and seven are red. Two discs are removed at random from the box.

By drawing a tree diagram, or otherwise, calculate the probability that

(a) both discs will be red   (b) at least one disc will be red.

7 When there are no darts in the 20 sector of the dartboard, a darts player estimates the probability of hitting the 20 sector as being 0.3. When she hits the 20 sector with one of her darts, her estimate of the probability of hitting the 20 sector changes to 0.15.

She throws three darts one after the other. Calculate an estimate of the probability that she will miss the 20 sector with her first dart, hit the 20 sector with her second dart and hit the 20 sector with her third dart.

Mixed exercise 29

1 Seven cards numbered 1 to 7 are shuffled thoroughly. The top two cards are turned face up on a table. Draw a probability tree diagram and use it to calculate the probability that the numbers will

(a) both be even   (b) have an odd sum.

2 There are two red sweets and seven green sweets in a bag. Two sweets are taken at random, one at a time, from the bag (the first sweet is not put back in the bag).
Find the probability that the sweets will be

(a) both red   (b) of different colours   (c) the same colour.

3 Arnold is throwing stones at a can. The probability that he will hit the can with his first throw is 0.4. If he hits the can with his first throw, the probability that he will hit the can with his second throw is 0.7. If he misses the can with his first throw, the probability that he will hit the can with his second throw is 0.5. Arnold throws two stones, one at a time, at the can.
Calculate the probability that he will hit the can with

(a) both stones

(b) exactly one stone

(c) at least one stone.

4 The probability that it will rain today is $\frac{5}{8}$. If it does not rain today, the probability that it will rain tomorrow is $\frac{2}{9}$.
Calculate the probability that it will rain today or tomorrow.

5 In a group of 20 students, five do not have a mobile phone. If three students are selected at random from the group, work out the probability that exactly one of these three students does not have a mobile phone.

6 There are 15 plugs in a box, of which six are known to be faulty.
Three plugs are taken at random from the box.
Work out the probability that

(a) all three plugs are faulty

(b) none of the three plugs is faulty

(c) at least one of the three plugs is faulty.

7 80% of Dr Blunt's patients have a flu injection.
The probability of catching flu after having the injection is $\frac{1}{40}$.
The probability of catching flu after not having the injection is $\frac{13}{20}$.
Calculate the probability that any one of Dr Blunt's patients,
selected at random, will catch flu.

8 A doctor diagnoses that a patient has a virus. She does not
know which type of the virus, type A, B or C, the patient has.
The probability of having each type of the virus is shown
in the table.

Type A	Type B	Type C
$\frac{1}{2}$	$\frac{3}{8}$	$\frac{1}{8}$

The probabilities that the patient will recover from each of the
types of the virus, A, B and C, are $\frac{1}{7}$, $\frac{4}{7}$ and $\frac{2}{7}$ respectively.
Work out the probability that the patient recovers from the virus.

9 A bag contains four strawberry, three orange and two lemon
flavoured sweets.
Three sweets are taken at random from the bag.
Find the probability that they will be

(a) all strawberry flavoured   (b) all the same flavour

(c) one of each flavour.

10 A school is divided into two parts, Upper school and Lower school.
Upper school has 300 boys and 200 girls.
Lower school has 400 boys and 300 girls.
A student is selected at random from the school.
If the first student comes from the Lower school, a second
student is selected at random from the Upper school; if the first
student comes from the Upper school, a second student is
selected at random from the Lower school.
Find the probability that the second student will be a girl.

Summary of key points

1 **Conditional probability** is the name for the probability of an event that is dependent on a previous event. The probability depends on the conditions before the event.

2 You only need to draw paths through a tree diagram which are needed to solve the problem set. Make sure you include all the paths you need.

3 It may be quicker to calculate the probability of a mutually exclusive event than to draw a tree diagram.

Examination practice paper

Non-calculator

1   Four coins are made from an alloy using these materials:

      60 g zinc,   30 g iron,   10 g copper

    Work out how much of each material is needed to
make six coins.   **(3 marks)**

2   A television is advertised at £500 plus 17.5% VAT.
Calculate the total cost of the television.   **(3 marks)**

3   (a)   Simplify $4p \times 3q$ **(1 mark)**

    (b)   Expand $y(y + 2)$ **(1 mark)**

    (c)   Factorise $x^2 - 5x$ **(2 marks)**

4   (a)   £600 is divided between Amy and Beth in the ratio 1 : 4.
How much does Beth receive?   **(2 marks)**

    (b)   What percentage of the £600 is Beth's share?   **(2 marks)**

5   Draw a plan, front elevation and side elevation of this solid.

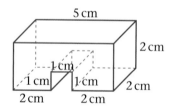

    Show the hidden details.   **(3 marks)**

6   (a)   Copy the diagram and make the axes go up to 6 in each
direction. Rotate **A** through 90° anticlockwise about (0, 0).
Label the image **B**.   **(2 marks)**

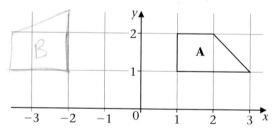

    (b)   Translate **A** by $\begin{pmatrix} 2 \\ 4 \end{pmatrix}$. Label the image **C**.   **(1 mark)**

    (c)   Fully describe the transformation which maps
B onto **C**.   **(3 marks)**

7 (a) Construct a stem and leaf diagram for the data given.
   Height of seedlings, in mm, after 3 weeks:

   8, 12, 27, 11, 33, 19, 21, 10, 15, 24, 10, 12, 23, 16, 31,
   7, 24, 19, 13, 32, 30, 26, 17, 20, 15, 11, 21, 10, 7, 21

   (3 marks)

   (b) What is the median value? **(1 mark)**

8 Two fair spinners can each score 1, 2, 3 or 4.
   They are spun at the same time.
   What is the probability that their sum will be 6? **(3 marks)**

9 (a) Solve the inequality $3(x + 2) \leqslant 7$. **(3 marks)**

   (b) Write the value of the greatest integer which satisfies
   this inequality. **(1 mark)**

10 Work out an estimate for the value of $\dfrac{23.2 \times 57.6}{0.43}$ **(2 marks)**

11 The diagram shows a rectangle with length $x + 2$ and width $2x - 7$.

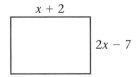

   All measurements are given in centimetres.
   The perimeter of the rectangle is 17 cm.
   Find the value of x. **(3 marks)**

12 (a) Work out $\frac{5}{8} \div \frac{2}{3}$
   Give your answer in its simplest form. **(2 marks)**

   (b) Work out $3\frac{3}{4} \times 2\frac{2}{5}$
   Give your answer in its simplest form. **(3 marks)**

13 A prism is 20 cm long. Its cross-section is a right-angled triangle
   ABC, in which $AB = 6$ cm and $AC = 10$ cm.

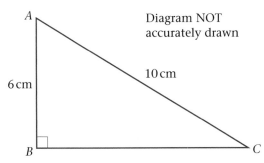

Diagram NOT
accurately drawn

   Calculate the total surface area of the prism. **(6 marks)**

14 The cumulative frequency graph gives information about the prices of 50 houses.

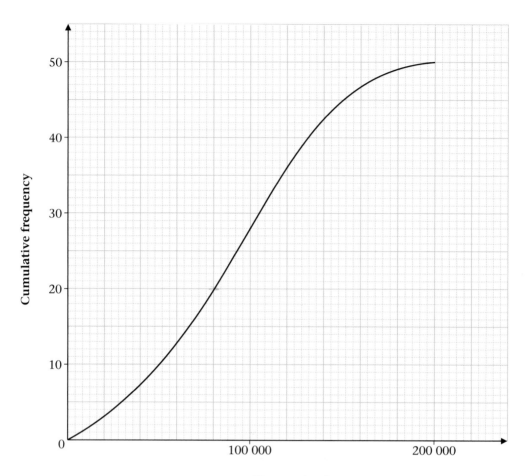

House price £

(a) Find the number of houses priced **below** £80 000.   **(1 mark)**

(b) Find the median house price.
     Show clearly how you found your answer.   **(1 mark)**

(c) Find the interquartile range of the house prices.   **(2 marks)**

15 A computer performs 400 million calculations per second.

(a) Write 400 million in standard form.   **(1 mark)**

(b) Write out the number of calculations the computer performs in 10 minutes.
     Give your answer in standard form.   **(2 marks)**

16 The diagram shows five points A, B, C, D and E on the circumference of a circle.

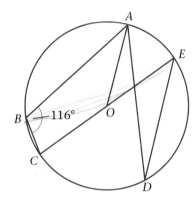

Diagram NOT
accurately drawn

The centre of the circle is O.
CE is a diameter.
Angle $ABC = 116°$.

(a) State the size of angle CBE. Give a reason. **(2 marks)**

(b) Calculate the size of angle ADE. **(2 marks)**

(c) Calculate the size of angle AOE. **(2 marks)**

17 (a) Copy and complete the table of values for
$y = x^3 - 2x^2 - 5x$.

x	-2	-1	0	1	2	3	4
y		2					

(2 marks)

(b) On graph paper plot the graph of $y = x^3 - 2x^2 - 5x$
for $-2 \leqslant x \leqslant 4$. **(2 marks)**

(c) Use your graph to solve the equation $x^3 - 2x^2 - 5x = 0$.
Where necessary, give your answers correct to 1 decimal
place. **(2 marks)**

18 (a) Evaluate

(i) 7^0

(ii) 4^{-3}

(iii) $64^{\frac{1}{3}}$ **(3 marks)**

(b) Express $\dfrac{6}{\sqrt{12}}$ in the form $a\sqrt{b}$, where a and b are integers.

(2 marks)

19 100 car owners were asked the distances, in miles, their cars had travelled.
The unfinished histogram and table show this information.

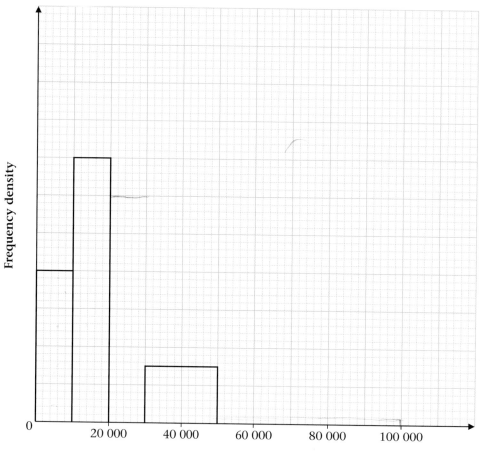

Distance (*d*) in miles

Distance (*d*) in miles	Frequency
$0 \leqslant d < 10\,000$	20
$10\,000 \leqslant d < 20\,000$	35
$20\,000 \leqslant d < 30\,000$	25
$30\,000 \leqslant d < 50\,000$	15
$50\,000 \leqslant d < 100\,000$	5

(a) Use the information in the histogram to complete the table.
(2 marks)

(b) Make a copy of the histogram. Use the information in the table to complete it.
(2 marks)

20 *y* is inversely proportional to the square of *x*. $y = 4$ when $x = 5$.

(a) Find a formula for *y* in terms of *x*.
(2 marks)

(b) Find the value of *x* when $y = 400$.
(2 marks)

21 *A* is the point with coordinates (2, 8) and *B* is the point with coordinates (14, 4).
A straight line *L* is perpendicular to the line *AB* and passes through *A*.
Find the equation of the straight line *L*. **(3 marks)**

22 Prove algebraically that the sum of two consecutive square numbers is always an odd number. **(3 marks)**

23 The diagram shows two squares *QRST* and *PMLR*.
PS and *LQ* are straight lines.

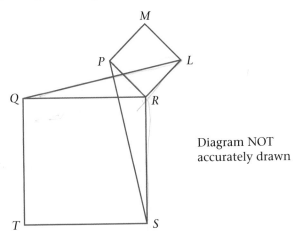

Diagram NOT
accurately drawn

Prove that triangle *QLR* and triangle *PRS* are congruent.
(3 marks)

24 Rearrange $\dfrac{a}{x + b} = \dfrac{b}{x - a}$ to make *x* the subject. **(4 marks)**

25 The diagram shows the net of a cone.

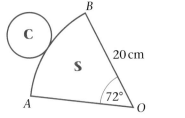

Diagram NOT
accurately drawn

The net is made from a circle, **C**, and a sector, **S**, of a circle centre *O*.
Angle *AOB* = 72°.

(a) Find the length of the arc *AB*.
Give your answer as a multiple of π. **(2 marks)**

(b) The sector is folded to make a cone.
Find the total surface area of the cone.
Give your answer as a multiple of π. **(4 marks)**

Examination practice paper

Calculator

1 (a) Use your calculator to find the value of $\sqrt{\dfrac{3.21 \times 5.42}{1.41 \times 2.74}}$

   Write down all the digits on your calculator display.

   (3 marks)

2 A car tyre is advertised at £80. There is a discount of 15%.
   Calculate the cost of buying **two** car tyres with this discount.

   (2 marks)

3 The cost of 8 metres of wire fence is £3.28.
   What is the cost of 5 metres of this wire fence?   **(2 marks)**

4 (a) The velocity of a particle is given by the formula
       $v^2 = u^2 + 2as$.
       Calculate the velocity v when $u = 3$, $a = 0.8$ and $s = 4.35$.

       (3 marks)

   (b) Solve $3(2x - 5) = 30$.   **(3 marks)**

5 Here are the first five numbers of an arithmetic sequence.

       4   9   14   19   24

   Write down, in terms of n, an expression for the nth term of the
   sequence.   **(2 marks)**

6 The diagram shows the floor plan of a workshop.
   Work out the area of the floor.   **(3 marks)**

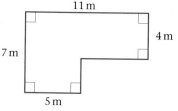

7 In the diagram BE is parallel to CD, angle $ABE = 105°$ and angle
   $BEC = 54°$.

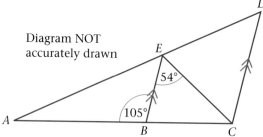

Diagram NOT
accurately drawn

   (a) Calculate the size of
       (i) angle EBC (ii) angle ECD.   **(3 marks)**
   (b) Given that $AE = 10$ cm, $ED = 5$ cm and
       $CD = 6$ cm, calculate the length of EB.   **(2 marks)**

8   In a bank account an amount of £500 appreciates at the rate of
    5% compound interest per annum.
    Calculate the amount of money that will be in the account at
    the end of three years.   **(3 marks)**

9   Find the size of the angle x.

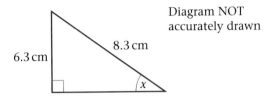

Diagram NOT
accurately drawn

8.3 cm

6.3 cm

x

                                                               (3 marks)

10 The table and the scatter graph show the number of units of
    electricity used in heating a house on ten different days and the
    average temperature for each day.

Average temperature (°C)	6	2	0	6	3	5	10	8	9	12
Units of electricity used	30	39	41	34	33	31	22	25	23	22

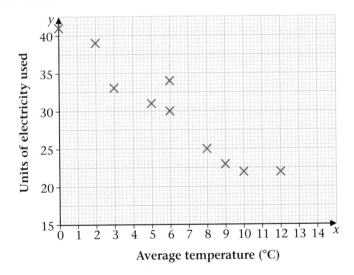

(a) Using the values in the table above, draw the scatter graph
     and mark on it a line of best fit.   **(1 mark)**

(b) Use your line of best fit to estimate
     (i)   the average temperature when 35 units of electricity
           are used,
     (ii) the units of electricity used when the average
           temperature is 7° C.   **(2 marks)**

11 The equation $x^3 - 7x = 81$ has a solution in the range $4 < x < 5$.
    Use a method of trial and improvement to obtain this solution
    correct to two decimal places.   **(4 marks)**

12 A year ago, Mrs Ford bought a new car. Its value has fallen by
12% since then. Its value now is £8624.

   (a) Calculate the value of the car when it was new.   **(2 marks)**

   (b) The value of the car will continue to fall by 12% each year.
        Calculate the value of the car when it is
        five years old.   **(3 marks)**

13 The grouped frequency table shows the
distribution of weekly rainfall at Heathrow
Airport in 2005.
Calculate an estimate for the mean weekly rainfall
in 2005.
Give your answer correct to the nearest
millimetre.   **(4 marks)**

Weekly rainfall (r mm)	Frequency
$0 < r \leqslant 10$	32
$10 < r \leqslant 20$	13
$20 < r \leqslant 30$	4
$30 < r \leqslant 40$	1
$40 < r \leqslant 50$	1
$50 < r \leqslant 60$	0
$60 < r \leqslant 70$	1

14 A factory employs 100 people.
76 of these people are men. The other 24 are women.
The mean weekly wage of all 100 people is £407.
The mean weekly wage of the men is £395.
Work out the mean weekly wage of the 24 women.   **(4 marks)**

15 (a) Solve $\dfrac{4}{y} - 3 = 7$ **(3 marks)**

   (b) Express $\dfrac{2}{3x - 1} + \dfrac{1}{x + 6}$ as a single algebraic fraction.
                                                    (3 marks)

   (c) Factorise $2x^2 + 3x - 14$ **(3 marks)**

16 The diagram represents the cross-section of a church door.
The cross-section consists of a rectangle with a semicircular top.
The door has a uniform thickness of 6 cm.
The door is made of metal of density 7.2 g per cm³.
Work out the mass of the door.   **(6 marks)**

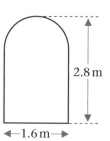

2.8 m

←—1.6 m—→

17 The diagram shows a square-based rectangular box.
Each side of its base is x cm long. The box has **no lid** and its
height is h cm.

   (a) Show that the surface area, A cm², of the box is given by
        the formula

        $A = x^2 + 4hx$ **(2 marks)**

   (b) Make h the subject of the formula.   **(2 marks)**

   (c) A lidless square-based box 2 cm high has a surface area of
        48 cm². Find the length of the sides of its base.   **(2 marks)**

   (d) $A = 100$ and $h = 3$. Find the value of x.
        Give your answer correct to 3 significant figures.   **(2 marks)**

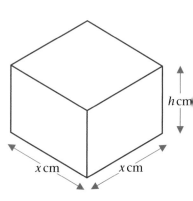

h cm

x cm x cm

18 (a) Sketch graphs of
(i) $y = \cos x$ (ii) $y = \sin 3x$
for all values of x in the range $0° \leqslant x \leqslant 180°$.   **(4 marks)**

(b) Solve the equation $\sin 3x = \dfrac{-2}{3}$
for all values of x in the range $0° \leqslant x \leqslant 180°$.   **(3 marks)**

19 The diagram represents a side view of a cylindrical tin with three tennis balls in it. The diameter of each tennis ball is 6.5 cm. Calculate, correct to 2 significant figures:

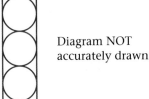

Diagram NOT accurately drawn

(a) the volume of the tin   **(2 marks)**

(b) the curved surface area of the tin   **(2 marks)**

(c) the total surface area of the three balls.   **(2 marks)**

20 A rectangle is 68 cm long and 8.4 cm wide.
Each measurement is correct to 2 significant figures.
Calculate the greatest lower bound for the area of the rectangle, **giving with an explanation** your answer to an appropriate degree of accuracy.   **(5 marks)**

21 The equation of a straight line is

$y = ax + b$, where a and b are constants.

The equation of a circle is:

$x^2 + y^2 = 64$

The straight line is a tangent to the circle.

Prove that $a^2 + 1 = \dfrac{b^2}{64}$ **(6 marks)**

22 A box contains three red pens, four blue pens and five black pens.
A pen is selected at random from the box, and is not replaced.
A second pen is then selected from the box.
Calculate the probability that the two pens selected will be the same colour.   **(4 marks)**

Formulae sheet: Higher tier

Volume of a prism = area of cross section × length

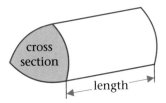

Volume of sphere = $\frac{4}{3}\pi r^3$

Surface of a sphere = $4\pi r^2$

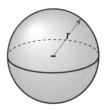

Volume of cone = $\frac{1}{3}\pi r^2 h$

Curved surface area of cone = $\pi r l$

In any triangle ABC

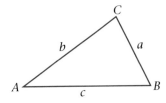

Sine rule $\dfrac{a}{\sin A} = \dfrac{b}{\sin B} = \dfrac{c}{\sin C}$

Cosine rule $a^2 = b^2 + c^2 - 2bc\cos A$

Area of a triangle = $\frac{1}{2}ab\sin C$

The quadratic equation
The solutions of $ax^2 + bx + c = 0$
where a = 0, are given by

$$x = \frac{-b \pm \sqrt{(b^2 - 4ac)}}{2a}$$

Answers

Chapter 1 Exploring numbers 1

Exercise 1A

1 (a) 1, 2, 3, 4, 6, 8, 12, 24
(b) 1, 2, 3, 4, 6, 9, 12, 18, 36
(c) 1, 2, 4, 7, 11, 14, 22, 28, 44, 77, 154, 308
(d) 1, 7, 11, 13, 77, 91, 143, 1001
(e) 1, 2, 4, 5, 7, 8, 10, 14, 20, 25, 28, 35, 40, 50, 56, 70, 100, 140, 175, 200, 280, 350, 700, 1400
(f) 1, 53
2 2, 3, 5, 7, 11, 13, 17, 19, 23, 29, 31, 37, 41, 43, 47, 53, 59, 61, 67, 71, 73, 79, 83, 89, 97
3 Yes $2 + 3 = 5$, $2 + 5 = 7$, $2 + 11 = 13$ etc.
One of the primes *must* be 2.
4 (a) $2^3 \times 3$ (b) 2^5 (c) 2×3^2 (d) 13
(e) $2^3 \times 3^2$ (f) 2×5^2 (g) $3^2 \times 11$ (h) $2^3 \times 3 \times 5$
5 (a) 4   (b) 3   (c) 6   (d) 12   (e) 36   (f) 28
6 (a) 12   (b) 12   (c) 14   (d) 30   (e) 144   (f) 850
7 Every 120 seconds (2 minutes)
8 (a) (i) 60   (ii) 124   (iii) 217   (iv) 511   (v) 63
(b) $p + 1$

(d) $1 + p + p^2 + \ldots + p^n = \left(\dfrac{p^{n+1} - 1}{p - 1}\right)$

Exercise 1B

1 (a) 64   (b) 343   (c) 144   (d) 1000
(e) 1, 4, 9, 16, 25, 36, 49, 64, 81, 100, 121, 144
(f) 1, 8, 27, 64, 125, 216, 343, 512
2 (a) (i) 1, 9, 49, 64   (ii) 1, 8, 64
(b) (i) 4, 16   (ii) 27, 125
(c) (i) 64, 81, 144   (ii) 64, 125
(d) (i) 81, 100, 169   (ii) 125, 216

Exercise 1C

1 (a) 169   (b) 12.25   (c) 1600   (d) 75.69
(e) 384.16   (f) 3294.76
2 (a) 216   (b) 13.824   (c) 8000   (d) −2.197
(e) 2406.104   (f) 47 437.928
3 (a) 11   (b) 15   (c) 130   (d) 1.7
(e) 0.7   (f) 5.8
4 (a) 15.91   (b) 48.38   (c) 4.29   (d) 92.07
(e) 5.43   (f) 1351.28
5 (a) 4.08   (b) 15.4   (c) 2.98   (d) −2.10
(e) 9.59   (f) 187

Exercise 1D

1 (a) 64   (b) 100 000   (c) 625   (d) 343
(e) 20 736 (f) 0.590 49
2 (a) 10^3 (b) 5^3 (c) 2^9 (d) 7^4
(e) 5^4 (f) 3^9
3 $64 = 2 \times 2 \times 2 \times 2 \times 2 \times 2 = 2^6 = 4 \times 4 \times 4 = 4^3$
4 2^5, by 7

Exercise 1E

1 (a) $5^6 = 15\,625$ (b) $5^8 = 390\,625$ (c) $10^2 = 100$
(d) $12^3 = 1728$ (e) $3^7 = 2187$ (f) 1
(g) 1   (h) $5^{-2} = 0.04$ (i) 6
(j) 150   (k) 14   (l) $4^6 = 2^{12} = 4096$
(m) $5^9 = 1\,953\,125$ (n) $2^{10} = 1024$ (o) $3^{-4} = \frac{1}{81}$
(p) $4^5 = 1024$ (q) $6^{-1} = \frac{1}{6}$ (r) $5^5 = 3125$

(s) $2^2 = 4$ (t) 4   (u) $5^{-6} = \frac{1}{15\,625}$
(v) $3^{-4} = \frac{1}{81}$ (w) $2^{-4} = \frac{1}{16}$ (x) $3^2 = 9$
2 $2^4 = 4^2$

Exercise 1F

1 (a) 2   (b) $\frac{1}{2}$ (c) 8   (d) $\frac{1}{5}$ (e) 2   (f) $\frac{1}{2}$
(g) $\frac{1}{8}$ (h) $\frac{1}{5}$ (i) $\frac{1}{10}$ (j) 10   (k) 125   (l) $\frac{1}{4}$
2 (a) $2^{10} = 1024$ (b) $3^2 = 9$ (c) 1
(d) $5^0 = 1$ (e) 40   (f) $4^{-6} = \frac{1}{4096}$
(g) $10^{-2} = \frac{1}{100}$ (h) $5^4 = 625$
3 (a) 3   (b) 5   (c) $\frac{1}{7}$ (d) 4   (e) 8   (f) $\frac{1}{100}$

Exercise 1G

1 (a) 8   (b) $\frac{1}{4}$ (c) 0.0001   (d) 10 000 000 000
(e) 800 000   (f) 400   (g) 0.008   (h) 3 200 000
(i) 0.000 001 (j) $\frac{1}{256}$
2 1000   3

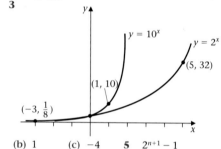

4 (a) $\frac{4}{5}$ (b) 1   (c) −4   5 $2^{n+1} - 1$

Mixed exercise 1

1 $2^2 \times 3 \times 13$
2 (a) 6, 12   (b) 6, 90
3 (a) 2^3 (b) 5^{-6}
4 (a) 9   (b) 6   (c) $\frac{1}{6}$
5 (b) $n + 1$ (c) $\tau(n \times m) = \tau(n) \times t(m)$ if n, m have no common factors except 1.
6 nth term = $(n - 1)$th term + $(n - 3)$rd term
7 $1^3 + 2^3 + 3^3 + \ldots + n^3 = (1 + 2 + \ldots + n)^2$
8 (a) LD(square number) = 0, 1, 4, 5, 6, 9 only
(b) LD$(n \times m)$ = LD [LD$(n) \times$ LD(m)]
(c) Cannot because $10n + 7$ has LD = 7
9 (a) $\frac{12}{5}$ (b) $-\frac{4}{3}$ (c) $-\frac{5}{2}$ 10 $\dfrac{5^{n+1} - 1}{4}$

Chapter 2 Essential algebra

Exercise 2A

1 (a) 13   (b) 17   (c) −1   (d) 2.5
(e) 4   (f) 14   (g) $3\frac{2}{5}$ (h) 1.5
2 (a) 16   (b) 15   (c) 32
(d) 9   (e) 15   (f) 14
3 (a) 35   (b) 8   (c) 6 ·
(d) 10   (e) 4   (f) −5
4 (a) 3   (b) 4   (c) 6   (d) 3   (e) 5

Exercise 2B

1 (a) 101   (b) 47 (c) 14   (d) 147 (e) 23 (f) 6
2 (a) 19   (b) 33 (c) 27   (d) 20 (e) 16 (f) 7
3 (a) 44   (b) 10 (c) 16   (d) 20 (e) 30 (f) 3

Exercise 2C

1 (a) a^4 (b) b^5 (c) c^6 (d) y^4
2 (a) $6a^2b^4$ (b) $10p^6q^4$ (c) $9x^4y^4$ (d) $3s^2t^5$

Exercise 2D

1 (a) $2n + 6$ (b) $12p + 8$ (c) $2y^2 + 3y$
   (d) $2x^2 + 10x$ (e) $3n - 6$ (f) $2p - 3$
2 (a) $4n + 12$ (b) $2p - 8$ (c) $5a + 5$
   (d) $6x - 2$ (e) $20q - 12$ (f) $6r + 3s$
   (g) $ab + 2a$ (h) $p^2 + pq$ (i) $2p^2 - p$
3 (a) $3x + 18$ (b) $4y - 8$ (c) $7t + 7$
   (d) $12x + 8$ (e) $15y - 20$ (f) $6p + 2q$
   (g) $x^2 + 3x$ (h) $x^2 + xy$ (i) $3y^2 - 2y$
4 (a) $2m + 10$ (b) $9n - 36$ (c) $8u - 8$
   (d) $35x + 15$ (e) $4y^2 - 24y$ (f) $6x^4 + 15x^2$
   (g) $20t^5 - 12t^3$ (h) $12x^5 + 28x^2y$ (i) $15y^4 - 20yx^2$

Exercise 2E

1 (a) $8r + 2$ (b) $2p - 6$ (c) $10x + y$ (d) $5m$
2 (a) $14a + 2b$ (b) $5p + q$ (c) $10r - 2s$ (d) $9p$
3 (a) $7x - y$ (b) $10m - 7n$ (c) $4p - 4q$ (d) $6p$
4 (a) $3a - 3b$ (b) $r^2 + r$ (c) $2mn$ (d) $3m + n$
5 (a) $3n + 2$ (b) $3n - 2$ (c) $2n - 4$
   (d) $3p + 5r$ (e) $r - 2q$ (f) $2r - s$

Exercise 2F

1 (a) $a + 3b$ (b) $p + 5q$ (c) $2n - 1$ (d) $4s - 4r$
   (e) $6y$ (f) $rs - 3r$
2 (a) $x - 2y$ (b) $11x - 3y$ (c) $2x^2 - x$
   (d) $20y^2 - 23y + 6$ (e) $2x^2 - 9x + 10$ (f) $16y^4 + 9y^3$

Exercise 2G

1   $5(x - 4)$ 2 $8(x + 3)$ 3 $6(2x + 3)$
4   $5(4x - 5)$ 5 $8(x^2 - 3)$ 6 $9(x^2 + 4)$
7   $5(2x^2 - 3)$ 8 $x(x + 6)$ 9 $x(x - 1)$
10   $p(x - 3)$ 11 $q(x + q)$ 12 $5x(x - 3)$
13   $3x(2x + 3)$ 14 $5x(3x - 7)$ 15 $4x(2 - 3x)$
16   $ax(x + 1)$ 17 $2a(2x - 3)$ 18 $3ax(2x + 5)$

Exercise 2H

1 (a) $x^2 + 3x + 2$

   (b) $x^2 + 10x + 16$

   (c) $x^2 + 6x + 9$

   (d) $x^2 + ax + bx + ab$

2 (a) $ab + 3a + 4b + 12$ (b) $cd + 4c + 5d + 20$
   (c) $xy + 6x + 3y + 18$ (d) $a^2 + 11a + 24$
   (e) $b^2 + 11b + 28$ (f) $x^2 + 8x + 12$

3 (a) $2ab + 8a + 3b + 12$ (b) $3bc + 9b + 2c + 6$
   (c) $4cd + 24c + 3d + 18$ (d) $2a^2 + 11a + 15$
   (e) $3b^2 + 10b + 8$ (f) $4c^2 + 13c + 10$
4 (a) $ab - 3a - 4b + 12$ (b) $cd - 4c - 5d + 20$
   (c) $xy - 6x - 3y + 18$ (d) $a^2 - 11a + 24$
   (e) $b^2 - 11b + 28$ (f) $x^2 - 8x + 12$
5 (a) $ab + 3a - 4b - 12$ (b) $cd - 4c + 5d - 20$
   (c) $xy + 6x - 3y - 18$ (d) $a^2 + 5a - 24$
   (e) $b^2 - 3b - 28$ (f) $x^2 + 4x - 12$
6 (a) $2ab + 10a - 5b - 25$ (b) $3b^2 - 14b - 24$
   (c) $3c^2 - 11c + 6$ (d) $3a^2 - 19a + 20$
   (e) $3x^2 + 11x - 20$ (f) $2a^2 - 9a - 18$
7 (a) $2x^2 + 5xy + 3y^2$ (b) $3x^2 - xy - 4y^2$
   (c) $5x^2 - 8xy + 3y^2$ (d) $6x^2 + 23xy + 20y^2$
   (e) $6x^2 - 5xy - 25y^2$ (f) $12x^2 - 28xy + 15y^2$
8 (a) $x^2 + 6x + 9$ (b) $y^2 - 10y + 25$
   (c) $4x^2 + 4x + 1$ (d) $9y^2 - 24y + 16$
   (e) $9x^2 + 30xy + 25y^2$ (f) $36x^2 - 84xy + 49y^2$
9 (a) $3a^2 - 5a - 2$ (b) $12p^2 + 5p - 3$
   (c) $3a^2 + 5a + 2$ (d) $12p^2 + 14p + 4$
   (e) $2b^2 - 5b + 3$ (f) $8b^2 + 2b - 1$
10 (a) $2a + b + 2$ (b) $pq - p + q - 2$
   (c) $3r^2 - 5r - sr + 2s$ (d) $4p^2$
11 (a) $9x^2 - 1$ (b) $1 - 16x^2$
   (c) $4x^2 - 25$ (d) $49x^2 - 4$
12 (a) $5x^2 - 20x - 60$ (b) $4x^2 - 4$
   (c) $30x^2 + 4x - 2$ (d) $36x^2 - 3x - 18$

Exercise 2I

1   $x^2 + 10x + 25$ 2 $x^2 - 6x + 9$ 3 $x^2 + 18x + 81$
4   $x^2 - 4x + 4$ 5 $p^2 + 2pq + q^2$ 6 $p^2 - 2pq + q^2$
7   $x^2 + 6x + 9$ 8 $x^2 - 14x + 49$ 9 $x^2 - 2ax + a^2$
10 (a) $(x + 4)^2 = x^2 + 8x + 16$
   (b) $(x - 10)^2 = x^2 - 20x + 100$
   (c) $(x + 1)^2 = x^2 + 2x + 1$
   (d) $(x - 12)^2 = x^2 - 24x + 144$

Exercise 2J

1   5 2 4 3 $2\frac{1}{2}$ 4 $2\frac{3}{5}$ 5 $\frac{3}{7}$
6   0 7 $\frac{5}{6}$ 8 $\frac{7}{9}$ 9 -3 10 -2
11   $-2\frac{1}{2}$ 12 $-2\frac{2}{5}$ 13 $-\frac{5}{8}$ 14 $-\frac{1}{2}$ 15 $-\frac{2}{5}$
16   1 17 -2 18 $\frac{1}{6}$ 19 -4 20 -6
21   $-2\frac{3}{5}$ 22 $1\frac{1}{2}$ 23 $\frac{1}{6}$ 24 $-\frac{1}{4}$ 25 $-1\frac{4}{5}$

Mixed exercise 2

1 (a) 19 (b) 9 (c) 54 (d) 90 (e) 100
   (f) 243 (g) 6 (h) 154 (i) 4 (j) $4\frac{1}{2}$
2 (a) $9a + 3b$ (b) $5ab$ (c) $4a - 3a^2$
   (d) $5p - 5q$ (e) $4d - c$
3 (a) $5p + 16$ (b) $12p + 5$ (c) $7a - 7b$ (d) -11
4   $pq - p^3$
5 (a) $y^2 + 2y - 15$ (b) $12x^2 + 11x + 2$
   (c) $25y^2 - 30y + 9$ (d) $9a^2 - 12ab + 4b^2$
6 (a) 24 (b) 4 (c) 12 (d) $\frac{3}{5}$ (e) -5 (f) 5
7 (a) $2(2p - 3)$ (b) $3(3q + 1)$ (c) $2(3a + 2b)$
   (d) $a(b + a)$ (e) $2a(b + c)$ (f) $2a(4a - 3)$

Chapter 3 Shapes

Exercise 3A

1 (a) $a = 48°$ (b) $b = 106°$, $c = 74°$, $d = 74°$
   (c) $e = 111°$ (d) $f = 60°$, $g = 120°$ (e) $h = 51\frac{3}{7}°$
2 (a) $a = 35°$, $b = 50°$, $c = 50°$, $d = 95°$
   (b) $e = 68°$, $f = 68°$ (c) $g = 149°$, $h = 125°$ $i = 55°$

Exercise 3B

1 $a = 35°$ (interior angles of a triangle)
2 $b = 35°$ (interior angles of a triangle)
3 $c = 70°$ (interior angles of a triangle)
   $d = 110°$ (angles on a straight line)
4 $e = 52°$ (angles on a straight line)
   $f = 85°$ (interior angles of a triangle)
5 $g = 47°$ (corresponding angles)
   $h = 101°$ (interior angles of a triangle)
   $i = 101°$ (corresponding angles)

Exercise 3C

1 (a) isosceles triangle   (b) parallelogram
   (c) parallelogram   (d) square
2 (a) 1440°   (b) 144°

Exercise 3D

$a + b + c = d + e + f = 180°$
$a + b + c + d + e + f = 360°$

Exercise 3E

1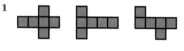

2
3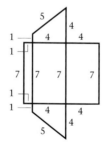

4 (a) plan   front   side   (b) plan   front   side

5 (a) (b) (c)

6
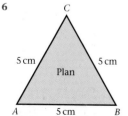
C

5 cm   5 cm
    Plan

A   5 cm   B

Front
elevation

5 cm

7 (a)
Side
elevation

(b)

8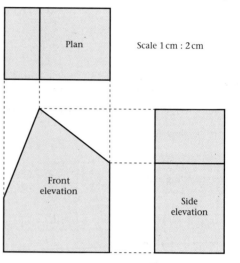

Plan   Scale 1 cm : 2 cm

Front
elevation

Side
elevation

Exercise 3F

1 (a) No. Only the angles are equal. The corresponding sides
        may not be the same length.
   (b) Yes. $PQRS = TUVW$
        Corresponding angles and sides are equal.
   (c) Yes. $XYZ = NLM$ (RHS)
2 (a) ABC, YZX (SAS)   (b) ABC, EFD (AAS)
   (c) XYZ, QRP (RHS)
3 $AB = CD$ (opposite sides of a parallelogram)
   $AD = BC$ (opposite sides of a parallelogram)
   $BD = BD$ (common)
   Triangles ABD, CDB are congruent (SSS)
4 $BC = BD - CD$
   $DE = CE - CD$
   But $BD = CE$
   $\therefore BC = DE$
   $CA = AD$ (given)
   $\angle BCA = 180° - \angle ACD$
   $\angle ADE = 180° - \angle ADC$
   But $\angle ACD = \angle ADC$ (base angles of an isosceles $\triangle$)
   $\therefore \angle BCA = \angle ADE$
   Triangles ABC, ADE are congruent (SAS)
5 $\angle ABC = \angle ACB$ (base angles of isosceles $\triangle$)
   $\angle ABC = \angle ADE$ (corresponding angles)
   $\angle ACB = \angle AED$
   $\angle ADE = \angle AED$ and ADE is isosceles
   $\angle AD = \angle AE$
   Triangles ACD and ABE are congruent (SAS)
6 $\angle ABE = 180 - 2 \times \angle AEB$ (3rd angle in an isosceles $\triangle$)
   $\angle DBC = 180 - 2 \times \angle BDC$
   But $\angle AEB = \angle BDC$
   $\therefore \angle ABE = \angle DBC$
   $\angle ABD = \angle ABE + \angle EBD$
   $\angle EBC = \angle DBC + \angle EBD$
   Therefore $\angle ABD = \angle EBC$
            $AB = EB$ (given)
            $BD = BC$ (given)
   Triangles ABD, EBC are congruent (SAS)
7 Yes, $ABC = FED$ (RHS)
   $AB = 4$ cm (Pythagoras' theorem)

Activity

For example:

Exercise 3G

1 (a) 4   (b) 2.34   (c) 19.8   (d) $d = 6.72$, $e = 4.8$
2 (a) $x = 1.5$, $y = 3$ (b) 12.8
  (c) $x = 3.75$, $y = 5.25$, $z = 6.75$
3 A and C   B and E   D and F
4 squares   circles   equilateral triangles   regular hexagons
5 cubes   spheres   tetrahedrons   octahedrons
6 (a) (ii)   (b) (iii)   (c) (iii)
7 (a) Two pairs of sides in same ratio, included angle equal.
    $x = 4.72$
  (b) All corresponding angles equal.
    $y = 1.8$
  (c) All corresponding angles equal.
    $a = 7.68$, $b = 9.6$
8 (a) All corresponding angles equal.
  (b) $x = 3.2$, $y = 2.625$

Exercise 3H

1 (a) 4   (b) 9   (c) infinite
2

3 (a)

  (b) 3 planes of this type: 6 planes of this type:

The cube and regular octahedron both have 9 planes of symmetry.

Mixed exercise 3

1 (a) (i) 29°   (ii) 122°
  (b) (i) 30°   (ii) 75°   (iii) 75°   (iv) 150°

2 (a)

Cuboid	8	6	12
Triangular prism	6	5	9
Hexagonal prism	12	8	18
Octagonal prism	16	10	24
Triangular-based pyramid	4	4	6
Square-based pyramid	5	5	8

  (b) $V + F = E + 2$

3 (a)

Tetrahedron	4	4	6
Cube	8	6	12
Octahedron	6	8	12
Dodecahedron	20	12	30
Icosahedron	12	20	30

  (b) 3, 4, 5
  (c) Six regular triangular faces meeting at a vertex have a total of 360°, which creates a surface not a vertex.
  (d) Regular polyhedra require 3 or more faces to create a vertex. 3 interior angles of regular hexagons = $3 \times 120° = 360°$, creating a surface not a vertex.
4 AF, BD, CH, EG
5 (a) $y = 2.7$
  (b) $q = 4.2$, $r = 5.88$ (c) $x = 6$ (d) $p = 25.2$
6 (a) Alternate angles and vertically opposite angles are equal so interior angles all match.
  (b) $x = 1.6$, $y = 4.48$
7 (a) AXY and ABC
  (b) One interior angle is shared, the other two are corresponding angles.
  (c) 1.9125
  (d) 1.788
8 $PQ = QR$ (equal sides of a kite)
  $PS = RS$ (equal sides of a kite)
  QS is common to both. Therefore the two triangles are congruent. (SSS)
9 Area of triangle = $\frac{1}{2} \times$ base $\times$ height

  Area $ABC = \frac{1}{2} \times BC \times h$

  Area $ACD = \frac{1}{2} \times CD \times h$

  $$\frac{\text{Area } ABC}{\text{Area } ACD} = \frac{BC}{CD}$$

10 $\dfrac{\text{Area } YXW}{\text{Area } WXZ} = \dfrac{YW}{WZ}$ (from 9)

  $Y\hat{X}W = W\hat{X}Z$.   Label these angles α

  $$\frac{\text{Area } YXW}{\text{Area } WXZ} = \frac{\frac{1}{2}XY \times XW \sin \alpha}{\frac{1}{2}XW \times XZ \sin \alpha} = \frac{XY}{XZ}$$

  $\therefore \quad \dfrac{XY}{XZ} = \dfrac{YW}{WZ}$

Chapter 4 Fractions and decimals

Exercise 4A

1 (a) $\frac{3}{4} = \frac{6}{8} = \frac{9}{12} = \frac{12}{16} = \frac{15}{20} = \frac{18}{24}$ (b) $\frac{2}{7} = \frac{4}{14} = \frac{6}{21} = \frac{8}{28} = \frac{10}{35} = \frac{12}{42}$
  (c) $\frac{4}{5} = \frac{8}{10} = \frac{12}{15} = \frac{16}{20} = \frac{20}{25} = \frac{24}{30}$ (d) $\frac{3}{9} = \frac{6}{18} = \frac{9}{27} = \frac{12}{36} = \frac{15}{45} = \frac{18}{54}$
2 (a) $\frac{1}{6} = \frac{3}{18}$ (b) $\frac{3}{7} = \frac{6}{14}$ (c) $\frac{3}{8} = \frac{18}{48}$ (d) $\frac{4}{7} = \frac{12}{21}$
  (e) $\frac{5}{6} = \frac{30}{36}$ (f) $\frac{2}{3} = \frac{6}{9}$ (g) $\frac{4}{9} = \frac{24}{54}$ (h) $\frac{5}{7} = \frac{40}{56}$
3 (a) $\frac{9}{10} = \frac{90}{100}$ (b) $\frac{3}{5} = \frac{9}{15}$ (c) $\frac{2}{5} = \frac{8}{20}$ (d) $\frac{5}{8} = \frac{25}{40}$
  (e) $\frac{8}{9} = \frac{40}{45}$ (f) $\frac{7}{12} = \frac{84}{144}$ (g) $\frac{7}{8} = \frac{49}{56}$ (h) $\frac{2}{9} = \frac{18}{81}$

Exercise 4B

1 $\frac{5}{12}, \frac{1}{2}, \frac{2}{3}$ 2 $\frac{2}{5}, \frac{3}{7}, \frac{5}{8}$ 3 $\frac{8}{12}, \frac{3}{4}, \frac{7}{8}$ 4 $\frac{2}{15}, \frac{1}{5}, \frac{1}{3}$ 5 $\frac{13}{16}, \frac{5}{6}, \frac{7}{8}$

Exercise 4C

1

	Tens	Units	.	tenths	hundredths	thousandths	ten thousandths	hundred thousandths
(a)		0	.	2	6			
(b)		1	.	7	9			
(c)	3	8	.	7	2	4		
(d)	1	4	.	8				
(e)		0	.	0	1	0	1	
(f)		0	.	0	3	3	5	
(g)		2	.	6	4	3	5	
(h)	8	0	.	9	3	4		
(i)		4	.	3	9	8	5	2
(j)		0	.	0	0	1	1	
(k)		5	.	1	0	7		
(l)		9	.	8	3	0	2	
(m)		7	.	4	5	7		
(n)	1	3	.	0	3	0	6	
(o)	5	4	.	7	0	5		
(p)	1	0	.	5	0	3	0	1

2 (a) 4.101 m, 4.009 m, 4.0059 m
  (b) 1.25 kg, 0.55 kg, 0.525 kg, $\frac{1}{8}$ kg
  (c) 5.306 km, 5.305 km, 5.204 km, 5.202 km
  (d) 9.99 t, 9.904 t, $9\frac{9}{10}$ t, 9.804 t
  (e) 6.556 s, 6.554 s, $5\frac{3}{4}$ s, 5.623 s
  (f) 2.02 cl, 2.0 cl, $\frac{1}{5}$ cl, 0.022 cl
  (g) 6.306 km, 6.305 km, 6.204 km, 6.202 km
  (h) 6.643 t, 6.448 t, 6.443 t, 4.512 t
  (i) $10\frac{4}{5}$ cm, 9.8 cm, 9.08 cm, 1.08 cm
  (j) 8.88 m, 8.801 m, 8.8 m, 8.701 m

Exercise 4D

1 13.1   2 27.9   3 1.7   4 28.8   5 4.9   6 6.4
7 10.6   8 27.6   9 1.2   10 1.3   11 6.3   12 5.1
13 (a) 11.242   (b) 154 000

Exercise 4E

1 (a) 8   (b) 40   (c) 2.4
2 (a) 17.1   (b) 29.2   (c) 26
3 (a) 26   (b) 14.5   (c) 415
   (d) 4520   (e) 521   (f) 56.2

Exercise 4F

1 (a) 0.8   (b) 0.25   (c) 1.125   (d) 0.19
   (e) 3.6   (f) 0.52   (g) 0.625   (h) 3.425
2 (a) 0.14   (b) 4.1875   (c) 3.15   (d) 4.3125
   (e) 0.007   (f) 1.28   (g) 15.9375   (h) 2.35

Exercise 4G

1 (a) $\frac{12}{25}$ (b) $\frac{1}{4}$ (c) $1\frac{7}{10}$ (d) $3\frac{203}{500}$ (e) $4\frac{3}{1000}$
2 (a) $2\frac{1}{40}$ (b) $\frac{49}{1000}$ (c) $4\frac{7}{8}$ (d) $3\frac{3}{4}$ (e) $10\frac{101}{1000}$
3 (a) $\frac{5}{8}$ (b) $2\frac{64}{125}$ (c) $\frac{13}{16}$ (d) $14\frac{7}{50}$ (e) $9\frac{3}{16}$

Exercise 4H

1 (i) 0.833 333 (ii) $0.8\dot{3}$ 2 (i) 1.222 222 (ii) $1.\dot{2}$
3 (i) 3.166 666 (ii) $3.1\dot{6}$ 4 (i) 0.916 666 (ii) $0.91\dot{6}$
5 (i) 5.555 555 (ii) $5.\dot{5}$ 6 (i) 4.818 181 (ii) $4.\dot{8}\dot{1}$
7 (i) 0.068 181 (ii) $0.68\dot{1}$ 8 (i) 2.636 363 (ii) $2.\dot{6}\dot{3}$
9 (i) 9.954 545 (ii) $9.9\dot{5}\dot{4}$ 10 (i) 0.833 333 (ii) $0.8\dot{3}$

Exercise 4I

1 (a) $\frac{1}{2}$ (b) $\frac{6}{7}$ (c) $1\frac{1}{5}$ (d) $1\frac{3}{5}$
   (e) $3\frac{2}{9}$ (f) $2\frac{2}{3}$ (g) $1\frac{3}{4}$ (h) $1\frac{7}{8}$
2 (a) $\frac{7}{8}$ (b) $3\frac{5}{8}$ (c) $1\frac{3}{8}$ (d) $6\frac{5}{8}$
   (e) $4\frac{1}{16}$ (f) $4\frac{1}{8}$ (g) $1\frac{1}{16}$ (h) $4\frac{3}{16}$
3 (a) $\frac{23}{40}$ (b) $\frac{11}{30}$ (c) $2\frac{29}{30}$ (d) $\frac{20}{21}$
   (e) $3\frac{19}{42}$ (f) $3\frac{41}{42}$ (g) $5\frac{13}{15}$ (h) $2\frac{8}{9}$
4 $7\frac{1}{12}$ miles   5 $17\frac{1}{2}$ in   6 $3\frac{11}{16}$ in   7 $\frac{17}{30}$ 8 $1\frac{22}{35}$ lb

Exercise 4J

1 (a) $\frac{1}{2}$ (b) $\frac{1}{4}$ (c) $\frac{1}{2}$ (d) $\frac{3}{7}$
2 (a) $\frac{1}{8}$ (b) $\frac{3}{8}$ (c) $\frac{1}{8}$ (d) $2\frac{3}{8}$
3 (a) $2\frac{3}{20}$ (b) $2\frac{18}{35}$ (c) $5\frac{4}{15}$ (d) $3\frac{29}{72}$
4 (a) $3\frac{5}{24}$ (b) $2\frac{4}{9}$ (c) $3\frac{17}{40}$ (d) $3\frac{6}{35}$
5 $4\frac{5}{8}$ kg   6 $3\frac{1}{8}$ pints   7 $2\frac{1}{8}$ ft   8 $\frac{1}{12}$ 9 $2\frac{1}{8}$ km

Exercise 4K

1 (a) 9   (b) 30   (c) 28 litres   (d) 21 pints
2 (a) £15   (b) 3.704 kg (c) £2.64   (d) 62.05 litres
3 (a) 27 kg   (b) £2.22   (c) 654.5 km   (d) 280 people
4 (a) £17.00 (b) 168 toys (c) 720 cars   (d) £37.15
5 140 days   6 21
7 (a) $\frac{3}{16}$ (b) $\frac{8}{15}$ (c) $\frac{4}{21}$ (d) $\frac{4}{15}$
   (e) $\frac{2}{35}$ (f) $\frac{10}{21}$ (g) $\frac{3}{8}$ (h) $\frac{1}{5}$
8 (a) $\frac{2}{7}$ (b) $\frac{5}{14}$ (c) $\frac{2}{5}$ (d) $\frac{1}{6}$
   (e) $\frac{1}{7}$ (f) $\frac{2}{5}$ (g) $3\frac{1}{2}$ (h) $\frac{13}{20}$
9 (a) $\frac{8}{9}$ (b) $\frac{14}{15}$ (c) $\frac{3}{8}$ (d) $3\frac{3}{4}$
   (e) $\frac{5}{8}$ (f) $1\frac{13}{15}$ (g) 16   (h) $\frac{3}{8}$
10 £32.65   11 £1.08   12 $20\frac{5}{8}$ lb   13 $14\frac{5}{8}$ min
14 $24\frac{1}{6}$ m²

Exercise 4L

1 (a) $\frac{4}{9}$ (b) $\frac{8}{15}$ (c) $\frac{9}{16}$ (d) 2
2 (a) $\frac{2}{9}$ (b) $\frac{1}{9}$ (c) $1\frac{1}{6}$ (d) $1\frac{3}{4}$
3 (a) $\frac{8}{9}$ (b) $7\frac{1}{2}$ (c) $6\frac{9}{10}$ (d) $1\frac{13}{20}$
   (e) $\frac{3}{5}$ (f) $1\frac{1}{9}$ (g) $\frac{4}{9}$ (h) $\frac{9}{10}$
4 16   5 36   6 $11\frac{2}{11}$ days

Exercise 4M

1 4.95 t   2 289 cars   3 $\frac{2}{9}$ 4 $\frac{17}{45}$
5 $1\frac{3}{5}$ miles   6 605   7 $\frac{9}{16}$ 8 $\frac{27}{55}$

Mixed exercise 4

1 (a) $\frac{5}{9} = \frac{10}{18} = \frac{15}{27} = \frac{20}{36} = \frac{25}{45}$ (b) $\frac{2}{5} = \frac{4}{10}$
   (c) $\frac{3}{8} = \frac{24}{64}$ (d) $\frac{8}{7} = \frac{40}{35}$
2 $\frac{2}{5}, \frac{7}{10}, \frac{3}{4}, \frac{7}{8}$
3 (a) 0.875 (b) 2.75 (c) 4.36 (d) $0.\dot{7}$ (e) $2.8\dot{3}$
4 (a) $\frac{1}{25}$ (b) $2\frac{29}{40}$ (c) $5\frac{17}{25}$ (d) $\frac{3}{10}$ (e) $\frac{9}{20}$
5 $\frac{4}{5}$ kg
6 $8\frac{3}{8}$ ft
7 (a) 85.1 (b) 24.5 (c) 5.7 (d) 1.2
   (e) 75 (f) 712
8 $4\frac{11}{16}$ ft
9 16 bags
10 $\frac{8}{9}$
11 $4\frac{11}{12}$ in
12 $\frac{1}{8}$
13 £18
14 3.91 km, 3.9 km, 3.671 km, 3.506 km, 3.451 km, 3.008 km

Chapter 5 Collecting and recording data

Exercise 5A

Qualitative: (b), (g), (h).
Quantitative, discrete: (e), (i), (j).
Quantitative, continuous: (a), (c), (d), (f), (k).
(f) and (k) are continuous variables but they are usually treated as if they were discrete.

Exercise 5B

1 (a) random (b) biased (c) random
 (d) random (e) random
2 (a) (i) 25 (ii) 8.5
 (b) She can't sample 0.5. She can double the sample size or decide, by tossing a coin, to sample 8 or 9.
 (c) 7.25
 (d) She can't sample fractions of a student, so she will need to round to the nearest whole numbers.
3 (a) 800 (b) 1020 (c) 567
4 Every 50th person in the directory starting at a random name from first to fiftieth.

Exercise 5C

1 (a) X (b) Y (c) Y (d) X
 (e) Y (f) X (g) X (h) X
2 (a) The question is biased.

(b) 'Terrible', 'quite good' and 'OK' mean different things to different people.
(c) The question is biased.
(d) The question cannot be answered if you have never played football.
(e) The question does not specify a period (e.g. each day/week).
(f) Everyone may not have the same local library.
3 (a) less than £5, £5 to £9.99,
 £10 to £14.99, £15 to £19.99
 (Other answers are possible.)

Exercise 5E

1 (a) secondary (b) secondary (c) primary
 (d) primary (e) secondary

Exercise 5F

1 1–10 4; 11–20 5; 21–30 5; 31–40 8
2 (a) 1–5 2; 6–10 5; 11–15 7; 16–20 8; 21–25 13; 26–30 5
 (b) 1–3 1; 4–6 1; 7–9 4; 10–12 4; 13–15 4; 16–18 4; 19–21 6; 22–24 5; 25–27 10; 28–30 1.
 (c) Six intervals are best. There is no useful extra information from the ten intervals.
3 61.0–61.9 6; 62.0–62.9 9; 63.0–63.9 8; 64.0–64.9 5; 65.0–65.9 6; 66.0 and over 2.

Exercise 5G

1

		\multicolumn{4}{c}{Height (cm)}
		$160 \leqslant h < 170$	$170 \leqslant h < 180$	$180 \leqslant h < 190$	$190 \leqslant h < 200$
Waist (cm)	$65 \leqslant x < 70$				
	$70 \leqslant x < 75$				
	$75 \leqslant x < 80$				
	$80 \leqslant x < 85$				
	$85 \leqslant x < 90$				
	$90 \leqslant x < 95$				
	$95 \leqslant x < 100$				

2

		\multicolumn{6}{c}{Average speed of journey (km/h)}
		$30 \leqslant s < 35$	$35 \leqslant s < 40$	$40 \leqslant s < 45$	$45 \leqslant s < 50$	$50 \leqslant s < 55$	$55 \leqslant s < 60$
Length of journey (km)	$10 \leqslant x < 20$						
	$20 \leqslant x < 30$						
	$30 \leqslant x < 40$						
	$40 \leqslant x < 50$						
	$50 \leqslant x < 60$						
	$60 \leqslant x < 70$						
	$70 \leqslant x < 80$						
	$80 \leqslant x < 90$						
	$95 \leqslant x < 100$						

3

		\multicolumn{4}{c}{Height of nest (cm)}
		$0 \leqslant h < 100$	$100 \leqslant h < 200$	$200 \leqslant h < 300$	$300 \leqslant h < 400$
No. of eggs	1				
	2				
	3				
	4				
	5				
	6				
	7				

4 (a)

	Sd	S	O	Total
Boys	12	3	1	16
Girls	8	4	2	14
Total	20	7	3	30

(b) 4

Exercise 5H

1 (a)

Length (cm)		Frequency
1 up to but not including 2		0
2	3	2
3	4	6
4	5	8
5	6	8
6	7	4
7	8	2

(b)

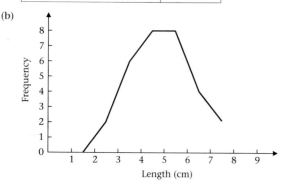

2 (a) You can't tell at what point the two boys gave up – it could have been quite quickly. You cannot 'make up' data for them so in this case it is best to ignore the two stars and consider the data for the 23 boys who completed the task.

(b)

Time (s)		Frequency
10 up to but not including 20		0
20	30	3
30	40	3
40	50	7
50	60	4
60	70	3
70	80	2
80	90	1

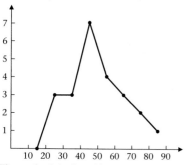

(c) The peak of the boys' frequency polygon is in the same interval as the girls. However, the boys' frequency polygon is extended towards longer times – especially if the two who gave up are included.

Mixed exercise 5

1 Make the question specific and comparative, e.g. 'Which food do you eat most often in the school canteen?'

2 Give a time frame and a numerical quantifier, e.g. 'How many hours of television did you watch last Saturday?'

3

	Tally
Comedy	
Horror	
Drama	
Fantasy	
Documentary	
Romance	

4

	Hot sweet	Cold sweet	Hot drink	Cold drink
Salad				
Sandwiches				
Hot meal				

5 Number each student from 0–999. Select students using random numbers.

6 (a) Not random, but unlikely to be biased.
(b) Random (unless the survey concerns possible effects of season of birth).
(c) Biased.
(d) Random (as long as the survey is not concerned with sport).

7 (a) His friends may have similar interests.
(b) Take a random sample from the school. A bigger sample (30 or more).

8 (i) **Random sample:** Number each student in the school and use random numbers to select the 60 students.
(ii) **Systematic sample:** From a list of the students, use random numbers to select a student from the first 20 students. Start from this student and select every 20th student from the list.
(iii) **Stratified sample:** Stratify the school according to Upper school and Lower school – 400 and 800 students (say). Use random numbers to select 20 students from Upper school and 40 students from Lower school.

9 7

10

	Tea bags	Packet tea	Instant tea	Total
50 g	2	0	5	7
100 g	35	20	5	60
200 g	15	5	13	33
Total	52	25	23	100

11 (a)

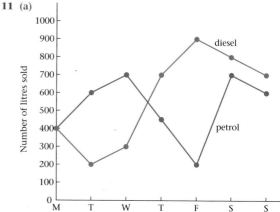

(b) Large sales of petrol correspond to smaller sales of diesel, and vice versa. Total sales were largest on Saturday and Sunday.

12

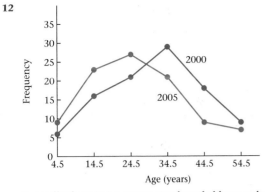

Generally there was a greater number of older members (and smaller number of younger members) in 2000.

Chapter 6 Solving equations and inequalities

Exercise 6A

1 $x = -5$ **2** $x = -2$ **3** $x = 2\frac{1}{2}$ **4** $x = -5\frac{1}{2}$
5 $a = -3$ **6** $b = 1.6$ **7** $c = -3\frac{1}{2}$ **8** $p = 3$
9 $q = 4$ **10** $s = -4$

Exercise 6B

1 18   **2** -6 **3** 24   **4** $9\frac{1}{3}$ **5** $4\frac{1}{2}$
6 14   **7** -6 **8** -12 **9** 6   **10** $-6\frac{2}{3}$

Exercise 6C

1 4   **2** 3   **3** $\frac{1}{2}$ **4** $\frac{4}{5}$ **5** -2
6 -1 **7** $1\frac{1}{2}$ **8** $3\frac{2}{3}$ **9** $-1\frac{1}{3}$ **10** $-2\frac{1}{5}$

Exercise 6D

1 1   **2** $\frac{1}{2}$ **3** $-1\frac{1}{2}$ **4** -3 **5** $\frac{3}{5}$
6 $-\frac{3}{4}$ **7** $1\frac{1}{5}$ **8** 1   **9** 0   **10** $-2\frac{1}{3}$

Exercise 6E

1 4   **2** -3 **3** $2\frac{1}{6}$ **4** $3\frac{1}{3}$ **5** $\frac{4}{5}$
6 $6\frac{1}{2}$ **7** -1 **8** 1   **9** $1\frac{1}{2}$ **10** 2
11 -3 **12** $-2\frac{1}{2}$ **13** 2   **14** -2 **15** 2
16 4   **17** $\frac{1}{2}$ **18** 15   **19** -14 **20** 0
21 $\frac{1}{2}$

Exercise 6F

1 60   **2** 60   **3** 28   **4** 12   **5** 8
6 15   **7** 24   **8** $6\frac{5}{8}$ **9** 33   **10** $3\frac{1}{9}$

Exercise 6G

1 18   **2** 7
3 (a) $8 + 2p = 28 - 6p$ (b) $p = 2\frac{1}{2}$ (c) 13
4 (a) $30 + x = 45 - 2x$ (b) $x = 5$ (c) 35
5 (a) $15 - 3c = 12 - 2c$ (b) £3
6 (a) $70 - 60g = 60 - 20g$ (b) $\frac{1}{4}$ litre

Exercise 6H

1 $x > 1$ **2** $x \leq -2$ **3** $-3 \leq x < 2$ **4** $0 \leq x < 3$
5

![number line 5]

6

![number line 6]

7

![number line 7]

8

![number line 8]

9

![number line 9]

10

![number line 10]

Exercise 6I

1 $x \leq 3$ **2** $x > \frac{2}{5}$ **3** $x \leq 1\frac{2}{3}$ **4** $x < -3$
5 $a \geq 1$ **6** $b < -1$ **7** $c > -4$ **8** $d \leq 1\frac{1}{2}$
9 $e \leq -4$ **10** $f < -28$ **11** $g \leq 2$ **12** $h \geq \frac{1}{2}$
13 $j > -1$ **14** $k \leq 1\frac{2}{3}$ **15** $m \geq -3\frac{1}{2}$ **16** $n > -1\frac{1}{2}$

Exercise 6J

1 $1 < p \leq 6$ **2** $5 < q < 7\frac{1}{2}$
3 $-1 \leq r \leq 2$ **4** $-6 < s < 4$

Exercise 6K

1 (a) $-2, -1, 0, 1$ (b) $1, 2, 3, 4$ (c) $-3, -2, -1$
    (d) $0, 1, 2, 3$
2 (a) e.g. $2 \leq x \leq 5$ (b) e.g. $0 \leq x \leq 3$
    (c) e.g. $-3 \leq x \leq -1$ (d) e.g. $-2 \leq x \leq 2$
3 (a) 2   (b) 1   (c) -1 (d) -3
4 (a) -1 (b) 2   (c) 0   (d) 2
5 (a) $2, 3$ (b) $2, 3$ (c) $-2, -1, 0, 1, 2$
    (d) $0, 1$

Exercise 6L

1 $x > 11\frac{2}{3}$ **2** $x > 4$ **3** $x \leq -3\frac{1}{2}$ **4** $x \geq 2$
5 $x > -\frac{1}{2}$ **6** $x \leq -1$ **7** $x > 2\frac{1}{2}$ **8** $x < -1$
9 $x \leq 3\frac{1}{5}$ **10** $x < 0$ **11** $x < -1\frac{1}{3}$ **12** $x \geq -\frac{1}{2}$
13 $x < -8$ **14** $x < -2$

Mixed exercise 6

1 3   **2** $1\frac{2}{5}$ **3** -1 **4** 4
5 -6 **6** $\frac{2}{3}$ **7** -2 **8** $-2\frac{1}{2}$
9 $\frac{3}{5}$ **10** 8   **11** -8 **12** 6
13 2   **14** $2\frac{1}{2}$ **15** 7   **16** 27p
17 (a) 5 cm
    (b) 2 cm
18 (a)   (b)

![number line 18a]   ![number line 18b]

    (c)   (d)

![number line 18c]   ![number line 18d]

19 (a) $x < 2$ (b) $x \leq -2$ (c) $x < 1\frac{1}{3}$
20 (a) (i) 3   (ii) 3
    (b) (i) 3   (ii) 0
21 $P = 53$
22 $x \leq 3\frac{2}{5}$
23 $y = 1\frac{6}{7}$

Chapter 7 Transformations, loci and constructions

Exercise 7A

1 (a) $\begin{pmatrix} 2 \\ -2 \end{pmatrix}$

(b) $\begin{pmatrix} -2 \\ 2 \end{pmatrix}$

(c) $\begin{pmatrix} 5 \\ 0 \end{pmatrix}$

(d) $\begin{pmatrix} -7 \\ 2 \end{pmatrix}$

(e) $\begin{pmatrix} -6 \\ -3 \end{pmatrix}$

2 (a)

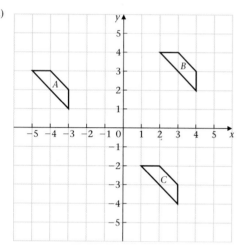

(c) $\begin{pmatrix} 6 \\ -5 \end{pmatrix}$ (d) $\begin{pmatrix} -6 \\ 5 \end{pmatrix}$

3

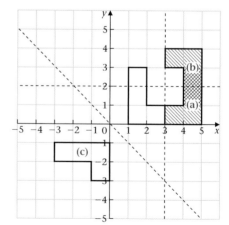

(a) A (3, 3), B (6, 4)   (b) A (6, 5), B (9, 6)
(c) $\begin{pmatrix} 6 \\ 4 \end{pmatrix}$ (d) A (1, 4), B (4, 5)
(e) A (1, 4), B (4, 5)   (f) They are the same.

Exercise 7B

1

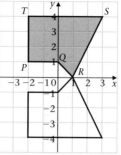

2   A reflection in the line $x = 2$

3

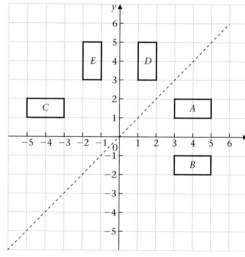

(a) (ii)   (3, −1) (3, −2) (5, −2) (5, −1)
     (iii) x coordinate remained the same
            y coordinate changed sign
(b) (ii)   (−3, 1) (−3, 2) (−5, 2) (−5, 1)
     (iii) x coordinate changed sign
            y coordinate remained the same
(e) Reflection in line $y = -x$

4

Exercise 7C

1

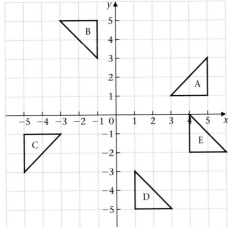

(a) $(-1, 3) (-1, 5) (-3, 5)$
(b) $(-3, -1) (-5, -1) (-5, -3)$
(c) Rotate 180° about the origin
(d) $(1, -3) (1, -5) (3, -5)$
(e) Rotate 180° about the origin

2 A rotation through 180° about the point (0, 0). (A half turn about (0, 0).)

Exercise 7D

1 (a)

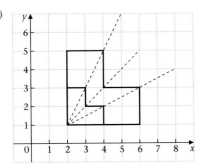

(b)

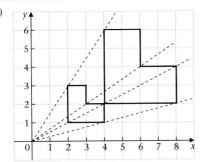

(c)

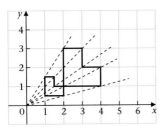

(d)

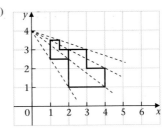

2 Enlargement, scale factor $\frac{1}{2}$, centre (7, 1)

3

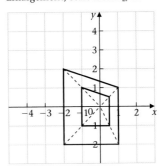

4

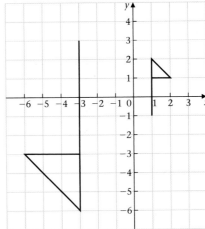

5 (a) (1, 1) scale factor 2
(b) (3, 3) scale factor 3
(c) (3, 3) scale factor $\frac{1}{3}$
(d) (2, 0) scale factor -1

6 Repeat 4 [FD 45 RT 90]

7 (a) Scale factor $\frac{3}{2}$, centre $(-1, -1)$
(b) Scale factor $\frac{2}{3}$, centre $(-1, -1)$
(c) Each is the inverse of the other.

8

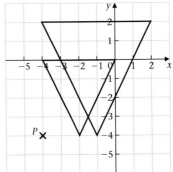

9 Enlargement, scale factor $-1\frac{1}{2}$, centre (1, 1)

Exercise 7E

1

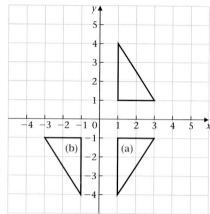

(c)   Rotation 180° about origin.

2

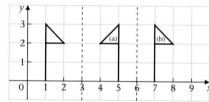

(c)   Translation $\begin{pmatrix} 6 \\ 0 \end{pmatrix}$

3

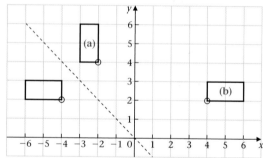

(c)   Reflection in $x = 0$ (y-axis)

4

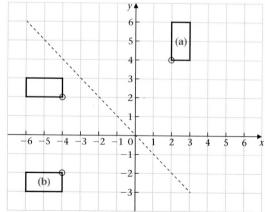

(c)   Reflection in $y = 0$ (x-axis)

5

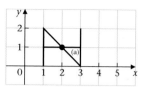

(b)   Rotation 180° about point (2, 1)

6

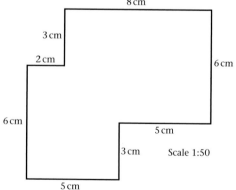

7 (a)   Enlargement, scale factor 2, centre (0, 0)
   (b)

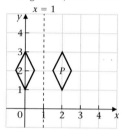

Exercise 7F

1 (a)   7.6 m   (b)   4.3 m
2 (a)   80 m   (b)   1.5 m
3 (a)   14 km   (b)   31 km   (c)   2.7 cm
4 (a)   1:200
   (b)   3 cm by 2 cm
   (c)   6 m by 4 m
   (d)   2 m
5 The scale drawing should have the marked dimensions.

Exercise 7G

6 60°
8 (a) $x = 120°$ (b) $y = 30°$
10 8.77 cm

Exercise 7H

1 Pupils should draw the perpendicular bisector of *AB*.

2

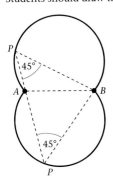

2.5 cm
2.5 cm
4 cm

3 The locus will consist of 4 quarter circles with centres on a straight line. The radii depend on the size of the box and position of the point. If the point is on one of the edges the box is tipped about there will be only 3 quarter circles.)

4 (a)

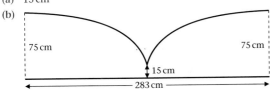

22.3 m
150 m

(b) 440 m to 3 s.f.

5 (a) 15 cm

(b)

75 cm   75 cm
15 cm
283 cm

6

1.8 m
0.3 m
1.2 m
0.8 m   1.4 m

7 Daniel could attend Ayleton or Bankbury

8 Students should draw the bisector of the angle at the gate.

9

P
45°
A   B
45°
P

10

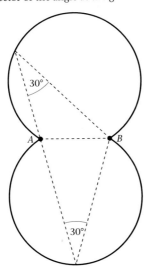

30°
A   B
30°

Exercise 7I

1 (a) (b) (c) (d) (e)

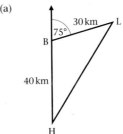

2 (a) 063°   (b) 170°   (c) 332°
3 (a) 235°   (b) 115°   (c) 295°
4 240°
5 325°
6 (a) 180° + bearing of *B* from *A*
   (b) bearing of *B* from *A* − 180°

Mixed exercise 7

1 (a)

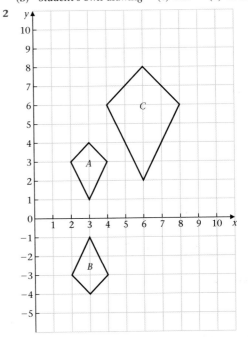

30 km   L
75°
B
40 km

H

(b) Student's own drawing   (c) 031°   (d) 56 km

2

3 (a)

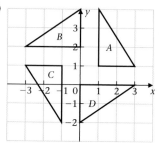

(b)   Rotation, 180° about (0, 1)

4

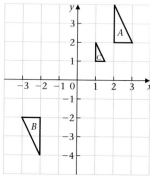

5 (a)

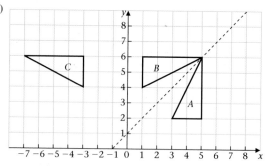

(b) $y = x + 1$
(c)   90° anticlockwise rotation around $(-1, 0)$

6 (a)   3 km
   (b)

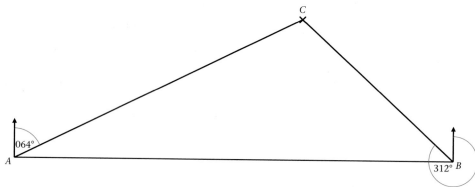

7

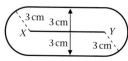

8

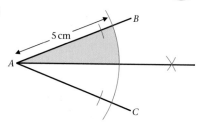

9

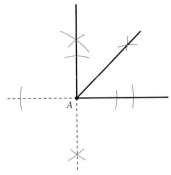

A

11 81°

Chapter 8 Using basic number skills

Exercise 8A

1 (a) 112% (b) 117% (c) 130%
   (d) 111% (e) 108.5%
2 (a) 1.24 (b) £13 454
3 (a) 1.4 (b) £21.56
4 £1525.15
5 1.04
6 (a) £144 (b) 70 kg (c) 2.784 m
   (d) £1370.20 (e) 128.52 cm

Exercise 8B

1 (a) 84% (b) 89%
   (c) 95% (d) 78%
2 (a) 0.66 (b) £3.47 to 2 d.p.
3 77.9 kg
4 (a) £96 (b) 42 kg (c) 2.016 m
   (d) £1109.80 (e) 123.48 cm
5 (a) 1.06 (b) £106 (c)

Date	Amount
1/1/99	£100
1/1/00	£106
1/1/01	£112
1/1/02	£119
1/1/03	£126

6 (a) £12 320 (b)

Original cost	£14 000
value after 1 year	£12 320
value after 2 years	£10 102
value after 3 years	£9597
value after 4 years	£9117
value after 5 years	£8662

7 (a) 1.1 (b) 1.188

Exercise 8C

1 (a) $\frac{7}{9}$ (b) 1 (c) 77.7% (d) 100%
2 (a) 50% (b) 60% (c) 16.73% (d) 20%
3 (a) 8 km/litre (b) 10.6 km/litre (c) 32.5%
4 (a) £5460 (b) 58%
5 (a) 7.5 litres (b) 25%

Exercise 8D

1 (a) £260 (b) 32.5% (c) 20.6% to nearest 0.1%
2 (a) 33% (b) 27% (c) 12% (d) 8% (e) 55%
3 (a) 20%, 31.2%, 40%, 47.2% (b) 12.8%

Exercise 8E

1 £428.40 **2** £54 538 **3** 13.04
4 (a) 42.86% (b) 33.3% (c) 4.76% decrease

Exercise 8F

1 (a) 90% (b) pre-sale price $\xrightarrow{\times 0.9}$ £540 (c) £600
2 £2.78
3 £104.11
4 (a) 117.5% (b) pre-VAT price $\xrightarrow{\times 1.175}$ £320 (c) £272.34
5 (a) 175 (b) 14
6 (a) £13 333.33
   (b) £14 814.82

Exercise 8G

1 (a) £14.63 (b) 5.852%
2

Year	Amount at start of year	Interest	Total amount at year end
6	£334.56	£334.56 × 0.06	£354.63
7	£354.63	£354.63 × 0.06	£375.91
8	£375.91	£375.91 × 0.06	£398.46
9	£398.46	£398.46 × 0.06	£422.37
10	£422.37	£422.37 × 0.06	£447.71

3 (a) £47.71 (b) 19.1% (c) 7.9%
4 (a) approx 12 years (b) 16 years 8 months
5 (a) £1082 (b) £1370.59 (c) £1878.53
6 If an annual interest rate of 50% is paid, then after the first year the value of the investment increases to 150% of its original value, and during the second year the value of the investment increases by 50% of its value *at the end of the first year*, giving an increase of 125%.
7 (a) £7440 (b) £24.80 per month
   (c) £644.80 (d) £754.32
8 (a) £10 800 (b) £9720 (c) £7085.88
   (d) 6 years 7 months approx
9 (a) 1.602 (b) 2.027 (c) 3.247
10 (a) £550.16 (b) £1755.16 (c) £973.36

Exercise 8H

1 (a) 87.17 km/h (b) 54 mph
2 25.71 mph
3 (a) 1.25 h (b) 0.4 h (c) 0.1 h (d) 0.42 h
   (e) 0.75 h (f) 0.17 h
4 (a) 162 min (b) 2.7 h (c) 18.52 km/h
5 (a) Shannon by 4 minutes (b) 1.63 h or 1 h 38 min
6 (a) 24 min (b) 5 h 18 min (c) 3 h 15 min
   (d) 6 h 17.4 min (e) 3 h 54 min (f) 9 h 52.8 min
7 (a) 5 hours (b) 5 km/h
8 (a) 456 km (b) 6.85 h (c) 66.57 km/h

Exercise 8I

1 14.5 km/l
2 (a) (i) 42 litres (ii) 14 km/litre
   (b) 84 km
3 (a) (i) regular 0.8125 p/h longlife 0.78 p/h
       (ii) regular 123 h/£ longlife 128 h/£
   (b) Longlife cheaper per hour or more hours/£
4 (a) 1.424 francs/dollar (b) 0.702 dollars/franc
5 (a) Cold 184 litres/min Hot 115 litres/min
   (b) 3 min 4.6 sec
6 (a) (i) 0.16 litres/km (ii) 6.4 litres/hour
   (b) (i) 10 km (ii) 1.6 litres

Exercise 8J

1 100 hours
2 (a) (i) $0.00026\,\text{m}^3/\text{sec}$ (ii) $15.6\,\text{litres/min}$
   (b) 1 hr 4 min 6 sec
   (c) 51 min 17 sec

Exercise 8K

1 (a) $32000\,\text{cm}^3$ (b) 32 m
2 (a) $2800\,\text{cm}^3$ (b) $75\,\text{g/cm}^3$
3 (a) 6475 kg (b) $0.463\,\text{m}^3$
4 (a) $546\,\text{cm}^3$ (b) $6.41\,\text{g/cm}^3$

Exercise 8L

1 (a) (i) 2 (ii) 4 (iii) 6
   (b) (i) $\frac{1}{6}$ (ii) $\frac{1}{3}$ (iii) $\frac{1}{2}$
2 (a) Sally $\frac{1}{5}$ Brian $\frac{3}{10}$ Mark $\frac{1}{2}$
   (b) Sally £8 Brian £12 Mark £20
3 (a) 8 parts (b) $\frac{3}{8}$ (c) £101.25 (d) $\frac{5}{8}$
   (e) £168.75
4 (a) 5:2 (b) Danny $\frac{5}{7}$ (Melissa $\frac{2}{7}$) (c) £19
   (d) £60 (e) £10
5 (a) 4:1 (b) 18 marks
6 (a) $\frac{2}{7}$ (b) 68 boxes (c) £32.64
7 (a) £120 (b) 2:1 (c) £282
8 (a) 540 (b) 180 (c) 324 (d) 216
9 (a) 86 (b) 86:5 (c) 86 (d) 80

Exercise 8M

1 (a) $\frac{5}{13}$ (b) $\frac{8}{13}$ (c) 60 (d) 195
2 9.37 km/litre
3 (a) 5:8 (b) No, 56 km/h (c) 112 km/h
4 (a) 50 (b) 37.5 (c) 1:1.75
5 (a) 27 days (b) 21 days (c) 6 days (d) 75 days
6 (a) 6:3:1 (b) 81 goals (c) 135 goals
7 60 years, 30 years, 20 years, 12 years

Exercise 8N

1 (a) 10^2 (b) 10^3 (c) 10^6 (d) 10^{10}
2 (a) 6 million (b) 3.4 million (c) 7.8 million
   (d) 5.5 million (e) 2.65 million (f) 7.642 million
3 (a) 3.1×10^6 (b) 4.3×10^6 (c) 5×10^5
   (d) 2.4×10^6 (e) 7.8×10^6 (f) 8.6×10^6
   (g) 4×10^6 (h) 9×10^6 (i) 4×10^5
4 (a) 10^3 (b) 10^4 (c) 10^1 (d) 10^6 (e) 10^5 (f) 10^3
5

Standard form
7.4×10^4
2.6×10^2
6.8×10^5
4.5×10^1
9.9×10^6
6.2×10^1
8×10^0

6 (a) 1.6×10 (b) 4.3×10^3
   (c) 6.5×10^5 (d) 8.7×10^7
   (e) 6.7×10^2 (f) 8.65×10^2
   (g) 9.87×10^6 (h) 9.85×10^4
   (i) 8.05×10^{11}

Exercise 8O

1 (a) 420 (b) 67000 (c) 5500 (d) 7500000
   (e) 620000 (f) 73000 (g) 24000000 (h) 11
   (i) 7.25
2 (a) 9×10^2 (b) 9.6×10^3 (c) 4.05×10^4
   (d) 3.91×10^5 (to 2 d.p.)

Exercise 8P

1

Standard form
2.4×10^{-3}
2×10^{-1}
6×10^{-5}
1.5×10^{-1}
7×10^{-3}
4.5×10^{-4}
3.46×10^{-2}
1.25×10^{-3}

2 (a) 2×10^{-3} (b) 1.5×10^{-1}
   (c) 4×10^{-4} (d) 5.4×10^{-2}
   (e) 8×10^{-6} (f) 6.8×10^{-11}
   (g) 3.46×10^{-1} (h) 9×10^{-2}
   (i) 5.6×10^{-3}

3 (a) 0.35 (b) 0.06 (c) 0.00072
   (d) 0.0022 (e) 0.0000135 (f) 0.00000533
   (g) 0.00000000088 (h) 0.00000044 (i) 0.4999

Exercise 8Q

1 (a) 9.6×10^{13} (b) 1.722×10^4 (c) 8.4×10^{-3}
   (d) 1.476×10^{-12} (e) 2.91×10^{-1} (f) 1.47×10^{-9}
   (g) 2.83×10^{30} (h) 1.0×10^3 (i) 7.84×10^6
   (j) 1.0648×10^{-5}
2 (a) 2.1×10^8 (b) 2.4×10^{-5} (c) 2.133×10^{-1}
   (d) 6.378×10^1 (e) 5.2 (f) 4.416×10^{-3}
   (g) 2.684×10^4 (h) 1.463×10^{-9}
3 Approximately 500 secs or 8 min
4 (a) Earth
   (b) 5.91×10^3 km (5910 km)
   (c) 1.949×10^4
   (d) 6.543

Mixed exercise 8

1 (a) £3150 (b) £7200
2 £24
3 £224.97
4 8.1 miles per litre
5 £1111.11
6 (a) £114.89 (b) £177.87 (c) £41.70
7 £1820
8 (a) 1.4×10^8 (b) 1.94×10^6
9 $£1.40 \times 10^7$
10 362.88 kg
11 3.38×10^8

Chapter 9 Functions, lines, simultaneous equations and regions

Exercise 9A

1 (a) $1 \to 0$ (b) $1 \to 5$ (c) $1 \to 8$
        $2 \to 2$ $2 \to 8$ $2 \to 10$
        $3 \to 4$ $3 \to 11$ $3 \to 12$
        $10 \to 18$ $10 \to 32$ $10 \to 26$
   (d) $1 \to 0$ (e) $1 \to 1.5$ (f) $1 \to 2$
        $2 \to 3$ $2 \to 2$ $2 \to 2.5$
        $3 \to 6$ $3 \to 2.5$ $3 \to 3$
        $10 \to 27$ $10 \to 6$ $10 \to 6.5$

Exercise 9B

1 (a) $y = x + 2$ (b)

x	0	1	2	3	4	5
y	2	3	4	5	6	7

(c)

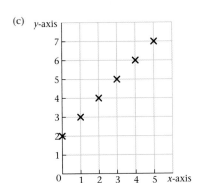

2 (a) $y = x + 4$

x	0	1	2	3	4	5
y	4	5	6	7	8	9

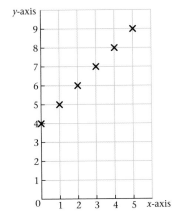

(b) $y = 2x + 3$

x	0	1	2	3	4	5
y	3	5	7	9	11	13

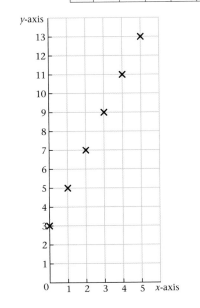

(c) $y = 3x + 2$

x	0	1	2	3	4	5
y	2	5	8	11	14	17

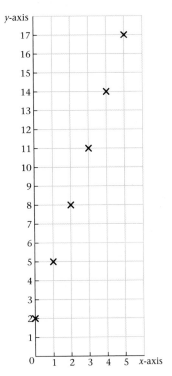

(d) $y = 2x$

x	0	1	2	3	4	5
y	0	2	4	6	8	10

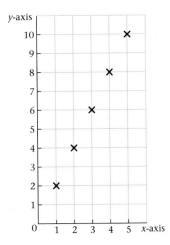

3 (a)

x	1	2	3	4	5
y	4	7	10	13	16

(b)

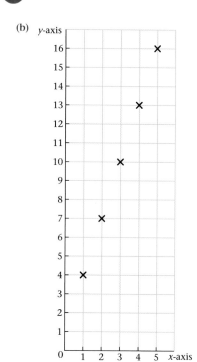

(c) $y = 3x + 1$

4 (a)

x	1	2	3	4	5
y	3	5	7	9	11

(b)

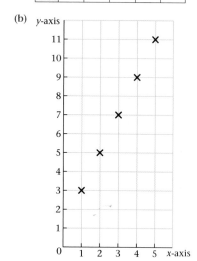

(c) $y = 2x + 1$

5 (a) (i)

x	1	2	3	4	5
y	3	5	7	9	11

(ii)

x	1	2	3	4	5
y	2	4	6	8	10

(b) (i) $y = 2x + 1$ **(ii)** $y = 2x$

Exercise 9C

1 (a)

x	0	1	2	3	4
y	0	4	8	12	16

(b)

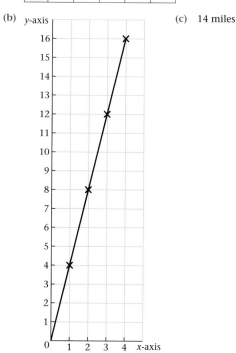

(c) 14 miles

2 (a)

x	0	5	10	15	20	25
y	0	8	16	24	32	40

(b)

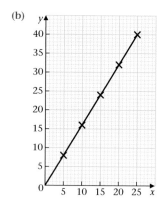

(c) 21 (20.8 to be exact)
(d) 9 (8.75 to be exact)

3 (a)

x	0	1	2	3	4	5	6
y	0	3	6	9	12	15	18

(b)

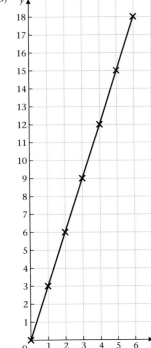

(c) 5 ($4\frac{2}{3}$ to be exact)

4 (a)

x	$\frac{1}{2}$	1	$1\frac{1}{2}$	2	$2\frac{1}{2}$	3	$3\frac{1}{2}$
y	1	$1\frac{1}{2}$	2	$2\frac{1}{2}$	3	$3\frac{1}{2}$	4

(b) A joint of lamb cannot weigh 0 kg.

(c)

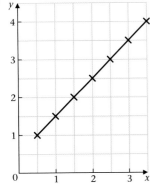

(d) About $2\frac{3}{4}$ hours

Exercise 9D

1 (a)

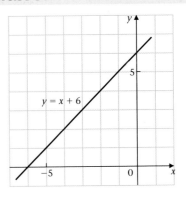

$y = x + 6$

(b)

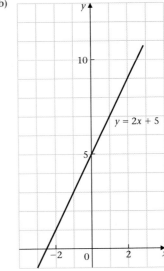

$y = 2x + 5$

(c)

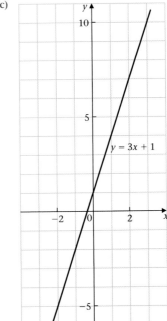

$y = 3x + 1$

(d)

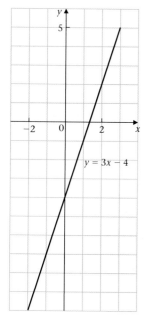

(e)

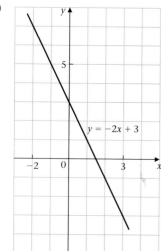

(f)

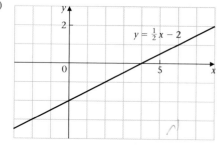

2 $y = 3x + 1$ and $y = 3x - 4$
3 (a) $y = \frac{1}{4}x + 1$ and $y = \frac{1}{4}x - 1$
   (b) Coefficients of x equal
4 (a) $(0, 5)$ (b) $(0, -2)$
   (c) $(0, -5)$ (d) $(0, 4)$
5 $y = 4x + 3$ **6** $y = 5x - 3$
7 $y = 3x + 5$ **8** $y = mx + c$
9 $c = 4$; $(0, 4)$ **10** $c = -3$; $(0, -3)$
11 (a) $(0, 1)$ (b) 5
12 (a) $y = \frac{1}{3}x - 4$ (b) $(0, -4)$

13 (a)

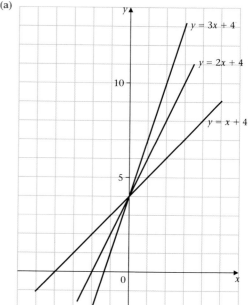

(b) all $(0, 4)$
(c) Lines with higher coefficients are steeper.
14 (a) 3   (b) 3   **15** (a) $\frac{1}{2}$ (b) $\frac{1}{2}$

Exercise 9E

1 (a) (i) 5   (ii) $(0, 4)$ (b) (i) 2   (ii) $(0, -7)$
   (c) (i) $\frac{1}{4}$ (ii) $(0, 9)$ (d) (i) -4 (ii) $(0, -3)$
   (e) (i) 9   (ii) $(0, 8)$ (f) (i) -2 (ii) $(0, 7)$
2 (a) (i) -2 (ii) $(0, 3)$ (b) (i) 3   (ii) $(0, -5)$
   (c) (i) $\frac{1}{2}$ (ii) $(0, -4)$ (d) (i) $-\frac{1}{3}$ (ii) $(0, 2)$
   (e) (i) $\frac{3}{5}$ (ii) $(0, -2)$ (f) (i) $-\frac{3}{4}$ (ii) $(0, 3)$
3 $y = 6x + 7$
4 (a) $y = 2x + 5$
   (b) $y = \frac{1}{2}x - 1$
   (c) $x + y = 8$ or equivalent
   (d) $5x + 2y + 10 = 0$
5 (a) 3   (b) $y = 3x + 2$
6 $y = 4x + 7$ **7** $y = -3x + 11$
8 (a) 4   (b) $y = 4x + 3$

Exercise 9F

1 (a) $-\frac{1}{5}$ (b) $-\frac{1}{2}$ (c) -1 (d) $-\frac{4}{3}$ (e) $\frac{2}{5}$ (f) $\frac{7}{2}$
2 (b)   **3** $y = -\frac{1}{3}x$ **4** $y = -x + 6$ **5** $y = \frac{5}{2}x - 9$
6 (c)   **7** $y = -\frac{3}{2}x + 10$ **8** $y = \frac{3}{4}x - 5$

Exercise 9G

1 (a) $(2, 2\frac{1}{2})$ (b) $(3\frac{1}{2}, 5)$ (c) $(3, 6\frac{1}{2})$
   (d) $(3, 4)$ (e) $(6, 8\frac{1}{2})$

Exercise 9H

1 e.g. $(1, 9)$ **2** (b), (c), (d)   **3** e.g. $(5, 2)$
4 (b), (c), (d)   **5** $x = 3, y = 1$
6 (a) $x = 2, y = 1$ (b) $x = 4, y = 2$ (c) $x = 4, y = 1$
   (d) $x = -3, y = 4$ (e) $x = 3, y = -1$ (f) $x = 4, y = 3$

Exercise 9I

1 $x = 3, y = 2$　　**2** $x = 4, y = 1$　　**3** $x = 2, y = -1$

4 $x = -2, y = -3$　**5** $x = \frac{1}{2}, y = 1\frac{1}{2}$　**6** $x = 2\frac{1}{2}, y = -1$

7 $x = 7, y = -1\frac{1}{2}$　**8** $x = \frac{3}{4}, y = -3$　**9** $x = -1, y = -2$

Exercise 9J

1 $x = 4, y = 1$　　**2** $x = 3, y = -2$　　**3** $x = -3, y = -5$

4 $x = 3, y = 2$　　**5** $x = \frac{1}{2}, y = -4$　　**6** $x = 5, y = 3$

7 $x = -7, y = -4$　**8** $x = 3, y = -5$　　**9** $x = 2\frac{1}{2}, y = 1$

10 $x = 1\frac{1}{3}, y = -1$

Exercise 9K

1 $a = 2.5, b = 4.5$　**2** £5 per hour　　**3** 29 and 37

4 £9.30　　　　　**5** 18　　　　　　**6** $a = 5, b = -5$

7 £3　　　　　　**8** 11p　　　　　**9** 300 kg

10 (a)　$a = 7, b = -2$

　　　(b)　(i)　$3\frac{1}{2}$　　(ii)　$(0, -5\frac{1}{2})$

Exercise 9L

1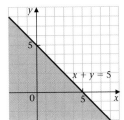

2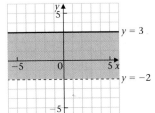

3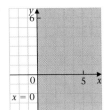

4

5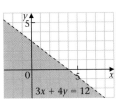

6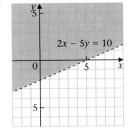

7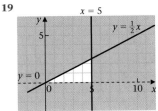

8

9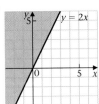

10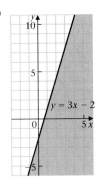

11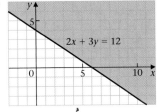

12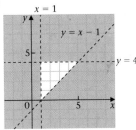

13 $x < 1$　　**14** $x + y \geqslant 3$　　**15** $y \geqslant x$　　**16** $2x + y \geqslant 4$

17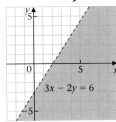

18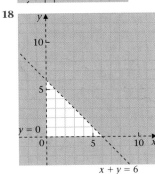

19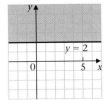

20

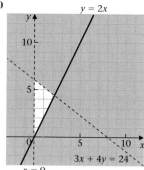

21

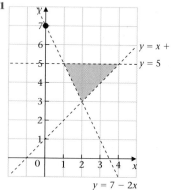

22

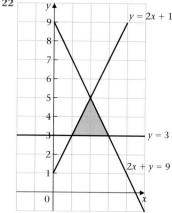

23 $y \leqslant 5$, $x \leqslant 4$, $y \geqslant -\dfrac{5x}{4} + 5$

Mixed exercise 9

1 17, 21

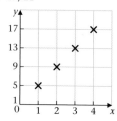

$y = 4x + 1$

2 (a)

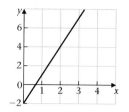

(b)

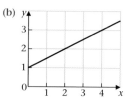

(c)

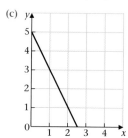

3 (a) $y = 3x + c$ ($c \neq -2$) (b) $y = \frac{1}{2}x + k$ ($k \neq 1$)
  (c) $y = a - 2x$ ($a \neq 5$)
4 $(\frac{1}{2}, 5)$
5 (a) $y = 9 - \frac{1}{3}x$ (b) $y = 6 - \frac{1}{3}x$
6 (a) 7, $(0, 3\frac{1}{2})$ (b) $-1\frac{1}{2}$, $(0, 3)$
  (c) 2, $(0, 5)$ (d) -1, $(0, -2)$
7 (a) $p = 4$, $q = -1$ (b) $x = 4\frac{1}{2}$, $y = -3$
8 (a) $y = 2x + 6$ (b) $y = 6 - \frac{1}{2}x$
9 (a)

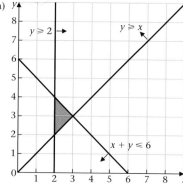

(b) $(2, 2)$, $(2, 3)$, $(2, 4)$, $(3, 3)$
10 (a) $y = \frac{1}{6}(15 - 5x)$ (b) $k = 20$
  (c) (i)

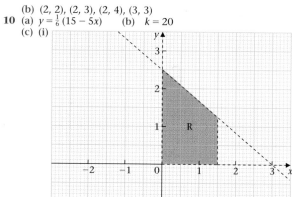

  (ii) $(1, 1)$

Chapter 10 Presenting and analysing data

Exercise 10A

1. | 4 | 5, 9 |
    | 5 | 3, 4, 5, 5, 6, 7, 7, 8, 8 |
    | 6 | 0, 0, 0, 2, 2, 3, 4, 4, 5, 6, 7, 8 |
    | 7 | 0, 3, 4, 5, 7 |
    | 8 | 2, 3   Key: 5│3 means 53 years |

    mode = 60, median = 62, mean = 63.07

2. | 2 | 5, 8, 8, |
    | 3 | 0, 1, 4, 7, 7, 7, 8, 9 |
    | 4 | 1, 2, 3, 5, 5, 6, 6, 6, 7, 7, 9 |
    | 5 | 0, 2   Key: 2│5 means 2.5 kg |

    mean = 4.01 kg, median = 4.15 kg, mode = 3.7 kg and 4.6 kg

3. | 44 | 7, 8, 8, 9, 9, 9 |
    | 45 | 0, 0, 1, 2, 3, 3, 3, 4, 5, 6, 7, 7, 7, 9 |
    | 46 | 0, 1, 2, 2, 3, 3   Key: 44│7 means 44.7 s |

    mode = 44.9 s, 45.3 s and 45.7 s
    median = 45.3 s

Exercise 10B

1. (a) (i) 34   (ii) 28.5   (iii) 26.4
    (b) Mean is lowest – the '4' depresses its value compared to the mode and median.
    (c) 31, 25.5, 23.4
2. (a) 32.6̇
    (b) No – because the ages are so spread out
3. 'Average' could be mean, median or mode.
    The mean will be distorted by the small number of very high earners.

Exercise 10C

1. (a) 287   (b) 258   (c) 319
2. (a) 19   (b) 11   (c) 40.5
3. (a) 51.06 g   (b) 50.89 g   (c) 51.19 g

Exercise 10D

1. 0, 20, 26, 12, 8, total f 45, total fx 66. Mean 1.47, Mode 1, Median 1
2. Mode 0, Median 0, Mean 0.5
3. (a) 2 goals   (b) 3 goals   (c) 3.2 goals

Exercise 10E

1. (a) 7   (b) 2
2. (a) £9   (b) £4
3. (a) $4\frac{1}{2}$ (b) $2\frac{1}{2}$

Exercise 10F

1. (a)

Class interval	Frequency f	Middle value	$f \times x$
1–5	16	3	48
6–10	28	8	224
11–15	26	13	338
16–20	14	18	252
21–25	10	23	230
26–30	3	28	84
31–35	1	33	33
36–40	0	38	0
41–45	2	43	86
Total:	100	Total:	1295

(b) 13 words
2. 11.8 years
3. (a) 61–100   (b) 61–100   (c) 69.3

Exercise 10G

1. Not necessarily. Drapeway Ltd may have a few very high wage earners which will push up the value of the mean.
2. On average students did better on paper A (higher median). There was a bigger spread of marks on paper A, with the spread of the middle 50% of candidates being over $2\frac{1}{2}$ times that on paper B.
3. Club 2 has an older median but a smaller spread. Generally there are more older people in Club 2 but because the spread in Club 1 is larger, it may have some members who are as old as the Club 2 members.

Exercise 10H

1. (a), (b)

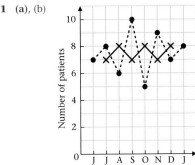

(b) No, the 3 month average is steady at 7 or 8.

2. (a)

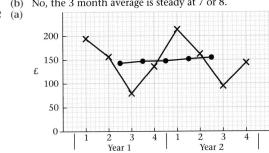

(b) The gas bills are rising slowly.

3.

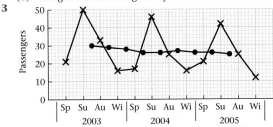

The number of passengers fell slowly between the beginning of 2003 and the end of 2005.

Exercise 10I

1.

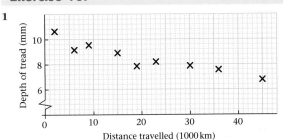

The greater the distance travelled the smaller the depth of tread.

2 (a)

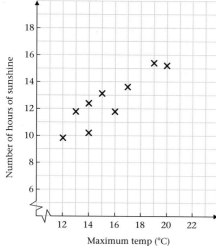

(b) As the number of hours of sunshine increases, the maximum temperature increases.

Exercise 10J

1 (a)

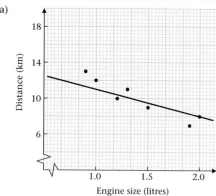

(b) High negative correlation   **(c)** 9 km

2 (a)

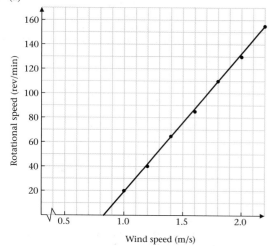

(b) There is strong positive correlation between rotational speed and wind speed
(c) 76 rev/min
(d) 4 m/s is far outside the range of data given, so the relationship may have changed.
3 (b) (i) 14 to 15 hours   **(ii)** 14°C to 15°C

Mixed exercise 10

1 (a)

7	6 7 7 7 8 8 9
8	0 0 0 0 2 3 5 6 7 7 8
9	0 0 5 5 5 5 5 5 5 8
10	0 1 5 6 6 8 8 Key: 7│6 means 76 kg

(b) (i) 95 kg   **(ii)** 88 kg   **(iii)** 89.6 kg
   (iv) 32 kg   **(v)** 15 kg
2 19
3 (a) 162 kg, 167.1 kg
   (b) (i) median **(ii)** Yes. Not affected by the larger values.
4 (a) 30   **(b)** 6, 2
   (c) 2   **(d)** No. The mode is better.
5 (a) $300 < w \leqslant 500$
   (b) 18 800 g
   (c) You do not know where the weights fall in each interval. The midpoint of each class interval is used to represent the class interval.
   (d) 470 g
6 (a) £67.50, £65.25, £63.75, £62.75, £62, £60.50, £59.75, £59.25, £58.50
   (b) Generally decreasing.
7 (a) 12, $13\frac{2}{3}$, 15, 18, 16, 10
   (b)

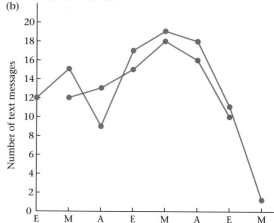

(c) Increasing trend until Sunday morning, than decreasing trend.
8 (a)

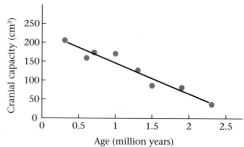

(b) Strong negative correlation. Points close to a straight line.
(c) 90 cm³

9 (a)

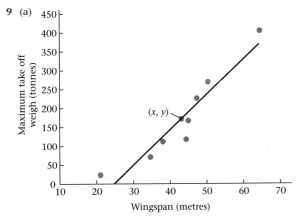

(b) (i) 43.2 m, 169.8 tonnes
(c) About 24 to 65 metres

10 (a)

	Median	IQR
Male	22	10
Female	25	8

(b) Smaller median for males, but larger IQR.
So not enough evidence to conclude a significant difference.

Chapter 11 Estimation and approximation

Exercise 11A

1 (a)

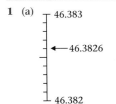

(b) 46.383

2 (a)

Name	Time
R Grey	1 min 45.48 sec
M Hobson	1 min 45.48 sec
T Knight	1 min 45.48 sec

(b) M. Hobson, T. Knight, R. Grey
3 (a) 137.3 (b) 0.6738 (c) 5.0 (d) 9.00
(e) 17.994 (f) 2.0097
4 (a) 95.047 (b) 9.760 (c) 10.91 (d) 176.0

Exercise 11B

1 (a) 3.17 (b) 965 (c) 16 (d) 10
(e) 55.764 (f) 55.90
2 (a) 196.161 (b) 196 (c) 196.16 (d) 200
(e) 196 (f) 200 (g) 200 (h) 196.1614
3 (a) 43 670 (b) 44 000 (c) 40 000
4 (a) 1940 (b) 4450 (c) 9800 (d) 86 000
(e) 1000 (f) 60 000
5 (a) 0.003 (b) 0.000 88 (c) 0.0090 (d) 0.0009
(e) 0.01 (f) 0.0004
6 (a) 0.007 (b) 0.066 67 (c) 0.000 0029
7 (a) 0.39 (b) 0.039 (c) 39.4 (d) 40 000

Exercise 11C

(Note: there are alternatives to questions 1 (i) and (ii), and 3.)
1 (a) (i) 9×10 (ii) 90 (iii) 82.7896
(b) (i) $4 \times (2 - 0)$ (ii) 8 (iii) 7.3875
(c) (i) 40×4 (ii) 160 (iii) 160.7296
(d) (i) $(2000 \times 300) + 400$ (ii) 600 400
(iii) 633 086
(e) (i) $\dfrac{200 \times 75}{50}$ (ii) 300 (iii) 294.831 3152
(f) (i) $\dfrac{0.6 \times 0.02}{0.04}$ (ii) 0.3 (iii) 0.262 550 588
2 (a) 8.5 (b) 40.488 (c) 4.6706 (d) 134.3632
(e) 76.704 (f) 8.0736
3 (a) 7 or 8 (b) 6 (c) 9 or 10 (d) 3
(e) 7 or 8 (f) 10 (g) 7 or 8 (h) 8
(i) 7 (j) 3 (k) 10 (l) 3 or 4
(m) 13 (n) 175 (o) 3 or 4

Exercise 11D

1 (a) 3.85 cm–3.95 cm (b) 17.6195 m–17.6205 m
(c) 52.95 cm–53.05 cm (d) 1.0295 m–1.0305 m
(e) 28.5 m–29.5 mm
2

School	Attendance to the nearest 100	Lowest value of range	Greatest value of range
Southpark	1400	1350	1450
Westown	400	350	450
Eastgrove	800	750	850
Northbury	1200	1150	1250
Southone	2100	2050	2150

3 4 min 7.555 sec to 4 min 7.565 sec
4 16.5 miles to 17.5 miles

Exercise 11E

1 (a) (i) 0.2335 km (ii) 0.2345 km
(b) (i) 0.3185 g (ii) 0.3195 s
(c) (i) 0.5445 kg (ii) 0.5455 kg
(d) (i) 0.7945 km (ii) 0.7955 km
(e) (i) 9.4165 km (ii) 9.4175 s
(f) (i) 3.4585 g (ii) 3.4595 g
2 (a) 29.5 cm (b) 18.45 cm, 18.35 cm
3 (a) 3.65 cm (b) 15.0 cm, 14.6 cm
(c) 3 s.f. ($4 \times 3.7 = 14.8$ lies midway between the bounds of the perimeter)
(d) Yes, 3 s.f.
4 (a) 99.5 m
(b) (i) 14.75 seconds (ii) 14.85 seconds
(c) (i) 6.700 336 7 m/s (ii) 6.813 559 3 m/s
(d) (i) 6.76 m/s
(ii) Midway betweeen the answers in (c)
5 (a) 5.65 cm, 5.55 cm
(b) 120.5 cm 119.5 cm
(c) 6042.32 cm³ 5781.94 cm³
6 (a) 8 450 000 km²–8 550 000 km²
(b) 21.83 people/km², 21.46 people/km²
7 (a) (i) 815 (ii) 805
(b) (i) 2.935 (ii) 2.925
(c) (i) 278.632 478 6 (ii) 274.275 979 6
(d) 276.5 to 4 s.f.
(There are alternatives; the quoted answer is the mean of the upper and lower bounds.)

Mixed exercise 11

1 (a) 9.5 (b) 0.046 (c) 7.70 (d) 7.43
2 (a) 424 (b) 0.2147 (c) 11 000 (d) 90 000

3 (a) $60 \div 1$ (b) 3×2
  (c) $110 - 80$ (d $60 + 20$
  (e) $(20 + 4) \div (3 + 5)$ (f) $8 \times 4 \div (4 + 5)$
4 (a) 15.25 km   (b) 15.15 km
5 (a) 29.25 cm²   (b) 0.538
6 (a) r : 1.75, 1.65;
    R : 31.05, 30.95
  (b) 29.2

Chapter 12 Sequences and formulae

Exercise 12A

1 2, 4, 6, 8, 10   **2** 3, 4, 5, 6, 7
3 10, 13, 16, 19, 22   **4** 13, 11, 9, 7, 5
5 6, 9, 14, 21, 30   **6** 2, 11, 26, 47, 74
7 4, 9, 16, 25, 36   **8** 1, 8, 27, 64, 125
9 2, 4, 8, 16, 32   **10** $-2, 4, -8, 16, -32$
11 $\frac{1}{2}, \frac{1}{4}, \frac{1}{8}, \frac{1}{16}, \frac{1}{32}$ **12** 1, 3, 6, 10, 15
13 $1, \frac{1}{2}, \frac{1}{3}, \frac{1}{4}, \frac{1}{5}$ **14** $\frac{1}{2}, \frac{2}{3}, \frac{3}{4}, \frac{4}{5}, \frac{5}{6}$
15 17

Exercise 12B

1 8   **2** 12   **3** 17   **4** 7   **5** 12
6 9   **7** 21   **8** 8   **9** 7   **10** 20

Exercise 12C

1 $n + 2$ **2** $2n + 3$ **3** $4n + 3$ **4** $10 - n$
5 $14 - 2n$ **6** $10 - 3n$

Exercise 12D

1 -11 **2** $53\frac{1}{8}$ **3** 30.2   **4** 26.3   **5** 5
6 47.8   **7** 223   **8** -45 **9** 100.152   **10** -43
11 47.4   **12** 195.5   **13** -65 **14** 82   **15** 9.09
16 10.8   **17** 3.44   **18** 1.30   **19** 9.71   **20** 23

Exercise 12E

1 18   **2** 5.4   **3** 22   **4** 0.8   **5** 4
6 4.5   **7** -8 **8** $-1\frac{1}{4}$ **9** 6   **10** $7\frac{1}{2}$
11 5   **12** 12.6   **13** 7 or -7 **14** 3.99   **15** 4.2
16 27.6   **17** 201   **18** 15.3   **19** 1.5   **20** 4.8

Exercise 12F

1 $x = q - p$ **2** $x = t - 7$ **3** $x = b + a$
4 $x = \frac{1}{5}w$ **5** $x = \frac{p}{n}$ **6** $x = 3f$
7 $x = md$ **8** $x = \frac{bc}{a}$ **9** $x = \frac{p + q}{n}$
10 $x = \frac{5 - b}{a}$ **11** $x = \frac{c - h}{d}$ **12** $x = \frac{p - q}{5}$
13 $x = \frac{b - a}{3}$ **14** $x = m(p - n)$

Exercise 12G

1 $I = \frac{V}{R}$ **2** $l = \frac{A}{w}$ **3** $d = \frac{C}{\pi}$
4 $w = \frac{V}{lh}$ **5** $u = v - at$ **6** $t = \frac{v - u}{a}$
7 $h = \frac{C}{2\pi r}$ **8** $R = \frac{100I}{PT}$ **9** $L = \frac{P - 2w}{2}$

10 $h = \frac{A - 2\pi r^2}{2\pi r}$ **11** $x = \frac{y}{3}$ **12** $T = \frac{PV}{k}$
13 $x = \sqrt{\frac{y}{5}}$ **14** $x = \sqrt{\frac{E}{k}}$ **15** $r = \sqrt{\frac{A}{\pi}}$
16 $n = \frac{2S}{a + d}$ **17** $a = \frac{2S - nd}{n}$ **18** $x = \sqrt{\frac{y}{k}}$
19 $h = \frac{3V}{\pi r^2}$ **20** $r = \sqrt{\frac{3V}{\pi h}}$ **21** $a = \sqrt{c^2 - b^2}$
22 $b = \sqrt{\frac{a^2 + 8}{4}}$ **23** $s = \frac{gt^2}{2}$ **24** $L = \frac{gT^2}{4\pi^2}$
25 $y = \sqrt{1 - x^2}$ **26** $u = \sqrt{v^2 - 2as}$ **27** $x = \sqrt{\frac{a^2w^2 - v^2}{w^2}}$
28 $x = \sqrt{\frac{9 - y^2}{9}}$ **29** $x = \frac{y - 5}{y + 1}$ **30** $u = \frac{vf}{(v - f)}$

Exercise 12H

1 $P = \frac{V^2}{R}$ **2** $s = \frac{1}{2}gt^2$ **3** $A = \frac{\pi d^2}{4}$
4 $a = rw^2$ **5** $A = \frac{P^2}{16}, P = 4\sqrt{A}$ **6** $x = \frac{y^2}{4a}$
7 $A = 2\pi r^2 + \frac{2V}{r}$ **8** $F = \frac{2A}{l} + l$

Mixed exercise 12

1 (a) 108, 147   (b) $3n^2$
2 (a) 27, 31   (b) 15, 12
  (c) 32, 64   (d) 28, 36
3 $t = \frac{(v - 30)}{7}$ **4** $g = \frac{4\pi^2a}{T^2}$ **5** $r = \sqrt{\left(\frac{A}{4\pi}\right)}$
6 $y = 2x + 5$ **7** $v = \frac{fu}{(u - f)}$ **8** $u = \frac{(D - kt^2)}{t}$
9 (a) (i) $g = 10, t = 20$ (ii) 2000
  (b 2748.12
10 (a) $r = 7, h = 11, \pi = 3$; 1617 cm³
  (b) 1860 cm³
  (c) 1891.88 cm³
11 (a) $f = 4(v - 55)$ (b) $v = \frac{1}{4}f + 55$

Chapter 13 Measure and mensuration

Exercise 13A

1 (a) 3.35 m, 3.45 m   (b) 4.55 m, 4.65 m
  (c) 15.24 m², 16.04 m²
2 (a) 194.46 km   (b) 190.55 km

Exercise 13B

1 (a) 52.5 cm², 31 cm
  (b) 80 cm², 48 cm
2 (a) $(6x + 8)$ cm   (b) 7   (c) 136 cm²
3 (a) 4 cm   (b) 16 cm²
4 62 500 m²
5 (a) 30 cm²   (b) 6 cm²   (c) 45 cm²
6 (a) $4x - 5$ (b) 8.5
7 (a) 40 cm²   (b) 36 cm²   (c) 30 cm²
8 (a) 70 cm²   (b) 8.75 cm
9 9.49
10 (a) 105 cm²   (b) 14 cm²   (c) 66 cm²
11 58.4 cm²   **12** 2.4 cm

Exercise 13C

1 (a) 31.42 cm, 78.54 cm²
  (b) 50.27 cm, 201.06 cm²
  (c) 37.70 cm, 113.10 cm²

2 (a) 15.71 m   (b) 19.63 m²
3 $\pi r^2 = 2\pi r$, $r^2 = 2r$, $r = 2$ (or 0)
4 (a) 2.39 cm   (b) 17.94 cm²   **5** 37.85 cm
6 (a) Area   = 39.27 cm²
           Perimeter = 25.71 cm
  (b) Area $= \dfrac{25\pi}{2}$ cm²
           Perimeter = $(10 + 5\pi)$ cm
7 (a) 181.70 cm (b) 1666.19 cm²

Exercise 13D

1 (a) 125 cm³   (b) 1728 cm³   (c) 54.872 cm³
2 (a) 10 cm   (b) 600 cm²
3 (a) 180 cm³   (b) 480.24 cm³
4 (a) 4 cm   (b) 352 cm²
5 (a) 504 cm³   (b) 396 cm²
6 240 cm³   **7** 12 cm   **8** 576 cm³

Exercise 13E

1 (a) 2-D   (b) 3-D   (c) 2-D   (d) 3-D
2 (2, 1, 3), (4, 3, 1), (5, 2, 12), (27, 8, 11)
3 O (0, 0, 0), A (4, 0, 0), B (4, 0, 3), C (0, 0, 3), D (0, 2, 3),
    E (0, 2, 0), F (4, 2, 0), G (4, 2, 3)
4 (a) O (0, 0, 0), A (4, 0, 0), B (4, 0, −3), C (0, 0, −3),
        D (0, −2, 0), E (4, −2, 0), F (4, −2, −3), G (0, −2, −3)
  (b) (i) $(0, -2, -1\frac{1}{2})$ (ii) $(4, -1, -1\frac{1}{2})$

Exercise 13F

1 $4\pi r^2$, dimension is 2
2 $n + m$ must equal 3
    so $n = 0, m = 3$ or $n = 1, m = 2$ or $n = 2, m = 1$ or $n = 3, m = 0$
3 $x + y + x$, length
    xyz, volume
    $xy + yz + xz$, area
4 $\pi a^2 b$, volume
    $\pi b^2 + 2h$, none
    $2ah$, area

Mixed exercise 13

1 1.2 m²
2 3 400 000 cm³
3 (a) 86.625 m³   (b) 119.5 m²
4 (a) 317.08 m   (b) 5963.5 m²
5 50 cm²
6 154.06 cm²
7 600 cm³, 660 cm²
8 (a) From $x^2 h = 50$ (b) 3.54   (c) 2
9 (a) A (−4, 0, 0), B (−4, 0, −2), C (−1, 0, −2), D (−1, 0, 0)
        E (−4, 2, −2), F (−4, 2, 0), G (−1, 2, 0), H (−1, 2, −2)
  (b) (i) $(-2\frac{1}{2}, 0, 0)$ (ii) $(-2\frac{1}{2}, 1, -1)$
10 Volume, as there are 3 variable dimensions.
11 (a) 4.35 m, 4.25 m; 3.75 m, 3.65 m
  (b) 16.3 m², 15.5 m²

Chapter 14 Simplifying algebraic expressions

Exercise 14A

1 x^8 **2** y^8 **3** $6x^7$ **4** $5a^4$
5 $24a^6$ **6** $48y^5$ **7** $10x^5 y^2$ **8** $24a^5 b^5$
9 $a^2 b^2 c^2$ **10** $18a^2 b^3 c^5$ **11** $6x^2 y^3$ **12** x^8
13 $9a^4$ **14** $8x^6$ **15** $2a^{20}$ **16** $2x^4 y^6$
17 $63x^8$ **18** $14a^{13} b^6$ **19** $36x^2$ **20** $27x^6$

Exercise 14B

1 x^3 **2** $3a^3$ **3** $6xy^3$ **4** x^{-2}
5 a^6 **6** $\frac{3}{2} a^3 x^{-1}$ **7** 4   **8** x^2
9 $5q^2$ **10** $2r^4$ **11** $2e^4$ **12** $3p^5$

Exercise 14C

1 5   **2** 2   **3** 4   **4** $\frac{1}{7}$
5 $\frac{1}{3}$ **6** 6   **7** $\frac{1}{5}$ **8** $\frac{1}{10}$
9 $\frac{1}{2}$ **10** 0.4

Exercise 14D

1 (a) 2   (b) $\frac{1}{100}$ (c) 4   (d) 32   (e) 81
2 (a) 125   (b) 32   (c) 32   (d) $\frac{1}{4}$
  (e) 32   (f) 25   (g) $\frac{1}{4}$ (h) $\frac{1}{32}$
  (i) 6   (j) $\frac{9}{4}$ (k) $\frac{64}{49}$ (l) $\frac{1}{125}$
3 (a) 9   (b) a^2 (c) $x^{\frac{3}{2}} y^{\frac{5}{2}}$ (d) $3x$ (e) $\frac{2}{x}$
  (f) $\frac{a^2}{5}$ (g) $729x^3$ (h) 16   (i) $24y^2$

Exercise 14E

1 (a) $\frac{1}{2}$ (b) 0   (c) −3   (d) $\frac{2}{3}$ (e) $\frac{7}{2}$
  (f) $\frac{9}{2}$ (g) 5   (h) 4   (i) 3   (j) $\frac{1}{2}$
  (k) $\frac{3}{2}$ (l) 3

Exercise 14F

1 $\dfrac{x}{4}$ **2** $\dfrac{b}{d}$ **3** $4x$ **4** $\dfrac{4t}{3r}$
5 $3ab$ **6** 2   **7** $\dfrac{(x + 2)}{2x}$ **8** $\dfrac{3}{(x - 4)}$
9 $\dfrac{x^3 y^2 (y - 2)}{(y - 6)}$

Exercise 14G

1 (a) $\dfrac{3x}{5}$ (b) $\dfrac{y + x}{xy}$ (c) $\dfrac{13}{5x}$
  (d) $\dfrac{11x}{12}$ (e) $\dfrac{7a + 2}{12}$ (f) $\dfrac{x + 17}{10}$
  (g) $\dfrac{(x + 3)}{(x + 1)(x + 2)}$ (h) $\dfrac{2}{x(2 - x)}$ (i) $\dfrac{y - 1}{2(y - 5)}$
2 (a) 18   (b) $x^2 (x - 1)$ (c) $12(x - 1)(x + 1)$
  (d) $x^3 (1 - x)(1 + x)$
3 (a) $\dfrac{5a}{18}$ (b) $\dfrac{(x - 2)}{x^2 (x - 1)}$
  (c) $\dfrac{(x + 2)}{12(x - 1)(x + 1)}$ (d) $\dfrac{x^2 + x + 2}{x^3 (1 - x)(1 + x)}$
4 $\dfrac{2(x + 10)}{(x - 2)(x + 2)}$

Exercise 14H

1 (a) $4x(2x + 1)$ (b) $3p(2p + 1)$ (c) $3x(2x - 1)$
  (d) $3b(b - 3)$ (e) $3a(4 + a)$ (f) $5c(3 - 2c)$
  (g) $7x^3 (3x + 2)$ (h) $4y^2 (4y - 3)$ (i) $2d^2 (3d^2 - 2)$
2 (a) $ax(x + 1)$ (b) $pr(r - 1)$ (c) $ab(b - 1)$
  (d) $q(r^2 - q)$ (e) $ax(a + x)$ (f) $by(b - y)$
  (g) $3a^2 (2a - 3)$ (h) $4x^3 (2 - x)$ (i) $6x^3 (3 + 2x^2)$
3 (a) $6ab(2a + 3b)$ (b) $2xy(2x - y)$
  (c) $4ab(a + 2b + 3)$ (d) $2xy(2x + 3y - 1)$
  (e) $3ax(4x + 2a - 1)$ (f) $abc(a + b + c)$

Exercise 14I

1 $(p + q)(x + y)$ 2 $(x + y)(3 - z)$ 3 $(x - y)(5 - x)$
4 $2(a - b)(5 + x)$ 5 $(y + 2)(x + z)$ 6 $(a + 1)(y + 2b)$
7 $(x - b)(1 - x)$ 8 $(x - y)(x - z)$ 9 $2(b + 1)(a - 5)$
10 $3(a - 2c)(1 - b)$

Exercise 14J

1 (a) $x(x + 3)$ (b) $8(x + 3)$ (c) $(x + 3)(x + 8)$
   (d) $x(x + 6)$ (e) $2(x + 6)$ (f) $(x + 6)(x + 2)$
   (g) $x(x - 5)$ (h) $8(x - 5)$ (i) $(x - 5)(x + 8)$
   (j) $(x + 5)(x + 3)$ (k) $(x + 4)(x - 8)$ (l) $(x - 6)(x - 2)$
   (m) $(x + 4)(x - 3)$ (n) $(x + 3)(x + 5)$ (o) $(x - 7)(x - 9)$
2 $x^2 - 5x - 6$
3 (a) $-2, -3$ (b) $1, 6$ (c) $-2, -6$
   (d) $4, 15$ (e) $4, -6$ (f) $-2, 12$
   (g) $3, -16$ (h) $-4, 12$
4 (a) $(x - 4)(x - 1)$ (b) $(x + 5)(x + 2)$ (c) $(x - 5)(x + 3)$
   (d) $(x + 3)(x + 2)$ (e) $(x - 10)(x - 1)$ (f) $(y - 4)(y + 3)$
5 (a) $(2x - 3)(x - 1)$ (b) $6(x + 2)(x + 5)$ (c) $(2y - 5)(y + 2)$
   (d) $2(3a + 2)(a + 1)$ (e) $(4x + 3)(x - 2)$ (f) $5(5 + 4x)(3 - x)$
   (g) $(5x - 1)(x - 3)$ (h) $(x^2 + 2)(x^2 + 3)$ (i) $2(x^2 + 4)(x^2 + 3)$
6 $(x^n + 1)(x^n + 8)$
7 $x(x^{2n} + 4)(x^{2n} + 2)$

Exercise 14K

1 (a) $(x - 1)(x + 1)$ (b) $(x - 5)(x + 5)$
   (c) $(2x - 3)(2x + 3)$ (d) $(2x + 3)(4x - 3)$
   (e) $2(p - 7)(p + 7)$ (f) $2(x - 2y)(x + 2y)$
   (g) $(x + 2)^2$ (h) $(x - 3)^2$ (i) $2(x + 7)^2$
2 (a) $(x - 2)(x + 2)$ (b) $x(x - 3)$ (c) $(x + 1)(x - 4)$
   (d) $4(x - 3)(x + 3)$ (e) $2x(2x - 35)$ (f) $2(2x + 1)(x - 18)$
   (g) $(x + 1)^2$ (h) $x(x + 2)$ (i) $2(4x - 7)(x + 2)$
3 (a) $(x - 4)(x - 1)$ (b) $(x + 12)(x - 3)$ (c) $(x + 8)(x - 4)$
   (d) $(x - 1)(2x - 5)$ (e) $(4x - 7)(x + 1)$ (f) $(5x + 3)(x - 2)$
   (g) $(x - 1)(x - 3)$ (h) $(y + 2)(y + 5)$ (i) $(x + 5)(x - 3)$
   (j) $(x - 7)(x + 3)$ (k) $(y + 4)^2$ (l) $(2x - 5)^2$
   (m) $(5x - 3)(x - 2)$ (n) $(4x + 5)(x - 1)$ (o) $(5x - 3)(x + 2)$

Exercise 14L

1 $\dfrac{x + 2}{2}$ 2 $\dfrac{x - 1}{x + 4}$ 3 $\dfrac{(x + 2)}{(x - 2)^2}$ 4 $\dfrac{(x + 9)}{(x + 1)}$

5 $\dfrac{(x + 2)}{x}$ 6 $\dfrac{4}{(x + 1)(x + 2)}$ 7 2

Mixed exercise 14

1 (a) x^4 (b) y^{12} (c) m^8 (d) $6r^3t^6$
2 (a) $2p(p - 2q)$ (b) $(p + q)(p + q + 5)$ (c) $(x + 1)(x + 6)$
3 (a) $3x(x + 2y)$ (b) $(x - 5)(2x - 7)$
4 (a) $\dfrac{3}{(y + 2)}$ (b) $\dfrac{3}{(x + 1)}$

5 $\dfrac{(2x + 3)}{(x - 1)}$

6 $\dfrac{(5x - 4)}{(x^2 - 1)}$

7 (a) p^9 (b) x^5 (c) y^2
8 (a) $\frac{1}{9}$ (b) 6   (c) 9   (d) $3\frac{3}{8}$

9 (a) $(3x - 1)^2$ (b) $\dfrac{(2x + 3)}{(3x - 1)}$

10 $\dfrac{5(x - 3)}{(x - 2)(x - 1)}$

11 $\dfrac{(2x - 5)}{(2x - 1)}$

12 $p = 3$ or -3, $q = -6$ or 30, $r = 9$
13 (a) $7p$ (b) (i) xy (ii) x^2y^2
14 (a) 3   (b) 7   (c) -2

Chapter 15 Pythagoras' theorem

Exercise 15B

1 (a) 13 cm   (b) 15 cm   (c) 12.5 cm   (d) 7.5 cm
2 (a) 11.4 cm   (b) 35.5 cm   (c) 7.2 m   (d) 20.8 cm
3 26 cm   4 14.14 cm

Exercise 15C

1 (a) 9 cm   (b) 24 cm   (c) 7.5 cm   (d) 10.5 cm
2 (a) 6.3 m   (b) 35.5 cm   (c) 8.7 cm   (d) 21.2 cm
3 24.4 cm   4 12.4 m
5 9.80 m   6 129.7 km

Exercise 15D

1 (a) $x = 13$ cm $y = 19.85$ cm
   (b) $x = 6$ cm $y = 12.04$ cm
   (c) $x = 9$ cm $y = 50.8$ cm

Exercise 15E

1 13 units
2 (a) 5   (b) 4.47   (c) 11.31   (d) 14.87
   (e) 5   (f) 12.08
3 5.83 units

Exercise 15F

2 (a) obtuse   (b) acute   (c) obtuse   (d) right
   (e) obtuse   (f) obtuse

Exercise 15G

2 (a) Yes   (b) Yes   (c) Yes   (d) No
   (e) Yes   (f) No

Exercise 15H

1 15.26 cm   2 18.31 mm
3 (a) (i) 13 cm   (ii) 17.72 cm   (iii) 20.81 cm   (b) acute

Exercise 15I

1 (a) $x^2 + y^2 = 16$ (b) $x^2 + y^2 = 49$
2 (a) $x^2 + y^2 = 25$ (b) $x^2 + y^2 = 144$ (c) $x^2 + y^2 = 225$
3 $x^2 + y^2 = 169$
4 Any eight of: $(5, 0), (0, 5), (-5, 0), (0, -5)$
                $(3, 4), (3, -4), (-3, 4), (-3, -4)$
                $(4, 3), (4, -3), (-4, 3), (-4, -3)$

Mixed exercise 15

1 12.4 cm   2 6.24 m   3 30.8 km
4 $x = 26$ cm, $y = 15.0$ cm
5 13.6
6 $7^2 + 8^2 < 11^2$, so obtuse
7 Only (b) is a Pythagorean triple, as $10^2 + 24^2 = 26^2$
8 20.8 cm
9 (a) 13 cm   (b) 23.3 cm   (c) 23.9 cm
10 5   11 $x^2 + y^2 = 16$ 12 $x^2 + y^2 = 289$
13 No. For example, they can be similar triangles or semi-circles.

Chapter 16 Basic trigonometry

Exercise 16A

1 (a) cosine   (b) cosine   (c) tangent
   (d) sine   (e) sine   (f) tangent
   (g) sine   (h) tangent   (i) cosine
2 About 9.1 m

Exercise 16B

1 (a) 0.7314   (b) 0.5299   (c) 0.3839
   (d) 8.1443   (e) 0.9063   (f) 0.8660
   (g) −3.4874   (h) −0.0523   (i) 0.7431
2 (a) 20.6°   (b) 82.2°   (c) 54.7°
   (d) 23.6°   (e) 9.8°   (f) 62.3°

Exercise 16C

1 44.4°   **2** 38.7°   **3** 48.2°
4 21.8°   **5** 65.4°   **6** 30°
7 18.4°   **8** 24.3°   **9** 52.4°

Exercise 16D

1 (a) 7.66   (b) 7.71   (c) 12.99
   (d) 5.34   (e) 7.52   (f) 6.75
   (g) 6.36   (h) 12.12
2 328° (to the nearest degree)

Exercise 16E

For example:
1 (b) $\sin 410° = \sin 50°$ (c) $\sin 50° = \sin 130°$
2 (a) $\cos 110° = -\cos 70°$ (b) $\cos (-85°) = \cos 85°$
   (c) $\cos 405° = \cos 45°$
3 (a) $\tan (-45°) = -\tan 45°$ (b) $\tan 130° = -\tan 50°$
   (c) $\tan 390° = \tan 30°$

Exercise 16F

3 (a)

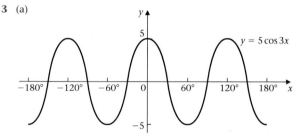

   (b) 120°   (c) Max of 5 at 0°, 120°, −120°
                 Min of −5 at 60°, −60°, 180°, −180°
4 (a)

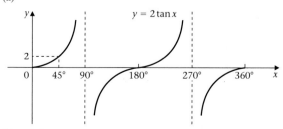

   (b) 180°

5 (a)

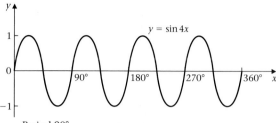

Period 90°
   (b)

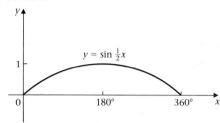

Period 720°
6 Period $\dfrac{360°}{B}$, maximim A, minimum $-A$
7 For sine and cosine functions, maximum is A, minimum is
   $-A$ and period is $\dfrac{360°}{B}$
   For tangent, the period is $\dfrac{180°}{B}$

Exercise 16G

1 60°, −60°, 300°, −300°
2 (b) 23.6°, 156.4°, 383.6°, 516.4°
3 63.4°, −116.6°
4 (a)

   (b) 70.5°, 289.5°
5 (b) 17.71°, 42.29°, 137.71°, 162.29°, −102.29°, −77.71°
6 (a) $y = 2 \cos \theta$ (b) $y = 10 \cos \theta$
   (c) $y = 3 \sin \tfrac{1}{2}\theta$ (d) $y = 3 \sin 2\theta$

Mixed exercise 16

1 (a) 12.37 m   (b) 72.1°
2 (a) 8.15 m   (b) 38.7°
3 7.71 m   **4** 021.6°   **5** $\sin 35° \neq 0.7$
6 (a) $h = 1247.47$ m   (b) 5868.89 m
7 (a) 26.4°   (b) 125.9 m
8 (a)   (b) 120°, 240°

9 (a) 13 m   (b) 11.1 hours
10 (a) 41.8°, 138.2°   (b) 113.6°, 246.4°   (c) 70.5°, 289.5°

Chapter 17 Graphs and equations

Exercise 17A

1

x	−4	−3	−2	−1	0	1	2	3	4
y	21	14	9	6	5	6	9	14	21

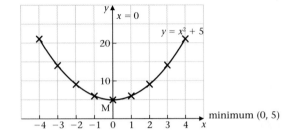

minimum (0, 5)

2

x	−4	−3	−2	−1	0	1	2	3	4
y	6	−1	−6	−9	−10	−9	−6	−1	6

$x = 0$, minimum $(0, -10)$

3

x	−4	−3	−2	−1	0	1	2	3	4
y	48	27	12	3	0	3	12	27	48

$x = 0$, minimum $(0, 0)$

4

x	−4	−3	−2	−1	0	1	2	3	4
y	8	4.5	2	0.5	0	0.5	2	4.5	8

$x = 0$, minimum $(0, 0)$

5

x	−4	−3	−2	−1	0	1	2	3	4
y	−16	−9	−4	−1	0	−1	−4	−9	−16

$x = 0$, maximum $(0, 0)$

6

x	−4	−3	−2	−1	0	1	2	3	4
y	−32	−18	−8	−2	0	−2	−8	−18	−32

$x = 0$, maximum $(0, 0)$

7

x	−4	−3	−2	−1	0	1	2	3	4
y	8	3	0	−1	0	3	8	15	24

$x = -1$, minimum $(-1, -1)$

8

x	−4	−3	−2	−1	0	1	2	3	4
y	4	0	−2	−2	0	4	10	18	28

$x = -1\frac{1}{2}$, minimum $(-1\frac{1}{2}, -2\frac{1}{4})$

9

x	−4	−3	−2	−1	0	1	2	3	4
y	9	4	1	0	1	4	9	16	25

$x = -1$, minimum $(-1, 0)$

10

x	−4	−3	−2	−1	0	1	2	3	4
y	36	25	16	9	4	1	0	1	4

$x = 2$, minimum $(2, 0)$

11

x	−4	−3	−2	−1	0	1	2	3	4
y	31	20	11	4	−1	−4	−5	−4	−1

$x = 2$, minimum $(2, -5)$

12

x	−4	−3	−2	−1	0	1	2	3	4
y	23	13	5	−1	−5	−7	−7	−5	−1

$x = 1\frac{1}{2}$, minimum $(1\frac{1}{2}, -7\frac{1}{4})$

13

x	−4	−3	−2	−1	0	1	2	3	4
y	24	16	10	6	4	4	6	10	16

$x = \frac{1}{2}$, minimum $(\frac{1}{2}, 3\frac{3}{4})$

14

x	−4	−3	−2	−1	0	1	2	3	4
y	15	4	−3	−6	−5	0	9	22	39

$x = -\frac{3}{4}$, minimum $(-\frac{3}{4}, -6\frac{1}{8})$

15

x	−4	−3	−2	−1	0	1	2	3	4
y	66	41	22	9	2	1	6	17	34

$x = \frac{2}{3}$, minimum $(\frac{2}{3}, \frac{2}{3})$

Exercise 17B

1

x	−3	−2	−1	0	1	2	3
y	−22	−3	4	5	6	13	32

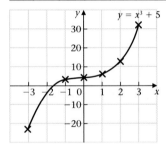

2

x	−3	−2	−1	0	1	2	3
y	−37	−18	−11	−10	−9	−2	17

3

x	−3	−2	−1	0	1	2	3
y	−54	−16	−2	0	2	16	54

4

x	−3	−2	−1	0	1	2	3
y	−13.5	−4	−0.5	0	0.5	4	13.5

5

x	−3	−2	−1	0	1	2	3
y	27	8	1	0	−1	−8	−27

6

x	−3	−2	−1	0	1	2	3
y	54	16	2	0	−2	−16	−54

7

x	−3	−2	−1	0	1	2	3
y	−125	−64	−27	−8	−1	0	1

8

x	−3	−2	−1	0	1	2	3
y	−8	−1	0	1	8	27	64

9

x	−3	−2	−1	0	1	2	3
y	−45	−16	−3	0	−1	0	9

10

x	−3	−2	−1	0	1	2	3
y	−18	−4	0	0	2	12	36

11

x	−3	−2	−1	0	1	2	3
y	−42	−18	−6	0	6	18	42

12

x	−3	−2	−1	0	1	2	3
y	−12	2	4	0	−4	−2	12

13

x	−3	−2	−1	0	1	2	3
y	6	12	8	0	−6	−4	12

14

x	−3	−2	−1	0	1	2	3
y	−38	−16	−6	−2	2	12	34

15

x	−3	−2	−1	0	1	2	3
y	−43	−14	−1	2	1	2	11

Exercise 17C

1

x	−3	−2	−1	−0.5	−0.2	0.2	0.5	1	2	3
y	−0.7	−1	−2	−4	−10	10	4	2	1	0.7

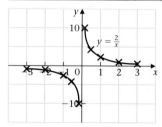

Asymptotes: $x = 0$
$y = 0$

2

x	−3	−2	−1	−0.5	−0.2	0.2	0.5	1	2	3
y	0.3	0.5	1	2	5	−5	−2	−1	−0.5	−0.3

Asymptotes: $x = 0$, $y = 0$

3

x	−3	−2	−1	−0.5	−0.2	0.2	0.5	1	2	3
y	1	1.5	3	6	15	−15	−6	−3	−1.5	−1

Asymptotes: $x = 0$, $y = 0$

4

x	−3	−2	−1	−0.5	−0.2	0.2	0.5	1	2	3
y	3.7	3.5	3	2	−1	9	6	5	4.5	4.3

Asymptotes: $x = 0$, $y = 4$

5

x	−3	−2	−1	−0.5	−0.2	0.2	0.5	1	2	3
y	6.3	7	9	13	25	−15	−3	1	3	3.7

Asymptotes: $x = 0$, $y = 5$

6

x	−3	−2	−1	−0.5	−0.2	0.2	0.5	1	2	3
y	−4.3	−5	−7	−11	−23	17	5	1	−1	−1.7

Asymptotes: $x = 0$, $y = −3$

Exercise 17D

1

x	−1	0	1	2	3	4
y	6	1	−2	−3	−2	1

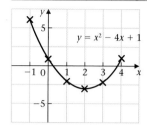

$y = x^2 - 4x + 1$

2

x	−3	−2	−1	0	1	2
y	−12	−3	−2	−3	0	13

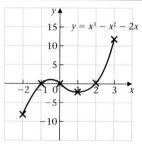

$y = x^3 + 2x^2 - 3$

3

x	−2	−1	0	1	2	3
y	−8	0	0	−2	0	12

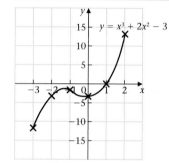

$y = x^3 - x^2 - 2x$

4

x	−3	−2	−1	0	1	2	3
y	−10	4	6	2	−2	0	14

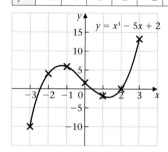

$y = x^3 - 5x + 2$

5

x	−2	−1	0	1	2	3	4
y	−7	−2	1	2	1	−2	−7

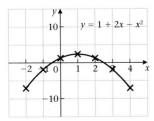

6

x	−3	−2	−1	−0.5	−0.2	0.2	0.5	1	2	3
y	−6.3	−4.5	−3	−3	−5.4	5.4	3	3	4.5	6.3

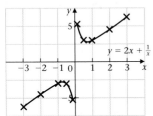

7

x	−3	−2	−1	−0.5	−0.2	0.2	0.5	1	2	3
y	9.3	4.5	2	2.3	5.0	−5.0	−1.8	0	3.5	8.7

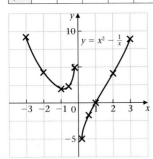

8

x	0.1	0.2	0.5	1	2	3	4	5
y	22.1	12.2	6.5	5	5	5.7	6.5	7.4

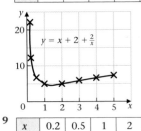

9

x	0.2	0.5	1	2	3
y	−5.0	−1.9	0	7.5	26.7

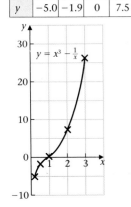

10

x	−3	−2	−1	−0.5	−0.2	0.2	0.5	1	2	3
y	11.3	5	0	−3.3	−9.8	9.8	3.8	2	3	6.7

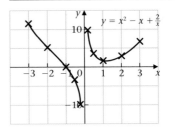

Exercise 17E

1 5.6 or 1.4	**2** 2.3 or −1.3	**3** 2.8 or −1.3
4 1.3 or 0.3	**5** −2.1, 0.3 or 1.9	**6** −2.3, 0 or 1.3
7 2.2	**8** 0 or 2	**9** 2.2
10 0.7 or 2.9		

Exercise 17F

1 1.3	**2** 2.1	**3** 2.1	**4** 2.47	**5** 6.6
6 8.56	**7** 5.6	**8** 6.89	**9** 4.35	**10** 2.4

Exercise 17G

1 4.64 cm	**2** 2.89 cm	**3** 71.6°C

Exercise 17H

1 (a) Sharon.
(b) Tracey overtakes Sharon.
(c) Pass each other travelling in opposite directions.

2

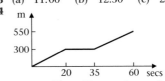

3 (a) 11:00 (b) 12:30 (c) 2 hrs (d) 60 mph

4

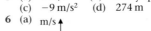

5 (a) 7 m/s² (b) steady speed of 36 m/s
(c) −9 m/s² (d) 274 m

6 (a) (b) 486 m

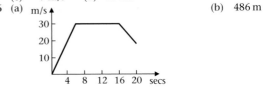

7 (a) E (b) A (c) B (d) C

8 (a) (b) (c) (d)

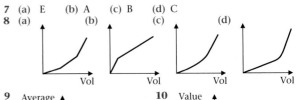

9   **10**

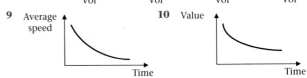

11 (a) Starts slowly then speeds up. Has a break. Continues at a steady speed more slowly than before.
  (b) **(i)** 09:00   **(ii)** Post Office closed
12 (a) B   **(b)** D

Mixed exercise 17

1 4.4   **2** 4.2

3

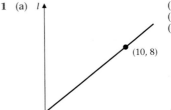

$x = 0$

4

$y = x^3 - 3x^2 + 14$
$y = 4\frac{1}{2}x + 8$
$y = 3x$

 (a) −1.7   **(b)** −1, 2   **(c)** −2   **(d)** 0.9, −1.7

5 (a)

x	0	1	2	3	4	5	6
y	10	5	2	1	2	5	10

(b)

(c) 1 (at $x = 3$)   **(d)** 1.4, 3.6

6 A
7 (a)

x	−2	−1	0	1	2	3	4
y	−18	−2	2	0	−2	2	18

(b)

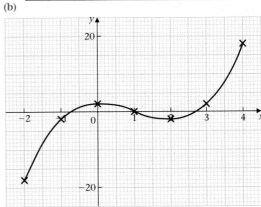

(c) **(i)** −0.7, 1, 2.7   **(ii)** 3.4
8 (a) About 6 m/s²   **(b)** About 100 m

Chapter 18 Proportion

Exercise 18A

1 (a) Yes   **(b)** No   **(c)** No
  (d) Yes   **(e)** Yes   **(f)** No
2 (a) 12   **(b)** 4
3 (a) 24   **(b)** $2\frac{1}{2}$
4 (a) 12   **(b)** 54

Exercise 18B

1 (a)

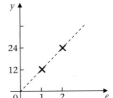

  (10, 8)

  (b) $l = 0.8h$
  (c) 15.625 = 15.63 (2 d.p.)
  (d) 12

2 C
3 (a)

  (14, 42)

  (b) $A = 3l$
  (c) 97.2 cm²
  (d) 8.2

4 (a)

  (b) Student's explanation is 2 → 24 so 1 → 12 etc.
  (c) 84 ÷ 12 = 7

Exercise 18C

1 (a) $w = 1.05\,h$, 12.6   (b) $s = \frac{41}{24}\,h$, 28.7
   (c) $p = \frac{39}{34}\,l$, 31.7   (d) $p = \frac{9}{4}\,a$, 248.4

2 (a) 9.36 volts, 5.2 amps   (b) $V = \frac{9}{5}\,I$
   (c) 6.4 amps   (d) 4.428 volts

3 (b) 4.6 should be 5.6   (c) $V = 2.8\,I$
   (d) 9.8   (e) 7.321

4 (a) 1.67   (b) 16.53

5 (a) $C = 3.6\,A$ (b) 360p or £3.60

6 (a) $d = 38\,t$ (b) 190 miles   (c) 11 hours

Exercise 18D

1 $12\frac{1}{3}$, $86\frac{1}{3}$ **2** $\frac{2}{3}$, $4\frac{2}{3}$ **3** $z = k\,x$, 8.4   **4** $z = k\,w$, 4725

5 (a) 15   (b) 39   (c) $3\frac{1}{3}$

6 (a) 225   (b) 1250   (c) 72

7 $y = 2.8$, $p = 1.43$ (2 d.p.)   **8** 5.71 cm³ (2 d.p.)   **9** 40%

10 (a) $h = 7.5d$
   (b) 187.5 cm
   (c) $42\frac{2}{3}$ cm

Exercise 18E

1 5, 80

2 (a) 6.25   (b) 156.25   (c) 1.265

3 (a) 2   (b) 128   (c) 4

4 (a) $l = k\,m^3$ (b) (i) 25.6   (ii) 0.952

5 25 000 N

6 (a) 45   (b) 101.25

7 $p = 0.75$, $q = 3$ **8** 0.002

9 (a) 1 m   (b) 1.5625 m

10 4   **11** $z = 2$, $w = 1.5$ **12** 125%   **13** 4.14 m

Exercise 18F

1 (a) 72   (b) 7.2

2 (a) 4.8   (b) 9.6

3 (a) 512   (b) 128

4 (a) 2.5   (b) 40

5 (a) 233   (b) 291

6 (a) $l \propto \dfrac{1}{w}$, $l = \dfrac{k}{w}$ (b) area of the rectangle

7 4 units

8 $f = \dfrac{33\,024}{w}$ (b) 384 hertz   (c) 96 cm

Mixed exercise 18

1 9.6 units

2 5.56 kg

3 28 800 km

4 (a) $h = \frac{35}{64}\,s^2$ (b) 315 m

5 18

6 (a) 31.25 m   (b) 45 m

7 1 hour 48 minutes

8 0.576 units

9 50

10 (a) $S = \dfrac{8000}{f^2}$ (b) 500

11 (a) D   (b) A   (c) B   (d) C

Chapter 19 Quadratic equations

Exercise 19A

1 (a) $-2, 1$ (b) 0, 3   (c) $-\frac{1}{2}, 4$ (d) $0, 1\frac{1}{2}$
   (e) $\frac{1}{2}, 1\frac{1}{3}$ (f) 2, 5

2 (a) 0, 3   (b) ± 4 (c) $-1, 4$ (d) $-7, -4$
   (e) $-3, \frac{1}{2}$ (f) $-\frac{3}{4}, 1\frac{1}{3}$ (g) $\frac{3}{4}, 2$ (h) 0, 9
   (i) $-\frac{1}{2}, 2\frac{2}{3}$

3 (a) ± 5 (b) $0, \frac{1}{3}$ (c) $-2, \frac{1}{3}$ (d) ± 2
   (e) $0, 2\frac{1}{2}$ (f) $-7, 5$ (g) $-3, 4$ (h) 1, 3
   (i) $-2\frac{1}{3}, \frac{1}{2}$ (j) $\frac{1}{2}, 1$

4 Both sides have been divided by $6y$ and $y = 0$ is a solution.
$y = 0$, $y = 2$

Exercise 19B

1 (a) $(x + 2)^2 - 4$ (b) $(x - 7)^2 - 49$ (c) $(x + \frac{3}{2})^2 - \frac{9}{4}$
   (d) $(x + \frac{1}{2})^2 - \frac{1}{4}$ (e) $(x - \frac{1}{2})^2 - \frac{1}{4}$ (f) $(x - 2)^2 - 4$
   (g) $(x + 3.5)^2 - 12.25$
   (h) $(x - 5)^2 - 25$

2 (a) $2(x + 4)^2 - 32$ (b) $3(x - 2)^2 - 12$ (c) $2(x + \frac{1}{4})^2 - \frac{1}{8}$
   (d) $5(x - \frac{3}{2})^2 - \frac{45}{4}$ (e) $2(x + \frac{1}{4})^2 - \frac{1}{2}$ (f) $4(x - 1)^2 - 4$
   (g) $3(x - 2.5)^2 - 18.75$ (h) $7(x - 2)^2 - 28$

Exercise 19C

1 (a) $-5 \pm \sqrt{22}$ (b) $4 \pm \sqrt{18}$ (c) $\dfrac{-9 \pm \sqrt{69}}{2}$
   (d) $1 \pm \sqrt{\dfrac{2}{3}}$ (e) $\dfrac{3 \pm \sqrt{41}}{4}$ (f) $1 \pm \sqrt{\dfrac{9}{2}}$
   (g) $\dfrac{-5 \pm \sqrt{61}}{6}$ (h) $\dfrac{7 \pm \sqrt{65}}{4}$ (i) $\dfrac{4 \pm \sqrt{19}}{3}$

2 (a) $-0.65, 4.65$ (b) $0.35, 5.65$ (c) $-3.12, 1.12$
   (d) $-0.69, 2.19$ (e) $-0.15, 2.15$ (f) $-1.29, 1.54$
   (g) $-0.65, 4.65$ (h) $-0.84, 0.24$ (i) $-0.47, 2.14$

Exercise 19D

1 (a) 5   (b) 8   (c) 44   (d) 288
   (e) 41   (f) 28

2 (a) $-2.62, -0.38$ (b) $-0.41, 2.41$ (c) $-3.16, 0.16$
   (d) ± 1.06 (e) $-2.35, 0.85$ (f) $-0.82, 1.82$

3 (a) $0.27, 3.73$ (b) $0.21, 4.79$ (c) $-2.13, -0.12$
   (d) $-0.65, 1.15$ (e) $0.13, 7.87$ (f) $-5.65, -0.35$

4 $9 - \sqrt{60}$ and $9 + \sqrt{60}$

5 $3 + \sqrt{24}$ and $-3 + \sqrt{24}$

Exercise 19E

1 $-\frac{4}{7}, 3$ **2** $0.22, 2.28$ **3** $-\frac{2}{3}, 2$
4 $-0.11, 1.11$ **5** $-7, 4$ **6** $-1.30, 2.30$

Exercise 19F

1 4, 5   **2** 5   **3** $-8, 3$ **4** 2.47 m, 247 cm

5 (a) $£\dfrac{400}{x}$ (c) £50

6 4.7

7 (a) $\dfrac{84}{x}$ km/h   (c) 7

Exercise 19G

1 (a) $x = 4$, $y = 16$; $x = -4$, $y = 16$
   (b) $x = 6$, $y = 36$
   (c) $x = 7$, $y = 49$; $x = -5$, $y = 25$
   (d) $x = 9$, $y = 81$; $x = -2$, $y = 4$
   (e) $x = \frac{3}{2}$, $y = \frac{9}{2}$; $x = -1$, $y = 2$
   (f) $x = -\frac{2}{3}$, $y = \frac{4}{3}$; $x = 3$, $y = 27$
   (g) $x = 2$, $y = -2$; $x = -3$, $y = -7$
   (h) $x = \frac{3}{4}$, $y = \frac{7}{4}$; $x = -\frac{4}{3}$; $y = \frac{49}{3}$

2 (a) $(1, 1)$, $(-3, 9)$ (b) $(2, 16)$, $(-\frac{5}{4}, \frac{25}{4})$
   (c) $(5, 10)$, $(-2, 3)$ (d) $(0, 7)$, $(-\frac{1}{2}, 5\frac{1}{2})$

3 (a) none:
(b) one: (5, 25)
(c) two: $(-1, 2)$ and $(\frac{7}{2}, \frac{49}{2})$
(d) none:
(e) two: $(3, 45)$ and $(-\frac{1}{5}, \frac{1}{5})$
(f) two: $(3, 36)$ and $(-\frac{1}{2}, 1)$
(g) one: $(2, 2)$
(h) two: $(\frac{5}{4}, \frac{25}{2})$ and $(-\frac{1}{3}, 3)$

Exercise 19H

1 (a) $x = 4, y = 3$; $x = -4, y = 3$ meet in 2 points
(b) $x = 0, y = 5$; $x = -5, y = 0$ meet in 2 points
(c) $x = -3, y = 4$; $x = -\frac{24}{5}, y = -\frac{7}{5}$ meet in 2 points
(d) $x = 3, y = -4$; $x = -4, y = 3$ meet in 2 points
(e) $x = 5, y = -5$ meet in 1 point, line is a tangent
(f) $x = 1, y = -7$; $x = -\frac{49}{65}, y = \frac{457}{65}$ meet in 2 points
(g) $x = 3, y = -1$ meet in 1 point, line is a tangent
(h) $x = 2, y = 1$ meet in 1 point, line is a tangent
(i) $x = -\frac{7}{5}, y = -\frac{1}{5}$ meet in 1 point, line is a tangent
(j) $x = 0, y = 2$; $x = \frac{48}{25}, y = -\frac{14}{25}$ meet in 2 points
3 (a) $y = x + 4$ touches circle at $x = -2$ only
(b) (i) $y = x - 4$; $(2, -2)$
(ii) $4\sqrt{2}$ or $2\sqrt{8}$
(c) $2r$

Exercise 19I

1 (a) (b)

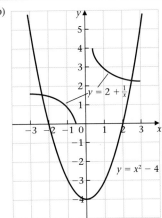

(c) (i) $-1.4, 1.4$ (ii) -0.8 (iii) $-2.4, 2.5$
2 (a) (i) $y = 1$ (ii) $y = -2$ (iii) $y = 2x + 1$
(iv) $y = -2x + 4$
(b) $x^3 - 2x = 4$ ($y = 4 - 2x$ meets the curve only one point, $(2, 0)$)
(c) $x^4 - 4x^2 = 1$
3 (a)

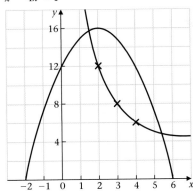

(b) 1.5 m, 5.4 m

4 (a)

x	-3	-2	-1	0	1	2
y	-8	5	6	1	-4	-3

(b) (c) (i)

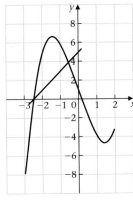

(ii) $-2.5, -0.5$
(d) $x^3 - 8x - 4 = 0$ (e) $y = 2x + 1$ (f) $0, \pm\sqrt{8}$

Mixed exercise 19

1 2 cm
2 $4\frac{1}{2}$ cm
3 $2 + \sqrt{13}$
4 $-3 \pm \sqrt{13}$
5 (a) $(2x - 7)(x - 14)$ (b) $x = 3.5, 14$
6 $x = 2, y = 5$; $x = -5, y = -2$
7 (a) $1.30, -2.30$
(b) Untrue for $x = 11$, as 11 is a factor
8 (a) Simplified from Pythagoras: $(x + 8)^2 = x^2 + (x + 5)^2$
(b) 9.93 cm
9 (a)

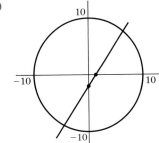

(b) $(6.4, 7.7), (-4.6, -8.9)$
(c) 9
(d) $x = 3, y = 10$

Chapter 20 Presenting and analysing data 2

Exercise 20A

1 (a)

Time watching TV (hours)	Cumulative frequency
$0 \leqslant t < 3.5$	3
$0 \leqslant t < 7.5$	8
$0 \leqslant t < 11.5$	16
$0 \leqslant t < 15.5$	19
$0 \leqslant t < 18.5$	20

(b)

Number on bus	Cumulative frequency
0–5	8
0–10	15
0–15	24
0–20	31
0–25	40

(c)

Age (years)	Cumulative frequency
$16 \leqslant a < 20.5$	3
$16 \leqslant a < 25.5$	9
$16 \leqslant a < 30.5$	26
$16 \leqslant a < 35.5$	52
$16 \leqslant a < 40.5$	63
$16 \leqslant a < 50.5$	65

(d)

Temperature (°C)	Cumulative frequency
$-10 \leqslant t < 0$	12
$-10 \leqslant t < 10$	98
$-10 \leqslant t < 20$	283
$-10 \leqslant t < 30$	362
$-10 \leqslant t < 40$	365

2

Weight of baby (kg)	Frequency
$1 \leqslant w < 2$	5
$2 \leqslant w < 3$	12
$3 \leqslant w < 4$	24
$4 \leqslant w < 5$	8
$5 \leqslant w < 6$	1

2 (a)

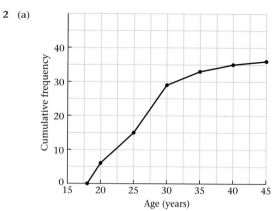

(b) (i) 26 years   (ii) 8 years
3 (a) 36.75°C   (b) 0.65°C   (c) 86 people

Exercise 20C

1 (a)

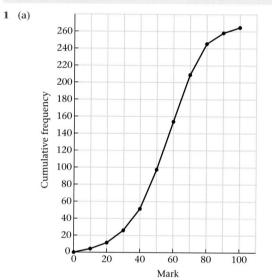

(b) 70   (c) 52
2 (a)

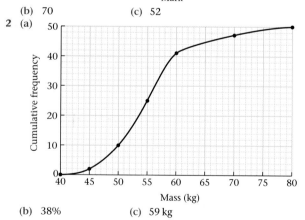

(b) 38%   (c) 59 kg

Exercise 20B

1 (a)

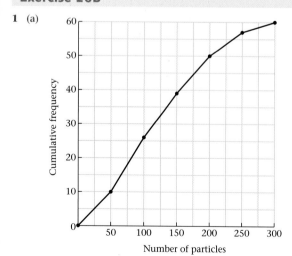

(b) (i) 115   (ii) 66   (iii) 177   (iv) 111

3 (a)

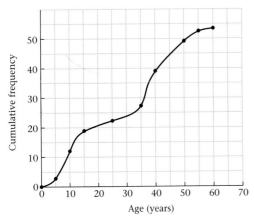

(b)   34 years
(c)   4

Exercise 20D

1

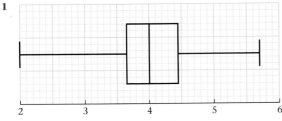

Weight (kg)

2

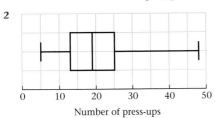

Number of press-ups

Exercise 20E

1   Class B had a slightly higher median but their results were spread over a wider range. Class A was more consistent.

2   (a)

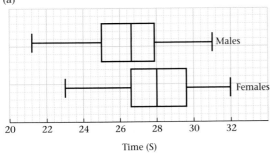

Time (S)

(b)   The male median time was quicker. The box plots are alike with the male times lower than the female times.

Exercise 20F

1

Class widths (s)	Frequency density
30	1
30	3
20	5
10	8
10	6
50	2.4

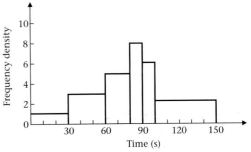

Time (s)

2

Class widths (g)	Frequency density
30	2.5
15	7
15	11
15	8
30	4.5

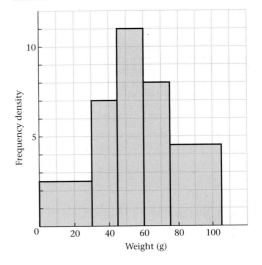

Weight (g)

3

Distance (m)	Frequency density
0–9	$18 \div 9.5 = 1.9$
10–14	$14 \div 5 = 2.8$
15–19	$16 \div 5 = 3.2$
20–24	$15 \div 5 = 3$
25–34	$17 \div 10 = 1.7$
35–50	$10 \div 16 = 0.6$

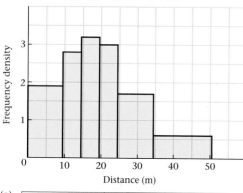

4 (a)

Hand length (cm)	Frequency	Frequency density
$12.5 \leqslant h < 17.5$	5	1
$17.5 \leqslant h < 19.5$	12	6
$19.5 \leqslant h < 21.5$	10	5
$21.5 \leqslant h < 25.5$	15	3.75

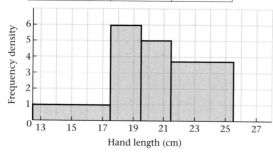

Exercise 20G

1

Estimated length (cm)	Frequency
$10 < l \leqslant 20$	14
$20 < l \leqslant 24$	12
$24 < l \leqslant 28$	20
$28 < l \leqslant 32$	16
$32 < l \leqslant 36$	12
$36 < l \leqslant 50$	7

2

Lifetime (hours)	Frequency
$100 < t \leqslant 400$	30
$400 < t \leqslant 600$	32
$600 < t \leqslant 700$	36
$700 < t \leqslant 900$	108
$900 < t \leqslant 400$	36

3 (a)

Weight (grams)	Frequency
$8 \leqslant w < 24$	24
$24 \leqslant w < 32$	41
$32 \leqslant w < 40$	32
$40 \leqslant w < 56$	30
$56 \leqslant w < 72$	18
$72 \leqslant w < 112$	15

(b) 160
(c) 39.4%

4 (a)

Time (minutes)	Frequency
$10 \leqslant t < 25$	15
$25 \leqslant t < 35$	30
$35 \leqslant t < 40$	20
$40 \leqslant t < 45$	40
$45 \leqslant t < 60$	75

(b) 45

Mixed exercise 20

1

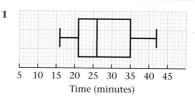

2 (a)

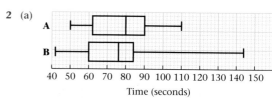

(b) 142 is an extreme value- minimum, lower quartile, median, upper quartile are all less than Assistant B's values.

3 (a) 5, 23, 35, 39, 40

(b)

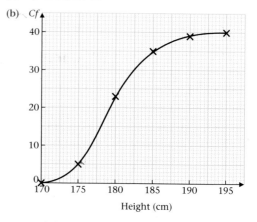

(c) 179 cm

4 (a) 32 seconds

(b)

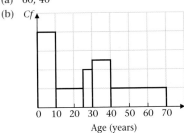

Time (seconds)

(c) Boys have: bigger median, wider range, wider interquartile range, bigger maximum value, smaller minimum value.

5 (a) 60, 40

(b)

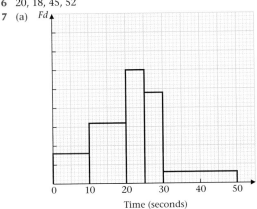

Age (years)

6 20, 18, 45, 52

7 (a)

Time (seconds)

(b) 10, 18, 14, 10, 8

8 (a)

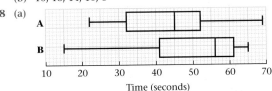

Time (seconds)

(b) Forest A has the oldest tree but generally the trees in forest B are older.

Chapter 21 Advanced trigonometry

Exercise 21A

(All answers correct to 2 d.p.)
1 (a) 17.32 cm²   (b) 29.60 cm²   (c) 76.31 cm²
    (d) 32.14 cm²   (e) 40.26 cm²   (f) 111.54 cm²
2 492.61 m²   **3** 73.54 cm²   **4** 53.13°   **5** 123.56°

Exercise 21C

1 (a) 9.92 cm   (b) 11.43 cm   (c) 14.95 cm
   (d) 8.28 cm   (e) 13.32 cm   (f) 6.24 cm
2 (a) 33.27°   (b) 23.71°   (c) 13.72°
3 (a) 61.90 km   (b) 37.82 km

Exercise 21D

$P\hat{Q}R = 63.73°$ or $P\hat{Q}R = 20.27°$

Exercise 21F

1 (a) 6.22 cm   (b) 6.25 cm   (c) 11.53 cm   (d) 24.70 cm
2 (a) 36.34°   (b) 101.54°   (c) 100.95°   (d) 98.25°
3 146.46 m
4 145.26 km

Exercise 21F

1 (a) 13 cm   (b) 23.85 cm   (c) 23.32 cm
   (d) 22.62°   (e) 12.10°   (f) 17.75°
2 (a) 14.14 cm   (b) 14.35 cm   (c) 63.77°
   (d) 26.23°
3 (b) (i) 38.42 cm   (ii) 30.81 cm
   (c) 6.72 cm   (d) 77.37°

Mixed exercise 21

1 (a) 15.76 cm²   (b) 6.30 cm   (c) 38.71°
2 (a) 56.4 cm²   (b) 7.84 cm
3 18.3 cm
4 (a) 183.97 km   (b) 320.64°   (c) 28.54 km
5 (a) 9.94 km   (b) 151.18°   (c) 3.19 km
6 (a) 17.4 cm   (b) 17.9 cm   (c) 54.47°

Chapter 22 Advanced mensuration

(All answers correct to 2 d.p. for Exercises 22A to 22D.)

Exercise 22A

1 (a) 9.77 cm   (b) 10.47 cm   (c) 7.38 cm
   (d) 39.10 cm   (e) 47.12 cm   (f) 60.21 cm
2 (a) 50.13°   (b) 80.21°   (c) 47.75°
   (d) 222.82°   (e) 73.52°   (f) 39.14°

Exercise 22B

1 (a) 22.34 cm²   (b) 61.09 cm²   (c) 30.54 cm²
   (d) 45.95 cm²   (e) 139.63 cm²   (f) 279.25 cm²
2 72.15°   **3** 10.06 cm

Exercise 22C

1 (a) 9.06 cm²   (b) 10.98 cm²   (c) 0.51 cm²
   (d) 3.22 cm²   (e) 308.83 cm²   (f) 192.04 cm²

Exercise 22D

1 (a) 68.75°   (b) 135 cm²
2 (a) 82.08 m   (b) 83.78 m   (c) 5026.54 m²
   (d) 398.38 m²
3 (a) 7.16 m   (b) 2.99 m²

Exercise 22E

1 314 cm³, 283 cm²   **2** 156 cm³, 211 cm²
3 0.942 m³   **4** 11972 cm³
5 (a) 8311 cm³
   (b) 2354 cm²

Exercise 22F

1 960 cm³   **2** 140 cm³   **3** 6.18 cm
4 (a) 301.59 cm³   (b) 6.71
5 7.08

Exercise 22G

1 (a) 2144.7 cm³, 804.25 cm²   (b) 1563.5 cm³, 651.4 cm²
  (c) 3591.4 cm³, 1134.1 cm²   (d) $\frac{1}{6}\pi x^3$, πx^2
2 10.61 cm
3 9.67

Exercise 22H

1 23.4 cm
2 $\left(\frac{12.5}{12}\right) = 1.13$ i.e. 13% increase
3 (a) 4:9   (b) 8:27   **4** $\frac{1}{50^3} = \frac{1}{125000}$
5 (a) 12.60 cm   (b) 1:1.31

Exercise 22I

1 618.6 cm³   **2** 112.72 cm³
3 496.37 cm³   **4** 45389.33 cm³

Mixed exercise 22

1 (a) 6.76 cm   (b) 6.98 cm   (c) 27.93 cm²   (d) 3.41 cm²
2 9048 cm³, 2413 cm²
3 268.08 cm³
4 $\frac{4}{3}\pi r^3 = r$, $r^2 = \frac{3}{4\pi}$, $r = \sqrt{\frac{3}{4\pi}}$
5 1120 cm³
6 9366.4 cm³
7 26.66 cm
8 2165.60 cm³
9 3200 cm³
8 270 cm³

Chapter 23 Exploring numbers 2

Exercise 23A

1 0.75, 0.6, 0.428 571…, 0.375, 0.3, 0.2727…
  $\frac{3}{7}$ and $\frac{3}{11}$ are recurring decimals
2 $\frac{5}{8}, \frac{5}{16}$ are terminating decimals because the denominators divide exactly into powers of 10. The others are recurring.
3 0.142 857…, 0.285 714…, 0.428 571…, 0.571 428…, 0.714 285…, 0.857 142…
  Same digits in the same order.
4 16
5 $\frac{1}{13}, \frac{3}{13}, \frac{4}{13}, \frac{9}{13}, \frac{10}{13}, \frac{12}{13}$ use the same digits in the same order
  $\frac{2}{13}, \frac{5}{13}, \frac{6}{13}, \frac{7}{13}, \frac{8}{13}, \frac{11}{13}$ use the same digits in the same order.
  Doubling the decimal form of the first set of fractions gives the pattern of recurring digits in the second set.
6 (a) Decimal recurs because 17 does not divide exactly into any power of 10.
  (b) When dividing by 17 there can only be 16 different remainders and 16 different subtraction sums.

Exercise 23B

1 (a) $\frac{2}{3}$ (b) $\frac{7}{9}$ (c) $\frac{34}{99}$ (d) $\frac{91}{99}$
  (e) $\frac{2}{11}$ (f) $\frac{125}{999}$ (g) $\frac{19}{37}$ (h) $\frac{91}{909}$
  (i) $\frac{1279}{9999}$ (j) $\frac{9}{101}$ (k) $1\frac{31}{75}$ (l) $4\frac{119}{165}$
  (m) $2\frac{19}{55}$ (m) $5\frac{69}{110}$
2 1; There is no number between 0.9999… and 1.

Exercise 23C

1 $\pm\sqrt{30}$
2 $\sqrt{40} = 2\sqrt{10}$
3 (a) Perimeter = 12 units
  (b) Area = 4 (units)²

4 (a) $\frac{\sqrt{7}}{7}$ (b) $\frac{3\sqrt{5}}{5}$ (c) $\frac{\sqrt{17}}{17}$
  (d) $\frac{\sqrt{11}}{11}$ (e) 4   (f) $\frac{3\sqrt{5}-5}{4}$
  (g) $-\frac{2\sqrt{7}+7}{3}$ (h) $-2\sqrt{3}+3$ (i) $-\frac{7\sqrt{11}+11}{38}$
5 (a) $x = 3 \pm \sqrt{7}$ (b) $-5 \pm \sqrt{11}$
7 $\sqrt{11}$ units
8 (a) $\frac{3}{2}$ (units)²   (b) $\sqrt{22}$ units
9 $\frac{\sqrt{6}}{2}$

Exercise 23D

1 3.5, 4.5; 13.5, 14.5; 103.5, 104.5; 9.5, 10.5; 99.5, 100.5
2 25, 35; 45, 55; 175, 185; 3015, 3025; 5, 15; 95, 105
3 4.65, 4.75; 2.85, 2.95; 13.55, 13.65; 0.25, 0.35; 157.45, 157.55; 9.95, 10.05; 99.95, 100.05
4 36.5, 37.5; 49.5, 50.5; 175, 185; 3.15, 3.25; 9.45, 9.55; 9350, 9450; 9.5, 10.5; 95, 105
5 4.25, 4.75; 7.25, 7.75; 16.25, 16.75; 2.75, 3.25; 15.25, 15.75; 9.75, 10.25; 99.75, 100.25
6 3.3, 3.5; 3.1, 3.3; 3.9, 4.1; 9.3, 9.5; 12.1, 12.3; 24.5, 24.7; 9.9, 10.1; 99.9, 100.1
7 3.625, 3.875; 3.375, 3.625; 4.125, 4.375; 5.875, 6.125; 15.375, 15.625; 9.875, 10.125; 99.875, 100.125
8 (a) 7.5 m, 6.5 m; 6.5 m, 5.5 m
  (b) 735 cm, 725 cm; 595 cm, 585 cm
  (c) 7.325 m, 7.315 m; 5.945 m, 5.935 m
  (d) Yes, the range of (a) contains the range of (b) which contains the range of (c).
9 (a) 74.365 s, 74.355 s
  (b) It is not necessarily the fastest lap time.

Exercise 23E

Answers are either exact or given to 4 s.f.
1 55.25 cm², 41.25 cm²   **2** 30 cm, 26 cm
3 57.0025 cm², 55.5025 cm²   **4** 30.2 cm, 29.8 cm
5 38.48 cm², 19.63 cm²   **6** 2.168 m, 2.105 m
7 5411 mm², 5153 mm²   **8** 27.625 cm², 20.625 cm²
9 24.35 cm², 23.65 cm²   **10** 161.5 cm², 127.5 cm²
11 6.297 cm², 4.984 cm²   **12** 23.56 cm, 20.42 cm
13 (a) 22 450 cm³, 8181 cm³   (b) 15 599 cm³, 12 770 cm³
  (c) 14 279 cm³, 13 996 cm³

Exercise 23F

Answers are either exact or given to 4 s.f.
1 1 cm, 0.8 cm   **2** 4 min, 2 min   **3** 20.5 cm, 9.5 cm
4 (a) 6.826 m/s, 6.780 m/s   (b) 6.860 m/s, 6.746 m/s
5 (a) 9.571 miles/l, 8.784 miles/l   (b) 43.45 mpg, 39.88 mpg
6 26.10 cm, 23.68 cm   **7** 3.857 g/cm³, 2.778 g/cm³
8 54.77°, 39.59°

Mixed exercise 23

1 (a) 0.875 T   (b) 0.$\dot{7}$ R   (c) 0.7 T
  (d) 0.6$\dot{3}$ R   (e) 0.58$\dot{3}$ R   (f) 0.$\dot{5}$3846$\dot{1}$ R
2 (a) $13\frac{74}{99}$ (b) $6\frac{25082}{33300}$
3 (a) $\frac{\sqrt{5}}{5}$ (b) $\frac{5\sqrt{7}}{7}$ (c) $\frac{\sqrt{29}}{29}$
  (d) 3   (e) $2\sqrt{6}$ (f) $2\sqrt{2}$
4 (a) 255 000, 245 000   (b) 6.5×10^6, 5.5×10^6
5 4 cm is correct to nearest cm
  4.0 cm is correct to 1 d.p.
  4.00 cm is correct to 2 d.p.
6 75.398 cm², 25.133 cm²   **7** 3.81 mm, 3.79 mm
8 (a) 4   (b) $4\sqrt{10}$ (c) 83.3%
9 (a) 9.719, 11.710   (b) 10 to 1 s.f.

Chapter 24 Probability

(Any correct fraction, decimal or percentage equivalents are acceptable unless otherwise stated.)

Exercise 24A

1 (a) $\frac{4}{9}$ (b) $\frac{2}{9}$
2 (a) $\frac{3}{10}$ (b) $\frac{8}{10}$
3 (a) $\frac{1}{6}$ (b) $\frac{3}{6}$ (c) $\frac{3}{6}$ (d) $\frac{5}{6}$
4 (a) $\frac{3}{8}$ (b) $\frac{2}{8}$ (c) $\frac{5}{8}$
5 (a) $\frac{6}{20}$ (b) $\frac{10}{20}$ (c) $\frac{16}{20}$ (d) $\frac{16}{20}$
   (e) Add the answers to (a) and (b)
   (c) Subtract p from 1

Exercise 24B

1 (a)

+	1	3	5	7
2	3	5	7	9
4	5	7	9	11
6	7	9	11	13
8	9	11	13	15

(b) $\frac{3}{16}$

2 (Only correct fraction equivalents are acceptable.)
   (a) $\frac{1}{6}$ (b) $\frac{9}{36}$ (c) $\frac{6}{36}$ (d) $\frac{15}{36}$
3 H H H H H T T T H H T T H T T T H T T T H T T T T T T T T T T T
   H H H T H H H T T H H T T H T T H T T T H T T T H T T T T T T T
   H H T H H H T T H H T T H T T H T H T H T T H T T T H T T T H T
   H T H H H H H T T H T H T T H H T T T T H T T T T T H T T T H T
   (a) $\frac{1}{16}$ (b) $\frac{6}{16}$ (c) $\frac{15}{16}$
4

Red	Red	Red	Red	Red	Red	Red	Red	Red
Bro	Bro	Bro	Gre	Gre	Gre	Yel	Yel	Yel
Blu	Pin	Bla	Blu	Pin	Bla	Blu	Pin	Bla
Whi	Whi	Whi	Whi	Whi	Whi	Whi	Whi	Whi
Bro	Bro	Bro	Gre	Gre	Gre	Yel	Yel	Yel
Blu	Pin	Bla	Blu	Pin	Bla	Blu	Pin	Bla

$\frac{3}{18}$
5 (a) $\frac{15}{36}$ (b) $\frac{18}{36}$ (c) $\frac{27}{36}$ (d) $\frac{22}{36}$
6 (a) $\frac{1}{6}$ (b) $\frac{1}{8}$ (c) $\frac{4}{48}$ (d) $\frac{6}{48}$
   (e) $\frac{6}{48}$ (f) $\frac{3}{48}$

Exercise 24D

1 (a) $\frac{1}{12}$ (b) $\frac{1}{10}$ (c) $\frac{26}{120}$
2 (a) $\frac{1}{5}$ (b) (i) $\frac{1}{25}$ (ii) $\frac{1}{5}$

Exercise 24E

1 (a) $\frac{1}{60}$ (b) $\frac{24}{60}$ (c) $\frac{9}{60}$ (d) $\frac{10}{60}$
2 (a) $\frac{2}{60}$ (b) $\frac{13}{60}$ (c) $\frac{45}{60}$

Mixed exercise 24

1 (H, H, H, H), (H, H, H, T), (H, H, T, H), (H, T, H, H),
   (T, H, H, H), (H, T, H, T), (T, H, T, H), (H, T, T, H),
   (T, H, H, T), (H, H, T, T), (T, T, H, H), (H, T, T, T),
   (T, H, T, T), (T, T, H, T), (T, T, T, H), (T, T, T, T)
   (a) $\frac{6}{16}$ (b) $\frac{5}{16}$
2 (a) $\frac{1}{13}$ (b) $\frac{3}{13}$ (c) $\frac{5}{13}$
3 120
4 (a) $\frac{x}{8}$ (b) $1 - \frac{x}{8}$ (c) 2 5 $\frac{140}{360}$

6

+	1	2	3	4
1	2	3	4	5
2	3	4	5	6
3	4	5	6	7

(a) $\frac{3}{12}$
(b) $\frac{3}{12}$
(c) $\frac{6}{12}$

7 $\frac{10}{24}$
8 (a) $\frac{4}{9}$ (b) $\frac{34}{63}$
9 $\frac{1}{72}$ 10 $\frac{32}{100}$
11 (a) $\frac{11}{17}$ (b) No, he may improve with practice.
12 (a) (i) $\frac{165}{340}$ (ii) $\frac{32}{340}$ (b) $\frac{137}{165}$
13 (a)

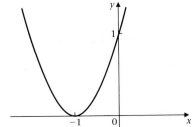

(b) 0.6 14 $\frac{\pi}{4}$

Chapter 25 Transformations of graphs

Exercise 25A

1 (a) 2 (b) 14 (c) -2 (d) -1
   (e) 7 (f) $-1\frac{15}{16}$
2 (a) x^2 (b) $9x^2$ (c) $x^2 + 4x + 4$
   (d) $x^2 + 2x + 4$ (e) $\frac{1}{4}x^2 - x - 3$ (f) $5 - 4x^2$
   (g) $k^2x^2 + 2akx + a^2 + b$
3 (a) (i) 5 (ii) -3 (iii) -4 (iv) -3 (b) $-2, 2$

Exercise 25B

1 $y = x^2 - 4$
2 $y = x^2 + k$ for any three values of k between 0 and 8
3 $-15 < k < 0$

Exercise 25C

1 (a) A: 4 units vertically in the negative y-direction.
      B: 4 units vertically in the positive y-direction.
      C: 6 units vertically in the positive y-direction.
   (b) A: 5 units vertically in the positive y-direction.
      B: 12 units vertically in the positive y-direction
      C: 8 units vertically in the negative y-direction.
   (c) A: 10 units vertically in the negative y-direction.
      B: 2.5 units vertically in the positive y-direction.
      C: 6 units vertically in the positive y-direction.
   (d) A: 2 units vertically in the positive y-direction.
   (e) A: 3 units vertically in the negative y-direction.
2 (c) Translation 2 units vertically in the positive y-direction.

Exercise 25D

1 $y = (x - 1)^2$
2 (a)

(b)

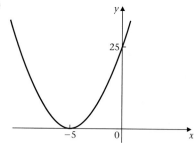

3

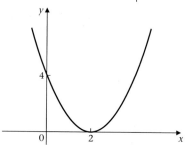

Exercise 25E

1 A horizontal translation of 1 unit in the positive *x*-direction.
2 A horizontal translation of 1 unit in the positive *x*-direction.
3 A vertical translation of 7 units in the positive *y*-direction.
4 $y = x^3 + 2$
5 $y = (x - 4)(x - 5)(x + 1)$

Exercise 25F

1 (a)   (b)

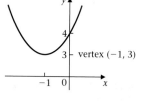 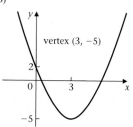

(c)   (d)

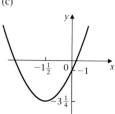 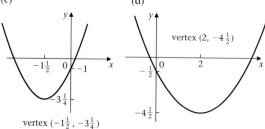

2 (a) A horizontal translation of 6 units in the negative
       x-direction.
   (b) A horizontal translation of 4 units in the positive
       x-direction followed by a vertical translation of 4 units
       in the negative *y*-direction.

3 (a)   (b)

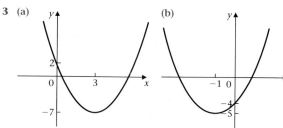

A horizontal translation of 4 units in the negative
x-direction followed by a vertical translation of 2 units
in the positive *y*-direction.

4 $y = x^2 - 18x + 14$
5

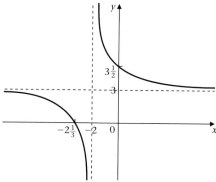

Exercise 25G

1 (a) $f(-x) = x^2 + 2x$
   (b)   (c)

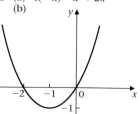 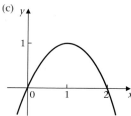

   (d) $(1,1)$
2 (a) $y = -x^3 - 2x^2$
   (b) $y = -x^3 + 2x^2$
3 (b)

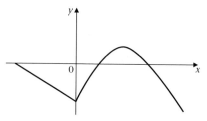

   (c) Reflection in the *x*-axis.
4 (b)

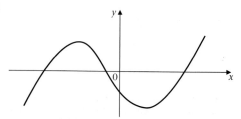

   (c) Reflection in the *y*-axis.

Exercise 25H

1 (a) Reflection in *x*-axis then a vertical translation of 6 units in the positive *y*-direction.

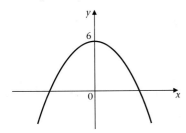

(b) Horizontal translation of 4 units in the positive *x*-direction then a vertical translation of 2 units in the positive *y*-direction.

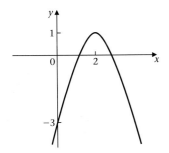

(c) Horizontal translation of 2 units in the positive *x*-direction then a reflection in *x*-axis then a vertical translation of 1 unit in the positive *y*-direction.

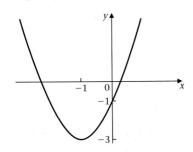

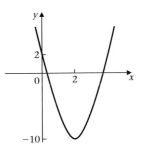

(c) Horizontal translation of 2 units in the positive *x*-direction then a vertical translation of 2 units in the negative *y*-direction

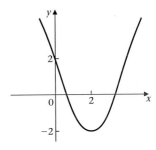

2 (a) Horizontal translation of 2 units in the negative *x*-direction then a stretch scale factor 4 parallel to the *y*-axis.

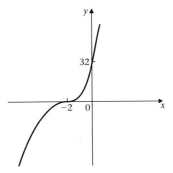

(b) Reflection in the *x*-axis then a vertical translation of 8 units in the positive *y*-direction.

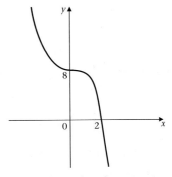

Exercise 25H

1 (a) Horizontal translation of 1 unit in the negative *x*-direction then a stretch scale factor 2 parallel to the *y*-axis then a vertical translation of 3 units in the negative *y*-direction.

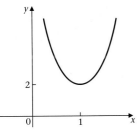

(b) Horizontal translation of 2 units in the positive *x*-direction then a stretch scale factor 3 parallel to the *y*-axis then a vertical translation of 10 units in the negative *y*-direction.

(c) Horizontal translation of 1 unit in the positive
x-direction then a vertical translation of 5 units in the
positive y-direction.

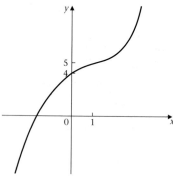

(d) Stretch scale factor 0.5 parallel to the x-axis then a
vertical translation of 8 units in the negative y-direction.

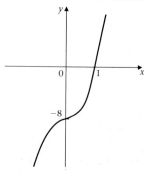

3 (a) Horizontal translation of 1 unit in the negative
x-direction.
(b) Horizontal translation of 4 units in the negative
x-direction, or a reflection in the y-axis.
(c) Reflection in the x-axis then a vertical translation of
2 units in the positive y-direction.

Exercise 25J

1 $\sin (x + 90°) = \cos x$
2 (a) Stretch scale factor 0.5 parallel to the x-axis.
(b)

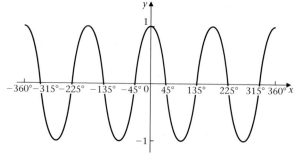

(c) 8
3 (a) Stretch scale factor 2 parallel to the y-axis.

(b)

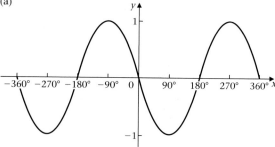

(c) 4
(d) 8
4 (a)

(b) Same graph for both $y = f(-x)$ and $y = -f(x)$.
5 (a) greatest: 5, least: -5
(b) greatest: 1, least: -1
(c) greatest: 1, least: -1
(d) greatest: 2, least: -2
(e) greatest: 9, least: 3

Mixed exercise 25

1 (a) (i) $y = \sin x + 1$ (ii) $y = 2\sin x$
(b) Horizontal stretch with scale factor $\frac{1}{2}$,
vertical stretch with scale factor 3
2 (a) $a = 2, b = -1$
(b) $0°, 233°, 360°, 593°, 720°$
(c) 3
3 (a) (i) $(5, -4)$ (ii) $(2, -9)$ (iii) $(2, 4)$ (iv) $(1, -4)$
(b) $f(x) = x^2 - 4x$
4 (a) $(1, 6)$ (b) $(3, 10)$ (c) $(3, -6)$
5 (a) (i) $(4, -25)$ (ii) $(2, -3)$ (iii) $(1, -25)$
(b) $a = 1, b = 9$

Chapter 26 Circle theorems

Exercise 26A

1 38°
2 (a) 66°   (b) 24°
3 (a) 34°   (b) 17°

Exercise 26B

1 (a) 56°   (b) 34°
2 (a) 34°   (b) 56°   (c) 68°   (d) 68°   (e) 22°
3 (a) $a = 36°, b = 36°, c = 108°, d = 36°$
4 (a) $P\hat{Q}R = 90°$ (b) $Q\hat{R}P = 47°$
5 (a) $A\hat{D}E = 78°$ (b) $A\hat{B}C = 78°$
(c) $D\hat{C}B = 78°$ (d) $B\hat{A}D = 102°$
6 25°, 27°   7 8.49 cm, 4.24 cm
8 82°, 8°   9 8 cm, 120 cm²
10 32°   11 112°, 22°
12 $90° - x°, x°$ 13 24 cm²

14 70°, 70°, 40°   **15** 128°, 29°

16 10.6 cm   **17** 40°

18 6.5 m

19 The sides are chords of the circle. So, the perpendicular bisectors pass through the centre. Hence, their point of intersection is the centre.

Exercise 26C

1 angle TCB = angle OAB (alternate segment)
angle OBA = angle OAB (isosceles triangle)
angle TBC = angle TCB (equal tangents)
Since two angles are equal then the third must also be equal. Hence triangles AOB, BTC are similar.

2 (a) P lies on the perpendicular bisection of RY
∴ $PR = PY$ and triangle PRY is isosceles
∴ angle YPX = angle XPR.
But angle XPR = angle XSQ (angles in the same segment)
∴ angle YPX = angle XSQ

(b) angle XSZ + angle XSQ = 180° (angles on a straight line)
angle XSZ + angle YPX = 180° (from part (a))
∴ angle PZS = 360° − 180° − angle PXS = 90°

3 (a) angle BAD = angle ADC (alternate angles)
angle ABC = angle ADC (angles in the same segment)
Hence, triangle ABE is isosceles.
∴ angle AEB = 180° − 2 × angle ABE
Also, angle AEB = 180° − angle AEC (angles on a straight line)
∴ angle AEC = 2 × angle ABC

(b) Since triangle ABE is isosceles, $AE = BE$

4 Let angle CXE = x° = angle AXE and angle XCD = y°
The angle CFX = 180° − x° − y° (angle sum of a triangle)
angle DFE = 180° − x° − y° (vertically opposite angles)
angle XAB = y° (cyclic quadrilateral)
In triangle AXE, angle XEA = 180° − x° − y° (angle sum of a triangle)
∴ angle DFE = angle FEA

5 angle ADC = 180° − angle DAB (interior angles of parallel lines)
angle BCD = 180° − angle DAB (opposite angles of a cyclic quadrilateral)
∴ angle ADC = angle BCD

6 Join D to B and A to C.
Triangle ABD is isosceles. ∴ angle ABD = angle ADB
angle ABD = angle ACD (angles in the same segment)
angle ADB = angle ACB (angles in the same segment)
∴ angle ACD = angle ACB
∴ AC bisects angle BCD.

7 Join A to E
angle AEC + angle ABC = 180° (cyclic quadrilateral)
angle AEC + angle AED = 180° (angles on a straight line)
∴ angle ABC = angle AED.
But angle ABC = angle ADC (opposite angles of a parallelogram)
∴ angle ADC = angle AED
∴ $AE = AD$ (isosceles triangle)

8 Let angle CDF = angle FDE = x°
angle CBF = angle CDF = x° (angles in the same segment)
angle CBA = 180° − angle CDA (cyclic quadrilateral)
angle CDE = 180° − angle CDA (straight line)
∴ angle CBA = 2x°
∴ FB bisects angle ABC

9 Join F to D. Let angle DAF = angle FAB = x°
Then angle AED = angle EAB = x° (alternate angles)
angle AFD = 90° (angle in a semi-circle)
Hence FD is a line symmetry of triangle AED ∴ $AF = FE$

10 angle COA = 2 × angle CBA (angle at the centre)
= 4 × angle TBA
But angle TBA = angle ATP (alternate segment)
∴ angle COA = 4 × angle ATP

Mixed exercise 26

1 (a) 8 cm   (b) 10.4 cm

2 $A\hat{D}C$, $B\hat{C}D$, $D\hat{A}B$

3 103°

4 (a) 90°   (b) 35°   (c) 125°

5 (a) 21°   (b) 53°   (c) 68°

6 26°   **7** 41°, 49°   **8** 30 cm, 36 cm

9 (a) 86°   (b) 52°   (c) 52°

10 $(1 + \sqrt{2})r$ **11** 101°, 63°

12 (a) 60°   (b) 90°   (c) 30°   (d) 90°
(e) 30° AC = 6.93 cm

13 (a) 47°   (b) 43°   (c) 25°

14 (a) 70°   (b) 35°   radius = 2.11 cm

15 (a) Two tangents from the same point are equal, so $PY = PA$ and $PX = PA$. So $PY = PX$, meaning that P is the midpoint of XY and PA is a bisector.
(b) As PX, PY and PA are all equal, they are all radii of a circle centre P. XY is a diameter of this circle, so angle XAY must equal 90°.

16 (a) 52°   (b) 47°   (c) 81°
(d) $D\hat{A}B$ = 180° − 47° − 52° (angles on a straight line)
= 81°
∴$D\hat{A}B$ = $C\hat{B}A$ and EAB is isosceles
∴$EA = EB$

17 53°

18 (a) 70°, 40°   (b) $B\hat{A}C$ = $B\hat{C}A$ = 70°
(c) In △PBR and △ACR
$P\hat{B}R$ = $C\hat{A}R$ (angles in same segment)
$B\hat{P}R$ = $R\hat{C}A$ (angles in same segment)
$P\hat{R}B$ = $C\hat{R}A$ (virtually opposite)
△PBR is similar to △CAR

19 1.23 cm

20 (a) The altitudes meet at a point.
(b) Angle BHD = 90° − y. Join B to C so that BC cuts AD at K. BC subtends 90° at F and at E, so is the diameter of a circle through $FBCE$.
Angle EFC = y (angles in the same segment: EH)
Angle EBC = y (angles in the same segment: EC)
From △ BHK:
Angle BKH = 180° − angle BHD − angle HBK
= 180° − (90° − y) − y
= 90°
(c) BE, CF are altitudes of ABC, meeting at H. AK also goes through H so is the third altitude. So AK is perpendicular to BC.

Chapter 27 Vectors

Exercise 27A

1 (a) $\begin{pmatrix} -2 \\ -1 \end{pmatrix}$ (b) $\begin{pmatrix} -3 \\ -5 \end{pmatrix}$ (c) $\begin{pmatrix} 5 \\ 6 \end{pmatrix}$ (d) $\begin{pmatrix} -5 \\ -6 \end{pmatrix}$

2 $(0, -2)$

3 Translation by $\begin{pmatrix} 6 \\ 0 \end{pmatrix}$; translation by $\begin{pmatrix} 2b - 2a \\ 0 \end{pmatrix}$

4 (a) $\begin{pmatrix} 4 \\ -2 \end{pmatrix}$ (b) $(6, 1)$

(c) $\overrightarrow{AB} = \begin{pmatrix} 5-2 \\ 7-3 \end{pmatrix} = \begin{pmatrix} 3 \\ 4 \end{pmatrix}$ $\overrightarrow{DC} = \begin{pmatrix} 9-6 \\ 5-1 \end{pmatrix} = \begin{pmatrix} 3 \\ 4 \end{pmatrix}$

Exercise 27B

1 (a) $\begin{pmatrix} 9 \\ 3 \end{pmatrix}$ (b) $\begin{pmatrix} 15 \\ 5 \end{pmatrix}$ (c) $\begin{pmatrix} -6 \\ -2 \end{pmatrix}$ (d) $\begin{pmatrix} -3 \\ -1 \end{pmatrix}$ (e) $\begin{pmatrix} 3k \\ k \end{pmatrix}$

2 (a) 2**b**   (b) **a** + **b**   (c) **a** + 2**b**

3 $\overrightarrow{AC} = \begin{pmatrix} 6 \\ 1 \end{pmatrix}$

4 (a) $\overrightarrow{AD} = \begin{pmatrix} -2 \\ 3 \end{pmatrix}$

(b) $\overrightarrow{AC} = \begin{pmatrix} 1+2 \\ 4-3 \end{pmatrix} = \begin{pmatrix} 3 \\ 1 \end{pmatrix}$

(c) $\overrightarrow{DB} = \begin{pmatrix} --2+1 \\ -3+4 \end{pmatrix} = \begin{pmatrix} 3 \\ 1 \end{pmatrix}$

5 $\frac{3}{2}\mathbf{a} + 3\mathbf{b}$

Exercise 27C

1 (i) $\begin{pmatrix} -2 \\ 4 \end{pmatrix}$ (ii) $\begin{pmatrix} 4 \\ 2 \end{pmatrix}$ (iii) $\begin{pmatrix} 16 \\ -7 \end{pmatrix}$ (iv) $\begin{pmatrix} 12 \\ -4 \end{pmatrix}$

for example, $c = \begin{pmatrix} 2 \\ 3 \end{pmatrix}$ or $\begin{pmatrix} -1 \\ 0 \end{pmatrix}$

3 (4, 3)

4 $\begin{pmatrix} -2 \\ 2 \end{pmatrix}$ **5** $\begin{pmatrix} 1 \\ -1 \end{pmatrix}$ **6** $\begin{pmatrix} 10 \\ -1 \end{pmatrix}$

Exercise 27D

1 (a) 5 (b) 25 (c) 10 (d) $\sqrt{170} = 13.0$
2 (a) $\sqrt{106} = 10.3$ (b) $\sqrt{18} = 4.24$
   (c) $\sqrt{208} = 14.4$ (d) $\sqrt{40} = 6.32$
3 (a) $\sqrt{13} = 3.61$ (b) $2\sqrt{13} = 7.21$
   (c) 10 (d) $\sqrt{26} = 5.10$

Exercise 27E

1 $p = 2$
2 $q = -2$
3 $x = 0, y = -2$
4 $p = 2.4, q = -2.2$
5 (2, 6), (3, 7), (4, 8), (5, 9) (6, 10) (7, 11). A straight line, $y = x + 4$.
6 (a) $\begin{pmatrix} 6 \\ 3 \end{pmatrix}$ (b) $\begin{pmatrix} 6 \\ 3 \end{pmatrix} + k\begin{pmatrix} -2 \\ 2 \end{pmatrix} = \begin{pmatrix} 6-2k \\ 3+2k \end{pmatrix}$
   (c) $k = 3$, $D = (2, 10)$

Exercise 27F

1 $\begin{pmatrix} 2 \\ 4 \end{pmatrix}$ **2** (4, 0)

3 (5, 9)   **4** $\begin{pmatrix} 3 \\ 5 \end{pmatrix}$ $\begin{pmatrix} 4 \\ 4 \end{pmatrix}$

Exercise 27G

1 $\mathbf{y} - \mathbf{x}$. XY is parallel to AB and $\frac{2}{3}$ of the length of AB.
2 $\mathbf{b} - \mathbf{a}$. $3\mathbf{b} - 3\mathbf{a}$. CD is parallel to AB and 3 times the length of AB.
3 (a) $\frac{1}{3}(\mathbf{a} + \mathbf{b})$ (c) $\frac{1}{3}(\mathbf{a} + \mathbf{b})$. G and H are the same point.
   (c) $\frac{1}{2}\mathbf{b}$.
4 (c) $k = \frac{1}{3}, m = \frac{1}{3}, \frac{4}{3}\mathbf{a} + \frac{1}{3}\mathbf{b}$
5 e.g. $\overrightarrow{OA} = 3\mathbf{a}$, $\overrightarrow{OC} = 3\mathbf{b}$. $\overrightarrow{PQ} = 2\mathbf{a} + \mathbf{b}$, $\overrightarrow{SR} = \mathbf{b} + 2\mathbf{a}$. $PQRS$ is a parallelogram. No longer true.
6 (a) $\mathbf{b} - \mathbf{a}$ (b) $-\mathbf{b}$ (c) $-\mathbf{b} - \mathbf{a}$
   (b) Midpoints are $\frac{1}{2}(\mathbf{a} + \mathbf{b})$, $\frac{1}{2}(2\mathbf{a} - \mathbf{b})$, $\frac{1}{2}(\mathbf{a} - 2\mathbf{b})$, $-\frac{1}{2}(\mathbf{a} + \mathbf{b})$, $\frac{1}{2}(\mathbf{b} - 2\mathbf{a})$, $\frac{1}{2}(2\mathbf{b} - \mathbf{a})$
   The third hexagon is an enlargement, scale factor 0.75, centre O, of the first hexagon.

Mixed exercise 27

1 (a) $\begin{pmatrix} 5 \\ 1 \end{pmatrix}$ (b) $\begin{pmatrix} -5 \\ -1 \end{pmatrix}$ (c) $\begin{pmatrix} -1 \\ -5 \end{pmatrix}$ (d) $\begin{pmatrix} 1 \\ 5 \end{pmatrix}$

2 (a) $\begin{pmatrix} 5 \\ 1 \end{pmatrix}$ (b) $\begin{pmatrix} 2 \\ 3 \end{pmatrix}$ (c) $\begin{pmatrix} 7 \\ 4 \end{pmatrix}$
   (d) $\begin{pmatrix} 3 \\ -2 \end{pmatrix}$ (e) $\begin{pmatrix} 10 \\ 2 \end{pmatrix}$ (f) $\begin{pmatrix} -3 \\ 2 \end{pmatrix}$

3 $\begin{pmatrix} 3 \\ -2 \end{pmatrix}, \begin{pmatrix} 1.5 \\ -1 \end{pmatrix}$ **4** $\begin{pmatrix} 3 \\ 5 \end{pmatrix}$ **5** $\begin{pmatrix} -1 \\ -7 \end{pmatrix}$

6 (a) $\begin{pmatrix} 3 \\ 0 \end{pmatrix}$ (b) $\begin{pmatrix} 1 \\ 2 \end{pmatrix}$

7 $\begin{pmatrix} 10 \\ 1 \end{pmatrix}$ **8** 3, 3

9 $a = 1, b = 1$ **10** $p = 1, q = \frac{1}{2}$
11 (a) $\mathbf{a} + \mathbf{b}$ (b) $-\mathbf{b}$ (c) $\mathbf{a} - \mathbf{b}$
12 (a) $\frac{3}{5}\mathbf{b}$ (b) $\mathbf{a} - \mathbf{b}$ (c) $\frac{3}{5}\mathbf{a} - \frac{3}{5}\mathbf{b}$ (d) $\frac{3}{5}\mathbf{a}$
   CD is parallel to OA and $\frac{3}{5}$ of the length of OA.
13 (a) $\mathbf{a} + \mathbf{b}$ (b) $\mathbf{a} + \mathbf{b} + \mathbf{c}$ (c) $2\mathbf{b}, \mathbf{a} + \mathbf{c} = \mathbf{b}$
14 $\overrightarrow{DA} = 4\mathbf{a} - 2\mathbf{b}$
   $\overrightarrow{PQ} = -\mathbf{a} + 2\mathbf{b} - (2\mathbf{a} - \mathbf{b}) = 2\mathbf{a}$
   $\overrightarrow{SR} = 2\mathbf{a} - \mathbf{b} + \mathbf{b} = 2\mathbf{a}$
   Thus PQ has equal length and is parallel to SR and $PQRS$ is a parallelogram.
15 (a) $\mathbf{b} - \mathbf{a}$ (b) $\frac{1}{2}\mathbf{b}$ (c) $\mathbf{c} - \mathbf{a}$
   (d) $\frac{1}{2}(\mathbf{c} - \mathbf{a})$ (e) $\frac{1}{2}(\mathbf{a} + \mathbf{c})$ (f) $\frac{1}{2}(\mathbf{a} + \mathbf{c}) - \frac{1}{2}\mathbf{b}$
16 (a) $\frac{3}{2}\mathbf{a}$ (b) $3\mathbf{b}$ (c) $\frac{3}{2}\mathbf{a} - 3\mathbf{b}$
   (d) $\overrightarrow{OZ} = \overrightarrow{OY} + p\overrightarrow{YX} = 3\mathbf{b} + p(\frac{3}{2}\mathbf{a} - 3\mathbf{b})$
   $\overrightarrow{OZ} = \overrightarrow{OB} + q\overrightarrow{BA} = \mathbf{b} + q(\mathbf{a} - \mathbf{b})$
   $\overrightarrow{OZ} = 2\mathbf{a} - \mathbf{b}$
17 $x = 4, y = -3$

Chapter 28 Introducing modelling

Exercise 28A

1 (a) £1500(1.06)^t (b) 16 years
2 (a) $8(0.6)^n$ metres (b) 23.36 metres
3 (a) 48 days (b) 96 days
4 (a) (i) £32 000 (ii) £16 384 (iii) £40 000(0.8)^n
   (b) 27 years

Exercise 28B

1 (a) (i) 6.5 m (ii) 5.25 m (iii) 4 m (iv) 1.5 m
   (v) 6.5 m
   (b) 1.14 pm (c)

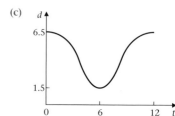

2 (a) (i) 15 cm (ii) 20 cm (iii) 10 cm
   (b)

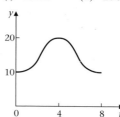

3 (a) (i) 7.5 m   (ii) 6 m
  (b) 3 m   (c) 3 am, 3 pm
  (d)

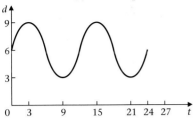

4 (a) $p = 9$, $q = 8$
  (b) $3\frac{1}{3}$ sec, $16\frac{2}{3}$ sec, $23\frac{1}{3}$ sec
  (c) $6\frac{2}{3}$ sec, $13\frac{1}{3}$ sec, $26\frac{2}{3}$ sec
  (d)

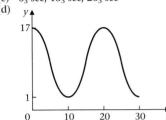

Exercise 28C

1 (a) $y = 0.25x + 2.5$ (b) 14.5
2 (a) $y = -0.25x + 0.75$ (b) -3.75
3 (a) $y = \frac{14}{15}x + \frac{46}{3}$ (b) $85\frac{1}{3}$

Exercise 28D

(The following answers are estimates from a line of best fit. Your answers may therefore differ slightly and still be correct.)
1 $a = 0.28$, $b = 5$
2 $a = 2$, $b = 1$
3 $a = -0.2$, $b = 7.2$
4 $a = 2$, $b = 0.2$
5 $a = -0.5$, $b = 3$
6 $a = -0.5$, $b = 6$

Exercise 28E

1 $p = 2$, $q = 4$, $k = 32$
2 $p = -3$, $q = 2$, $k = 6$
3 $a = 3$, $b = 4$, $k = 67$
4 $\frac{1}{3}$
5 $\frac{2}{3}$ and $-\frac{2}{3}$
6 $a = 4$, $b = 3$, $k = \frac{37}{9}$
7 $\frac{1}{2}$
8 (a) $p = 2$, $q = 4$
  (b) 3

Chapter 29 Conditional probability

Exercise 29A

1 (a) 0   (b) $\frac{2}{7}$ (c) $\frac{5}{7}$ (d) $\frac{1}{7}$
2 (a) $\frac{5}{42}$ (b) $\frac{1}{21}$ (c) $\frac{20}{21}$
3 $\frac{17}{24}$
4 (a) $\frac{89}{110}$ (b) $\frac{21}{110}$
5 (a) $\frac{1}{12}$ (b) $\frac{1}{12}$ (c) $\frac{1}{4}$

Exercise 29B

1 $\frac{421}{455}$
2 (a) $\frac{1197}{2990}$ (b) $\frac{1793}{2990}$
3 (a) (i) $\frac{2}{35}$ (ii) $\frac{2}{35}$ (iii) $\frac{2}{35}$ (b) $\frac{12}{35}$
4 $\frac{3}{10}$
5 (a) $\frac{1}{5}$ (b) $\frac{2}{5}$ (c) $\frac{3}{5}$

Exercise 29C

1 (a) (i) 0.00015   (ii) 0.99985
  (b) The probabilities of a dart landing in treble 20 when the first or second miss are unknown.
2 0.999   **3** 0.79
4 (a) 0.56   (b) 0.18   (c) 0.26
5 0.7625
6 (a) $\frac{7}{22}$ (b) $\frac{28}{33}$
7 0.0315

Mixed exercise 29

1 (a) $\frac{1}{7}$ (b) $\frac{4}{7}$
2 (a) $\frac{1}{36}$ (b) $\frac{7}{18}$ (c) $\frac{11}{18}$
3 (a) 0.28   (b) 0.42   (c) 0.7
4 $\frac{17}{24}$
5 $\frac{35}{76}$
6 (a) $\frac{4}{91}$ (b) $\frac{12}{65}$ (c) $\frac{53}{65}$
7 (a) $\frac{3}{20}$
8 (a) $\frac{9}{28}$
9 (a) $\frac{1}{21}$ (b) $\frac{5}{84}$ (c) $\frac{1}{7}$
10 $\frac{173}{420}$

Examination practice paper: Non-calculator

1 90 g zinc, 45 g iron, 15 g copper
2 £587.50
3 (a) $12pq$ (b) $y^2 + 2y$ (c) $x(x - 5)$
4 (a) £480   (b) 80%
5

6 (a)
  (b)

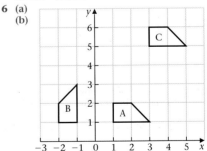

  (c) Rotation, 90° clockwise about (3, 1)
7 (a)

```
0 | 7, 7, 8
1 | 0, 0, 0, 0, 1, 1, 2, 2, 3, 5, 5, 6, 7, 9, 9
2 | 0, 1, 1, 1, 3, 4, 4, 6, 7
3 | 0, 1, 2, 3            Key: 2 | 1 means 21
```

  (b) 18 marks
8 $\frac{3}{16}$

9 (a) $x \leqslant \frac{1}{3}$ (b) 0

10 $\dfrac{20 \times 60}{0.5} = 2400$

11 4.5

12 (a) $\frac{15}{16}$ (b) 9

13 528 cm²

14 (a) 19   (b) £93 000   (c) £65 000

15 (a) 4×10^8 (b) 2.4×10^{11}

16 (a) 90°   (b) 26°   (c) 52°

17 (a)

x	-2	-1	0	1	2	3	4
y	-6	2	0	-6	-10	-6	12

(b)

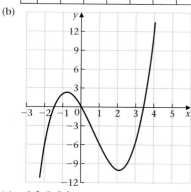

(c) $-1.5, 0, 3.5$

18 (a) (i) 1   (ii) $\frac{1}{64}$ (iii) 4
  (b) $\sqrt{3}$

19 (a) Frequency: 20, 35, 25, 15, 5
  (b)

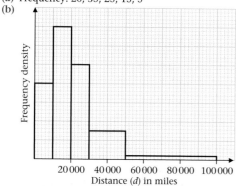

20 (a) $y = \dfrac{100}{x^2}$ (b) $\frac{1}{2}$

21 $y = 3x + 2$

22 $x^2 + (x+1)^2 = x^2 + x^2 + 2x + 1$
$\qquad\qquad\qquad\quad = 2(x^2 + x) + 1$
1 more than an even number is an odd number.

23 $LR = PR$ (sides of a square)
$\quad QR = SR$ (sides of a square)
$\quad L\hat{R}Q = 90° + P\hat{R}Q$
$\quad P\hat{R}S = 90° + P\hat{R}Q = L\hat{R}Q$
$\quad LQR$ and PRS are congruent (SAS)

24 $x = \dfrac{a^2 + b^2}{a - b}$

25 (a) 8π cm   (b) 96π cm²

Examination practice paper: Calculator

1 2.122 107 214

2 £136

3 £2.05

4 (a) 3.995   (b) 7.5

5 $5n - 1$

6 59 m²

7 (a) (i) 75°   (ii) 54°   (b) 4 cm

8 £578.81

9 49.38°

10 (a)

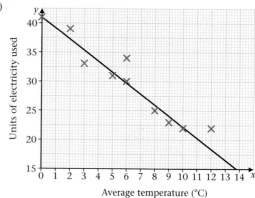

Average temperature (°C)

(b) (i) 3°C   (ii) 27.5 units

11 4.86

12 (a) £9800
  (b) £5171.77

13 12 mm

14 £445

15 (a) $y = 0.4$ (b) $\dfrac{5x + 11}{(3x - 1)(x + 6)}$
  (c) $(2x + 7)(x - 2)$

16 1817 kg or 1.817 tonnes

17 (a) A = area of base + area of 4 sides
$\qquad = x^2 + 4hx$

(b) $h = \dfrac{A - x^2}{4x}$

(c) 4 cm

(d) 5.66

18 (a)

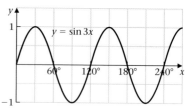

(b) $x = 80°$ or $x = 100°$

19 (a) 650 cm³   (b) 400 cm²   (c) 400 cm²

20 The lower bound for the length is 67.5 cm and for the width is 8.35 cm
So the lower bound for the area is 67.5×8.35 cm²
$\qquad\qquad\qquad\qquad\qquad\qquad = 563.625$ cm²

21 If $y = ax + b$ is a tangent to the circle, then $x^2 + (ax + b)^2 = 64$ has only one solution, so is a perfect square.
$(a^2 + 1)x^2 + 2abx + (b^2 - 64) = 0$ is a perfect square
So $\sqrt{(a^2 + 1)} \times \sqrt{(b^2 - 64)} = ab$
$\qquad \therefore (a^2 + 1)(b^2 - 64) = a^2 b^2$
$\qquad a^2 b^2 - 64a^2 + b^2 - 64 = a^2 b^2$
Rearranging gives $b^2 = 64(a^2 + 1)$
$\qquad\qquad a^2 + 1 = \dfrac{b^2}{64}$

22 $\frac{19}{66}$

Index

A

acute angles 29
addition
  algebraic fractions 262–4
  fractions 60–1
  lower bounds 440
  upper bounds 440
  vectors 504
algebra basics, summary of key points 27–8
algebraic expressions
  collecting like terms 20
  definition 16
  evaluating 16
  expanding the brackets 19
  factorising 22
  multiplying bracketed expressions 23
  negative numbers involved 21
  removing brackets 19
  simple equations 25–6
  simplifying 20
  squares involved, evaluating 17–18
  squaring 25
  summary of key points 275
algebraic formulae
  basics 228–9
  changing the subject 232–4
  manipulating 230–1
  two formulae 235
algebraic fractions 261–4
  complex 272–3
  quadratic equations 354
allied angles 30
alternate angles 30
altitude 500
angles
  acute 29
  allied 30
  alternate 30
  bisecting 120
  co-interior 30
  complete turn 29
  corresponding 30
  meeting at a point 29
  obtuse 29
  perpendicular lines 29
  polygons, exterior angles 33
  polygons, interior angles 33–4
  quadrilaterals, interior angles 33
  reflex 29
  right 29
  supplementary 29
  triangles, exterior angles 35
  triangles, interior angles 31–2
  trigonometric ratios 295
  vertically opposite 29
approximating
  checking 218
  rounding 214–16
  standard form 156
  summary of key points 224
arc 412–13
area
  circles 243
  quadrilaterals 239–41
  similar shapes 424–6
  triangles 239–41, 399–400
arithmetic sequences 227
arrowhead 32

asymptote 313
averages
  appropriate 186–8
  extreme values 187
  four–point moving 202
  frequency distributions 192–3
  grouped data 195–7
  mean 184–5, 192–3
  mean, estimate of 196–7
  median 184–5, 188–90, 192–3, 195
  modal class 196
  mode 184–5, 192–3
  moving 201–2
axiom 485

B

back-bearing 127
bar 339
bearings 126–7
best fit, lines of 206–7
biased sampling 73
BIDMAS 17
box plots 379–80, 382
box and whisker diagrams 380

C

Celsius 319
Centigrade 319
centre of rotation 107–8
certain events 446
checking by estimating 218
checking by number for x 264
chord 412
circle based pyramid 36
circle theorems
  1. lengths of two tangents 485–6, 492
  2. perpendicular to a chord 486, 493
  3. angle subtended at centre 487, 493
  4. right angle in a semicircle 488, 494
  5. angles in same segment 488, 494
  6. opposite angles of a cyclic quadrilateral 488–9, 494
  7. angle between tangent and a chord 489, 494
  summary of key points 500
circles
  arc length 412–13
  area 243, 412
  basics 412
  circumference 243, 412
  equation 286–7
  sector area 414–15
  segment area 416–17
  trigonometric ratios 298–9
circular prism 36
class intervals 383–6
co-interior angles 30
coefficient 165
collecting like terms 20
column vectors 501–2
complete turn 29
complex numbers 360
compound interest 140–1
compound measures 143–7
conditional probability
  basics 534, 535
  summary of key points 541

cone 36
  volume 421
congruent shapes 38–40
constructions
  bisecting an angle 120
  hexagon 118
  perpendicular 118–20
  right angle 118–20
  triangles 117
continuous data 71, 238, 386–7
correlation 206–7
corresponding angles 30
cos x 292–3, 301
cosine rule 405–6
cube root 6
cube [number] 5, 6
cube [shape] 35, 46, 244
cubic functions, graphs 311–12
cubic proportionality 335–7
cuboid 35, 45, 244–5, 285–6
cumulative frequency 370–1
cumulative frequency graphs 371–8
cylinder
  circular prism 36
  surface area 231, 418
  volume 419

D

data
  bias, avoiding 77
  box plots for comparison 382
  class intervals 82
  collection 76–9
  collection, reasons 72
  conclusions from 199–200
  continuous 71, 238, 386–7
  database 81
  discrete 71, 184–5, 238
  frequency polygons 86
  grouped 82–3, 85, 195–7
  histograms 383–7, 389–91
  information 72
  internet as source 80–1
  market research 72
  measurement 79
  observation 79
  pilot survey 77
  primary, collection 76–9
  qualitative 71
  quality control 72, 79
  quantitative 71
  questionnaire 76–7
  sampling 73–5
  secondary 72, 80–1
  two-way tables 85
data collection and recording
  summary of key points 89–90
data presentation and analysis
  summary of key points 212–3, 398
database 81
decimals
  basics 54–5
  division, written methods 57
  fractions 58, 434–5
  mental calculations 56–7
  ordering 54–5
  percentages 133
  recurring 59–60, 432–5
  rounding 214–15
  summary of key points 69–70
  terminating 432–3
demonstration vs proof 34
denominator 52
density 147

dependent events 534–5
difference of two squares 271
dimension theory 248–51
dimensionless 250
direct proportion 328–30
directed line segments 503
discrete data 71, 184–5, 238
division
  algebraic fractions 261–2
  decimals, written methods 57
  expressions with indices 257
  fractions 65–6
  indexes 9
  inequalities 98
  lower bounds 441–2
  mixed numbers 66
  upper bounds 441–2
dodecahedron 46

E

elevation 36–7
enlargement 109–10
equation vs formula 228
equations
  algebraic fractions 95
  brackets involved 93–4
  fractional terms 91
  graphical solutions 315–16
  indices 260
  negative coefficients 93
  problem solving 95
  rearranging 91, 92
  reduction to linear form 527–30
  solving simple 91–5
  solving simple algebraic 25–6
  summary of key points 103
  trial and improvement methods 317–18
  unknown on both sides 92, 93
  $y = px^2 + q$ 527–9
  $y = px^2 + qx$ 529–30
equivalent fractions 52–3
estimating
  checking 218
  probability using experience 453
  rounding 214–16
  standard form 156
  summary of key points 224
Euclid 29
Euler's Theorem 46, 47
examination
  formulae sheet 552
  practice paper – calculator 548–51
  practice paper – non–calculator 542–7
expanding the brackets 19
exponential functions 519–21, 531–2

F

factor 1
factor tree 2
factorising 22, 265–71
Fahrenheit 319
flow measure 146
formula vs equation 228
formulae, summary of key points 237
four-point moving averages 202
fractional indices 10–11, 258–9
fractions
  addition 60–1
  algebraic 261–4
  decimals 58, 434–5
  denominator 52
  division 65–6
  equivalent 52–3

improper 52
inverting 65
mixed numbers 52, 66
multiplication 63–4
numerator 52
ordering 54
problems 67
subtraction 62
summary of key points 69–70
frequency distributions
averages 192–3
interquartile range 194
spread measures 194
frequency polygons 86
fuel consumption 145
function 16
function notation 460–2
functions, definition 160
functions, summary of key points 183

G

general term of sequences 227
gradient 167–8
graphs
combined transformations 474–5
cubic functions 311–12
direct proportion 329–30
distance between two points 282
distance–time 320
double translations 470–1
equation solving 315–16
gradient 167–8
horizontal translations 467–9
intercepts 165, 167
linear functions 161–3
mid-point of line segment 172
parallel lines 165
perpendicular lines 170–1
quadratic equations 315–16
quadratic functions 310
reciprocal functions 312–13
reflections 472–3
region of an inequality 178–9
simultaneous equations 172–3
speed–time 320–1
stretches 475–8
summary of key points 327
transformations 460–84
trigonometric functions 479–80
trigonometric ratios 300–3
vertical translations 463–5
water levels 321–2
$y = ax^3 + bx^2 + cx + d + \frac{e}{x}$ 314–15
$y = mx + c$ 167–8
greatest lower bound 439
grouped data 82–3, 85, 195–7

H

HCF (highest common factor) 2, 22, 265
hexagon construction 118
hexagonal prism 36
hexagonal pyramid 421
highest common factor (HCF) 2, 22, 265
histograms 83, 383–7, 389–91
hypotenuse 278

I

icosahedron 46
identities 23
impossible events 446
improper fraction 52
independent events 453–4, 534–5
index, fractional 10–11

index laws 8–10
index numbers 7–8
indices
basics 18–19, 255–7
combining 259
division 257
equations 260
fractional 258–9
multiplication 255–6
negative 258–9
notation 255
zero 257
inequalities
brackets involved 100–1
integer solutions 99–100
number line 97
region on graph 178–9
solving 98–9
summary of key points 103
two-sided 99
intercepts 165, 167
interquartile range 191, 194, 375–6
intersection of line and circle 361–2
inverse proportion 339–41
inverting fractions 65
isosceles trapezium 32

K

kite 32

L

large numbers 152–6
LCM (lowest common multiple) 3–4, 263–4
least upper bound 439
line of best fit 525–6
line of symmetry 105–6, 120
linear functions, graphs 161–3
lines of best fit 206–7
lines, summary of key points 183
locus
around a fixed line 123
around a fixed point 122–3
definition 122
point equidistant from two lines 124
point equidistant from two points 124
summary of key points 131–2
lower bounds 220–1, 439–42
lower quartile 189–91, 374–6
lowest common multiple (LCM) 3–4, 263–4

M

major segment 412
market research 72
mathematical modelling
basics 519
constants in exponential relationships 531–2
equations reduced to linear form 527–30
exponential functions 519–21
line of best fit 525–6
scatter diagrams 525–6
summary of key points 533
trigonometric functions 522–4
mean 184–5, 192–3
mean, estimate of 196–7
measurement
accuracy 219–20
continuous data 238–9
data 79
data, definitions 238
summary of key points 254
time 238
median 184–5, 188–90, 192–3, 195, 374–6
mid-point of line segment 172

minor segment 412
mirror image 105–6
mirror line 106
mixed numbers 52, 66
mnemonics
    ASA 39
    BIDMAS 17
    DMV triangle 246
    DST triangle 143
    MDV triangle 147
    RHS 39
    SAS 39
    SSS 39
modal class 196
mode 184–5, 192–3
moving averages 201–2
multiple 1
multiplication
    algebraic fractions 261–2
    bracketed expressions 23
    expressions with indices 255–6
    fractions 63–4
    indexes 8
    inequalities 98
    lower bounds 440
    two fractions 64
    upper bounds 440
    vectors by scalars 505
mutually exclusive events 448–9

N

negative correlation 206–7
negative indices 258–9
negative powers 9
nets 36
notation
    cube root 6, 10, 258
    decimal places 214
    index form 7, 8
    indices 18–19, 255
    line boundary of inequality region 178–9
    not equal to 256
    nth term of sequence 225
    positive or negative root 347
    powers 2, 255
    probability of an event 447
    proportion 328
    recurring decimals 59–60
    significant figures 216
    square root 6
    surds 350
    vectors 501, 503
number skills, summary of key points 159
numbers explored, summaries of key points 15, 444–5
numerator 52

O

observation 79
obtuse angles 29
octahedron 46
1-D line, dimension 248–9, 248–9

P

parabola 310
parallel lines, graphs 165
parallelogram 32, 240
pascal 339
percentages
    compound interest 140–1
    decimals 133
    decreases 134, 137
    fractions 133
    increases 133, 136
    mixed increases and decreases 138
    problem calculations 139
    summary of key points 159
perimeter
    quadrilaterals 239–41
    triangles 239–41
perpendicular
    constructions 118–20
    lines 29
    lines, graphs 170–1
pilot survey 77
plan 36–7
Plato 46
Platonic solids 46
polygons
    exterior angles 33
    interior angles 33–4
polyhedron 35
population 73
position vectors 511–13
positive correlation 206–7
powers
    0 (zero) 9, 257
    basics 18–19
    definition 2, 7
    negative 9
    of 10 12–13, 152–6
    of 2 12–13
    raised to power 9
    summing series 13
prime factor form 2
prime factors 1–2, 263–4
prime number 1
prism 36, 245–7
    surface area 419–20
    volume 419–20
probability
    basics 446–7
    conditional, basics 534, 535
    definition 446
    dependent events 534–5
    estimating from experience 453
    human behaviour 538
    independent events 453–4, 534–5
    lists and tables 449–50
    mutually exclusive events 448–9
    relative frequency 452–3
    summary of key points 459, 541
    tree diagrams 536–7
probability trees 455
proof vs demonstration 34
proportion
    cubic 335–7
    direct 328–30
    formulae 333
    inverse 339–41
    ratio and rules 331–2
    square 335–7
    summary of key points 344–5
proportionality statement 333
pyramid 35
    volume 421–2
Pythagoras 276
Pythagoras' theorem
    applying twice 281
    confirmed 277
    distance between two points 282
    lengths, finding 277–8
    shorter side 279
    stated 276
    summary of key points 289–90
    three dimensions 285–6, 407–8
    triangles that are not right–angled 283
Pythagorean triples 284

Q

quadratic equations
    algebraic fractions 354
    basics 346
    graphical solutions 363–5
    graphs 315–16
    intersection of line and circle 361–2
    problem solving 355–7
    simultaneous linear equations 358–60
    solving by completing the square 349–50
    solving by factorising 346–7
    solving by formula 351–3
    solving $y^2 = k$ 348
    summary of key points 368
quadratic expressions, factorising 267–9
quadratic functions
    definition 310
    graphs 310
quadrilaterals
    area 239–41
    interior angles 33
    overview 32–3
    perimeter 239–41
qualitative data 71
quality control 72, 79
quantitative data 71
quartiles 188–91, 374–6
questionnaire 76–7

R

random number tables 75
random process 73
random sampling 73–4
range 191
ratios 148, 150
reciprocal functions, graphs 312–13
rectangle 32
rectangular prism 245–7
recurring decimals 432–5
reflection 105–6
reflex angles 29
    region of an inequality 178–9
summary of key points 183
relative frequency 452–3
repeated reflection 106
rhombus 32
right angle, constructions 118–20
rotation 107–8
rounding
    decimal places 214–15, 438
    significant figures 216

S

sampling
    biased 73
    population 73
    random 73–4
    random process 73
    stratified 74
    systematic 75
scale drawings 115–16
scale factor 42–3, 109–10, 424–6
scale models 115–16
scatter diagrams 203–5, 525–6
secondary data 72, 80–1
sector 412, 414–15
segment 412, 416–17
sequences
    arithmetic 227
    general term 227
    pattern 160
    summary of key points 237
    terms 225–6

series of powers 13
shapes, summary of key points 49–51, 431
side elevation 36–7
sigma function 4
similar shapes 42–3
    area 424–6
    volume 424–6
simultaneous equations
    algebraic solutions 174–6
    graphical solutions 172–3
    problem solving 177
    summary of key points 183
simultaneous linear and quadratic equations 358–60
$\sin x$ 291–3, 300
sine rule 402–5
slant height 422
small numbers 154
space diagonal 285
speed, averages 143
sphere 36
    surface area 423
    volume 423
spread measures 191, 194, 199–200
square proportionality 335–7
square root 6
square [number] 4–5
square [shape] 32
square–based pyramid 35, 421
standard form 152–6
stem and leaf diagrams 185–6, 380
stratified sampling 74
subtraction
    algebraic fractions 262–4
    fractions 62
    lower bounds 441–2
    upper bounds 441–2
    vectors 505
summaries of key points
    algebra basics 27–8
    algebraic expressions 27–8, 275
    approximating 224
    circle theorems 500
    conditional probability 541
    data collection and recording 89–90
    data presentation and analysis 212–3, 398
    decimals 69–70
    equations 103
    estimating 224
    formulae 237
    fractions 69–70
    functions 183
    graphs 327
    inequalities 103
    lines 183
    locus/loci 131–2
    mathematical modelling 533
    measurement 254
    number skills 159
    numbers explored 15, 444–5
    percentages 159
    probability 459, 541
    proportion 344–5
    Pythagoras' theorem 289–90
    quadratic equations 368
    regions of an inequality 183
    sequences 237
    shapes 49–51, 431
    shapes, 2-D 49–51
    shapes, 3-D 49–51
    simultaneous equations 183
    transformations 131–2
    transformations of graphs 483–4
    trigonometric ratios 308–9
    trigonometry 411
    vectors 518

supplementary angles 29
surds 7, 278, 350, 351, 436–7
symmetry
    line of 105–6, 120
    plane of 45–6
systematic sampling 75

T

table of values 161
tan *x* 292–3, 302–3
terminating decimals 432–3
tetrahedron 46
3-D shapes
    dimensions 248–9
    overview 35–7
    summary of key points 49–51
    surface area 244–7
    volume 244–7
time 238
time series graph 201–2
transformations
    combined 113
    enlargement 109–10
    reflection 105–6
    rotation 107–8
    summary of key points 131–2
    translation 104
    vector 104
transformations of graphs
    basics 460
    combined transformations 474–5
    double translations 470–1
    horizontal translations 467–9
    reflections 472–3
    stretches 475–8
    summary of key points 483–4
    trigonometric functions 479–80
    vertical translations 463–5
translation 104
    vectors 501–2
trapezium 32, 240
tree diagrams 536–7
trial and improvement methods
    equations 317–18
    problems 318–19
triangle-based pyramid 35, 46, 421
triangles
    altitude 500
    area 239–41, 399–400
    basics 31
    constructions 117
    cosine rule 405–6
    exterior angles 35
    interior angles 31–2
    perimeter 239–41
    Pythagoras' theorem 276–90
    right-angled 276–90
    sine rule 402–5
triangular prism 36

trigonometric equations 304–5
trigonometric ratios
    advanced 298–9
    angles 295
    basics 291–3
    calculator 294–5
    graphs 300–3
    inverse ratios 295
    length of sides 296–7
    mathematical modelling 522–4
    summary of key points 308–9
    transformations of graphs 479–80
    unit circle 298–9
    values 294–5
trigonometry
    background 291
    summary of key points 411
    three dimensions 407–8
2-D shapes
    congruent 38–40
    dimensions 248–9
    overview 31–5
    similar 42–3
    summary of key points 49–51
two-way tables 85

U

upper bounds 220–1, 439–42
upper class boundary 369–70
upper quartile 189–91, 374–6

V

vectors
    addition 504
    algebra 506–7
    column 501–2
    directed line segments 503
    linear combinations 509–10
    magnitude 508
    multiplication by scalars 505
    position 511–13
    proving geometrical results 514
    subtraction 505
    summaries of key points 518
    transformations 104
    translations 501–2
vertically opposite angles 29
volume
    compound solids 427–8
    cone 421
    cube 244
    cuboid 244–5
    cylinder 419
    prism 419–20
    pyramid 421–2
    similar shapes 424–6
    sphere 423